The Palgrave Handbook of Sexual Ethics

David Boonin
Editor

The Palgrave Handbook of Sexual Ethics

palgrave
macmillan

Editor
David Boonin
Department of Philosophy
University of Colorado Boulder
Boulder, CO, USA

ISBN 978-3-030-87785-9 ISBN 978-3-030-87786-6 (eBook)
https://doi.org/10.1007/978-3-030-87786-6

This Palgrave Macmillan imprint is published by the registered company Springer Nature Switzerland AG.
The registered company address is: Gewerbestrasse 11, 6330 Cham, Switzerland

Contents

Notes on Contributors

Cheryl Abbate is Assistant Professor of Philosophy at the University of Nevada, Las Vegas, USA, and the co-president of the Society for the Study of Animal Ethics. She specializes in animal ethics, and she has a growing interest in the ethics of dating. Her recent publications include "A defense of free-roaming cats from a hedonist account of feline well-being" (*Acta Analytica*), "Meat eating and moral responsibility: Exploring the moral distinctions between meat eaters and puppy torturers" (*Utilitas*), and "Valuing animals as they are: Whether they feel it or not" (*European Journal of Philosophy*).

Ellie Anderson is Assistant Professor of Philosophy at Pomona College in Claremont, CA, USA. She has published articles on the phenomenology of love and intimacy in *Continental Philosophy Review*, *Philosophy Today*, and *Symposium: Journal of Canadian Continental Philosophy*. Ellie is also co-author of the *Stanford Encyclopedia of Philosophy* entry "Feminist Perspectives on the Self," and her *American Philosophical Association* blog post "The Limits of Consent in Sexual Ethics" (April 2019) has been among the website's most-read pieces since its publication. She co-hosts the philosophy podcast *Overthink*.

Peter Brian Barry is Professor of Philosophy and the Finkbeiner Endowed Professor in Ethics at Saginaw Valley State University, USA. He is the author of *Evil and Moral Psychology* (2013) and *The Fiction of Evil* (2016) and multiple papers in ethics and social and political philosophy. He is writing a book on the ethics of George Orwell.

Francis Joseph Beckwith is Professor of Philosophy and Church-State Studies, Affiliate Professor of Political Science, and Associate Director of the Graduate Program in Philosophy, at Baylor University (Waco, Texas, USA). His books include *Never Doubt Thomas: The Catholic Aquinas as Evangelical and Protestant* (2019), *Defending Life: A Moral and Legal Case Against Abortion Choice* (2007), and *Taking Rites Seriously: Law, Politics, and the Reasonableness of Faith* (2015), winner of the American Academy of Religion's

2016 Book Award for Excellence in the Study of Religion in the category of Constructive-Reflective Studies.

Sonu Bedi is the Joel Parker 1811 Professor in Law and Political Science and Professor of Government at Dartmouth College. Along with numerous articles, law reviews, and book chapters, he has published four books, including most recently *Private Racism* (2019). His research interests are in the areas of contemporary political theory, constitutional law and theory, and race, law, and identity. Additional information about his publications is available here: https://faculty-directory.dartmouth.edu/sonu-s-bedi.

David Benatar is Professor of Philosophy at the University of Cape Town, South Africa. His books include *Better Never to Have Been: The Harm of Coming into Existence* (2006), *The Second Sexism: Discrimination Against Men and Boys* (2012), and *The Human Predicament: A Candid Guide to Life's Biggest Questions* (2017).

Andria Bianchi received her PhD in Philosophy from the University of Waterloo, where her research considered dementia and sexual consent from an ethics perspective. She works as a bioethicist and clinician-scientist at the University Health Network in Toronto, Canada. She is affiliated with the Dalla Lana School of Public Health, University of Toronto, as an assistant professor (status-only) and is an affiliate scientist at the KITE research institute at Toronto Rehab. In addition to sex and dementia, she has published on various topics, including ethics and eating disorders, transgender women in sports, and deceased directed organ donation.

David Boonin is Professor of Philosophy at the University of Colorado Boulder. He is the author of *Thomas Hobbes and the Science of Moral Virtue* (1994), *A Defense of Abortion* (2003), *The Problem of Punishment* (2008), *Should Race Matter?* (2011), *The Non-Identity Problem and the Ethics of Future People* (2014), *Beyond Roe* (2019), and *Dead Wrong: The Ethics of Posthumous Harm* (2019) as well as a number of articles on subjects such as animal rights, euthanasia, same-sex marriage, and our moral obligations to past and future generations.

Luke Brunning is Lecturer in Ethics at the University of Birmingham. He writes on the philosophy of sex and love, and ethics broadly construed.

John Corvino is Professor of Philosophy and Dean of the Irvin D. Reid Honors College at Wayne State University in Detroit, Michigan, USA. He is the author or co-author of several books, including *Debating Same-Sex Marriage* (with Maggie Gallagher, 2012), *What's Wrong with Homosexuality?* (2013), and, most recently, *Debating Religious Liberty and Discrimination* (with Ryan T. Anderson and Sherif Girgis, 2017). In addition to his academic writing, he has contributed to *The New York Times*, the *Detroit Free Press*, *Slate*, and various other popular venues; he also has a YouTube channel. He is working on a book on the intersection of civility and free speech.

Dan Demetriou is Associate Professor of Philosophy at the University of Minnesota, Morris, USA. Demetriou recently co-edited *Honor in the Modern World: Interdisciplinary Perspectives* (Lexington Books) and has published a number of articles and chapters on honor ethics. He is writing essays on racist monuments, immigration, and low-trust ethics.

Tom Dougherty is Associate Professor of Philosophy at the University of North Carolina at Chapel Hill. He specializes in normative ethics. He has recently completed one project on the ethics of consent, culminating in a forthcoming book, *The Scope of Consent*, and is nearing completion of a second project on consent and coercion.

Edward Feser is Professor of Philosophy at Pasadena City College in Pasadena, California, USA. He is the author of many academic articles and books, including *Aquinas*, *Scholastic Metaphysics: A Contemporary Introduction*, and *Aristotle's Revenge: The Metaphysical Foundations of Physical and Biological Science*. Further information can be found at his personal website: www.edwardfeser.com.

Jessica Flanigan is the Richard L. Morrill Chair in Ethics and Democratic Values at the University of Richmond. Her research addresses the ethics of public policy, medicine, and business. In "Pharmaceutical Freedom" (2017), she defends rights of self-medication. In "Debating Sex Work" (2019), she defends the decriminalization of sex work.

Raja Halwani is Professor of Philosophy at the School of the Art Institute of Chicago, USA. He is the author of *Virtuous Liaison: Care, Love, Sex, and Virtue Ethics* (2003) and *Philosophy of Love, Sex, and Marriage: An Introduction* (2nd ed., 2018) and co-author of *The Israeli-Palestinian Conflict: Philosophical Essays on Self-Determination, Terrorism, and the One-State Solution* (2008). He is the editor of *Sex and Ethics: Essays on Sexuality, Virtue, and the Good Life* (2007) and the lead editor of *Queer Philosophy: Presentations of the Society for Lesbian and Gay Philosophy, 1998–2008* (2012) and of *The Philosophy of Sex: Contemporary Readings*, 7th ed. (2017).

Carol Hay is an associate professor in the Department of Philosophy at the University of Massachusetts Lowell. Her research interests focus primarily on issues in analytic feminism, liberal social and political philosophy, oppression studies, Kantian ethics, and the philosophy of sex and love. Her monograph, *Kantianism, Liberalism, & Feminism: Resisting Oppression* (Palgrave Macmillan, 2013), received the American Philosophical Association's Gregory Kavka/UCI Prize in Political Philosophy. Her most recent trade book, *Think Like a Feminist: The Philosophy Behind the Revolution* (2020), has been called "a crisp, well-informed primer on feminist theory" by *Publisher's Weekly* and "a winning mix of scholarship and irreverence" by *Kirkus Reviews*.

Christopher Kaczor (rhymes with razor) is Professor of Philosophy at Loyola Marymount University, USA. He graduated from the Honors Program of

Boston College and earned a PhD four years later from the University of Notre Dame. A Fulbright Scholar, Kaczor did post-doctoral work as a Federal Chancellor Fellow at the University of Cologne and as William E. Simon Visiting Fellow at Princeton University. He has written more than 100 scholarly articles and book chapters. An award-winning author, his 15 books include *Disputes in Bioethics*, *The Seven Big Myths about Marriage*, and *The Ethics of Abortion*.

Stephen Kershnar is a distinguished teaching professor in the philosophy department at the State University of New York at Fredonia, USA, and an attorney. Kershnar is the author of ten books, including *Desert Collapses: Why No One Deserves Anything* (forthcoming), *Total Collapse: The Case Against Morality and Responsibility* (2018), *Abortion, Hell, and Shooting Abortion-Doctors: Does the Pro-Life Worldview Make Sense?* (2017), and *Adult-Child Sex: A Philosophical Defense* (2015). He has also written more than 100 articles and book chapters on such diverse topics as abortion, affirmative action, capitalism, discrimination, equal opportunity, hell, punishment, sexual fantasies, slavery, and torture.

Alida Liberman received her PhD from the University of Southern California and is Assistant Professor of Philosophy at Southern Methodist University, USA. Her research focuses on theoretical and applied ethics and the places where they intersect. Her work on promises and other forms of commitment has been published in the *Journal of the American Philosophical Association*, *Oxford Studies in Normative Ethics*, *Pacific Philosophical Quarterly*, and the *Journal of Ethics and Social Philosophy*, among others. Her work in bioethics has been published in *Bioethics*, *Social Theory and Practice*, and the *Journal of Medical Ethics*.

Patricia Marino is Professor of Philosophy at the University of Waterloo in Canada, where she works in ethics, epistemology, philosophy of economics, and philosophy of sex and love. She is the author of *Moral Reasoning in a Pluralistic World* (2015) and *The Philosophy of Sex and Love: An Opinionated Introduction* (2019) as well as articles on moral dilemmas, ambivalence, sexual objectification, values in law and economics, and other topics. For more information, visit patriciamarino.org.

Neil McArthur is Professor of Philosophy and Director of the Centre for Professional and Applied Ethics at the University of Manitoba. He is the co-editor of *Robot Sex: Social and Ethical Implications* (2017). In addition to his academic publications, he has written about sexual ethics, alternative sexualities, and LGBTQ issues for publications including VICE, the Guardian, Time, and the Globe and Mail.

Natasha McKeever is Lecturer in Applied Ethics at the University of Leeds, United Kingdom. She received her PhD in Philosophy from the University of Sheffield in 2014, which she wrote on the topic of romantic love and monogamy. Her research interests are primarily in the philosophy of love and sex, and she has published articles on topics including rape, asexuality, prostitution, romantic love, sexual infidelity, and sexual exclusivity. She is co-editing two forthcoming anthologies: *The Philosophy of Sex: Contemporary Readings*, 8th edition, and *The Philosophy of Love in the Past, Present and Future.*

Shaun Miller is an instructor at Salt Lake Community College. He received his PhD in Philosophy from Marquette University which he wrote on the topic of moral assumptions of sex education classes in the USA. His research topics specialize in the philosophy of sex and love, and he has written articles on topics that range from sexual consent, positive male sexuality, and BDSM.

Kevin Mintz received his PhD in Political Science from Stanford University and is an affiliate faculty in the Department of Philosophy at George Mason University, USA. He also holds a doctorate degree in Human Sexuality from The Institute for Advanced Study of Human Sexuality, an MSc in Political Theory from The London School of Economics and Political Science, and an AB in Government from Harvard College. Born with cerebral palsy, his research focuses on disability ethics and sexual health ethics. His work has appeared in a variety of venues including *Pediatrics*, *Disability & Society*, and the *Los Angeles Times.*

Alastair Norcross is Professor of Philosophy at the University of Colorado Boulder, where he has taught since being allowed out of Texas for good behavior in 2007. He has published extensively on consequentialism, in particular defending a scalar version of the theory (see, e.g., *Morality by Degrees: Reasons without Demands*), and in applied ethics, including the widely reprinted "Puppies, Pigs, and People: Eating Meat and Marginal Cases" (*Philosophical Perspectives*, 2004). He also runs marathons, with somewhat less success than Eliud Kipchoge, and writes, directs, and acts in the theater, with somewhat less success than Kenneth Branagh.

Jordan Pascoe is Associate Professor of Philosophy at Manhattan College in New York City, where she also serves as the director of the Lasallian Women and Gender Resource Center. She has published on gender, race, sex, and domestic labor, as well as disaster epistemology, and ethics. She is the director of the Society for the Philosophy of Sex and Love.

James Rocha is Associate Professor of Philosophy at Fresno State. His books are *The Ethics of Hooking Up* (2019) and *Joss Whedon, Anarchist?* (with Mona Rocha; McFarland 2019). He has published in numerous journals, including *Ethical Theory & Moral Practice*, *Journal of Applied Philosophy*, *Social Theory and Practice*, and *Public Affairs Quarterly*. He coordinates the Social Justice and Social Change Certificate and is the Central Valley Scholars Law Pathway Liaison for Fresno State.

Laurie Shrage received her PhD in Philosophy from the University of California San Diego (1983) and is Professor of Philosophy at Florida

International University. Her books include *Abortion and Social Responsibility: Depolarizing the Debate* (2003), *Moral Dilemmas of Feminism* (1994), an edited collection *You've Changed: Sex Reassignment and Personal Identity* (2009), and the co-authored textbook *Philosophizing About Sex* (2015). She has published in numerous scholarly journals and served as co-editor of *Hypatia* from 1998 to 2003. She was a fellow-in-residence at the Edmond J. Safra Center for Ethics, Harvard University, 2015–2016 and was a Laurance S. Rockefeller Visiting Fellow, Princeton University Center for Human Values, 2011–2012. She has contributed several pieces to "The Stone" series in *The New York Times*.

Alan Soble, now Emeritus, was at the University of New Orleans, 1986–2006, eventually as Research Professor of Philosophy. He was later, 2007–2016, Professor at Drexel University. He founded the Society for the Philosophy of Sex and Love in 1977 and began teaching, writing, and lecturing about sexuality, love, friendship, and marriage in 1978. He has also published in epistemology, biomedical ethics, and history of philosophy. Soble's leisure activities include chess, philately, and mathematical logic; when he was younger, they had also included sexuality, love, friendship, and marriage. His heroes are Søren Kierkegaard, Karl Marx, Georg Cantor, Philip Roth, and Nero Wolfe.

Michael Tooley has written, in metaphysics, on laws of nature, causation, and the nature of time, in philosophy of religion, on the evidential argument from evil and the non-existence of God, and, in applied ethics, on abortion, voluntary euthanasia, animals' moral status, and cloning. He is the author of *Abortion and Infanticide* (1983), *Causation: A Realist Approach* (1987), *Time, Tense, and Causation* (1997), and *The Problem of Evil* (2019), and a co-author, with Alvin Plantinga, of *Knowledge of God* (2008), and, with Alison Jaggar, Philip Devine, and Celia Wolf-Devine, of *Abortion: Three Perspectives* (2009).

Helga Varden is Professor of Philosophy and of Gender and Women Studies at the University of Illinois at Urbana-Champaign. Her main research interests are Kant's practical philosophy, legal-political philosophy and its history, feminist philosophy, and the philosophy of sex and love. In addition to her *Sex, Love, and Gender: A Kantian Theory* (2020), Varden has published many articles on a range of classical philosophical issues including Kant's answer to the murderer at the door, private property, care relations, political obligations, and political legitimacy, as well as on applied issues such as privacy, poverty, nonhuman animals, and terrorism.

Caleb Ward is a postdoctoral researcher specializing in feminist philosophy at the University of Hamburg. He holds a PhD in Philosophy from Stony Brook University. His research focuses on feminist approaches to phenomenology, critical theory, and ethics. Ward has published work on Audre Lorde in the *Journal of the American Philosophical Association*, and he has co-edited two volumes on food ethics: *The Routledge Handbook of Food Ethics* (with Mary

C. Rawlinson, 2017) and *Global Food, Global Justice: Essays on Eating under Globalization* (with Mary C. Rawlinson, 2015).

Elise Woodard is a doctoral candidate at the University of Michigan, Ann Arbor. Her research focuses primarily on epistemology, ethics, social philosophy, and their intersections. Her recent work focuses primarily on norms governing further inquiry, including re-deliberation, evidence-gathering, and double-checking. She also has strong interests in sexual ethics and issues regarding consent, including the epistemology of consent.

CHAPTER 1

Introduction: Sex, Ethics, and Philosophy

David Boonin

Sex raises some of the oldest of ethical questions. What is my body for? Who can I share it with and under what conditions? It also raises some of the newest. Is it wrong to swipe left on Tinder solely because of a person's race? What uses of sex robots, if any, are morally impermissible? Thinking about such questions with the clarity and rigor contemporary philosophers aspire to can be challenging for a number of reasons. For one thing, sex strikes most people as a distinctly intimate and personal matter. This can make it difficult for them to talk about it at all, let alone to talk about it frankly and openly. Sexual desire and sexual activity also tend to generate powerful and primitive emotions. These can interfere with the attempt to think about the moral problems associated with sex calmly and dispassionately. And many of the most pressing and important questions in sexual ethics can't be answered without delving into debates on further philosophical issues that are themselves difficult to grapple with because of their depth and complexity: the nature of human autonomy, for example, what it means to treat a person with respect, why consent sometimes makes the difference between an act being permissible and its being impermissible. The very reasons that make it valuable to bring together some of the latest philosophical thinking about a variety of issues in sexual ethics in the way this Handbook seeks to do also make the tasks involved in thinking philosophically about such issues a daunting one.

The contributors to this volume respond to this challenge in a variety of ways. Some turn for philosophical assistance, at least in part, to the work of historical figures from antiquity, like Aristotle (Chaps. 2 and 3), the modern

D. Boonin (✉)
Department of Philosophy, University of Colorado Boulder, Boulder, CO, USA
e-mail: david.boonin@colorado.edu

D. Boonin (ed.), *The Palgrave Handbook of Sexual Ethics*,
https://doi.org/10.1007/978-3-030-87786-6_1

period, like Kant (Chaps. 5 and 9), or the previous century, like Simone de Beauvoir (Chap. 4) and Hannah Arendt (Chap. 5). Others engage exclusively with more contemporary literature or with none at all. Some largely ground their arguments for a general thesis in the intuitive reactions they expect their readers to have about specific cases (e.g., Chaps. 11, 14, and 22). Others focus more substantially on arguing in the opposite direction: from a general principle to a more particular conclusion (e.g., Chap. 20). Some rely heavily on empirical claims (e.g., Chaps. 8 and 10). Others rely more on conceptual analysis (e.g., Chaps. 3 and 15). Some focus their attention on relatively narrow questions (e.g., Chaps. 21, 26, and 30). Others address issues that are considerably broader or more general (e.g., Chaps. 2, 3, and 4). Some aim to uphold traditional values (e.g., Chaps. 10 and 11), some aim to upend them (e.g., Chap. 14), and some aim to upset common assumptions about their implications (Chap. 13). What they all have in common is the philosopher's attempt to bring the tools of critical analysis and reason to bear on questions whose sexual nature can threaten to render them resistant to such treatment.

This book is divided into five parts. Part I follows this introductory chapter with a set of seven substantive chapters each of which engage, in one way or another, with basic questions about, or features of, human nature. The first three of these seven chapters emerge from quite general and metaphysical questions concerning reality, supervenience, and the subject/object distinction. The remaining four focus on more specific features of human beings and human existence: our susceptibility to temptation and to jealousy and our vulnerability to sour grapes thinking and to various forms of imprudent irrationality. All seven chapters connect these issues about human nature to some general or more specific issue in sexual ethics.

In Chap. 2, Edward Feser contrasts two fundamentally different perspectives we might take on human beings and thus on human nature. One views human beings from the point of view of everyday experience and common sense. The other views them, or perhaps I should say views us, from the vantage point of modern science. Feser argues that there is a strong correlation between viewing human nature through the first lens and endorsing the edicts of traditional sexual morality and between viewing human nature through the second lens and rejecting traditional sexual morality in favor of the more liberal fruits of the sexual revolution. Feser argues, moreover, that recognizing the relationship between each metaphysical view of human nature and the view of sexual ethics that corresponds to it can help us see how these two very different views of sexual ethics are both understandable, and even reasonable, given the general views of human nature they fit with best. This doesn't lead Feser to endorse a relativistic conclusion on which the two views of sexual ethics are equally valid, but it does lead him to conclude that the clash between the two runs much deeper than a mere competition between superficial intuitions about sex and to suggest some ways in which recognizing this feature of the debate might lead people to engage with those on the other side in a more intellectually productive manner.

In Chap. 3, Raja Halwani explores a general question about the nature of sexual desire and its relation to sexual activity and sexual pleasure. On one view, the object of sexual desire is sexual pleasure and one engages in sexual activity merely as a means of satisfying the desire for this kind of pleasure. On a second view, the object of sexual desire is the sexual activity itself. What one wants, most fundamentally, is to engage in the activity and the pleasure brought about by the activity is merely a foreseeable consequence of getting what one wants: more like the icing on the cake than like the cake itself. In this chapter, Halwani argues in favor of the first view over the second, both by marshaling considerations in favor of the first view and by identifying problems with the second, and he develops a more precise and refined version of this first view. The view itself is a metaphysical view about a part of human nature rather than an ethical view about sex itself. But, as Feser does in Chap. 2, Halwani argues that the metaphysical view can have significant moral implications. In particular, Halwani suggests that this view about the relationship between sexual desire and sexual pleasure can shed light on the question of whether it's morally wrong to take pleasure from mere simulations of wrongful sexual activities, a question closely connected to the subject of John Corvino's discussion in Chap. 31, and that it can be used to help to solve a dilemma for both liberal and traditional views of sexual ethics first posed in an earlier article by David Benatar, a dilemma that Benatar in part returns to in Chap. 13 of this volume.

Contemporary philosophical discussions of sexual ethics focus heavily on questions of autonomy, agency, and the morally transformative power of consent. In doing so, they stress the importance of respecting human beings as moral subjects in their own right and of not treating people merely as sexual objects or as a mere means to one's own ends. This approach to sexual ethics can seem to presuppose a kind of exclusive dichotomy about the locus of value in human nature: human beings ought to be subjects, not objects. But as Caleb Ward and Ellie Anderson point out in Chap. 4, people are both subject and object in their encounters with other people, and perhaps especially so in their sexual encounters. The authors therefore argue for a revision to this familiar approach to sexual ethics, one that broadens the focus to include questions about the moral significance of being an erotic object. Drawing from the phenomenological tradition in general, and from the work of Simone de Beauvoir in particular, Ward and Anderson develop an approach that gives more weight than is typically given to the ambiguous situation of being simultaneously erotic subject and object in intimate encounters. In doing so, they make the case for the view that important moral features of intimacy are revealed through the erotic experiences both of being an object and of perceiving another as an object, features that are easily overlooked if we focus too exclusively on issues of moral agency and autonomy. While acknowledging the moral significance of such autonomy and agency, Ward and Anderson aim to help develop a more complete account of sexual ethics, one that does justice to how human intimacy entails being both subject and object.

One familiar feature of the human condition is that having a good sexual life can be challenging while settling for a bad one can be tempting. In Chap. 5, Carol Hay and Helga Varden consider why this is. In doing so, they draw on and build from ideas from a wide variety of sources, including work in feminist philosophy and the literature on the philosophy of sex and love in general, as well as the work of two figures in particular who might seem unlikely sources for such an enterprise and an unlikely pairing, too: Immanuel Kant and Hannah Arendt. More specifically, Hay and Varden propose that Kant's account of human nature, including both the good and the bad, when combined with some of Arendt's ideas about the problems with the Western philosophical tradition's treatment of our animality, provides a good starting point for exploring the nature and value of good sexual love and for understanding the challenges that confront those who seek it. Some of the sources of these challenges, they suggest, lie in the difficulties involved in trying to transform, develop, or integrate certain unruly emotions. Others involve barriers generated by inherited oppressive behaviors and feelings that make emotionally healthy, morally responsible realizations of sexuality difficult. Despite these difficulties, Hay and Varden conclude that striving for a satisfying sexual life can nonetheless be a meaningful and exciting part of a good human life.

Another familiar fact about human beings is that they get jealous. This is perhaps especially so in the case of sex and romance. If Bob and Carol are in a close romantic and sexual relationship and Carol has sex with Ted, it's likely that her doing so will make Bob jealous. This is a commonly observed feature of human existence, but it raises some significant questions that are not commonly raised, let alone seriously addressed. What, precisely, does it mean to say that Bob is jealous? Is his jealousy valuable in any way? What, if anything, should Bob or Carol do about the fact that Carol's behavior has made Bob jealous? In Chap. 6, Natasha McKeever and Luke Brunning address these questions. They begin by asking what jealousy is and answer that it's best understood as an emotional response to the threatened loss to a rival of love or attention that one believes one deserves. They then consider the relationship between romantic love and jealousy and argue that it can be consistent to feel jealousy toward someone we love. They next address the question of jealousy's value, arguing that claims made about its positive value must be balanced against a variety of potential harms it can cause. And finally, they assess two potential ways of managing jealousy, one that involves a policy of monogamy and one that doesn't, and they argue that the second approach should be taken more seriously than it typically is.

A third familiar feature of human nature is our tendency to convince ourselves that we didn't really want something that we tried but failed to get. This phenomenon has been recognized at least since the time of Aesop and his fable of the fox and the grapes, and in recent years philosophers have tended to discuss it in the context of "adaptive preferences": preferences that, roughly speaking, a person forms to help them cope with their non-ideal circumstances and that lead them to settle for less than they would have preferred under better conditions. In Chap. 7, Patricia Marino brings this feature of human nature to bear on an important question about the relationship between sexual consent and sexual objectification. It's plausible to suppose that it's wrong to treat a person merely as a sexual object, as nothing but a means to achieving sexual pleasure, and also plausible to suppose that this wrongness can be mitigated, and perhaps even eliminated entirely, if they consent to being treated in such a way. But what should we say if the person who consents to such treatment does so only because they've been conditioned into doing so by an unjust society? A sex worker, perhaps, or a porn starlet, who wants to have sex with men for money but who wouldn't have this desire if she lived in a less misogynistic and patriarchal society. In this chapter, Marino explores a variety of ways in which the ideas of social autonomy and adaptive preferences can be used to help us think more clearly about the choices people make that involve sexual objectification in different social contexts. Her discussion helps to illuminate some of the ways in which sexual objectification can be a positive thing for particular individuals in particular circumstances as well as some of the ways in which it can nonetheless represent a significant widespread social harm.

One final and distressingly familiar feature of human nature is the way it often leads people to do things that are self-destructive and harmful to others, perhaps especially so when it comes to sex. Michael Tooley addresses this topic in Chap. 8 and argues that, with the exception of sex that aims at reproduction, most human sexual activity is both irrational and immoral because it is dangerous to its participants in ways that can be easily avoided by pursuing sexual pleasure in other ways. In particular, Tooley argues that masturbation, either mutual or solo, can provide the same kind of pleasure as other forms of sexual activity without running any of the risks those other activities generate in terms of transmitting potentially serious diseases and producing unwelcome pregnancies. Tooley's claim that sexual intercourse is immoral unless it is aimed at procreation may be welcomed by sexual conservatives, but his robust defense of solo and mutual masturbation certainly won't be, nor will his rejection of much of the sexual morality associated with the Jewish and Christian traditions.

Part II gathers together six chapters under the heading of sex and traditional values. The first three can be read as offering a defense of some aspect of traditional sexual morality. The last three can be read as offering a critique. Three claims that are central to traditional views of sexual morality maintain that people should be sexually monogamous, that sexual partners should not live together before they get married, and that sex is special in a way that makes it importantly different from ordinary recreational activities. In Chap. 9, Dan Demetriou defends the first claim, in Chap. 10, Christopher Kaczor defends the second, and in Chap. 11, Francis Joseph Beckwith defends the third.

An "incel," as that term has come to be used, is an involuntarily celibate heterosexual man. Traditional sexual morality maintains that sexual relationships should be monogamous. If a hundred heterosexual men follow the traditional monogamous norm, they remove a hundred heterosexual women from the pool of women who might be available to have sexual relationships with other heterosexual men. If a hundred heterosexual men are each involved in a sexual relationship with three heterosexual women while each of those women is involved only with that one particular man, then the hundred men instead remove three hundred women from the pool, making it harder for the other heterosexual men to find partners and increasing the number of incels. This has led some people, including perhaps most prominently Jordan Peterson, to recommend "enforced monogamy" as a solution to the "incel problem." In Chap. 9, Dan Demetriou attempts to develop the strongest argument that can be made for this view and then subjects the argument to critical scrutiny. He concludes that, at least in some sense of the term, the argument may well succeed in justifying "enforced monogamy." And even though he also concludes that there is a stronger sense of the term in which the argument probably doesn't succeed, the result of the chapter as a whole nonetheless provides at least a partial defense of one fundamental component of traditional sexual morality.

In Chap. 10, Christopher Kaczor defends the claim that it's immoral for sexual partners to live together before they get married. Much of the chapter is devoted to presenting empirical evidence for the claim that such cohabitation on average produces worse consequences for the people involved than does waiting until marriage to live together. These include claims about increasing the chances of infidelity and divorce once the couple is married and about increasing the risk of drug abuse, physical violence, and child abuse within the marriage. While the chapter largely focuses on factual claims, as distinct from moral claims, Kaczor argues that the factual claims can be used to ground a moral case against cohabitation before marriage by appealing to a moral principle on which it's prima facie immoral to risk causing such harms. He also argues that cohabitation is morally objectionable on the grounds that it systematically disadvantages women who wish to marry because it typically leaves women worse off relative to their male cohabitors.

Chapter 11 defends a more general claim than those endorsed in Chaps. 9 and 10. In it, Francis Joseph Beckwith targets the popular contemporary view that sex is just like any other recreational activity and that, as with those other activities, it's always okay for adults to engage in it as long as they have freely and competently consented to doing so. While Beckwith's goal is to defend the very general and abstract claim that sex is special in the way that traditional sexual morality takes it to be, his approach to defending this claim largely takes place at a more particular level: he counts on the reader to agree that in specific, concrete cases, examples of sexual assault and of sexual harassment are, morally speaking, significantly worse than otherwise parallel examples of non-sexual assault and non-sexual harassment, and he argues that the traditional view of sex as special can easily account for these judgments while the competing contemporary view on which sex isn't special cannot.

An argument can be at odds with traditional sexual morality in a variety of ways. At a minimum, it might aim to go beyond traditional views about sex in a way that reveals those views to be incomplete. More critically, it might aim to show that an argument typically made by proponents of traditional sexual morality can be satisfactorily rebutted on grounds that proponents of those views already accept. And more critically still, it might aim to defend a conclusion that is diametrically opposed to a central tenet of traditional sexual morality. The last three chapters of Part II provide an example of each.

The scourge of sexual harassment in the workplace is old but the concept of sexual harassment is new. A victim or perpetrator would likely have had little understanding of the notion in the 1950s, let alone in earlier periods when most of our traditional beliefs about sexual morality were first generated and then refined. As a result, traditional approaches to sexual ethics have relatively little to teach us about what makes some forms of behavior forms of sexual harassment and what makes those forms of behavior morally wrong. In Chap. 12, Jessica Flanigan aims to fill this gap in our understanding by critically evaluating and rejecting three possible answers to this question and defending a fourth alternative. On the accounts that Flanigan rejects, sexual harassment violates the victim's rights because it causes them emotional distress, because it coerces them, or because it treats them unequally. On the account that she defends, sexual harassment violates the victim's rights because it involves a certain form of deception.

Traditional sexual morality maintains that sexual relationships between members of the same sex are morally wrong. A familiar argument for this view maintains that if same-sex sexual relationships (or same-sex marriages) are morally permissible than so is sex with (or marriage to) non-human animals as well as sex with corpses. On the assumption that it's clear that bestiality and necrophilia are immoral, this would show that homosexuality must be immoral, too. In Chap. 13, David Benatar argues that proponents of traditional sexual morality cannot consistently appeal to this argument because there are conceptual resources contained within their own tradition of thinking about sexual ethics that can help to show that same-sex sexual (and marital) relationships are relevantly different from those involving non-human animals and corpses. Benatar also argues that more liberal and permissive approaches to sexual ethics lack these conceptual resources and so his chapter can be seen as presenting one kind of challenge to sexual traditionalists and another kind to sexual liberals.

One of the most familiar tenets of traditional sexual morality maintains that it's wrong for people to have sex with each other before they get married. In Chap. 14, Alastair Norcross turns this piece of conventional thinking on its head, arguing not just that it isn't wrong for people to have sex with each other before they get married but that it's positively wrong for them not to have sex with each other before they get married. While the conclusion that Norcross defends in this chapter is as untraditional as any defended in this volume, however, the argument he offers in its defense is grounded in a quite traditional moral thought: that it's wrong to break a weighty promise and so wrong to make such a promise without first doing what one reasonably can to collect evidence about whether one will be able to keep the promise. Like the chapter by David Benatar that precedes it, then, Norcross's chapter can be read as maintaining that a component of traditional ethical thinking has implications that traditional ethical thinkers may be surprised, or even dismayed, by.

Questions about consent play a central role in a great deal of thinking about sexual ethics. Indeed, on one common view of the matter, they play virtually the only role. It's permissible to engage in any form of sexual activity, on this account, if and only if all the participants have validly consented to it. Part III therefore devotes a significant amount of space to issues involving sex and consent. It begins with two chapters about the nature of sexual consent itself. These are followed by three chapters that engage in one way or another with the question of whether consent is really all that's required for sexual interactions to be morally acceptable. This part of the book then concludes with four chapters that consider a variety of issues that can arise when consent is given under imperfect conditions, including cases where the person giving the consent might be deemed less than fully competent to do so, cases where one person might be seen as having undue influence over the person giving consent to have sex with them, and cases where a person has been deceived into consenting to sex.

In Chap. 15, Shaun Miller considers a very general question: what is the relationship between sexual consent and personal autonomy? Philosophers have sometimes distinguished between "thin" and "thick" accounts of personal autonomy, and Miller applies these categories to sexual autonomy in particular to see what they imply about sexual choices. These "thin" and "thick" accounts of sexual autonomy correspond to a "thin" and "thick" account of sexual consent. He begins by examining the "thin" accounts, what he calls procedural sexual autonomy and consensual minimalism, and considers the advantages and disadvantages of this position. He then examines the "thick" accounts: substantive sexual autonomy and consensual idealism, again considering advantages and disadvantages of the view. Finally, Miller attempts to navigate a middle ground between the "thin" and "thick" accounts with what he calls weak substantive sexual autonomy and consensual realism and concludes that this account better represents what sexual autonomy looks like and what counts as sexual consent.

Chapter 16 also considers a question about sexual consent in general: must a person's consent to sex be enthusiastic in order for their partner not to wrong them by having sex with them? In addressing this question, Tom Dougherty defends three substantive claims. First, that it's not the case that a person's consent to sex is fully valid only if they are eager, have a settled motivation for having sex, or endorse their motivation for having sex. Second, that to responsibly have sex with a person, that person must clearly communicate their consent to having sex. And third, that a person's consent to sex is fully valid only if they are not motivated by certain reasons. In addition to these matters of substance, Dougherty also defends a terminological claim: that we should not express Dougherty's second or third substantive claims by saying that consent is fully valid only if the consent is enthusiastic.

Chapters 17, 18, and 19 turn to the question of whether valid consent is enough to make a form of sexual activity morally acceptable. In Chap. 17, Alan Soble defends the claim that it is. He begins by clarifying what he calls the Principle of Informed Consent and by distinguishing between its two components: that valid agreement to sexual activity must be knowledgeable and that it must be given freely. He then defends the thesis that this principle provides not just necessary conditions for the morality of sexual activity but sufficient conditions as well. Soble considers an objection that might be raised against this view. The objection maintains that it might be wrong to have sex with someone even if they gave free and informed consent to having sex if the person's motive for having sex with the consenter is itself morally objectionable, such as a desire to humiliate the consenter. Soble acknowledges the potential significance of bad motives but argues that the Principle of Informed Consent can account for it by treating such motives as one of the things that should be included in the "knowledgeable" component of the principle. Not knowing that one's potential partner has a bad motive for proposing sex can invalidate one's consent to having sex with them, on Soble's account, but as long as the bad motive is disclosed, the bad motive itself can't make the sexual act wrong.

In Chap. 18, Elise Woodard offers a strikingly different response to the question of whether valid consent is enough to make a form of sexual activity morally acceptable. Indeed, she argues that there exists a broad range of cases in which sex can be both consensual and morally problematic. Woodard refers to cases that fall into this category as cases of "bad sex." And rather than trying to reconfigure our understanding of the nature of consent so that we could justify viewing such cases as cases of sex without true consent, she urges us to reconsider our assumptions about how much work it is reasonable for us to expect the mere fact of consent to sex to do in the first place. In addition, Woodard develops a useful typology of such cases, distinguishing between examples of bad sex that involve what she refers to as psychological pressure, social coercion, and epistemic risk. Finally, Woodard considers an objection on which at least some cases of bad sex should actually be treated as cases of rape. She responds by arguing that the considerations raised in support of this objection ultimately count against it.

In Chap. 19, Jordan Pascoe examines a different way that sex with consent and morally acceptable sex might come apart by exploring the relationship between consent and respect. Focusing specifically on cases involving sex between a man and a woman that take place under general conditions of patriarchy and misogyny, Pascoe offers a critical response to the view that the moral requirement to "respect women" can be fully satisfied simply by respecting a woman's "yes" as a yes and her "no" as a no. Drawing both on some feminist thought since the 1970s in general and on some feminist interpretations and developments of the thought of the philosopher Immanuel Kant in particular, Pascoe rejects this view as simplistic and instead develops and defends a view of sexual respect as requiring that one know and share one's partner's concrete and communicated sexual ends.

When one person coerces another into consenting to sex, it's clear that their consent isn't valid. The fact that the consenter said yes to having sex with the coercer doesn't make it okay for the coercer to have sex with them. But coercion isn't the only thing that can undermine the validity of sexual consent and Chaps. 20, 21, 22, and 23 each consider cases where consent is given voluntarily but where it can still seem unclear what we should say about it. One such case occurs when the person who says yes to sex is not a competent adult. This can happen either because the person is not yet an adult or because they are an adult but suffer from some kind of impairment. Chapter 20 considers a case of the first sort and Chap. 21 considers a case of the second sort.

Statutory rape occurs when at least one of the participants in a sexual act is post-pubescent but under the statutory age of consent. In many parts of the United States, and in many other parts of the world, this is not just a crime but a felony that brings with it a potentially lengthy prison sentence. In Chap. 20, Stephen Kershnar focuses on cases where the minors in question are no younger than 15 and considers the question of whether statutory rape in such cases should be a crime at all. His answer is no. Kershnar's argument for this position appeals to the following general claim: the government should punish an activity only if (1) the activity violates a moral right, (2) punishing someone for engaging in the activity satisfies a cost-benefit analysis, and (3) doing so satisfies a standard of intermediate scrutiny, where intermediate scrutiny permits the state to punish a type of act only if doing so directly advances a legitimate and important state interest and is the least restrictive alternative available to advancing that interest. Kershnar then presents a variety of reasons to doubt or deny that the practice of punishing people for committing acts of statutory rape satisfies all three of these conditions and concludes that statutory rape should not be a crime.

Dementia is a general term for the condition, most commonly associated with Alzheimer's disease and the aftermath of strokes, that involves a significant loss of memory and significant decline in various forms of cognitive functioning. People with dementia who want to have sex may be unable to provide consent to sex that meets the standards for valid consent that we tend to apply in the cases of adults who don't have dementia. This poses a problem. In Chap. 21, Andria Bianchi addresses this problem and considers whether, and if so, under what circumstances, it may be ethically permissible for people with dementia to have sex. Rejecting the view that their cognitive impairments make it morally impermissible for people with dementia to have sex, Bianchi instead proposes that we should enable people with dementia to pursue activities that promote their well-being, including sexual acts, even in some cases when they cannot clearly consent. In doing so, she defends a framework to apply to cases of sex and dementia that involves prioritizing considerations of well-being.

Cases of sexual consent that don't involve coercion can still generate a variety of problems even when they're limited to competent adults, and Chaps. 22 and 23 address two examples of such issues. In Chap. 22, David Boonin considers cases involving competent adults who are in a situation of positional inequality, focusing on a case where a psychologist asks her patient to have sex with her and where the patient says yes in response. In an important article called "Exploited Consent," David Archard argued that the patient's consent to sex in such cases should not be considered valid, appealing to the claim that patients are prone to develop feelings of deference, trust, and affection for their therapists, that these feelings would render them prone to say yes to their therapist's proposals when they might not otherwise do so, and that treating their consent as valid in such cases would therefore involve their therapist taking unfair advantage of them. Boonin offers a critical response to Archard's position, arguing that the considerations that Archard appeals to have unacceptable implications in other cases. He concludes that while there may well be good reason to prohibit sexual relationships between a psychologist and their patient, the claim that their professional relationship prevents the patient from giving valid consent to the sexual relationship is not among them.

Another kind of problem that can arise in cases of non-coerced consent among competent adults involves deception. In non-sexual contexts, it's generally agreed that deception can invalidate consent. If a customer says they'll only buy a used car if it has less than 20,000 miles on it and the salesman lies and says a particular car fits that description when it really doesn't, it's commonly accepted that if the customer is thereby tricked into agreeing to buy the car, they'd be entitled to get their money back because their agreement to buy the car wasn't really valid. But in sexual contexts, cases in which one adult deceives another into having sex with them are rarely treated as sex without valid consent, let alone as cases that might be understood as a form of rape. But if rape involves sex without valid consent and deceiving someone into consenting to something invalidates their consent to it, then why shouldn't sex that results from deception be viewed as on a par with, and perhaps simply a form of, rape? In Chap. 23, Laurie Shrage discusses this puzzle, sometimes referred to as "the problem of rape by fraud." In it, she investigates two proposals for reconceiving the moral purpose of anti-rape statutes: that rape violates a person's fundamental right to bodily possession or control and that rape violates a person's fundamental right to be treated with dignity and respect. She also considers an argument that defends the current consensus about the purpose of anti-rape law—that rape violates a person's fundamental right to sexual autonomy—but ultimately argues for a more pluralistic approach to anti-rape law, one that challenges the idea that there is just one basic human right that anti-rape statutes should protect.

Philosophers have become increasingly concerned in recent years with moral questions about discrimination and inclusion along such varying lines as race, gender, and disability status. While much of their work has focused on social and political issues, like those concerning police violence, racial gerrymandering, and access to health care, some has brought attention to concerns about discrimination and exclusion within the personal domain of sexual relationships. Part IV contains four chapters that provide examples of such work.

The first two chapters focus on race. In Chap. 24, Sonu Bedi extracts insights from Susan Okin's 1989 book *Justice, Gender and the Family* and uses them to help illuminate some fundamental questions about sexual racism. Unpacking the now familiar notion that "the personal is the political," Bedi first follows Okin in considering the various ways in which justice can be seen to apply not just to the public or political domain, but also within the more personal and intimate sphere of family relations. He then goes on to focus more specifically on the implications of racial justice, in particular, within the personal and intimate sphere of sexual relations in particular. Here Bedi distinguishes between two kinds of racial injustice that can arise—racial discrimination and racial stereotyping—and he shows how each corresponds to a distinct form of sexual racism. Drawing from and building on some of his previous work in this area, Bedi also considers the ways in which racism and sexism can intersect, making a suitably nuanced analysis of the various forms of injustice that can result from discrimination and stereotyping more complex and challenging.

Chapter 25 then dives more deeply into one particular question raised by the phenomenon of sexual racism: morally speaking, do people with racialized sexual preferences have the right to act on those preferences? Here, Cheryl Abbate considers an argument for the conclusion that the answer to this question is yes. The argument appeals to the claim that acting on such preferences doesn't harm other people and that even if it does, the harms can't count as wrongful harms because people can't really control what their sexual preferences are and they can't be held morally responsible for what they can't control. Abbate responds to this position by arguing that when white people act on anti-Black sexual preferences, they do in fact harm other people and that people can, at least to some degree, exert control over their sexual preferences. This leads Abbate to reject the argument under consideration and to conclude that white people have no moral right, not even a prima facie moral right, to act on anti-Black sexual preferences.

Chapter 26 also looks in detail at one particular question, but it's a question generated by a different form of sexual discrimination: the kind of discrimination against people with certain types of physical disabilities that leads them to have great difficulty in finding willing sexual partners and that is often referred to as sexual ableism. Some people have argued that hiring sex workers should be the primary means of mitigating the challenges posed by this form of discrimination. In this chapter, Kevin Mintz considers the advantages and limitations of the arguments for this view. He argues that sex workers can indeed play an important role in enabling some people with disabilities to have sex, but he concludes that focusing so heavily on access to sex work runs the risk of reinforcing sexual ableism. This is so, Mintz argues, because it can lead us to ignore important questions about how to empower people with impairments so that they can enjoy sexual freedoms that do not involve paying for sex.

This part of the book concludes with Chap. 27, which returns to a more general focus. In this chapter, Alida Liberman distinguishes between three kinds of sexual exclusion that are often conflated: (1) lack of access to sexual gratification or pleasure, (2) lack of access to partnered sex, and (3) lack of the kind of social or psychological validation that comes from being seen by others as a sexual being. Liberman offers proposals about what our responses to these harms should be and weighs in on debates about whether there are rights to various kinds of sexual goods. She concludes that we ought to provide mechanical assistance to those who are incapable of self-stimulation, enhance access to sexual education for everyone, and engage in a systematic effort to change the harmful social norms, stereotypes, and cultural ideals that drive exclusion from partnered sex and that can lead to social invalidation.

Part V brings this collection to a close with four chapters that, to one degree or other, connect issues in sexual ethics with some of the latest developments in digital technology. At a general level, technology can be good for people's sex lives and sexual relationships in a variety of ways. It can enable people to form connections they could not otherwise have formed, to maintain these connections over long distances, and to have experiences that are qualitatively new and different from those that were previously available to them. But the same technology can also present risks. It can exacerbate social divisions and inequalities. And it is vulnerable to various forms of restriction, control, and surveillance. In Chap. 28, Neil McArthur argues that we should welcome the fact that sexual technology has become central to the intimate lives of many people, and that we should equally welcome the development of new sexual technologies, such as sex robots and sexually explicit virtual reality environments. Such technology, McArthur argues, on balance increases both the overall happiness and the autonomy of its users. At the same time, though, McArthur argues that we must take an active role in managing the environment in which this technology operates, to ensure that the benefits are fully realized and are equally distributed, and to ensure that the technology is not ultimately controlled by governments and large corporations.

Several features of a typical college party environment can render it difficult to obtain meaningful consent to sex: loud music, for example, and vast amounts of alcohol. The situation can be made even more challenging by the existence of gendered double standards on which women have their autonomy disrespected in ways that men typically don't. In a so-called stoplight party, participants are supposed to indicate consent through the color of their clothing. Green means advance consent to sex, red means no consent, and yellow allows for uncertainty. Because binding sexual consent can't really be given in advance, these stoplight parties highlight either a deep confusion on the part of their participants about what consent requires or an insufficient concern on their part about securing consent. In Chap. 29, James Rocha argues that morality requires both avoiding such problematic party practices and developing better practices for obtaining consent. Connecting the concerns of this chapter to the latest developments in digital technology, Rocha explores the possibility that a specialized phone app designed for this purpose might help with this important work.

Doxing involves the deliberate public release of personally identifying information on the Internet by a third party, typically with the intent to humiliate, threaten, intimidate, or punish that person for some form of wrongdoing. In Chap. 30, Peter Brian Barry focuses specifically on the use of doxing to secure justice for the victims of sexual wrongdoing. Such doxing is often rejected as morally objectionable, but Barry defends the thesis that doxing sexual transgressors is sometimes a morally permissible response to their sexual misconduct. In doing so, he considers several rationales for the view that doxing sexual transgressors can be justified. These include the possibility that that doxing sexual transgressors is justified in virtue of warning prospective victims of sexual misconduct, and the related possibilities that sexual transgressors are liable to be punished in virtue of their misconduct and that doxing them can realize some of the goods that can be achieved by just and deserved punishment. Throughout the chapter, Barry also considers and responds to a variety of objections that can be raised against his position, and he offers some guidelines for identifying the kinds of circumstances in which the case for the permissibility of doxing sexual transgressors is particularly strong.

In 2002, John Corvino published an article called "Naughty Fantasies." In it, he considered the question of whether it's morally wrong to deliberately indulge in sexual fantasies that involve imagining doing things that it would be wrong to do in real life, things like raping someone or whipping a slave. While acknowledging some objections that can be raised against answering this question in the affirmative, Corvino's article tentatively endorsed the view that it's wrong to eroticize activities that are themselves wrong to engage in, that what he called "naughty fantasies" do eroticize activities that are themselves wrong to engage in (and don't, e.g., merely eroticize the simulation of such activities), and that it's therefore wrong to indulge in such fantasies. While all the other chapters in this collection consist of previously unpublished material, the volume concludes with a new edition of this article, one that Corvino has slightly modified to improve clarity, along with a new postscript that appears here for the first time. In this postscript, among other things, Corvino helps to connect the argument of his original article with one of the most disturbing developments in the area of digital sexual technology that has occurred since the paper's initial publication: sex robots with a so-called resistance setting that are specifically designed to help men act out the fantasy of raping a woman. The argument of Corvino's chapter, formulated before the creation of such devices, can be used to raise a potentially powerful objection to the production, distribution, and use of such devices.

The field of sexual ethics is large and rapidly expanding. Because of this, no single collection can credibly claim to offer a fully comprehensive representation of the kind of work currently being done in the area. And any editor will have their own views and blind spots about which are the questions most worth asking, approaches most worth pursuing, and positions most worth representing. So no collection can credibly pretend to impartiality and objectivity in its coverage, either. But what I hope this Handbook can do is provide a helpful and stimulating introduction to the field for those who are coming to it for the first time as well as a useful resource for those who have already been thinking in some detail about the questions it addresses. I hope you find the material contained in the chapters that follow to be provocative and challenging. And where you find gaps in the coverage, I hope the work contained here will inspire you to seek out equally strong work elsewhere that helps to fill those gaps or, even better, to create such work yourself. If this book can contribute something to the health and growth of the field in these ways, it will amply reward the efforts of the contributors whose thoughtful and dedicated work made it possible.

PART I

Sex and Human Nature

CHAPTER 2

The Metaphysical Foundations of Sexual Morality

Edward Feser

The Sexual Revolution has yielded the mainstreaming of extramarital sex, easy divorce, pornography, homosexuality, transgenderism, abortion, and other practices and attitudes diametrically at odds with traditional sexual morality. Earlier generations would have regarded this outcome as shockingly decadent, and many conservatives today share that judgment. By contrast, liberals and others sympathetic to these developments see in them a liberation from oppressive constraints, continued resistance to which is a sure mark of bigotry. Each side finds it difficult to credit the basic reasonableness and moral decency of the other.

I propose that the two sides are, implicitly if not always explicitly, operating from very different sets of basic metaphysical commitments and that their beliefs and attitudes about sex are perfectly intelligible given those commitments. Moreover, each set of commitments is defensible via arguments that contemporary philosophers could regard as respectable. To be sure, I am not saying that both sets of commitments are at the end of the day equally plausible. They are too far apart for that to be likely. But laying bare the nature and grounds of these different metaphysical pictures can at least facilitate understanding and rational debate between proponents of the Sexual Revolution and defenders of traditional sexual morality.

In particular, and borrowing Wilfrid Sellars's famous distinction, I would argue that these different views about sex reflect different estimations of "the manifest image of man-in-the-world" relative to "the scientific image of man."[1] The manifest image is the picture of the natural world and of the place of

E. Feser (✉)
Pasadena City College, Pasadena, CA, USA

D. Boonin (ed.), *The Palgrave Handbook of Sexual Ethics*,
https://doi.org/10.1007/978-3-030-87786-6_2

human beings within it that is afforded by everyday experience and common sense. The scientific image is the picture of nature and of human beings provided by modern science. What is traditionally (and by Sellars too) called the "perennial philosophy" is the style of metaphysics that upholds the essential soundness of the manifest image. It maintains that the manifest image is essentially correct as far as it goes, even if it doesn't go deep and gets certain details wrong. It is represented by views like Aristotelianism, "ordinary language" philosophy, and other examples of what P. F. Strawson called "descriptive metaphysics."[2] A very different tradition in the history of philosophy holds that the manifest image is largely or even wholly wrong and ought to be replaced by the scientific image. It is represented by ancient atomism, contemporary philosophical naturalism, and other examples of what Strawson called "revisionary metaphysics."

I am proposing that the more inclined one is to defend traditional views about sex, the more likely one is to be committed to the soundness of the manifest image, and that the more inclined one is to reject traditional views about sex, the more likely one is to favor a revisionary metaphysics that takes the manifest image to have been undermined by the scientific image. I do not claim that these are exceptionless correlations, only rough ones. More importantly, I claim that each of these views about sexual morality is *understandable* and indeed *reasonable*, given the metaphysical pictures each is correlated with. The clash between these views about sex is not a mere war of prejudices or intuitions, but something philosophically deeper and more interesting.

The best-known example of a defense of traditional sexual morality that reflects a commitment to the basic soundness of the manifest image is the one associated with the natural law theory of Thomas Aquinas and the Thomist tradition. The version of the "perennial philosophy" that Thomistic natural law theory regards as the best articulation of the metaphysics of the manifest image is Aristotelianism. The best-known example of a view that rejects traditional sexual morality as oppressive and bigoted is the liberal individualist position that whatever is agreed to by consenting adults is permissible. At least implicit in the typical expression of this view is the idea that the badness that traditional sexual morality purports to see in certain consensual sexual behaviors (such as extramarital sex or homosexual behavior) is not to be found in the scientific image of human beings.

I say that the two views reflect different metaphysical visions, but I don't mean to suggest that those who endorse the liberal position are all self-consciously committed to some positive metaphysical picture, in the way that natural law theorists are. I think that those who endorse the liberal position typically regard both its moral and metaphysical aspects as merely *negative* theses, to the effect that neither traditional sexual morality nor any metaphysical theory that seeks to make it intelligible has been given a convincing rational justification. However, *that* charge is in turn typically grounded in the idea that the metaphysics underlying natural law theory and other traditionalist moral argumentation does not sit well with the scientific image. Hence the liberal position is at least *implicitly* committed to a positive metaphysical thesis after all—namely the thesis that the scientific image tells us the whole truth about human beings, or at least provides the standard by which all metaphysical claims ought to be judged.

There are defenses of traditional sexual morality other than the Thomistic natural law approach, and criticisms of traditional sexual morality other than the liberal individualist one.[3] But I take natural law and liberal individualism to be the paradigmatic illustrations of my thesis, so I will develop it in terms of them. In what follows, I will first provide an exposition of the Thomistic natural law account of sexual morality and its foundation in an Aristotelian articulation of the manifest image. Then I will discuss how the scientific image—or, to be more precise, a certain metaphysical interpretation of the scientific image—displaced Aristotelianism beginning in the early modern period, and how increasingly liberal attitudes about sex were a natural consequence of this displacement. I end with some remarks about how a recognition of the divergent metaphysical assumptions underlying traditionalist and liberationist views about sex can allow their proponents to engage one another in a more intellectually fruitful way.

Natural Goodness

It is a commonplace that Thomistic natural law theory takes the content and imperative force of morality to be grounded in human nature. But less widely understood is the specific *conception* of nature (whether the nature of human beings or of anything else) that it operates with—a conception in terms of a metaphysics we might call *teleological essentialism*.

The label is fancy, but the basic idea is common sense. To borrow some examples from Steven Jensen, consider beavers and trees.[4] Beavers gnaw at tree trunks in order to get the trees to fall down. The point of doing that is to get the raw materials they need in order to construct their dams, and the point of the dams is to provide a stable environment in which to construct the lodges that beaver families live in. Beavers' teeth have a composition, structure, and growth pattern that facilitates this heavy duty use, and presumably they find pleasant or satisfying the use of their teeth for this purpose. Trees, meanwhile, produce nuts in order to reproduce themselves, sink roots in order to anchor themselves and draw in water and minerals, and grow leaves in order to carry out photosynthesis and bark in order to conserve water and to protect themselves against insects and other sources of damage.

Now, common sense would say that what we see here are purposes to which these organisms are as a matter of objective fact directed by nature. It is simply of the nature of trees that they produce nuts, sink roots, and grow leaves and bark and that these organs exist for purposes like reproduction, protection, taking in and conserving water, and producing food. It is simply of the nature of beavers that they like to use their teeth to gnaw on trees and that the point of their doing so is to facilitate their building dams and lodges and thereby sheltering their families. The thesis that natural kinds like trees and beavers have real essences or natures, essences or natures that we discover rather than invent, is the "essentialism" part of teleological essentialism. The thesis that having such a nature or essence involves a no less objective directedness toward distinctive ends or purposes is the "teleological" part.

This conception of nature and its claim to be capturing common sense go back at least to Aristotle's *Physics*. Aristotle says there that a thing that is natural in this sense "contains *within itself* a source of change and of stability" and in this respect differs from the products of artifice.[5] A watch is directed toward the end of telling time, and its changes (such as the movements of the hands across the watch face) and stability (such as the way that the hour hand will reliably point to "3" when it is three o'clock) reflect that end. But the end is not *intrinsic* to this particular kind of object. Human designers and users impose it from outside. By contrast, that a tree exhibits its characteristic changes (such as growing new leaves) and stability (such as reliably growing *leaves*, specifically, rather than something else) reflect an end that *is* intrinsic to it. Human beings have to do something to bits of metal to make them act in a watch-like way, but we don't have to do anything to a tree in order to make it act in a tree-like way. It does so *naturally*, just by virtue of being a tree, whereas the bits of metal function as a watch only *artificially*.

That there are things that are natural in this sense is just obvious, Aristotle says, and not in need of argument:

> It would be absurd, however, to try to *prove* that nature exists, since it is evident that there do exist many things of this sort. To rely on the non-obvious to establish the obvious is a sign of being incapable of distinguishing between what is and what is not intelligible in itself. … Inevitably, then, people in this kind of situation argue only at the verbal level, but do not understand anything.[6]

I would suggest that what Aristotle is saying here is that there is a kind of incoherence in denying the reality of things that are natural in the sense in question. For if *nothing* is natural in this sense, then everything is artificial. Yet the artificial itself presupposes the natural. For example, you can make watches (which have *extrinsically* imposed tendencies toward the changes and stability characteristic of watches) only because there already exists metal (which has an *intrinsic* tendency toward the changes and stability characteristic of metals). Things can be non-natural or artificial only because there are, more fundamentally, other things that are not—things that are just naturally a certain way. A skeptic might *verbally* contend otherwise, but his words would be without sense. The reality of nature in this sense is thus simply a bedrock part of the commonsense or manifest image of the world—and so too is the reality of essences and teleology, in terms of which the notion of nature is to be elaborated.

Common sense also takes the normative notion of *dysfunction* to be a corollary of the reality of essences and teleology, and takes the notions of proper function and dysfunction to yield in turn an objective standard of goodness and badness. For example, if some genetic or nutritional problem kept a particular beaver from developing enamel strong enough to cut into wood, we would say it had bad teeth, whereas if it had no such problems and thus enamel and other dental features typical of a healthy beaver, we would say that it had good teeth. By the same token, if a beaver for some reason had no inclination to gnaw on trees or build a dam or a lodge, we would say that there is something wrong with it. The goodness or badness has to do with whether an organ or behavioral disposition facilitates or frustrates the realization of the ends toward which beavers are by nature directed.

So far I am not yet talking about moral goodness or badness. There is nothing immoral about a beaver with bad teeth or even a lazy beaver, because beavers lack intelligence and free will and therefore can neither understand nor choose what is good for them. They simply either do it or do not do it. All the same, it remains objectively true that a beaver with weak enamel or lacking a disposition to gnaw trees is a dysfunctional beaver, and in that sense a bad beaver. That is to say, it is a bad or defective specimen of its kind. The normative notion operative here is what Philippa Foot has called "natural goodness."[7] Along with teleology and essentialism, it is part of the metaphysical inheritance from Plato and Aristotle that Aquinas and other natural law thinkers build on. But again, it is a metaphysics that merely articulates and systematizes common sense.

Now a few points of clarification to forestall some common misunderstandings. Note first that nothing I have said implies that an organ or behavioral disposition has only a single natural end or purpose. Naturally, a beaver's teeth have the function of chewing its food alongside the function of gnawing at trees. They also have both proximate and remote ends. The teeth can be said to have the proximate function of gnawing at trees, but also the remote function of felling the trees, and the even more remote function of facilitating the acquisition of materials by which to construct dams and lodges. It doesn't follow, however, that anything goes. There is no sense in which a beaver's teeth have the function of breaking off or of tapping out a little sequence of noises. Of course, sometimes they will in fact happen to break off, but that is not what the teeth are *for*, and indeed is positively contrary to what they are for. Tapping out a little sequence of noises is not positively contrary to what they are for, but neither does it have anything to do with what they are for. If a beaver happens to do that, that is incidental to the proper functioning of the teeth, rather than constitutive of it.

This example illustrates another point: not every trait or action that is *other than* the natural function of an organ or of a behavioral disposition counts as dysfunctional or bad, because not every such trait or action is *contrary to* that function. If some beaver for whatever reason developed a tendency occasionally to tap out a little noise with its teeth, this would not constitute a defect if it in no way frustrated the purposes of the teeth. It would not be comparable to the breaking off of the teeth, or to a disposition to avoid gnawing wood and eating, both of which would be contrary to those purposes.

These examples illustrate a third point: the fact that some trait or behavioral disposition is natural to a kind of thing does not entail that it will always in fact manifest, because a thing can be a defective instance of its kind. Aquinas distinguishes between the essence of a thing and the properties or proper accidents that flow from its essence. One of his stock examples of a proper accident of human beings is the capacity for laughter, which follows from our nature as rational animals. Though we all have this capacity, we don't all manifest it insofar as the manifestation can be blocked by a number of factors, from severe depression to brain damage. Similarly, a particular beaver might, due to injury or genetic defect, lack teeth or a disposition to gnaw trees. That does not entail that these traits are not, after all, proper accidents flowing from the nature of beavers as such, but only that the flow has in such an instance been blocked.

These points should elucidate what a natural law thinker like Aquinas means when he uses a term like "natural" in a normative way. What is natural in the normative sense is what is conducive to the realization of the ends toward which a thing is directed given its nature. What is unnatural or contrary to nature (*contra naturam*) is what frustrates the realization of those ends. Hence, simply being part of the natural world, or occurring in accordance with the laws of nature, does not suffice to make something natural in the normative sense. Deformations, injuries, and the like are part of the natural world and they occur in accordance with laws of nature, but they nevertheless make an organism dysfunctional and are therefore unnatural in the relevant sense. Furthermore, being "artificial" or man-made does not suffice to make something *un*natural in the normative sense, since a thing can be artificial and still consistent with the realization of a natural end. Indeed, it may even facilitate the realization of a natural end, either by correcting a dysfunction (as eyeglasses do) or by enhancing a natural function (as binoculars do).

Hence if some beaver liked to tap out sequences of noises with its teeth, this would not be unnatural in Aquinas's sense, as long as this tendency did not frustrate functions such as eating and gnawing at tree trunks. Similarly, if a beaver broke its leg and somehow figured out a way to put a splint on the leg in order to help it return to its normal activities, the fact that a splint is an artificial device would not make it unnatural in the relevant sense. On the contrary, it would be facilitating rather than frustrating natural function. However, if a beaver instead had a disposition to gnaw only at rocks and never at tree trunks, this would be unnatural, and it would also be contrary to nature for it to put splints on its legs in a way that would keep them from functioning as legs typically do.

Even if the inclination toward such odd behaviors were very deep-seated, and even if the beaver took great pleasure in them, this would not make these behaviors any less unnatural or dysfunctional. On the contrary, it would make them even more so. Behavioral inclinations and pleasure have natural functions no less than bodily organs do. An inclination to chew exists in beavers precisely in order to get them to chew tree trunks and food, specifically, not rocks. A tendency to take pleasure in chewing exists in beavers precisely to make it more likely that they will chew tree trunks and food, specifically, not rocks. If a beaver that was not able to chew tree trunks or food would be dysfunctional, a beaver that did not even want to do so, and took pleasure in things that prevented it from doing so, would be even more dysfunctional.

Rational Social Animals

For the natural law theorist, what is true of beavers and of trees is no less true of us. We have a nature or essence by virtue of which we are directed toward certain ends, and what is good for us is the realization of those ends. Now, our nature is that of a *rational animal*, where our animality is of the *social* kind. Like other animals, we take in nutrients, go through growth stages, reproduce ourselves, possess sensory organs and conscious awareness, have appetites inclining us toward what is pleasurable and away from what is painful, and can move about and interact with the world through our bodies. Like other social animals, we are highly dependent on one another. Unlike other animals, we can form concepts, put them together into propositions, and reason logically from one proposition to another.

These distinctively rational capacities transform our animal capacities. Our perceptual experiences, unlike those of other animals, possess a conceptual content that puts them into what Sellars called "the logical space of reasons" and thereby connects them up with the most abstract of our thoughts.[8] The signaling capacity other animals have is in us transformed into languages capable of expressing those thoughts and allowing us all thereby to tap into a common body of ideas and knowledge. Like other animals, we eat and copulate, but unlike other animals we attach norms and cultural conventions to these activities—culinary principles distinguishing a well-prepared meal from a poorly prepared one, rules of etiquette governing how a meal should be shared and consumed, romantic idealizations of sexual partners and the sexual act, wooing customs, and so on. Like other animals, we form social units, but unlike other animals we conceptualize them as families, religious congregations, ethnic groups, political orders, clubs, corporations, and so on, and either explicitly or implicitly expect other members of these units to share certain purposes and follow certain rules. Our rationality is thereby not merely superimposed on our animality but seeps down into and permeates it. It is also manifested, and most clearly, in practices that have no echo at all in the lives of other animals—in science, philosophy, religion, literature, music, and high culture generally.

Aquinas identifies three general classes of goods that follow upon our having this nature.[9] The first comprise those entailed by the need for self-preservation, which we have in common with all living things. Given our particular physiology, these include food, shelter, clothing, and so on. The second are those that follow upon our nature as animals, specifically. Significantly, the examples Aquinas gives are sexual intercourse and child-rearing. The third are those goods that follow upon the fact that we are animals of a distinctively *rational* kind. The examples Aquinas gives are participating in social orders like the ones referred to above, acquiring knowledge, and coming to know God in particular.

That ends like these are good for us and what is contrary to their realization bad is, natural law theory holds, as much a matter of objective fact as the facts that sinking roots is good for trees and building dams is good for beavers. It simply follows from the kind of thing we are. Moreover, on Aquinas's analysis, the intellect has as its natural end the knowledge of what is true and good, and the will has as its natural end the pursuit of what the intellect takes to be true and good. An intellect beholden to error and a will directed at what is in fact bad for us are as dysfunctional as beaver teeth incapable of gnawing tree trunks, or as a behavioral disposition in a beaver to gnaw rocks rather than tree trunks. For the Thomistic natural law theorist, pursuing what is true and good is therefore *constitutive* of rationality, so that skeptical doubts about whether we should care about what is true or about doing what is good are ultimately incoherent. The very entertaining of such skepticism involves taking skepticism itself to be true and adherence to it to be good. The skeptic who tries to use reason to cast doubt on the imperative to pursue the true and the good is like someone who tries to play chess without following the rules constitutive of chess.

It is with rational animals that the "natural goodness" we have been discussing becomes a distinctively *moral* kind of goodness. Moral goodness is simply the goodness characteristic of a properly functioning rational animal, manifest when the intellect grasps what is in fact good for an animal of that kind and the will aims at realizing it. Moral badness is a kind of dysfunction, manifest when a rational animal knowingly and willingly pursues what is contrary to what is good for it.

As this account indicates, morality for Thomistic natural law theory, as for Aristotelian ethics more generally, is about the self-perfection of the moral agent. Yet by no means does this exclude a social dimension to morality. On the contrary, precisely because human moral agents are animals of the social kind, realizing their own good as individuals is in part a matter of playing certain roles relative to each other and relative to the larger social orders of which they are parts—just as the health of a body part requires the health of the whole organism of which it is a part and requires that it play the role assigned to it in contributing to the health of the whole.

As Alasdair MacIntyre has emphasized, our social nature is reflected in part in the *dependence* on one another that follows upon our needing material provision and preparation for life while we are children, and our being subject to illness, injury, disability, economic distress, and the like even after we are adults.[10] But our social nature is manifest even when we are fully mature and healthy, and even in those aspects of our nature that are most distinctively human. For example, it is a commonplace of modern philosophy of language that linguistic and conceptual competences have an irreducibly social dimension,[11] and a commonplace of post-Kuhnian philosophy of science that science is an essentially social activity (acknowledgment of which, of course, does not require agreement with everything else said by Kuhn or post-Kuhnian philosophers of science).

The Essence and Teleology of Sex

This has all been foreplay, as it were, to a treatment of the natural law account of sex, so let's get to that at last. The fundamental way in which we are social animals is by being *familial* animals. And sex—both in the sense of there being two sexes, and in the sense of the sexual act—exists for the purpose of creating new families. In particular, the distinctive physiology and psychology of male human beings exists for the sake of making them fathers, and the distinctive physiology and psychology of female human beings exists for the sake of making them mothers. Of course, not all men and women actually become fathers and mothers, but the point is that that is what their being either men or women in the first place is *for*. If we did not reproduce in a way that required fathers and mothers, there would be no males and no females. Hence there would be no sex organs, no sexual arousal, and no sexual act.

Now, the most obvious respect in which sex has this teleology is that male sexual physiology and arousal have the biological function of getting semen into the vagina, whereas female sexual physiology and arousal have the biological function of facilitating reception of the semen, so as to get the sperm it contains into proximity with an ovum, so that pregnancy will result. It is often assumed that getting this plumbing right is the main concern of the natural law theorist. Nothing could be further from the truth. To be sure, the natural law theorist does insist on getting the plumbing right, but that is because the plumbing ultimately exists for the sake of something more important (just like beaver teeth ultimately exist for the sake of building shelters for beaver families, gnawing trees being an essential means to that end).

The *locus classicus* for Aquinas's treatment of these matters is the discussion in *Summa Contra Gentiles*, Book III, Part 2, Sections 122–126. There is a little bit there about emissions of semen and the like, but there is much, much more about what children and mothers need in order for family life to be possible, and how fathers have to provide it. That is to say, Aquinas's treatment of what it is to be a man or a woman goes well beyond having sex organs of a certain kind and using them in a certain way, and that is exactly what we should expect given that, as I have said, we are *social* animals, and *rational* social animals. Sex is for making you a father or a mother, with *all* that that entails given our social and rational nature, and any deliberate use of sex that positively frustrates that end with *all* that it entails given our rational and social nature is as contrary to what is good for us as breaking off teeth or gnawing only rocks rather than trees is contrary to what is good for beavers.

Now, one way this might happen is when a man sleeps with a woman to whom he has not committed himself in the way definitive of marriage. For any children that result from such acts, and the woman too who becomes a mother as a result, will be left helpless by such a man. Aquinas emphasizes several respects in which this is so. First, mother and children are in need of material provision, yet especially when the children are young it is very difficult for the mother to supply this herself. Second, it is not only material provision children need, for they are rational animals and thus need education as well, which takes a long time. Third, they need not only maternal nurturing but paternal discipline. All of this is the work of many years and thus requires the stable commitment of being a husband. Providing all of these things is no less part of the role of father than emitting semen is, and thus it is toward the fulfillment of *this whole paternal/husbandly function* that a man's sexual faculties point. A woman's sexual faculties point toward the fulfillment of *the whole maternal/wifely function* that complements the paternal one. In this way, there is for the natural law theorist a natural teleological connection between sex, marriage, and child-rearing, rather than a merely conventional one.

When natural law theorists say that sex has a *procreative* function, then, they don't mean merely that it has the function of generating new animals, but that it has the function of generating new animals *of the social and rational type*, with the long-term commitment that that entails. The making of a new rational social animal isn't completed with birth, but only when children have matured to the point that they are capable of leaving home and beginning families of their own. To have sex is to carry out an action that has *all of that* as its teleology, just as for a beaver to gnaw at a tree is to carry out an action that has the sheltering of the beaver's family as its teleology. And in both cases, this larger teleological context determines what counts as healthy or dysfunctional (and thus good or bad) behavior.

Of course, sex is pleasurable, but the pleasure of sex has its own teleology, just as the pleasure the beaver takes in gnawing trees or eating nuts and the like does. In both cases, the end or point of the pleasure is to draw the animal toward carrying out the action with which the pleasure is associated. But here too, it is the *whole* teleological picture that must be kept in view, not just the sexual act considered in isolation. And here as in every other aspect of our animal nature, our social and rational nature gives new significance to what in a non-human animal might be mere pleasurable sensations. Hence the pleasure of sex has as its natural end the drawing of the rational animal toward fatherhood or motherhood and the family life that that entails. And that is why, in rational animals, sexual desire comes to be associated with romantic fantasy, idealization of the sexual partner, a disposition toward playfulness and affection, and so on. What cognitive scientists call "theory of mind" plays a crucial role as well, insofar as sexual desire typically involves not just a desire to sleep with another person but also a desire that the other person want the same and feel a similar attraction. The pleasure looked forward to is not the mere release of one's own bodily tension but rather a shared pleasure in an essentially interpersonal activity. The perceptual and affective components of sexual arousal and pleasure are, in human beings, fused with an irreducible conceptual element. As Aquinas writes, "the lower powers follow the motion of the higher if that motion is more intense (as we see that a man's whole body is inflamed and set in motion at the sight of a woman he loves)."[12]

Thomistic natural law theorists thus hold that in addition to its procreative end, the sexual act has a *unitive* end, but that this second end is subsidiary to the first insofar as it exists in order to facilitate the first. Aquinas notes that "the greatest friendship between husband and wife" can be produced by their commitment to a common domestic project together with "the act of fleshly union, which produces a certain gentle association even among beasts."[13] But it is *because* sex is for creating new families that it also happens in this way to facilitate a bond between spouses. If there were no such procreative end, there would not be two different sexes, and thus no sexual act, and thus none of the pleasure and gentle association the sexual act produces. In short, the procreative end provides the larger teleological context within which the unitive end must be understood.

Unnatural Sexuality

Now, it is for the purpose of facilitating this unitive end, so that the procreative end might in turn be fulfilled, that sex involves "the *greatest* of pleasures … [which] absorb the mind more than any others."[14] The upside of this is that sexual pleasure can function as a kind of superglue that bonds a man and woman together long enough for a new family to get started and retains enough strength to help maintain a stable bond even after the initial intensity of romantic passion has subsided. The downside is that, precisely because sexual pleasure is the most intense of pleasures, it has the greatest tendency to cloud reason. In particular, when we take pleasure in what is contrary to the teleology of sex, and especially when we become habituated in doing so, it becomes harder for us to acknowledge that teleology, and easier to engage in rationalizations that blind us to it. And this can corrupt reason in general, insofar as the very idea of a natural order of things that implies that indulgence of some pleasures is dysfunctional, and therefore bad, becomes hateful to us. Accordingly, Aquinas identifies what he calls "blindness of mind" as the chief of the "daughters" of lust or sexual vice, and argues that sexual vices more than any other tend to erode "prudence" or the capacity for practical reason.[15]

Hence, consider some of the behaviors and habits that natural law theory condemns as contrary to the natural teleology of sex. Fornication tends to bring children into the world outside of the stable two-parent family unit they need for their full maturation. Hence while an act of fornication is not *per se* contrary to the *proximate* end of the sexual act (the climax which brings both insemination and emotional bonding), it is contrary to its *ultimate* end (the creation and maintenance of a stable marital-cum-family unit). You might say that such an act is directed toward the right sort of object, but in the wrong sort of context. Homosexual acts, though, are not even directed toward the right sort of object and are on the natural law analysis contrary to the proximate end as well as the remote end. If the fornicator is like a beaver who gnaws on trees but doesn't build dams, the person acting on homosexual desire is like the beaver who gnaws on rocks instead of trees.

Now, repeated indulgence in and rationalization of fornication dulls the intellect's capacity to see the natural end of sex and the will's capacity to pursue it, making sexual pleasure an end in itself rather than a facilitator of a larger purpose. Repeated indulgence in and rationalization of homosexual desire has an even greater tendency to dull the intellect and will in these ways, since it is not even directed toward the right sort of object. The intense pleasure associated with such behaviors "superglues" the mind onto ends other than the natural one, hardening one's orientation in an unnatural direction, like a kind of psychological crippling. Aristotle compares habituated homosexual desire to the compulsion to eat dirt or other nonnutritive substances, a disorder known as pica.[16] Just as pica would be no less dysfunctional even if it turned out to have a genetic basis, so too, for the natural law theorist, homosexual desire would be no less dysfunctional even if *it* turned out to have a genetic basis. That would entail, not the absence of psychological dysfunction, but rather the presence of both psychological and genetic dysfunction.

As habituated and rationalized sexual vice becomes more widespread, it inevitably takes a toll on the stability of the family, as individuals no longer see it as the end for which sexual desire exists. Instead of seeking to restrain and reform disordered sexual desire in a way that will be conducive to strengthening the institution of the family, they seek to alter the institution of the family in a way that will be conducive to indulging whatever disordered sexual desires they happen to have. The tail comes to wag the dog. The natural order of things becomes harder to see and people become less willing to see it. Increasing numbers of children come to lack the stability and discipline provided by parents who sacrifice their short-term desires for the good of the family and are neither encouraged nor prepared to form such stable and self-sacrificing unions themselves. In these ways, sexual vice strikes deep at both our rational nature and our social nature, which is why traditional moralists took it so seriously.

Of course, this raises all sorts of questions. For example, why would the teleology of sex entail that homosexual acts are bad but that sex between a sterile man and woman is not? The answer is that the two cases are different in just the way that there is a difference between a beaver who wants to gnaw on rocks rather than wood and a beaver who gnaws on wood with teeth that are too worn out to accomplish the task. The one act positively *aims away from* the natural end, whereas the other *aims at* the natural end but is simply prevented by circumstances from fulfilling it. But my point here is not to spell out all the implications and nuances of the natural law account of sexual morality.[17] Rather, it is to explain how Thomistic natural law theory takes traditional sexual morality to follow naturally from a certain metaphysical picture of the world.

Sex and the Scientific Image

It will not have escaped the reader's notice that Aristotelian essentialism and teleology go absolutely to the core of this conception of sexual morality. But they are precisely the parts of that conception that the scientific image most directly challenges. Or to be more precise, they are the parts most directly challenged by the house philosophy of the founders of the scientific revolution—a philosophy known as the "mechanical philosophy" or the "mechanical world picture."[18]

The idea of the world as *mechanical* or *machine-like* was precisely a rejection of the Aristotelian distinction between nature and artifice. Again, for the Aristotelian, a watch is artificial rather than natural insofar as its parts have no *intrinsic* tendency to behave in a time-telling way. *Objectively* there are just inert bits of metal there, and that the bits together make up something called a "watch" is true relative only to human purposes. What that entails is that the distinctive *essence* and *teleology* of a watch are not discovered in nature, but rather invented by us. This is what distinguishes man-made things like watches from natural objects like beavers and trees.

But for the mechanical philosophy, the distinctive essences and teleology of those things too are not really in nature. What exists in nature are just basic particles in motion, comparable to the bits of metal that make up a watch. Some of these particles make up beavers, some make up trees, some make up watches, and some make up yet other things. But with beavers and trees no less than with watches, the distinctive essences and teleology of these things are not *intrinsic* to the particles, but imposed from outside. In particular, they are imposed by God understood as a kind of artificer standing outside of nature and putting purposes into systems of particles in the way that a watchmaker puts a time-telling function into bits of metal. Or they are imposed by the human mind understood as a Cartesian immaterial *res cogitans* standing outside of nature, which classifies systems of particles in terms of essences and teleology it has invented and projected onto the world, rather than discovered there.[19] Science, on this model, proceeds by ignoring these externally imposed essences and teleology (also known as formal and final causes in Aristotelian philosophy) and instead identifying the mathematically specifiable laws that govern systems of particles as they push and pull against one another, gravitationally attract each other, and so forth (a focus on revamped versions of what Aristotelians call material and efficient causes).

This picture has undergone various alterations over the centuries. Particles have become less particle-like and their interactions less crude than the push-pull model of the early versions of the mechanical philosophy. The model's reductionism has been softened insofar as few people now believe that the facts about natural phenomena can be entirely captured by a description making reference only to the motions and configurations of basic particles. God and *res cogitans* are no longer part of the story for most people working with the model. But what survives is the idea that there are no purposes intrinsic to the physical world, nor the sharply demarcated kinds into which common sense would classify the middle-sized objects of everyday experience. There are at most configurations stable enough falsely to appear *as if* they fell into sharply demarcated kinds, and outcomes regular enough (due to natural selection) to make it falsely appear *as if* nature was aiming at certain purposes. But the essences and teleology the manifest image attributes to nature, and of which "perennial philosophies" like Aristotelianism give a systematic metaphysical articulation, do not feature in the scientific image of the world, which has been hammered out in "mechanistic" terms.

Now, you could take this mechanistic scientific image of the world to amount to a simplifying model of nature that is useful for certain purposes (such as technological purposes), but which doesn't capture all there is to nature. That is how modern Aristotelians would regard it. But you might also take it to be a complete metaphysical picture of the world, or at least the best approximation we have to a complete picture. That is how metaphysical naturalists regard it. And if you take it this latter way, then you are bound to conclude that the metaphysical presuppositions of Thomistic natural law theory are just false. As a corollary, you are bound to conclude that the natural law defense of traditional sexual morality fails.

It is important to emphasize that this by no means entails that metaphysical naturalists cannot agree with the natural law theorist on many points of fact, even if they would interpret the facts in a different way. For example, they could agree that sex is naturally procreative *in the sense that* the sexual organs and sexual desire were molded by natural selection in the way they have been precisely because this facilitated procreation. They could hold, as a thesis in evolutionary psychology or social science, that marriage developed precisely as a way of ensuring that mothers and children had necessary resources they could not have otherwise acquired for most of human history. What naturalists will deny is that these facts have, by themselves, any normative significance. They do not reveal nature's purposes, because there are no purposes in nature. They are simply contingent empirical circumstances which we can factor in when deciding what actions and policies we should favor, given the purposes *we* happen to have.

Now, for the liberal individualist, since these purposes are as diverse as the individuals whose purposes they are, different individuals can reasonably assign different weights to these empirical circumstances. For example, for some individuals, the well-being of children ranks high enough and personal sexual and romantic satisfaction low enough that they will judge it best to refrain from fornication and divorce, given the empirical facts about the conduciveness of stable marriages to the well-being of children. For other individuals, the well-being of children and personal sexual and romantic satisfaction rank equally, so that encouraging stable marriage as an ideal while permitting fornication and divorce for those who find its demands onerous seems an acceptable trade-off. Yet other individuals may have no interest at all in children and see in sexuality only a means of personal fulfillment, judging its procreative biological function as of merely scientific rather than moral relevance.

The liberal individualist self thus relates to the natural world as described by the scientific image in something like the way the Cartesian *res cogitans* does, albeit most liberals are not metaphysical dualists. Purposes are *in individual selves*, not in the world. If good actions are those conducive to fulfilling purposes, then good actions will be as multifarious as individual purposes are. There is, contra the natural law theorist, no good for human beings *as such*. There is only what is good for *this* human being, what is good for *that* one, and so forth. More to the point, there is only what this human being or that one is willing to *consent* to, given the purposes he or she happens to have.[20]

Talking Past One Another

Failure to see how deep are the differences between the metaphysical assumptions underlying traditional and liberal views about sex often leads to both sides talking past one another. Those with conservative attitudes about sex cannot understand why liberals are not more impressed by the obviously heterosexual biological functions of the sex organs, by the statistics regarding the consequences of fatherlessness, and so on. Liberals cannot understand how conservatives can be unmoved by evidence for a genetic basis for homosexuality, by the sexual frustration of those not ready to marry or stuck in unhappy marriages, and so forth. Each side fails to see that what the other is impressed or unimpressed by is perfectly intelligible and even reasonable *given* their respective metaphysical assumptions.

Hence it is not surprising that appeals to social scientific considerations and moral intuitions are so ineffective in settling disputes about sexual morality. For each side interprets the social scientific considerations and moral intuitions in light of very different metaphysical commitments. It is only by drawing out and debating these deeper commitments that each side can understand the other, much less hope to convince the other.

The idea that the scientific image captures all there is to human beings and the natural world has so permeated intellectual life and Western culture more generally that conservatives should be less surprised than they seem to be that they keep losing political and cultural battles over sexual morality. They have tried to conduct the debate as far as possible without adverting to the most controversial metaphysical issues, but as their failures and my analysis alike suggest, this is a naïve strategy.

On the other side, despite their cultural strength, liberals need to see just how challengeable are their implicit metaphysical assumptions. Recent decades have seen a revival in academic philosophy of Aristotelian metaphysical ideas—including essentialism and teleology—largely among thinkers who have no natural law axe to grind (and indeed who might be horrified if they realized their work might lend itself to a revitalization of traditional natural law arguments).[21] As philosophers like Alex Rosenberg have argued, it is also very difficult to regard the scientific image as an exhaustive metaphysical picture of reality (or at least an approximation to that) without falling into a radically eliminativist position regarding the self, intentionality and semantic meaning, free will, moral value, and so on (that is to say, a position that proposes flatly *eliminating* these notions altogether from our picture of reality, rather than trying to accommodate them to the scientific image).[22] And while Rosenberg and like-minded writers are willing to bite this bullet, it is notoriously difficult to make such a position coherent. Certainly it is far from obvious that, if scientism entails eliminativism, we should embrace eliminativism rather than taking it to be a *reductio ad absurdum* of scientism.[23]

Though each side sometimes pretends otherwise, conservatives and liberals alike clearly take sex to be a very big deal. I am arguing that it is an even bigger deal than they realize, raising as it does the deepest metaphysical questions about the nature of human beings and the world they inhabit.[24]

Notes

1. Wilfrid Sellars, "Philosophy and the Scientific Image of Man," in *Science, Perception, and Reality* (Atascadero, CA: Ridgeview Publishing Company, 1991).
2. P. F. Strawson, *Individuals: An Essay in Descriptive Metaphysics* (London: Methuen, 1959), p. 10.
3. For example, Roger Scruton's defense of traditional sexual morality in *Sexual Desire* (New York: The Free Press, 1986) reflects a respect for the manifest image, but as philosophically articulated by way of phenomenology rather than Aristotelian metaphysics. Friedrich Engels' critique of traditional sexual morality in *The Origins of the Family, Private Property and the State* (London: Penguin Books, 2010) reflects the perspective of the scientific image, but as interpreted through Marxism rather than liberalism. To complicate things further, in *Sex and Social Justice* (Oxford: Oxford University Press, 1999), Martha Nussbaum is critical of traditional sexual morality, despite being sympathetic to a broadly Aristotelian point of view. Whereas in "Why Homosexuality is Abnormal" (in *The Monist* 67 (2): 251–283, 1984), Michael Levin is sympathetic to traditional sexual morality, despite approaching the question more from the point of view of the scientific image than that of the manifest image. Again, I claim only *rough* correlations between the moral and metaphysical views I'm describing, not exceptionless ones.
4. Steven J. Jensen, *Knowing the Natural Law* (Washington, D.C.: Catholic University of America Press, 2015), pp. 72–73.
5. Aristotle, *Physics*, translated by Robin Waterfield (Oxford: Oxford University Press, 1996), p. 33. Emphasis added.
6. Ibid., p. 34.
7. Philippa Foot, *Natural Goodness* (Oxford: Clarendon Press, 2001).
8. Wilfrid Sellars, "Empiricism and the Philosophy of Mind," in *Science, Perception, and Reality*, at p. 169.
9. *Summa Theologiae* I-II.94.2.
10. Alasdair MacIntyre, *Dependent Rational Animals* (Chicago and La Salle, IL: Open Court, 1999).

11. Cf. Wittgenstein's critique of the notion of a private language, Hilary Putnam on the division of linguistic labor, Tyler Burge's anti-individualism, Donald Davidson on radical interpretation, and so on.
12. Thomas Aquinas, *On Love and Charity: Readings from the "Commentary on the Sentences of Peter Lombard,"* translated by Peter A. Kwasniewski, Thomas Bolin, O.S.B., and Joseph Bolin (Washington, D.C.: Catholic University of America Press, 2008), p. 34.
13. Thomas Aquinas, *Summa Contra Gentiles, Book Three: Providence, Part II*, translated by Vernon J. Bourke (Notre Dame: University of Notre Dame Press, 1975), Chapter 123, at p. 148.
14. *Summa Theologiae* II-II.46.3, in *Summa Theologica*, in five volumes, translated by the Fathers of the English Dominican Province (New York: Benziger Bros, 1948). Emphasis added.
15. *Summa Theologiae* II-II.153.5 and II-II.53.6.
16. *Nicomachean Ethics* 1148b 15–19a 20.
17. I have said more along those lines in "In Defense of the Perverted Faculty Argument," in *Neo-Scholastic Essays* (South Bend: St. Augustine's Press, 2015).
18. The latter phrase is from Tim Crane, *The Mechanical Mind*, Third edition (London: Routledge, 2016), p. 1. Crane's expression may have been inspired by the title of E. J. Dijksterhuis's *The Mechanization of the World Picture* (Oxford: Clarendon Press, 1961).
19. In Descartes' version of the mechanical philosophy, reality consists of two radically different kinds of substance: the mind or *res cogitans* ("thinking substance"), and the material world or *res extensa* ("extended substance"). On this Cartesian dualist picture, the material world is pure geometrical extension, utterly devoid of any of the properties associated with thought or consciousness and also of any teleology or purpose. Meanwhile, the mind just is pure thought or consciousness, standing altogether outside the desiccated, mechanical material world and projecting onto it whatever qualitative features it thinks it sees there. The mind relates to the material world thus conceived as a "ghost in the machine" (in Gilbert Ryle's famous phrase).
20. I am aware that "the liberal individualist self" is a loaded phrase, but my focus here is less on liberals' conception of the *self* than on their implicit conception of *nature*—even if, as my remarks indicate, the two topics cannot be entirely disentangled. Hence for present purposes I put to one side disputes about the nature of the self between liberals and communitarian critics like Michael Sandel, which prescind from the specific metaphysical issues I'm addressing here. Cf. Sandel's *Liberalism and the Limits of Justice* (Cambridge: Cambridge University Press, 1982).
21. I survey these developments in "Natural Law Ethics and the Revival of Aristotelian Metaphysics," in Tom Angier, ed., *The Cambridge Companion to Natural Law Ethics* (Cambridge: Cambridge University Press, 2019).
22. Alex Rosenberg, *The Atheist's Guide to Reality* (New York: W.W. Norton and Company, 2011).
23. I have developed a systematic articulation and defense of Aristotelian-Thomistic metaphysics and a critique of scientism in *Scholastic Metaphysics: A Contemporary Introduction* (Heusenstamm: Editiones Scholasticae, 2014) and *Aristotle's Revenge: The Metaphysical Foundations of Physical and Biological Science* (Neunkirchen-Seelscheid: Editiones Scholasticae, 2019).
24. I thank David Boonin for helpful comments on an earlier draft of this paper.

CHAPTER 3

The Ethics of Sexual Pleasure

Raja Halwani

The discussion of sexual desire and sexual pleasure has been sporadic in philosophy. In this chapter, I explain the "sexual pleasure view" of sexual desire (SPV), which basically states that sexual desire aims at sexual pleasure.[1] I then defend the claim that the moral evaluation of sexual pleasure depends on the moral evaluation of the sexual activity on which the pleasure supervenes. Thus, talk of sexual pleasures being ethical or not *simpliciter* is mistaken. Instead, talk of the ethics of sexual pleasure should be understood as shorthand for talk of the ethics of sexual acts. Similarly, I will also argue that whether a sexual desire is morally problematic depends on which sexual activities it seeks, even though those sexual activities are sought for their pleasures. However, I will also argue that the type of sexual pleasures sought help us evaluate the character of the person who seeks them. I conclude by showing how the SPV and this understanding of the ethics of sexual pleasure solve a dilemma (of sorts) in sexual ethics that David Benatar has raised.[2]

Three crucial concepts in this discussion are sexual desire, sexual activity, and sexual pleasure. Because sexual activity can be motivated by nonsexual desires (such as the desire to make money), I focus on those sexual activities motivated by sexual desire in order to keep the discussion squarely tied to that of sexual pleasure.

R. Halwani (✉)
School of the Art Institute of Chicago, Chicago, IL, USA
e-mail: rhalwa@saic.edu

D. Boonin (ed.), *The Palgrave Handbook of Sexual Ethics*,
https://doi.org/10.1007/978-3-030-87786-6_3

The Sexual Pleasure View (SPV)[3]

The SPV claims that the goal of sexual desire is sexual pleasure. Roughly put, when *X* experiences sexual desire, *X* experiences the desire for sexual pleasure.[4] Stated a bit more technically, *X* experiences sexual desire if, and only if, *X* experiences the first-order desire for sexual pleasure.[5] So, if *X* sexually desires sex with *Y*, *X* desires the sexual pleasures *X* would experience (usually) through sexual activity with *Y*. "Sexual activity" is to be cashed in depending on the tastes of the particular desirer, and no other person need be involved. If *X* sexually desires to masturbate using a sex toy, *X* desires the sexual pleasures *X* would experience through using the sex toy.

Note two other positions that compete with the SPV: one is the "Sexual Activity View" (SAV) and the other is the "Sexual Pleasure and Activity View" (SPAV). On the former, sexual desire is for sexual activity, while on the latter sexual desire is for both sexual pleasure and sexual activity. While on the SPV sexual pleasure is the basic goal of sexual desire, and sexual activity is the (usual) means to the goal, on the SAV sexual activity is the basic goal of sexual desire, and sexual pleasure is at most a foreseeable or expected consequence of achieving the goal. The SAV seems false, however, because sexual activity as such would not be desired if it were to be devoid of pleasure. So either the SPV or the SPAV is the more plausible option. For the purposes of this essay, I will not decide between the two, though for ease of reference I will proceed only with the SPV.[6]

Because on the SPV sexual pleasure is usually desired through sexual activity broadly understood (meaning that the activity need not involve more than the desirer themselves), we could restate it as follows: *X* experiences sexual desire if, and only if, *X* experiences the first-order desire for sexual pleasure attained through sexual activity. Possible inventions of advanced machines—such as Nozick's experience machine[7]—would allow individuals to desire sexual pleasure without the sexual activity, as long as the simulation is perfect or nearly perfect. For this reason, I prefer to leave out "through sexual activity." But this should not affect our discussion.

A *general* support for the SPV is our experience of sexual desire: human beings experience it as desire for sexual activity and sexual pleasure, though it can have other goals, such as reproduction, and can occur in various contexts, such as marriage. This experience strongly supports the SPV. Moreover, because in human beings sexual desire is typically intentional or conceptual,[8] because there is tremendous variety in terms of how sexual desire is experienced, with sexual pleasure as their common denominator, this also strongly supports the claim that the basic goal of sexual desire is pleasure, not other things such as reproduction or the expression of love.[9]

The SPV claims that sexual desire is for sexual pleasure attained through sexual activity. This implies that sexual pleasures will differ depending on the activity that gives rise to them and that there are various types of sexual pleasures given the various types of sexual activities, such as the pleasures of vaginal intercourse, of anal intercourse, of providing oral sex, and of receiving oral sex. Moreover, the pleasures differ in token, not just in type: even though *X* desires oral sex, *X*'s sexual pleasures will differ depending on the person with whom *X* has oral sex and how. Whether type or token, the supervenience of pleasure on the activity gives the SPV an Aristotelian aspect, given Aristotle's view that pleasures supervene on their activities: "Pleasure completes the activity ... as a sort of consequent end, like the bloom on youths."[10] This does not mean that the pleasure occurs at the end or completion of the activity (though orgasm can in many cases be described this way), but is experienced as the activity is ongoing.

To better understand the SPV, we need to say more about sexual pleasure. Pleasure in general admits of various forms.[11] There is pleasure-as-sensation, a type of pleasure felt in the body and that has a specific bodily location (e.g., scratching an itch); pleasure-as-enjoyment, in which one enjoys an activity or an experience, and which, depending on the activity or experience, might or might not involve bodily sensations (e.g., enjoying a swim vs. a game of chess); pleasure-as-feeling, "the warm glow one feels after a major achievement or unexpected fortuitous change for the better";[12] and pleasure-as-attitude, being pleased at a fact or a state of affairs (e.g., being pleased that I was accepted to compete on a TV show).

Note three things. First, pleasure-as-sensation is not intentional in that it typically has no object, whereas the other three typically are. I focus on pleasure-as-sensation and pleasure-as-enjoyment because they are paradigmatic of sexual pleasure. Second, one can experience some but not all these pleasures. For instance, I can be pleased at the idea of having been accepted to compete on a TV show, even though the prospect fills me with dread which I also physically feel. I can enjoy a rigorous jog even though it is physically exhausting. Third, all these pleasures have their opposite pains: obviously we experience pains as physical sensations, but we can also find an activity painful (e.g., taking a difficult exam, being humiliated), we can feel pain after having done something bad (the opposite of glow—e.g., feeling disgusted with oneself after having done something revolting), and we can be pained at the idea of something (e.g., at the prospect of a meeting with a dreadful co-worker).

Orgasm is the most obvious example of a sexual pleasure-as-sensation, but it is clearly not the only one: for instance, people enjoy the sensations of body-to-body contact; the sight, taste, and smell of someone else's body (oral sex is a common example); and the sounds of pleasure of sexual partners. However, with the exception (perhaps) of the sensation of orgasm, no bodily sensations are as such sexually pleasurable or painful. Depending on the context, any bodily sensation can be experienced as sexually pleasurable or non-pleasurable, even painful, disgusting, revolting, and so on: the sensation of one's ear being licked by *X* (to whom one is attracted) feels pleasurable, but it feels non-pleasurable if licked by *Y* (to whom one is not attracted). Moreover, there are cases when one experiences pleasurable sexual bodily sensations but does not enjoy the activity. One might have sex with another and enjoy the sensations of it (especially orgasm), but, not finding the other sexually attractive, the experience is not enjoyable. And there are cases in which one enjoys the activity but feels revulsion afterward at what one has done. One might have sex with one's student, enjoy it as sensation and as activity, but afterward feel disgusted at oneself for having done so.

Because the SPV understands "sexual desire" in terms of "sexual pleasure," distinguishing between pleasure-as-sensation and pleasure-as-enjoyment allows us to understand how sexual desire is both special and not-so-special in regard to pleasure and to see the centrality of pleasure-as-sensation for sexual desire.

Pleasure can be, and often is, the motive for many of our nonsexual activities. Indeed, with leisurely activities, pleasure, usually pleasure-as-enjoyment, is a principal motive for undertaking them: going to the beach, hiking, watching a movie, going shopping, reading, having coffee with a friend, and so on. If we remove the prospect of pleasure, few reasons are left for undertaking these activities. Sexual pleasure is similar in this respect: like other types of pleasure, it is a principal motive for satisfying the desire. It is thus not so special in this regard.

What does make sexual desire special is that pleasure-as-sensation is a crucial component of the pleasures that satisfy the desire. Indeed, between pleasure-as-sensation and pleasure-as-enjoyment, it is the former that is more basic because—and setting aside love and other emotions—enjoying sex is usually dependent on enjoying the sensations of sex, and thus sex-as-enjoyment is parasitic on sex-as-sensation. Sexual bodily pleasures are the central component of sexual activity, and this sets sexual desire apart from other desires for activities whose principal motive is pleasure.[13]

Sexual desire, however, is special in another way. In its grip, we find those body parts that are normally off-putting to be highly enticing. Soble puts the point nicely when discussing oral sex: "Your mouth is latched onto an organ, be it penis/foreskin/testicles or labia/clitoris/vagina, that even after bathing and even if powdered and perfumed, within an hour starts to drip and become caked with pungent excretions and fluids; and while your mouth is latched onto that organ, your eyes and nose and lips are located inches, perhaps only a few millimeters, away from the most taboo orifice in the human body, drenched in its own secretions and excretions"—an orifice that is itself highly sexually desirable in sexual contexts (though more so to men than to women).[14] Thus, when experiencing sexual desire, we gladly, joyfully, *lustfully* engage in activities that involve some of the most off-putting areas of the human body—off-putting, that is, in nonsexual contexts—and the pleasures we derive from this activity can be tremendous. This means that not only is sexual pleasure intense when it is good, but as importantly, "displeasure or disgust during sex is at the other extreme. It is unbelievably bad."[15]

Thus, the pleasures of sex can be very enticing while its pains can be very off-putting.[16] And this is because of the pleasures-as-sensation: the usual presence of the sexual body parts of another human being in close physical proximity to one's own body can tip one from pleasure to disgust, depending on the context.

Thus, because of the importance of the pleasures-or-pains-as-sensation, sexual pleasure is a principal motive for undertaking sexual activities, such that without the prospect of such pleasure the activity is likely to not be undertaken, and with the prospect of sexual *dis*pleasure, any additional reasons for having sex might very well be quashed. I might be willing to have sex with a famous celebrity even if I do not expect it to be pleasurable if I can post pictures of my date on Facebook. But if I come to realize that the sex will be un-pleasurable or nauseating, the motive of boasting about the date on Facebook might very well be overridden entirely. Contrast this case with one in which the date with the celebrity involves only dinner, but I come to know that the celebrity is not only boring, but also infuriatingly shallow. I might still go through with it so as to post the pictures on Facebook.

Because we normally avoid contact with others' body parts, especially sexual ones, and because we normally find them off-putting, without the prospect of sexual pleasure, sexual activity would not only have neutral value but also negative value. Although we are willing to engage in many activities without the prospect of pleasure, be it work, social events, or even leisurely activities that, for some reason or other, we have to go through, sex is special: without the prospect of sexual pleasure, we are hard-pressed to go through with the act (this is part of what makes prostitution a difficult job and explains the element of desperation in resorting to it, the exceptions notwithstanding).

Thus, as tempting as it is to consider the goal of sexual desire to be sexual activity,[17] especially since it seems to cohere with popular views about sexual desire—that if someone is sexually excited then he or she wants to have sex—the main goal is sexual pleasure, be it sensation or enjoyment, *through* the activity. This is another Aristotelian aspect to the SPV: one desires an activity according to Aristotle because one takes it to be good; so whenever we claim that we desire an activity, period, it is short for desiring an activity because of its perceived goodness.[18] In the case of sexual desire, the good in question is the pleasure one derives from sexual activity.

We can then uphold the view that sexual desire is for sexual pleasure. The pleasure is usually enjoyment-cum-sensation, and it is not an entity detachable and separately attained from the activity. It supervenes on the activity, and it can come in various forms depending on the specific sexual activity, by which I mean not only the type of activity—solo masturbation, paired sexual acts, orgies, anonymous sex, voyeurism, and so on—but also the token activity: masturbating while fantasizing about sex with *Y*, having sex with *Y*, having an orgy with a soccer team, and so on. In this respect, sexual pleasure and the desires for it can be as various as individual tastes. What unifies them is sexual pleasure.

The Ethics of Sexual Pleasure

The SPV is not a moral view, but a metaphysical one, concerning the nature of sexual desire. It is Aristotelian in taking sexual pleasure to supervene on the sexual activity and in countenancing various types and tokens of sexual pleasure, depending on the type and token of the sexual activity.

This raises the question of whether there are right and wrong sexual pleasures, and, if so, whether it is the pleasures or the activities on which they supervene that are morally primary. One way to put this question is as follows: granted that some sexual desires are morally problematic (e.g., sexual desires for children), are they problematic because of the pleasures that they seek? Or are they problematic for the sexual activities that they seek?

One reason in support of the claim that they are problematic for the sexual activities that they seek is that because pleasure supervenes on activity, then whether a pleasure is right or wrong depends on whether the activity is right or wrong. Although this is the position that I accept, it is worth noting that it does not follow from the SPV. That is, even if it is true that sexual pleasure supervenes on sexual activity, it is compatible with this truth to claim that whether a pleasure is right or wrong is the primary unit of evaluation, and that sexual acts are secondary. Just because the pleasures supervene on the activities, it does not follow that the activities are morally more basic.

On the other hand, one reason in support of the view that it is pleasures that are morally more basic is that on the SPV sexual pleasure is the goal of sexual activity, which might make pleasure the primary focus of moral evaluation: if the goal is unworthy, then so is the activity. One problem with this reasoning is that the goal of sexual pleasure (or any other pleasure, for that matter) is not an ordinary goal in that it cannot be arrived at separately from sexual activity.[19] Except in science fiction cases, such as plugging into an experience machine, the goal of experiencing sexual pleasure can be attained only through engaging in sexual activity. Indeed, this is what makes it plausible to claim that the goal of sexual desire is both pleasure and activity. Nonetheless, science fiction cases press on us the possibility that it is the pleasures, not the activities, that are the primary target of evaluation.

To address the question of whether it is sexual pleasure or sexual activity that is the primary target of moral evaluation, let us use the example of the wrongful sexual activity with prepubescent girls or boys whose age ranges from, say, eight to 12 years old (the sex need not be intercourse, but can be fondling or rubbing through the clothes). When *X*, for instance, fondles a prepubescent boy or girl (henceforth "child" or "children"), let's say that *X* engages in pedophilic activities, and let's call the pleasures that *X* derives from such fondling pedophilic pleasure.

Might Aristotle's view about which pleasures the virtuous person experiences help settle the question of whether it is pleasure or activity that is the basic target of evaluation? Aristotle, after discussing in Book X of the *Nicomachean Ethics* how pleasures differ among human beings, and how "the same things delight some people, and cause pain to others," comments,

> But in all such cases it seems that what is really so is what appears so to the excellent person. If this is right ... and virtue, i.e., the good person insofar as he is good, is the measure of each thing, then what appear pleasures to him will also really be pleasures, and what is pleasant is what he really enjoys. And if what he

> finds objectionable appears pleasant to someone, that is not at all surprising; for human beings suffer many sorts of corruption and damage. It is not pleasant, however, except to these people in these conditions. Clearly, then, we should say that the pleasures agreed to be shameful are not pleasures at all, except to corrupted people.[20]

Aristotle is not claiming in this passage that the corrupt do not enjoy their activities—he would not deny that pedophiles experience pleasure from pedophilic activities—but that some pleasures are rightfully enjoyed, whereas others are not.[21]

Aristotle makes similar claims in his discussion of temperance, which is the virtue concerned with bodily pleasures, stating that the temperate person "finds no pleasure in what most pleases the intemperate person, but finds it disagreeable." The temperate person would also desire "anything else that is pleasant, if it is no obstacle to health and fitness, does not deviate from the fine, and does not exceed his means."[22] Again, we find the idea that the virtuous person (in this case specifically the temperate one) would not pursue or enjoy shameful pleasures; she would enjoy those pleasures that do not "deviate from the fine."

Aristotle's discussion of pleasure might imply that Aristotle is advocating a direct evaluation of pleasures, and not the activities on which they supervene. This conclusion, however, would be hasty, because his discussion does not tell us whether pleasures can be *directly* ethically evaluated or whether their evaluation depends on an evaluation of the activities on which they supervene. For what Aristotle says in these passages is that the virtuous person neither finds pleasurable nor enjoys wrongful pleasures, and this, on its own, does not answer the question of whether it is pleasures or activities that are ethically primary (though it does tell us about the character of the agent, as I note below).[23] Put differently, and using the virtuous person as our guide, does the virtuous person reject pedophilic pleasures because they are pedophilic *pleasures* or because they are the pleasures of pedophilic *activity*? Aristotle's discussion does not answer this question.

John Stuart Mill's discussion in Chapter 2 of *Utilitarianism* of higher and lower pleasures is more helpful. His test for determining which pleasures are higher and which are lower, as problematic as it is,[24] attempts to establish this hierarchy by resorting to the experience of those familiar with the pleasures, arguing that of two pleasures, if those who are familiar with both pleasures favor one over the other, then the favored one is the higher pleasure. This test seems to directly appeal to the pleasures themselves. Mill's proof, however, goes only so far. First, we need a way to individuate sexual pleasures so as to know what to ask: should we ask whether people prefer oral sex over sexual intercourse or prefer oral sex in particular positions over intercourse in particular positions? Second, and more importantly, even if we do agree on a way of individuating the pleasures, the results won't tell us much: even if, say, 80% of men claimed that they prefer pleasure with adolescents over adult men or women that would not imply that such pleasures are morally superior.[25]

The best way to test our intuitions is to try to isolate the pleasures to see whether any sense can be made of evaluating them as such or whether any such evaluation would have to rely on activities. Let us consider two types of case. The first is the case in which someone, *X*, plugs into an experience machine to experience pedophilic pleasures. This case involves no pedophilic activity (and no actual sexual activity of any kind), only its pleasures. The second is the case in which someone, *Y*, who experiences pedophilic pleasures by (a) playing video games in which *Y*'s character has sex with children; (b) watching computer-generated pornography involving children; or (c) having sex with a child sex robot—a robot made to look and act like a child.[26] Like the first case, none of these three variations in the second case involves sexual pedophilic activity, even if they all involve the sexual activity of masturbation (though [c] does raise the question about whether it is a case of masturbation or something else).

Let us introduce a distinction between these two types of cases having to do with the reasons as to why *X* and *Y* do not have sex with children. Let's use the subscript "m" for moral reasons and "p" for pragmatic reasons, such that X_m and Y_m do not have sex with children for moral reasons (e.g., to not harm the children), so they use alternative ways of acting on their desires.[27] On the other hand, X_p and Y_p do not have sex with children for pragmatic reasons, because they are, say, afraid of getting caught and its implications. So, if they were assured that no such bad consequences would befall them, they would have sex with children if the opportunity arose (while X_m and Y_m would not).

Although we are likely to have a range of intuitions about *X* and *Y*, all the intuitions will probably condemn their sexual desires. That is, we are likely to consider their sexual desires to be morally compromised or problematic, precisely because they are desires for sex with children. I will set aside the cases of X_p and Y_p because not only are their desires morally compromised, but also because they would act on them if they were to have the chance. This disposition to act on their desires and engage in pedophilic activities indicates that our moral evaluation of their desires is activity-based: we deem their desires problematic (partly) because, if given the chance, they would act on them. This means that their cases are not going to shed much light on whether their sexual desires are morally problematic because they are primarily for wrong pleasures or because they are primarily for wrong acts.

This is not so for X_m and Y_m: given their strong moral fiber, they would not act on their desires even if given the chance, so whatever negative intuitions we have about them need to be explained by something other than their disposition to engage in pedophilic activity. In this vein, note that even in these cases of refraining from acting on pedophilic desires for moral reasons our intuitions remain negative: something is morally off with these men, even though they are not engaging in any pedophilic activities.

Why do we continue to view the two men with moral disapproval, given that none of them engages in pedophilic activities? In the second kind of case, one can argue that even though Y_m does not harm any children, what Y_m does is still morally problematic because it involves sexual activities, albeit not with real children. But this just pushes the question one step back: granted that Y_m engages in sexual activities, these activities are *not* pedophilic,[28] though they might mimic such activities (as in [c] and maybe also in [a]). So what is the problem exactly with Y_m's sexual desires if they are not for pedophilic pleasures or activities?

The above questions and remarks seem to support the following argument for the conclusion that it is sexual pleasure, not sexual activity, that is the primary locus of evaluation: (1) each of X_m and Y_m has morally compromised sexual desires; (2) each of X_m and Y_m desires pedophilic pleasure; (3) yet neither X_m nor Y_m desires pedophilic activity; (4) therefore, the only reason why their desires are morally compromised is because of sexual pleasure. Does this argument succeed in showing that it is pleasure, not activity, that is the locus of moral evaluation?

Not quite. What the argument shows is that because X_m and Y_m understand that pedophilic activity is wrong, they opt for alternative means to satisfy their desires. If anything, the argument underscores the idea that it is the activity, not the pleasure, that is the problem. But then why do we continue to think badly of X_m and Y_m? Or are our intuitions about them misguided? There is no way to address this issue other than by isolating sexual pleasure, and this we can do by addressing the case of X_m, who plugs into the experience machine. Note, in this regard, what is interesting about this case: typically, the experience machine is used as a *premise* in an argument. For example, if hedonism about our well-being were correct,[29] then plugging into the experience machine should not be problematic. Since it is problematic, then hedonism is wrong. Plugging into the experience machine is a moral datum in the argument. Not so in the case of X_m, because the question is precisely whether there is anything problematic about plugging into the experience machine to experience simulated pedophilic pleasure. Indeed, and to continue with the comparison, the experience machine is a problem for hedonism precisely because it shows that there is more to life than experiencing pleasure—we want to experience "the real thing." This response is blocked in our case: experiencing the real thing (pedophilic sex) is *more* problematic than not experiencing it. So what, if anything, is wrong with X_m?

Sexual pleasure is usually, and for lack of a better term, *conceptual* pleasure[30]: it is accompanied by, caused by, or sustained by perceptions, thoughts, or images that arouse one and that eventually cause one to experience the pleasure. Unlike the pleasures of scratching an itch, sexual pleasure is usually experienced in mental terms. Consider someone having sex with another, or masturbating while watching pornography, or masturbating while fantasizing about sex with, say, her next door neighbor. The experience of pleasure is accompanied and causally sustained by perceptions and/or thoughts[31]: in the case of having sex, the perceptions are caused by visual and non-visual sensations of one's sex partner and their activity; in the case of watching pornography, the perceptions are caused by visual and aural imagery; and in the case of masturbating solely by fantasizing, the mental images are caused by fantasizing (or that constitute the fantasizing). In short, perceptions of sexual activity are crucial for sexual pleasure. Whether we think of sexual pleasure as caused or constituted by perceptions, there is usually no such thing as sexual pleasure without the sustenance of perceptions.[32]

In X_m's case, what is morally problematic is the content of X_m's thoughts. X_m's pleasure is the pleasure sustained by perceptions of having sex with children, and what is problematic about these perceptions is the fact that they are of sex with children. Pedophilic activity seems to be central to the moral evaluation of X_m's pleasures because the pleasures that he experiences are pervaded with, or constituted by, perceptions of pedophilic activity. Thus, even if X_m does not and would not have sex with children, the very pleasures that X_m experiences through simulated sex are constituted by perceptions of pedophilic activity. The moral evaluation of X_m's pleasure is dependent on the activity. The activity is therefore primary in our moral evaluation of the pleasure.[33] In short, the sexual pleasures are morally troubling because their content is morally troubling.

Before turning to Y_m, I wish to be clear about what this argument is meant to show. It is not meant to show that pedophilic pleasures (or activities) are morally wrong.[34] Instead, it is meant to show that *if* pedophilic pleasures are wrong, they are wrong because of their connection to pedophilic activities—that the evaluation of their wrongness is dependent on the wrongness of their activities. The locus of moral evaluation is the sexual activity, not the pleasures that supervene on it. The argument also indicates that even though sexual pleasure is the goal of sexual desire, such a goal is typically experienced either through sexual activity or through perceptions of sexual activity, so its moral evaluation remains tied to the evaluation of the activity.

The argument is also not meant to assume, let alone show, that we do have intuitions about the X_m's (and the Y_m's) of the world to the effect that they are morally problematic people. Indeed, those intuitions might not be universal, and when they exist, they might be because of deep-seated doubts about whether X_m (and Y_m) really would not succumb to temptation if given the chance. Instead, the argument is meant to show that if such intuitions exist, the best way to explain them is by appealing to the wrongness of sexual activity, not merely the pleasures.

Something similar can be said about Y_m: his pleasures are also not the primary target of evaluation because they, too, are constituted by perceptions caused by what Y_m perceives as he watches pornography, plays a video game, or has sex with a sex robot. The differences between X_m and Y_m lie not here, but in whether one is morally better than the other. Here, and briefly, it is tempting to lump Y_m with X_m on the grounds that his pleasures are also constituted by imagery of pedophilic activities. However, one nuance has to do with whether the content of Y_m's pleasure is the object itself (e.g., the sex robot) or whether his imagination transforms the object into an image of a real-life child so that the pleasures he experiences by having sex with the robot are experienced as sex with (an imagined) child. The answer is likely to depend on the individual him or herself. Still, two crucial issues to address here (for another essay and perhaps by someone else) are the following: first, suppose that Y_m does enjoy the sex robot or the anime child pornography (not an imagined child), would this mean that the content of his pleasures are morally off the hook? Second, what implications would an analysis here have for other sexual activities that depend on imaginative transformations, such as BDSM activities that are consensual but that nonetheless can involve some tainted role-playing (master/slave, rapist/victim)?[35]

Of course, additional support for the claim that it is sexual activity, not sexual pleasure, that is the primary target of moral evaluation comes from the moral relevance of consequences: it is the activity of rape and pedophilia, not the pleasures that the rapist and the pedophile derive, that causes serious harm to their victims. But such considerations are plausible in ruling in sexual activity as a target of evaluation. They are not plausible in ruling out pleasure as a non-basic target. The argument I offer does not rule out sexual pleasure as a target of evaluation, but it does rule it in only as a non-basic target of evaluation, because of its conceptual dependency on sexual activity.

If the argument is convincing, it also explains what is morally problematic about pedophilic desires, namely, that they dispose the agent to either engage in pedophilic activities or partake of sexual pleasures that rely on pedophilic sexual perceptions.

Before concluding this section, there is one intriguing thought regarding the moral primacy of sexual pleasures: even if sexual pleasure is not the basic target of moral evaluation, sexual pleasure is nonetheless crucial in the moral evaluation of the *person* with the desires. Indeed, people who are willing to rape or have sex with children are worse for engaging in morally wrong activities for no more than the pursuit of the pleasures that they obtain from it, much like carnivores eat meat simply for the pleasure of it, even if that comes at the expense of abetting an industry that inflicts massive cruelty on animals. Put differently, it is their willingness to pursue particular (types or tokens of) pleasures that morally indict their characters.

Consider Z_m: Z_m has pedophilic desires but does not, for moral reasons, act on them at all. Z_m does not plug into an experience machine, have sex with child sex robots, watch simulated child pornography, or play video games during which Z_m has sex with children, let alone have sex with actual children. Z_m experiences neither pedophilic activities nor pleasures. Is there anything problematic about him? It depends. If Z_m is *continent*, he would constantly struggle with his desires for pedophilic pleasures—he struggles between doing what is right and acting on a base desire. What tempts Z_m is, clearly, the prospect of pedophilic sex, but this activity is tempting precisely because of its pleasures. If Z_m is virtuous, then he would not struggle in this way, and there would be no reason for these pleasures to taint Z_m's character. In these ways, then, sexual pleasures might be crucial for the moral evaluation of the characters of the people with the desires for the pleasures.[36]

There is much more to be said about this issue: the role of activity and pleasure in the moral evaluation of character, the importance of pleasure in our pursuits, and a comparison of various pursuits of pleasure and the means that are employed. They are thoughts worth pursuing.

If the above is correct, then the SPV as a metaphysical view about the supervenience of sexual pleasure on sexual activity has a correlative moral view about the relationship between the two, namely, that the moral evaluation of sexual pleasure is dependent on the moral evaluation of sexual acts. Crucially, though for certain purposes we might want to speak of sexual pleasure, period, and sexual activity, period, a proper analysis must at some point account for the metaphysical and moral varieties of sexual activities and their pleasures.

Concluding Remarks

I conclude by using the SPV and its moral correlate to solve a dilemma that David Benatar raised for sexual ethics.

Benatar argues that there are two views of sexual ethics, each of which is inconsistent with some of our beliefs about some sexual acts. The *casual view* of sex considers sexual pleasure to be insignificant, on a par with other types of pleasures, such as the pleasures of eating or of playing a game of tennis. Because the pleasures of sex are insignificant, the casual view is able to explain why casual and promiscuous sex are morally permissible: if sexual pleasure may be enjoyed like other pleasures, subject to the usual moral restrictions, then casual sex and promiscuous sex should not be singled out for criticism, and we may enjoy them as we see fit, much like we enjoy the pleasures of eating as we see fit.[37] The casual view is then consistent with our beliefs that casual sex and promiscuity are permissible.

Benatar does not explicitly state that on the casual view sexual activity is permissible *because* sexual pleasure is insignificant, but this is the argument that can be gleaned from his discussion. For instance, "Sexual pleasure, according to this view, is morally like any other pleasure and may be enjoyed subject only to the usual moral constraints." After making an analogy with the insignificance of gastronomic pleasures, Benatar states, "Similarly, according to the casual view, erotic pleasures may permissibly be *obtained from sex* with strangers or mere acquaintances."[38] The wording indicates that on the casual view sexual activity is permissible *because* sexual pleasures are insignificant.

But the casual view, according to Benatar, is inconsistent with another set of beliefs we have about sexual acts: it is inconsistent with the belief that pedophilia is wrong and that rape is more wrong than other bodily violations. The idea is that if sexual pleasure is insignificant, then sex between an adult and a child should also be insignificant: "If sex is (morally) just like other pleasurable activities and bears no special significance, why may it not be enjoyed with children?"[39] Of course, children might be traumatized by such encounters, but this, argues Benatar, might be a function of the social taboos about adult-child sex rather than a function of the wrongness of the sex itself (and these taboos are to be explained by factors other than such wrongness).[40] If one were to argue that pedophilia is wrong because children cannot genuinely consent, then, Benatar argues, it is not clear why consent is necessary if the activity is insignificant. Indeed, parents often do not care about children's consent when they force them to do all sorts of things that might be good for them.[41]

For similar reasons, Benatar argues that the casual view of sex cannot adequately explain the special wrongness of rape. If sexual pleasure is insignificant, then rape is wrong inasmuch as forcing someone to eat something is. One exception would be the rape of someone who considers sex to be significant: the wrong of raping her would be as severe as forcing a vegetarian or a religious Muslim or Jew to eat pork.[42] For the analogy to be apt, the force in question cannot simply be putting a gun to the vegetarian's head and making her eat the meat; the enforcer would have to shove the meat down the vegetarian's throat (this is important, as we will see). Thus, according to Benatar, while the casual view of sex can explain the moral permissibility of casual sex and promiscuity, it cannot offer an adequate explanation of the wrongness of pedophilia and the special wrongness of rape.

On the other hand, the significance view of sex considers sexual activity to be "acceptable only if it reflects the reciprocal love and affection of the parties to that union."[43] It can explain the wrongness of pedophilia and the special wrongness of rape. Having sex with a child is using him or her as a mere means because the child "may lack the capacity to understand how sex expresses a kind of love," even if the child is sexually aroused.[44] Raping someone is forcing that person "to engage in an activity that should be an expression of deep affection."[45] The significance view, however, cannot adequately explain the moral permissibility of casual sex and promiscuity; indeed, it condemns them as wrong because they are love-less or affection-less.[46]

Benatar himself does not argue for either the casual or the significance view. His point is to highlight an inconsistency between each view and our beliefs about some sexual acts—to wit, if we adopt the significance view, we can explain why pedophilia and rape are wrong but cannot account for the permissibility of promiscuity and casual sex; if we adopt the casual view, we can explain the permissibility of promiscuity and casual sex but not the wrongness of pedophilia and the special wrongness of rape. The assumption, which I believe is true and which, at any rate, I do not argue against, is that it is implausible to bite the bullet on either view by accepting that pedophilia is not wrong and that rape is not that wrong or by accepting that promiscuous or causal sex are impermissible.

Although Benatar does not argue for either the casual or the significance view, the criticism I offer rejects a unitary view of the moral permissibility of sexual activity.[47] To see this clearly, keep in mind the SPV and Aristotle's view of pleasure as supervening on an activity and his remarks about the temperate person, the crucial idea being that when the pleasures are wrong the temperate person would not enjoy them by partaking in them. And when they do not deviate from what is "fine," the temperate person might enjoy them ("might" because she does not have to enjoy them). Given the view that pleasure supervenes on the activity, and given the argument from the previous section, bad and good pleasures depend on whether the activity on which they supervene is itself good or bad.

Thus, we should reject the very bases of each of the casual and the significance views of sex—we should reject their principal premises that sexual pleasures are insignificant and significant, respectively. Given the arguments of this chapter, whether a pleasure is significant or not depends on the activity on which it supervenes. This takes us directly, therefore, to the question of whether rape and pedophilia are wrong. If they can be shown to be wrong on independent grounds, then we can resolve the inconsistency between the casual view and our beliefs by rejecting the view of sexual pleasure on which it is based: there is no inconsistency because there is no unitary view of sexual pleasure as casual. Whether a sexual pleasure is significant or casual depends on the type (or even token) of the act itself, not on whether sexual pleasures are, as such, significant or insignificant.

And surely rape is wrong for reasons having nothing to do with the pleasures and pains of sex in general. Much has been written on the wrongness of rape and how it is, as a violation of one's body, quite bad, and I won't repeat them here. But I want to emphasize two points. First, unlike having, by sheer accident, a metal stick pierce one's shoulder or a large wasp fly into one's mouth, events that are in themselves experienced as traumatic (to various degrees, depending on the event), someone using his genitalia to penetrate his victim is of a different order altogether. The close bodily proximity, the invasion of one's private and immediate space, the penetration by a person's genitalia, and the intentional use of force to do all this—four elements that are typical of cases of rape, though rape can occur with at least only one of them—cause the experience of the act to be worse than accidental events. Second, it is usually women who are the victims of rape, and it is not farfetched to argue that because of evolutionary biology, women experience rape in worse ways than men—that is, because when it comes to matters sexual men and women's psychologies generally differ from each other, with this difference explained crucially, though perhaps partially, in terms of evolution, women experience rape as very serious violation.[48] We can then agree that rape is a serious wrong and hence its pleasures and pains are significant indeed.

Similarly, we can argue that pedophilia is wrong and hence its pleasures are quite serious. I won't enter into such arguments because the philosophical debate is more contentious in that there is a question as to whether the harms of pedophilia are fully or partly a function of social attitudes. But the point remains that pedophilia's wrongness or rightness does not depend on the general significance (or lack thereof) of sexual pleasure, but on the type of act it is. Indeed, even if, *per impossibile*, *no* type of sexual activity turns out to be morally wrong, this would not be because sexual pleasure, as such, is insignificant, but because each type of sexual activity, after argument, turns out to be not wrong. And on the basis of those conclusions, we can further conclude, if we wish, that sexual pleasures are insignificant. Thus, the dilemma that Benatar raises for these two views of sex seems to rely on claims that should be the *conclusions* of our inquiry into the moral permissibility of types of sexual acts, not the premises for them.

The SPV is, obviously, a morally neutral view of sexual desire. But its acceptance of the claim that pleasure supervenes on the activity, coupled with a moral correlate that claims that it is sexual activity, not sexual pleasure, that is the primary target of moral evaluation, opens the door for a richer evaluation of sexual acts and their pleasures, and it helps resolve Benatar's dilemma.

Notes

1. I defend this account in Halwani (2020).
2. Benatar (2002).
3. The discussion in this section is an abbreviated version of parts of Halwani (2020).
4. An early formulation of the view is Alan Goldman's: sexual desire is "desire for contact with another person's body and for the pleasure which such contact produces" (1977, 268). See also Goldman (2016, 95).
5. The addition of "first-order" is to keep out cases in which *X* desires to feel what it is like to experience sexual desire.
6. I prefer the SPV because the SPAV excludes cases in which sexual pleasure can be sought through means other than sexual activity, such as by plugging into Nozick's experience machine (Halwani 2020). See the next paragraph.
7. The experience machine "would give you any experience you desired." It is operated by "superduper neuropsychologists" who can stimulate your brain as you float in a tank, in order to make you think and feel that you are having whichever experience you desire (Nozick 1974, 42). For instance, if you desire to experience walking down the red carpet for your anticipated movie premier, you can experience this, and it would feel as real as if it had actually occurred.
8. See Morgan (2003) and Halwani (2020).
9. Of course, all this is compatible with the claim that sexual desire can have other goals or that the sexual pleasure is desired with someone one loves or with whom one wants to have children.
10. *Nicomachean Ethics* (1999, 1174b32). *The Revised Oxford Translation* explicitly uses "supervene": "Pleasure completes the activity not as the inherent state does, but as an end which supervenes as the bloom of youth does on those in the flower of their age" (quoted in Broadie [1991], 336).
11. See Katz (2016) for a general discussion of pleasure. Bramble (2013) defends the first two conceptions of pleasure. Feldman (2004) defends pleasure as attitudinal.
12. Goldman (2016, 88).
13. Not fully apart because it shares this with desires for food and drink.
14. Soble (1996, 216). Soble's point is about the possible degrading nature of oral sex, given the positioning of someone's head—the locus of the mind—right into someone else's crotch—the locus of the animal.
15. Brogaard (2015, 189–190).

16. There might be gender differences in terms of how men and women experience the pleasures and pains of sex. As a colleague colorfully put it to me, "I have yet to hear of a woman willing to jeopardize her career because she wanted to get fucked." The non-philosophical literature on this is immense but a good philosophical review of the issues is Wertheimer (2003), ch.3.
17. Giles (2008, ch. 3).
18. Pearson (2012, 36).
19. Thus, the problem here is not that this is an instance of the end justifying the means.
20. Aristotle (1999, 1176a10–25).
21. For discussion of Aristotle on pleasure, see Sarah Broadie (1991, ch. 6).
22. Aristotle (1999, 1119a10–20).
23. For a discussion of Aristotle on sexual pleasure, see Halwani (2018a).
24. It is very problematic; see Hauskeller (2011).
25. Mill's proof might also be ambiguous as regards the question of appealing to pleasure directly or not for its moral evaluation. This is because Mill was not concerned with the distinction between pleasure and the activity on which it supervenes, and for all we know, he might have been indifferent to whether the people he surveys are referring to the pleasures themselves or the pleasurable activities.
26. On these issues, see Bartel (2012), Levy (2002), Luck (2009), and Strikwerda (2017).
27. I am assuming that plugging into the experience machine and experiencing sexual pleasure in forms (a), (b), or (c) are adequate substitutes for pedophilic activities. That is, they are not pedophilic activities, yet they come close to them in yielding a pleasure similar to what X_m and Y_m would have experienced had they engaged in pedophilic activity. The notion of substitution here, however, can lead to difficulties (see Soble [2013], 84).
28. A reminder that I use "pedophilic activity" to mean "sex with actual children or prepubescents."
29. Roughly, hedonism claims that pleasure is the only important element of what constitutes well-being (and pain is the only important element of what constitutes negative well-being or "ill-being").
30. See Morgan (2003) and Dent (1984).
31. For ease of reference, I henceforth use only "perceptions" instead of "perceptions or thoughts." I wish to be understood that some cases of pleasure include thoughts only, others perceptions only, while others both.
32. I say "usually" because there are cases of sexual pleasure caused by other things, such as women experiencing pleasure while horseback riding, or a man rubbing his penis to enjoy the sensations without any recourse to fantasizing about anything or anyone. Many asexuals also seem to enjoy sexual activity without the sustenance of perceptions (see Brunning and McKeever, 2021).
33. In cases in which the pleasures arise without any mental activity, then they have no connections to issues about pedophilic pleasure, so the question is moot.
34. Otherwise, it might show too much, such as the wrongness of sexual pleasures experienced by two consenting adults who are enacting, say, a master-slave BDSM scene. I return to this in the next paragraph.
35. On this, see Miller (2017) and Halwani (2018b, 310–318).

36. Of course, the pleasures are not the only factor we use in character evaluation. Those cases in which people act on their desires, the actions are as, if not more, crucial.
37. Benatar (2002, 192).
38. Benatar (2002, 192; my emphasis).
39. Benatar (2002, 194).
40. Benatar (2002, 194).
41. Benatar (2002, 195). Note that the phrase "subject to the usual moral constraints" that Benatar uses in formulating the casual sex view seems innocuous. However, given that the casual view goes on to deny the necessity of consent in the case of pedophilia, it is important to look into what "usual moral constraints" refers given that they typically include consent.
42. Benatar (2002, 196).
43. Benatar (2002, 192).
44. Benatar (2002, 193).
45. Benatar (2002, 193). To be clear, Benatar believes that the significance view *can* explain the wrongness of rape (because it is a violation of bodily autonomy), but it cannot explain why rape is *especially* wrong or more seriously wrong than forcing someone to eat something (hence my wording, "the special wrongness of rape").
46. Benatar (2002, 193). I doubt the adequacy of the explanation that the significance view of sex offers of the wrongness of pedophilia and the special wrongness of rape, because love is irrelevant: we can imagine cases of rape and pedophilia in which both parties can express their affection for each other during the act, but this would not make the act permissible. Indeed, the lack of deep affection is not something that comes to mind when thinking about the wrongness of rape and pedophilia.
47. For other criticisms of Benatar's dilemma, see Brogaard (2015, 187–190).
48. Wertheimer (2003, ch. 3).

Works Cited

Aristotle. 1999. *Nicomachean Ethics.* Translated by T. Irwin. Indianapolis, IN: Hackett.

Bartel, Christopher. 2012. Resolving the Gamer's Dilemma. *Ethics and Information Technology* 14 (1): 11–16.

Benatar, David. 2002. Two Views of Sexual Ethics: Promiscuity, Pedophilia, and Rape. *Public Affairs Quarterly* 16 (3): 191–201.

Bramble, Ben. 2013. The Distinctive Feeling of Pleasure. *Philosophical Studies* 162 (2): 201–217.

Broadie, Sarah. 1991. *Ethics with Aristotle.* New York and Oxford: Oxford University Press.

Brogaard, Berit. 2015. *On Romantic Love: Simple Truths about a Complex Emotion.* Oxford: Oxford University Press.

Brunning, Luke, and Natasha McKeever. 2021. Asexuality. *Journal of Applied Philosophy* 38 (3): 497–517.

Dent, N.J.H. 1984. *The Moral Psychology of the Virtues.* Cambridge: Cambridge University Press.

Feldman, Fred. 2004. *Pleasure and the Good Life: Concerning the Nature, Varieties, and Plausibility of Hedonism*. Oxford: Clarendon Press.
Giles, James. 2008. *The Nature of Sexual Desire*. Lanham, MD: University Press of America.
Goldman, Alan. 1977. Plain Sex. *Philosophy & Public Affairs* 6 (3): 267–287.
———. 2016. Pleasure. In *Philosophy: Sex and Love*, ed. J.M. Petrik and A. Zucker, 79–101. Farmington Hills, MI: Macmillan Reference USA.
Halwani, Raja. 2018a. Sexual Ethics. In *The Oxford Handbook of Virtue*, ed. N. Snow, 680–699. New York: Oxford University Press.
———. 2018b. *Philosophy of Love, Sex and Marriage: An Introduction*. 2nd ed. New York: Routledge.
———. 2020. The Sexual Pleasure View of Sexual Desire. *Philosophical Papers* 49 (1): 107–135.
Hauskeller, Michael. 2011. No Philosophy for Swine: John Stuart Mill on the Quality of Pleasures. *Utilitas* 23 (4): 428–446.
Katz, Leonard. 2016. Pleasure. In *The Stanford Encyclopedia of Philosophy* (Winter 2016 Edition), Ed. E. Zalta. https://plato.stanford.edu/archives/win2016/entries/pleasure/.
Levy, Neil. 2002. Virtual Child Pornography: The Eroticization of Inequality. *Ethics and Information Technology* 4 (4): 319–323.
Luck, Morgan. 2009. The Gamer's Dilemma: An Analysis of the Arguments for the Moral Distinction Between Virtual Murder and Virtual Pedophilia. *Ethics and Information Technology* 11 (1): 31–36.
Miller, Shaun. 2017. BDSM. In *The Philosophy of Sex: Contemporary Readings*, ed. R. Halwani, A. Soble, S. Hoffman, and J. Held, 8th ed., 421–436. Lanham, MD: Rowman & Littlefield.
Morgan, Seiriol. 2003. Sex in the Head. *Journal of Applied Philosophy* 20 (1): 1–16.
Nozick, Robert. 1974. *Anarchy, State, and Utopia*. New York: Basic Books.
Pearson, Giles. 2012. *Aristotle on Desire*. Cambridge: Cambridge University Press.
Soble, Alan. 1996. *Sexual Investigations*. New York and London: New York University Press.
———. 2013. On Jacking Off, Yet Again. In *The Philosophy of Sex: Contemporary Readings*, ed. N. Power, R. Halwani, and A. Soble, 6th ed., 77–99. Lanham, MD: Rowman & Littlefield.
Strikwerda, Litska. 2017. Legal and Moral Implications of Child Sex Robots. In *Robot Sex: Social and Ethical Implications*, ed. John Danaher and Neil MacArthur, 133–151. Cambridge, MA and London: The MIT Press.
Wertheimer, Alan. 2003. *Consent to Sexual Relations*. Cambridge: Cambridge University Press.

CHAPTER 4

The Ethical Significance of Being an Erotic Object

Caleb Ward and Ellie Anderson

Contemporary philosophical discussions of sexual ethics focus heavily on questions of autonomy, agency, and the moral power of consent. Since the late 1980s, moral and legal theorists have debated the conditions of moral validity for sexual consent, while feminist social and legal theorists have for even longer criticized social practices and institutional structures that undermine the agency of women to steer their sexual lives. Common across these discussions is the hope that respect for another's freedom, autonomy, or will can prevent the harm of *objectification*, that is, treating another merely as a sexual object or means. The task of sexual ethics, on the conventional view, becomes reconciling one's grasp of another as an object of erotic desire with the moral authority of their subjectivity. This generally involves placing their subjectivity *above* their status as object, so as to avoid the danger of objectification.

However, we are always both subject and object in our encounters with others. As a result, theorists disregard essential, morally relevant features of intimacy when they locate the moral significance of another only in that person's active subjectivity. In this chapter, we argue for a methodological adjustment to sexual ethics that widens the narrow focus on freedom, autonomy, and agency to address the moral significance of being an erotic object as well. Drawing on phenomenology, particularly the insights of Simone de Beauvoir, we propose

C. Ward (✉)
University of Hamburg, Hamburg, Germany

E. Anderson
Pomona College, Claremont, CA, USA
e-mail: ellie.anderson@pomona.edu

D. Boonin (ed.), *The Palgrave Handbook of Sexual Ethics*,
https://doi.org/10.1007/978-3-030-87786-6_4

an approach that recognizes a constitutive ambiguity between the freedom appropriate to subjectivity and the phenomenon of being an erotic object in intimate encounters. While we are committed to the feminist view that all people should have agency over their sexual lives, we reject the conflation of this commitment with the idea that the autonomous will is the sole source of another's moral authority in sexual intimacy. Important moral features of intimacy are disclosed through the erotic experiences of both being an object and perceiving another as object, and the ethical implications of objecthood remain unexamined when the division between moral subject and erotic object is too sharply defined. By engaging with the ethics of erotic objecthood, theories of sexual ethics can become more adequate phenomenologically and more sensitive to the complexity of intimacy, even as they work to capture the moral significance of agency and to acknowledge the wrongs of sexual violation.

We proceed in three sections. The first section, "Living One's Ambiguity and Being an Object for Others," examines how being an erotic object to another is experienced in sexual intimacy, particularly how it is intertwined with one's sense of active agency or subjectivity. We interpret this situation as an example of what Beauvoir describes as *ambiguity*, that is, the simultaneous sense of being a free subject and being exposed as a corporeal object to others. The second section, "Experiencing Another as an Erotic Object," articulates a distinction between the wrongs associated with the term 'objectification' in philosophical debates and the *other-as-object* valency of both perception and treatment of others in sexual intimacy. A person is encountered as an object unlike other objects, but not in the sense that they appear as a moral person endowed with a rational will. Instead, another person approaches one as a synthetic whole that discloses both subjectivity and bodily presence. The final section, "Ethical Implications," argues that theories of sexual ethics ought not to overlook the fact that being an object is inextricable from subjectivity or personhood. Being an erotic object is morally significant for two reasons. First, my erotic objecthood plays a central role in the explicit claims that I as an agential subject make on another, such as the claims I make by intentionally communicating sexual affirmation or refusal. Second, to grasp another as an erotic object is to be solicited morally by another, even absent the authoritative expressions of will familiar from standard accounts of sexual ethics.

Living One's Ambiguity and Being an Object for Others

Philosophical discussions of the ethics of sexual intimacy often focus on how and why a person wrongs another by failing to gain valid consent or to respect refusal. Often, these wrongs are articulated in terms of treating another as a mere object, or means, rather than as a person worthy of respect. This approach has been supported by feminist accounts of the experience of sexual violation that focus on how violation renders one passive or inert in the face of another's unwanted sexual behaviors.[1] However, feeling like an object need not entail feeling used or harmed; indeed, being an erotic object for another is a central feature of virtually all erotic encounters. In this section, we analyze the structure of being an erotic object using the phenomenological concept of *ambiguity*. We suggest that being an object plays a constitutive role in experiences of sexual selfhood and agency in general, which merits consideration in sexual ethics.

The idea that selfhood is ambiguous comes from phenomenology, a philosophical method that describes phenomena as they appear in lived experience—that is, in the first-person perspective of an embodied perceiver in the world. Simone de Beauvoir uses *ambiguity* to characterize an irreducibly dual character of human experience: as living in a world always shared with others, persons are unavoidably both subjects (for themselves) and objects (for others). Persons are ambiguous because they experience themselves as being a unique perspective on the world, with a sense of agency and in a process of becoming over time; and yet they also experience being objects within the world, affected by their environments and showing up in others' horizons of experience. That is, persons are both transcendent (or *free*) and immanent (or limited by being bound up in a situation with others to whom they are vulnerable). Beauvoir describes ambiguity as "that engaged freedom, that surging of the for-oneself which is immediately given for others."[2] Freedom can only exist engaged in a world, where it always depends on others both for its practice and for its products to have meaning.

As a free being, I transcend my situation; but, because I am also a material, living body, my transcendence can in turn be transcended by others. Maurice Merleau-Ponty describes the living body as a third kind "between the pure subject and the object"; the willing and perceiving subject *is* the body, and this body shows up for others as an object.[3] As body, I am sensitive to others, undergoing the affective drama of intimacy that includes pleasure, pain, violation, and other responses that arise from contact with others.

Ambiguity describes how the body is minded and the mind is embodied, calling into question the dualism between the two. And yet, the distinction between subject and object persists in experience, where subjectivity and objecthood—and, correspondingly, agency and passivity—are lived as being in tension. This indicates how, though mind–body dualism is untenable as a metaphysical claim, the subject/object distinction remains as what Nancy Bauer calls a "phenomenological dilemma," which is lived out in the perspectival character of embodied, subjective experience.[4] An individual is bound to a first-person perspective; I cannot directly perceive or know myself as an object, as we discuss below. And yet we *feel* that we are both subject and object. Ambiguity appears to us in experience as this felt duality, which evades our cognitive grasp.

Attention to ambiguity also reveals the intertwining of passivity and activity in everyday experience. As ambiguous, the person is not sometimes active and other times passive. Rather, as Merleau-Ponty puts it, "we are entirely active and entirely passive."[5] Touching and being touched, for instance, involve a simultaneous sense of being felt from within and being accessible from without.[6] The touched body is also a touching body: the activity of touching is situated in the passivity of being touched. Activity is indeed a taking-up of my situation and hence only transcends the situation by virtue of being rooted in it. Put differently, the body's vulnerable openness to others is intertwined with its agency. For the lived body, passivity and activity are not opposed; moreover, objecthood is not merely a passive state to be contrasted with the willful activity of subjectivity.

The intertwining of passivity and activity in embodied consciousness is not felicitously rendered in English, a language with a strong distinction between the passive and active voice. French phenomenologists frequently render it with the reflexive verb *se faire*. *Se faire* can be translated into English alternatively as 'to make oneself' or as 'to be made,' but French captures both senses at once, better illuminating the ambiguity. *Se faire* refers to the process of becoming that characterizes consciousness: self-creation happens ongoingly, in an environment with others. Notably for our purposes, this verb is frequently used in phenomenological descriptions of objecthood, as in the phrase *se faire objet* (to make oneself/to be made object) or, more broadly, *se faire être* (to make oneself be). This construction expresses how subjects actively take up their constitution as objects even as they do not 'cause' or 'control' their objecthood.[7] We simultaneously are vulnerable to others *and* take up our vulnerability in agential ways.

I first experience my own objecthood through the immediacy of relating to another person: this encounter makes me feel like an object. When looked at by another, for example, I feel known as an object by them. Sartre describes how the experience of the look of another is "of the affective order," which is distinct from cognition.[8] Only the other knows me as an object.[9] Unable to get around to the 'other side' of my subjectivity due to my limited perspective, I cannot know myself as the object that I am. I will often try to gain knowledge of myself-as-object by objectifying myself: I might look in a mirror and assess how another person would view the attractiveness of my form, or I might touch my skin and wonder how the other would experience it. Yet this project is doomed to failure: seeing or touching myself is phenomenologically different from seeing or touching another person. The best I can do is to feel that I am an object to another without cognitively grasping the character of this objecthood.

In erotic encounters, one may experience oneself as the subject of erotic intimacy—desiring, intending, willing, and experiencing pleasures or pains—and simultaneously as an erotic object that is perceived and acted on by another. Erotic encounters thus disclose ambiguity. Indeed, Beauvoir uses erotic experience as a paradigmatic example: the experience of erotic reciprocity in particular "most poignantly reveals to human beings their ambiguous condition; they experience it as flesh and as spirit, as the other and as subject."[10] One feels one's objecthood through being the object of the other's desirous activity, and this is intertwined with one's desire for the other.

In mutually erotic encounters, a crucial element of my eroticized self is this feeling of being an object for the other. I feel desired by the other in my body: the other is oriented toward me as attractive. Being taken as an object in this light is not experienced as a simple diminishment of my agency. On the contrary, I *fascinate* the other inasmuch as I am a desired object—I am the object that orients another's intentions and response. Thus, living one's own erotic objecthood need not be in tension with the experience of being an agent and a consciousness. Indeed, in being an erotic object to another, I can feel myself recognized in the flesh by them. Although, when unwelcome, such a feeling of being an object can strip me of my sense of agential efficacy, in other cases it can produce a new or different sense of myself, which might open onto new feelings and enhanced agential possibilities.

Beauvoir describes how the intimacy of positive erotic encounters permits one to feel one's objecthood as compatible with one's subjectivity.[11] When balanced with the recognition that one's partner acknowledges one's free subjectivity as well, one "recovers her essentialness at the moment she becomes object [*se fait objet*]."[12] Feeling like an erotic object can be liberating, because it offers one an entry point into a crucial side of one's ambiguity that is not otherwise accessible; the presence of another person is needed in order for me to experience myself as an object. Beauvoir's affirmation of erotic objecthood may in part be read as a response to Sartre's more pessimistic account of erotic encounters. Sartre suggests that the feeling of being the object of another's desire is experienced as a threat, to which one responds either by struggling to reassert one's subjectivity or by trying to embrace one's objecthood completely. For Sartre, the latter response is *seduction*, where I try to fascinate the other with my objecthood. He dismisses such behavior as bad faith, or self-deception, insofar as I pretend to abdicate my freedom of being a subject. Yet, Beauvoir shows us that erotic encounters can provide space for both my subjectivity and my objecthood and, reciprocally, for both the subjectivity and the objecthood of the desired other.

Talia Mae Bettcher makes a compatible point in her interactional account of "erotic structuralism," which also further clarifies how ambiguity functions to shape the erotic self. Bettcher suggests that accounts of erotic experience tend to overlook the role of the eroticized self by focusing exclusively on attraction to the other. Yet she argues that the content of erotic desire is not reducible to my attraction to an eroticized other (the source of attraction); it additionally includes my eroticized sense of self (the locus of attraction), which has an importantly gendered, embodied structure.[13] Furthermore, the interaction between my eroticized self and my eroticization of another is crucial to the dynamic of sexual encounters. It is not that I am *attracted to* myself as an object in such encounters but rather that I experience myself as the locus of my own attraction to the other and as the target of their attraction. I am excited by my exposure to an eroticized other.[14]

While Bettcher does not explicitly thematize the eroticized self as an *object*, we take her analysis to show how one's sense of erotic selfhood is both agential and receptive: in Beauvoir's terminology, it is ambiguous. While my attraction to another reflects a sense of my own eroticized body-self, that same sense of selfhood is open to the effects of feeling myself eroticized by another. Bettcher suggests that the eroticized self is not attracted *to itself* but is nonetheless a key element of erotic interest (she terms this the interest/attraction distinction). This distinction illuminates the importance of what we are describing as erotic objecthood, inasmuch as the eroticized self is the embodied target of the other's desire. "Some significant erotic content is not reducible to the 'source of attraction,'" Bettcher argues, and one's erotic self is part of this content.[15] Thus, the eroticized self—in part as object of the other—is key to subjectivity in sexual encounters, even as the other is the source of one's attraction.

The ambiguity of being both subject and object is an inherently difficult target for philosophical reflection. Philosophers have tended to overlook or minimize ambiguity's moral significance by thinking about it in terms of familiar distinctions between mind and body or between activity and passivity. In sexual ethics, theorists who set out from a Kantian moral framework tend to locate value in the autonomy of the will while taking feeling and the senses as secondary, contingent aspects of selfhood. Moral conduct on this view depends on an actor's ability to will that their actions respect the autonomous will of another—being an object for another is at most only a problem to be solved, not an integral aspect of the self. Even when a wider range of sexual goods are recognized as warranting respect, as in the work of Martha Nussbaum, those goods are often posited as goods *for a person's mind or subjectivity*, rather than as part of the integral body-self that is itself the origin of agency.[16]

Theories of 'morally transformative' consent depend on a similar simplification: they take certain conditions of the will to change another's obligations, while the phenomenon of being an erotic object drops out of moral consideration entirely.[17] The risk of such approaches is that erotic embodiment is treated as a good over which each person should have autonomy, rather than the unavoidable setting from which any possible practice of agency or autonomy must emerge. Feminist critical theorists have recently begun addressing the negative political effects of this failure to recognize ambiguity in theorizing consent. For example, overinflating the 'moral power' attributed to consenting can obscure the regularity with which women (and others) experience sexual violation within encounters to which they have given consent.[18] While consent is theorized primarily as a necessary condition of non-harmful sex, it does little to capture the complexities of embodied encounters between persons. Lived erotic encounters sometimes involve ambivalence or partial consent, and they are often shaped by pernicious social structures that, for certain subjects, limit choices and constrain the interpretive resources for understanding what is taking place.[19] In practice, putting a sharp notion of morally transformative consent at the center of sexual ethics ignores the moral significance of ambiguity.

A similar refusal to address ambiguity is also present in many poststructuralist discussions of sexual ethics. Theorists who ground their approach in the theory of Jacques Lacan, in a certain reading of Michel Foucault, or in the thought of Gilles Deleuze sometimes invert the valuation to elevate objecthood over subjectivity, but in doing so they retain the two terms' polarity. Some queer theorists in these traditions argue that sexual pleasure has a power to shatter subjectivity in a way that always exceeds the regime of rationality or the morally attuned will.[20] However, positing the senses as dissolving subjectivity fails to disturb the dualistic view of selfhood; feeling is here still taken to be opposed to rational thought or knowledge. Our analysis above suggests that being an erotic object does disrupt the rationalist's claim that subjectivity and the will are sovereign; in this it resonates with the work of queer feminists like Elizabeth Grosz, Lynne Huffer, and Lauren Berlant.[21] However, our phenomenologically informed rejection of dualism suggests that the sensing capacities of the body are constitutive of subjectivity, not opposed to it. To elevate the pursuit of pleasures and the materiality of the erotic body as valuable—over and against considerations of agency and subjectivity—is to impose an implicit, libertarian view of relationality and the ethical claims others make on our desires and intentions. Linda Martín Alcoff has diagnosed how libertarian readings of Foucault smuggle in an idealized notion of freedom while producing unacceptable results when used for ethical analysis of sexual violation.[22]

Ambiguity is the condition of possibility both for sexual violation to be experienced as a profound harm and for the joy of 'losing oneself' in erotic intimacy. Ann Cahill argues that the responsiveness of the body-self to others is the ground for both the profound injury of rape and the possibility of resilience. She writes, "the intersubjectivity necessitated by a bodily existence includes a vulnerability that rape exploits, but it also includes an openness, an ongoing process of development that limits the power of the rapist."[23] This is the openness of ambiguity, manifest in the possibility of ongoing narrativizing and meaning-making and in the ability of sexual subjectivity—one's sense of oneself as an agential, sexual being—to adapt and be remade.[24] Being an object for others is part of this meaning-making process: in shaping narratives, we locate ourselves in relation to others, taking up our sense of how those others treat us in dynamic interactions, including (but not limited to) interactions in which I am the object of the other's desire.

Experiencing Another as an Erotic Object

We have highlighted the significant role played by the feeling of being an erotic object and how being an object discloses one's embodied sense of ambiguity. It is a common view in sexual ethics, however, that treating another as an object in a sexual encounter is a paradigmatic moral wrong. Feminist theorists have drawn attention to how sexual objectification is a central component of women's subordination to men in many societies, and the project of adequately defining objectification (and distinguishing it from ethically unproblematic eroticism) has become an ongoing philosophical debate spanning several decades.[25] Objectification is often described as treating a person as a mere object rather than as an autonomous subject or moral person with their own will, interests, and integrity. This seems especially apt for explaining certain sexual harms, particularly those that violate one's agency or freedom. However, our discussion in the previous section suggests that the experience of being an erotic object is fundamental to eroticism and implicated in our subjectivity and agency. Being an erotic object to another makes one vulnerable to agential harms, but it does not of itself entail suffering those harms.

Thus, perceiving and treating another as an erotic object should not be conflated with the wrong of objectification. While the harms attributed to objectification can certainly be overlaid on perceiving the other as an object, some nontrivial form of treating others as objects is an integral feature of interpersonal perception and relationality.[26] I can—indeed, must—perceive others as objects, just as others also perceive me as an object. This should be approached as the starting point for moral reflection rather than as a problem to be remediated or a breakdown in morality that ought to be corrected.

A crucial insight from phenomenology is that other persons are grasped as objects of perception in a way that lays a claim on the perceiver. The body of another is not a mere thing in the world: rather, it is a *lived body*, and it is present to the perceiver as living. The other person presents themselves as a synthetic totality—a whole that cannot be replicated or understood by considering the particular qualities or parts that hang together in it.[27] In Sartre's analysis, "the body appears within the limits of the situation as a synthetic totality of *life* and *action*."[28] In this same vein, Merleau-Ponty describes the body of another as exemplifying a cohesion and depth of significance comparable to an artwork rather than a mere physical object.[29] While mere objects in the world reveal themselves as fungible instruments available for use or enjoyment, other persons show up within perception as integrated, non-fungible wholes. We do not natively experience another person in terms of potential use or enjoyment, but instead as demanding a relational response. The other's body is not merely an object, but rather the bearer of a behavior.[30] And their behavior solicits us to engage with them, that is, to change our own behaviors and intentions in response.

Whereas an object like a tool or a door handle solicits a response in the sense that it offers affordances for use—available to anyone with hands or other means of grasping—another person solicits in me a response of a different kind. Through their interactive, discursive presence, the solicitation of another is personally addressed *to me* and invested with ethical significance; in principle, how they perceive me and experience my response to them is an issue for me, even if I resolve to treat them with indifference or disdain.[31] As a living body, the other as an object of my experience implicates me: my perception of the other is dynamically related to them and intertwined with my ethical treatment of them. To say that the other is an object for me is not to say that I perceive them neutrally or 'objectively'; rather, the other is a phenomenal body that is part of my world, in proximity to me as a body-self as well.

Thus, the other's body is an object for me, but it is an object that materially gestures beyond my access and power over it. The other's perspective is 'over there,' implicated in their embodied behaviors but never accessible to me from within. My response can reach out toward the other's subjectivity but cannot grasp it. The other's body is expressive of agency, but their expressions—speech, movements, postures, and gestures—do not "consist in *giving* us the Other's interiority."[32] Instead, the other present to me in their expressive body opens up the field of interaction through discourse, touch, and other forms of relationality. To perceive someone as an object in this way is to perceive them not as a 'mere' body but as an affective totality whose movements express agency and intention. Such a body is not a *what* but a *who*.

Importantly, the other's body is not a signal that they additionally have a mind or a sign that points to a mental content; rather, the body is enminded in its presence to me as a body. Although I can never grasp the other's subjectivity directly as they experience it, I do receive a direct experience of their subjectivity *in the second person,* in relating to them as a body (object).[33] We see here the other side of the "phenomenological dilemma" of subjectivity and objecthood described earlier. Just as I can *feel* that I am an object, but cannot know the object that I am for the other (because I am a subject for myself), so too can I *feel* the other as a subject, but cannot know the subject that the other is for themselves (because they are an object for me). Feeling the other's subjectivity is direct and immediate, and I experience it by being solicited by the other, for whom I am an object. We co-constitute the situation of our encounter as subjects for ourselves and objects for each other. We are synthetic totalities, but we each live out the phenomenological dilemma of ambiguity. Merleau-Ponty writes that our experience of others, for this reason, is "not ... ultimately thinkable."[34] Yet it is directly felt.

Early phenomenologists such as Edith Stein describe this in terms of empathy: one *feels* the presence of another consciousness when perceiving the other-as-object, even as one cannot put oneself in the perspective of that consciousness. Empathy is a way of describing our feeling of a foreign consciousness as an embodied, synthetic totality. So-called inner experience and gestural expression are continuous; as Merleau-Ponty describes, the grimace on a person's face is a part of the anger that they feel, rather than a sign that points to anger as a content.[35] As a result, the expressiveness exhibited by another's body as an object of my perception is essentially integrated with their subjectivity. To see another's expression of anger in their face or body language is to sense their anger itself—not as my own anger, but as that of the other.[36]

Moreover, this uniquely relational mode of experiencing the other-as-object is heightened in erotic encounters. As Merleau-Ponty writes, the perception of another as an object is "inhabited by a more secret [perception]: the visible body is underpinned by a strictly individual sexual schema ... which is itself integrated into this affective totality."[37] The erotic other is not a collection of parts but an integrated totality to which I find myself *attracted* in a way that reorganizes my intentions. My desire for the other decenters my perception, drawing me out of myself as if magnetized by the other. Here, the other is an object—the object of my desire—but not as one over which I have a hold.

Rather, the other-as-object is the source of my attraction, which in a sense has a hold on me—not because they entice my mind but because they engage my senses. As noted in the first section, this attraction toward the other also implicates and facilitates my own subjectivity as a sexual being. Thus, experiencing the other as an erotic object not only involves sensing their subjectivity present through their living body, but it also involves experiencing how my own subjectivity and erotic objecthood depend on them. Erotic encounters are thus co-constituted rather than produced through an intentionality from the side of the desirer alone. How I treat the eroticized other who is the object of my erotic attention cannot be extricated from my own sense of subjectivity and my feeling of being an object to them.

Ethical Implications

In light of the foregoing phenomenological reflections about the role of ambiguity in experience and perception, objectifying treatment—that is, treating another as merely an object or means—indeed constitutes a moral wrong that reflects a disregard for the value-imbued nature of another human being. However, the wrongs of objectification need not arise from the particular failure identified by most theorists—that is, the failure to respect another as autonomous, or as having a will or mind worthy of moral regard.[38] While agency, the will, and other aspects of another's subjectivity are indeed of vital moral concern, an overinvestment in the ostensibly interior subjectivity as site of moral value leads us to neglect the ethical significance of being an erotic object and treating others as erotic objects—including how the other's bodily presence makes moral claims on me to change my behaviors and intentions.

We propose two insights that can help theories of sexual ethics take seriously the significance of erotic objecthood. First, claims on behalf of the value of sexual autonomy, freedom, or agency should acknowledge that the embodied feeling of *being an erotic object myself* plays a constitutive role in those practices. Second, normative claims about sexual ethics should recognize that valorizing another's subjectivity requires a certain orientation of responsiveness *toward another as an erotic object*. While the first insight calls for a methodological shift in how theorists approach objectification, the second has substantial ramifications for sexual ethics writ large, especially with respect to the normative significance of consent and other sexual communications.

The first point, that being an erotic object is constitutive of sexual agency, sets out from an acknowledgment that another's subjectivity—their agency or freedom, in particular—is an essential component of their value and moral appeal, what calls on us as responsible to them. Being denied agency in one's sexual encounters is often experienced as a profound violation, and there is a moral obligation to behave toward others in ways that acknowledge the significance of their agency and will. I wrong another personally when I prevent their agency from appropriately influencing our encounter, and the society-wide tendency to subject some people to this treatment—particularly women,

BIPOC, sex workers, incarcerated people, and disabled people of all genders—constitutes a widespread injustice. A commitment to agency or freedom is thus indispensible for critique of sexual violence and oppression. However, the moral imagination to embed this commitment within a nuanced sexual ethics is limited when theorists focus on agency as solely an aspect of active subjectivity, without recognizing the constitutive role played by passivity and being acted upon by others. Agency is always practiced within a situation of embodied ambiguity beyond one's control, which includes both the condition of being an object to others and the situation of being subjected to features of one's position within a social context.

Conventional discussions of sexual agency, and especially of what is described as sexual autonomy, often focus on the will, rational deliberation, and reasons for action as foundational elements. Our analysis suggests that the felt sense of oneself as an object for others should also be given sustained philosophical treatment. A complex phenomenology of sexual desire and self-relation can help provide a more nuanced account of agency in sexual intimacy. Further, an appreciation of the significance of being eroticized by another might help shed light on situations where agency breaks down, such as where a person is not coerced but is nonetheless unable to assert their boundaries and refuse unwanted touch. The notion of ambiguity provides a phenomenological rather than purely psychological explanation of such situations, which might interact productively with contemporary feminist analyses of power and the ideology of gender norms to further enhance their descriptive adequacy.[39]

As with *being an erotic object*, the ethical significance of *treating another as an erotic object* is also under-examined in sexual ethics, largely because the notion of objectification associates such treatment with moral harm. In particular, ethicists tend to focus on obligations to another in a sexual encounter in highly intellectual terms. They focus on whether a person's communications can be reasonably interpreted as consenting, whether behaviors or speech acts provide sufficient reasons for another to change their actions, or whether conditions of competence, information, or intentionality have been satisfied in a person's decision to consent to sex. Such a heavy investment in the value of the will as a mental process, taken purely as an 'interior' feature of subjectivity, leads to the view that speech is the only reliable medium for making moral claims on one another.

While speech and explicit communication are of undeniable importance for enacting agency, the agency or freedom that endows subjectivity with moral value is also enacted in our bodily comportment. The claim that the body-self is a synthetic totality, articulated in the second section, suggests that speech and bodily expressiveness should not be treated as mere indicators or signs to be decoded in order to sniff out the real will or intentions behind another's behaviors. Rather, attending to another's bodily expressiveness in gesture, posture, and movement should be recognized as an ethically significant practice in its own right. Another's bodily presence is a source of meaning or sense-making, often grasped on the affective level, that can be highly significant for the purposes of acting responsibly to them and acknowledging the moral significance of their agency.

This claim makes sense if we reflect on lived experience: people often either pursue or refuse sexual encounters through non-deliberative, habituated modes of expression—postures, gestures, nonverbal utterances, particular tones of voice, or cadences of speech—that fall short of the intentional communication which is the focus of most philosophical literature. Similar to more explicit, conventionally recognized expressions of consent or refusal, these oblique modes of sexual communication also solicit changes to another person's actions and intentions.

Perceiving the other-as-object in the way we have described is morally neutral. To grasp another as an object is to receive them as a synthetic totality, and in doing so I cannot help but become oriented toward ethical concerns regarding my treatment of them. The body's movement and expression, including the voice, bring another's agency to bear on me, enabling another to solicit my ethical response.[40] Because of the kind of object they are, I can in principle feel some aspect of the freedom that is their embodied situation. This is implicit in how I respond to any other person with whom I come into relation in daily life, including how I pick up on a wide range of subtle, contextually dependent meanings and affects that seamlessly shape our relations in everyday communication.

And yet, when grasping another as an erotic object, it is common for these implicit or embodied solicitations to fail to achieve the appropriate response. At least this is the case in the society the authors inhabit, which is shot through with pernicious social norms and gendered expectations around sex. Harms and moral wrongs occur when one refuses to take the other-as-object as a synthetic totality with their own bodily agency. In situations that might be described as objectification, another is treated not just as an object but as a particular, reductive kind of object: as a thing or instrument rather than a living body. We agree with the view that the kind of treatment described as harmful objectification is prima facie morally problematic because it denies the other's freedom or agency. In this sense, our phenomenological account coincides with mainstream feminist views. However, we disagree with the view that the kernel of this wrong is a lack of respect for another as a moral person with the dignity of a rational mind or will. Recognizing another's freedom in sexual intimacy also requires recognizing their erotic objecthood as a living body.

Given the phenomenological claim that we always grasp another as a synthetic whole, the moral failure of treating another as a thing requires explanation. This cannot simply be a question of instrumentalization or treating an ensouled person as merely an object. Rather, the moral failure in question here is that of bracketing the whole of another in favor of *pretending* that their body is an object in the same manner as non-persons are objects. It is a kind of bad faith, or renunciation of one's capacities, to pretend that the other is merely an object in the instrumental or impersonal sense—that is, a means to my own gratification or a mere aesthetic thing. To take on such a position is not a genuine belief about another or an orientation toward them, but a posture or pretense. When we objectify another person sexually—whether by representing them or treating them a certain way—we do not actually believe they are only a sexual object. Rather, we pretend that they are, and that pretending is propped up by discourses of male sexual entitlement and a libertarian view of one's sexuality and body as property that a person, as proprietor, has a right to dispose of as they see fit.[41]

The increasingly juridical understanding of consent feeds this tendency by encouraging the view that, as long as another person validly consents to a sexual act, their autonomy cannot be harmed by what takes place. The vulnerability and phenomenological significance of the erotic body drops out of ethical consideration in such a framing; we are encouraged to think it axiomatic to a person's autonomy that their consent has complete authority over how another treats their body. Most consent-based accounts lack nuanced consideration of the complex processes—including bodily influences—that shape the will.

Moreover, many moral theorists take culpability and responsibility in sex to hinge on the clarity of another's communications that might come to count as consent or refusal. The bad faith we have described in treating another's body as a thing is reinforced by this limited view of morally salient communication in consent-based sexual ethics, which ignores how another as an erotic object is always grasped as broadly and evidently communicative. If subjectivity is enacted in and through the body, as we claim, then freedom, agency, or the will can be expressed not only through verbal communication but also, importantly, through other cues—nonverbal utterances, bodily gestures, postures, and so forth.

Contemporary social science research has called into question the dominant narrative that sexual violation often results from miscommunication, exposing the role of bad faith in the way men in particular interpret the wider range of women's embodied sexual communications. Researchers have demonstrated that women (and others) often communicate sexual unwillingness, refusal, or ambivalence in subtle and implicit ways that are actually quite similar to the rich communication norms that govern other areas of life.[42] However, because of ready-made alibis provided by social discourses surrounding sexuality, men often ignore these embodied communications and discursive patterns in sexual encounters.[43] Here, it seems that a social position of privilege and entitlement enables the bad faith posture of denying one's ability to sense another's agency; the wider meaning of another's bodily expressiveness is ignored, often with grievous effects.

Most discussions of sexual ethics hesitate to valorize such corporeal expression as ethically decisive—that is, as producing claims on another that generate obligations—largely due to reasonable concerns about ambiguity between sexual desire and the will and the unreliability of interpreting the bodily cues of another. However, attending to bodily expression need not rely on the level of transparency expected of explicit communications of consent: agency is enacted through both bodily and verbal expression, and the two ought not to be considered in isolation. Indeed, ignoring bodily expressiveness encourages the manipulation of another's consent—the kind of maneuvering of another to a 'yes' that falls short of coercion but is nonetheless ethically problematic.

When another's bodily and verbal communication are apparently in contradiction, this might indicate a person's genuine ambivalence or the inadequacy of one's own interpretive abilities, and either possibility carries with it an ethical imperative toward hesitation, responsiveness, and ongoing conversation. Thus, this approach does not dismiss the relevance of explicit affirmation or the authority of verbal refusal for sexual ethics. It does, however, suggest that a person ought to have a certain receptiveness or attention to the bodily expression of another, particularly in cases where verbal communication is circumspect. In place of the eroticization of the other's body as passive flesh—which is encouraged by some pervasive, gendered sexual social norms—an ethical orientation calls for eroticizing the other's body insofar as it is receptive, as a body that is unified with an agential subjectivity. The receptive flesh actively receives one not as flesh alone but as a living body—that is, it generates possibilities for relationality rather than awaiting like a canvas.

Conclusion

We have suggested that responses to another's embodied presence and erotic involvement can be morally better or worse, but the moral valence of those responses does not follow only from how they map onto another's authority and dignity as an autonomous subject. Rather, a person's subjectivity in erotic intimacy is intertwined with being an object to another, and the way another appears as a bodily object of perception should be appreciated as morally significant as well. As an aspect of a synthetic totality, the other-as-object embodies the subjective dimensions of their agency, will, and desires. Another grasped as body, gesture, posture, movement, and so forth is contiguous with and inseparable from their subjectivity, and their moral claims solicit the perceiver through that totality. So, moral obligations to another in intimacy spiral out from another's presence, including the expressiveness of their living body, which is more than a bundle of signs that might indicate the content of that person's will or subjectivity. The ethical orientation toward another requires acknowledging this subject-object nature of another person—which they themselves live as ambiguity—and responding to how that sense of another demands changes to one's intentions and behaviors.

The fact that we experience others as objects creates the ineradicable moral risk of harming another through objectification, that is, through denying our own dependence on the other and pretending in bad faith that they are a thing. However, this risk cannot be eradicated through more explicit norms of sexual communication or a more airtight definition of valid consent. People must become more skillful at navigating the ambiguity of intimate encounters, including developing an embodied habituation and attunement to the wider range of expressions of others. While some movement in this direction may be possible through new forms of sex education, a large-scale shift will demand feminist transformation of disciplinary institutions, cultural practices, and discursive norms that limit the erotic possibilities of some while investing others with privilege and a sense of entitlement. A change to how we theorize sexual ethics can be a part of this shift if sexual ethics can valorize the body's agential expressiveness as a source of responsibility between people, while acknowledging the constitutive ambiguity of erotic intimacy.

Notes

1. While feminist analyses do not universally use the template of passivity to describe violation, it is prominent in radical feminist accounts of gender hierarchy, which remain perhaps the dominant lens among theorists examining sexual violence against women.
2. Beauvoir, *The Ethics of Ambiguity*, trans. Bernard Frechtman (New York: Citadel Press, 1976 [1947]), 8.
3. Merleau-Ponty, *Phenomenology of Perception*, trans. Donald A. Landes (New York: Routledge, 2012 [1945]), 366.
4. Bauer, "Beauvoir on the Allure of Self-Objectification," in *How to Do Things with Pornography* (Cambridge, MA: Harvard University Press, 2015), 47.
5. *Phenomenology of Perception*, 452.
6. See Merleau-Ponty, *The Visible and the Invisible*, trans. Alphonso Lingis (Evanston: Northwestern University Press, 1968 [1964]), 133.
7. The term *se faire objet* in Sartre and Beauvoir is often interpreted with a negative valence associated with bad faith and/or situations of oppression (see, for instance, Jennifer McWeeny, "Operative Intentionality," in *50 Concepts for a Critical Phenomenology*, ed. Gail Weiss et al. [Evanston: Northwestern University Press, 2020], 256–61). While this is certainly apt in many cases, *se faire* is also often used as a neutral description of the human condition, especially in its valence of facticity. It is this neutral sense we intend to capture; we discuss the case of bad faith in section three.
8. Sartre, *Being and Nothingness*, trans. Hazel E. Barnes (New York: Washington Square Press, 1984 [1943]), 288.
9. See Ibid., 350. For a more detailed explanation of the distinction between affect and knowledge in Sartre's account of being-for-others, see Ellie Anderson, "Sartre's Affective Turn: Shame as Recognition in 'The Look,'" *Philosophy Today* 65, no. 3 (2021): 709–726.
10. Beauvoir, *The Second Sex*, trans. Constance Borde and Sheila Malovany-Chevallier (New York: Doubleday, 2012 [1949]), 416.
11. Ibid.
12. Ibid., 415.
13. See Bettcher, "When Selves Have Sex: What the Phenomenology of Trans Sexuality Can Teach About Sexual Orientation," *Journal of Homosexuality* 61, no. 5 (2014): 606.
14. Ibid., 614.
15. Ibid., 609. A crucial upshot of Bettcher's argument, though outside our purview, is that this erotic self is gendered. Hence, Bettcher argues that gender is inseparable from the erotic structure of sexual encounters.

16. See Nussbaum, "Objectification," *Philosophy & Public Affairs* 24, no. 4 (1995): 249–291. We adopt the term "body-self" from Ann J. Cahill, *Rethinking Rape* (Ithaca: Cornell University Press, 2001).
17. Note that most accounts of morally transformative consent define consent as a performative or communicative act, not merely as a condition of the will. Nonetheless, where consent is understood performatively, the condition of the will is widely agreed to be decisive for the *moral validity* of consent.
18. See Nicola Gavey, *Just Sex?: The Cultural Scaffolding of Rape* (New York: Routledge, 2005); Ann J. Cahill, "Unjust Sex vs. Rape," *Hypatia* 31, no. 4 (2016): 746–761; and Elise Woodard, "Bad Sex and Consent," in this volume.
19. See Linda Martín Alcoff, *Rape and Resistance* (Cambridge: Polity Press, 2018). For a discussion of a variety of limitations of consent discourse, see Ellie Anderson, "Women in Philosophy: The Limits of Consent in Sexual Ethics," *Blog of the APA* (blog), April 24, 2019, https://blog.apaonline.org/2019/04/24/women-in-philosophy-the-limits-of-consent-in-sexual-ethics/.
20. See, for example, Lee Edelman's contribution to Lauren Berlant and Edelman, *Sex, or the Unbearable* (Durham: Duke University Press, 2013).
21. See Grosz, *Space, Time, and Perversion: Essays on the Politics of Bodies* (New York: Routledge, 1995); Huffer, *Are the Lips a Grave?: A Queer Feminist on the Ethics of Sex* (New York: Columbia University Press, 2013); and Berlant and Edelman, *Sex, or the Unbearable.*
22. See Alcoff, *Rape and Resistance*, 79–95. Alcoff specifically argues against Gayle Rubin, although her argument is equally well-suited as a critique of Janet Halley and others in the contemporary queer theoretical backlash to feminism. See also Rosanne Kennedy and Hannah McCann, "Splitting from Halley: Doing Justice to Race, Unwantedness, and Testimony in Campus Sexual Assault," *Signs* 46, no. 1 (2020): 79–102.
23. Cahill, *Rethinking Rape*, 138.
24. On sexual subjectivity as an agential activity of self-making, see Alcoff, *Rape and Resistance*, 110–47. On narrative self-making after trauma, see Susan J. Brison, *Aftermath: Violence and the Remaking of a Self* (Princeton: Princeton University Press, 2002).
25. See, for example, Nussbaum, "Objectification"; Rae Langton, *Sexual Solipsism: Philosophical Essays on Pornography and Objectification* (Oxford: Oxford University Press, 2009); and Sally Haslanger, *Resisting Reality: Social Construction and Social Critique* (New York: Oxford University Press, 2012).
26. See Ann J. Cahill, *Overcoming Objectification* (New York: Routledge, 2011).
27. Sartre, *Being and Nothingness*, 345–51.
28. Ibid., 346, emphasis original.
29. Merleau-Ponty, *Phenomenology of Perception*, 152.
30. Ibid., 364.
31. See Emmanuel Levinas, *Totality and Infinity*, trans. Alphonso Lingis (Pittsburgh: Duquesne University Press, 1979 [1961]), 187–240.
32. Ibid., 202, emphasis original.
33. On how we "co-inhabit others' perspectives" in this way, see Kym Maclaren, "Intimacy as Transgression and the Problem of Freedom," *Puncta: Journal of Critical Phenomenology* 1, no. 1 (2018): 22.
34. Merleau-Ponty, *Phenomenology of Perception*, 382.

35. Ibid., 190–92.
36. This suggests a phenomenological solution to the so-called problem of other minds. See Søren Overgaard, "Other People," in *The Oxford Handbook of Contemporary Phenomenology*, ed. Dan Zahavi (Oxford: Oxford University Press, 2012), 460–79.
37. Merleau-Ponty, *Phenomenology of Perception*, 158.
38. See, e.g., Langton, *Sexual Solipsism*; and the discussion of sexual objectification in *Evaluative Perception*, ed. Anna Bergqvist and Robert Cowan (Oxford: Oxford University Press, 2018).
39. This might support accounts such as those in Gavey, *Just Sex?*; and Kate Manne, *Down Girl: The Logic of Misogyny* (New York: Oxford University Press, 2018).
40. On the embodied basis for the "intervocality" of voice and speech, see Ann J. Cahill and Christine Hamel, *Sounding Bodies: Identity, Injustice, and the Voice* (London: Bloomsbury, 2021).
41. See Ngaire Naffine, "The Legal Structure of Self-Ownership: Or the Self-Possessed Man and the Woman Possessed," *Journal of Law and Society* 25, no. 2 (1998): 193–212.
42. Celia Kitzinger and Hannah Frith, "Just Say No? The Use of Conversation Analysis in Developing a Feminist Perspective on Sexual Refusal," *Discourse & Society* 10, no. 3 (1999): 293–316.
43. See Melanie Beres, "Sexual Miscommunication? Untangling Assumptions about Sexual Communication between Casual Sex Partners," *Culture, Health & Sexuality* 12, no. 1 (2010): 1–14; and Rachael O'Byrne, Susan Hansen, and Mark Rapley, "'If a Girl Doesn't Say "No"…': Young Men, Rape and Claims of 'Insufficient Knowledge,'" *Journal of Community & Applied Social Psychology* 18, no. 3 (2008): 168–93.

CHAPTER 5

Kant and Arendt on the Challenges of Good Sex and the Temptations of Bad Sex

Carol Hay and Helga Varden

Introduction[1]

A good philosophical account of human sexual love requires an account of our animalistic desires. Although certain aspects of sexual desires don't require such a reference—for example, some aspects can be analyzed by reference to our imaginative minds and social senses of self—a full account of human sexual desires is impossible without an account of our animality. After all, sexual desire is something we feel in a distinctly embodied, animalistic way: when I—that is, subjectively, from an internal, self-conscious first-personal point of view—feel sexual desire very strongly, I feel my body very strongly. When I feel my body in this way I strongly want or long for the attention, and typically touch, of another, along with some kind of intimacy and/or physical union with them.

Conversely, when I am experiencing sexual frustration, I may, for example, feel my body as a little in the way, as annoyingly and persistently demanding attention, or as betraying me by making me focus on something I don't really want to focus on, namely sexual satisfaction in general or with regard to some person(s) in particular. Alternatively, sexual frustration can manifest as the body being experienced as having an unfulfilled hunger or appetite, such as when you want to orgasm but can't achieve it—which is a little like having an itch

C. Hay (✉)
University of Massachusetts Lowell, Lowell, MA, USA
e-mail: carol_hay@uml.edu

H. Varden
University of Illinois at Urbana-Champaign, Champaign, IL, USA
e-mail: hvarden@illinois.edu

D. Boonin (ed.), *The Palgrave Handbook of Sexual Ethics*,
https://doi.org/10.1007/978-3-030-87786-6_5

that you can't scratch or being thirsty with no water available. Sexual frustration can also be the opposite: you might be able to bring yourself to orgasm, but you don't enjoy the process leading to it, so you sprint for a somewhat unsatisfying end. Another kind of frustrated sexual desire involves experiencing your body as betraying you because it makes you feel sexual desires that you do not know how to own or manage, or that make you feel ashamed, embarrassed, or vulnerable. For example, if you have been taught that the only "normal" sexual desires are heterosexual and procreative in nature, finding yourself in situations where you feel a desire that doesn't fit this conception of good sexual love can make you feel betrayed by your body.

What is it about our nature that makes us feel in these ways or have these desires? What is their value (if any)? How we can realize them well? Part of the challenge of articulating a philosophical account that can answer these questions is that having sexual desires is not, as such, distinctive of human being—non-human animals have these desires too, though somewhat differently. Good philosophical accounts of sexual love therefore need to capture both the similarities and the differences between human and other animals' sexual desires, including how only human animals are morally responsible for their sexual desires—for developing, transforming, and integrating them into their lives in good ways. In philosophical accounts of human nature, the exploration of these themes therefore non-accidentally involves an analysis of human animality, including how it is complemented by our distinctively human biological, social, aesthetic, rational, and temperamental aspects. In addition, such accounts must address our human tendency to engage in self-destruction, or the destruction of others, by disrespecting or attacking our sexually loving aspects.

Sexual love was not Hannah Arendt's intellectual focus, and many of Immanuel Kant's writings on the topic appear unpromising and hostile or even antithetical to any interesting theory of sexual love.[2] Despite this, in what follows we use philosophical resources they left us to explore these themes. In particular, we propose that Kant's account of human nature—good and bad—in conjunction with some of Arendt's thoughts on the problems with the Western philosophical tradition's engagement with animality, yield a good starting point for exploring the nature and value of good sexual love for many.

In Kant's philosophical framework, understanding sexual love requires an analysis of how we develop our "natural vital force" through our "animality" (our conscious strivings for self-preservation, sex, and affectionate community), which is distinct from and yet is also integrated, developed, and transformed by our "humanity" (our social sense of self and capacity to set ends of our own rationally), our "personality" (our moral feeling or moral vital force, or our ability to act as motivated by practical reasons), and our aesthetic-teleological abilities (our susceptivity to aesthetic pleasures and to engage the world through parts-and-wholes[3] rationality). Moreover, our analysis of sexual wrongdoing will use Kant's account of our propensity to evil, which includes an analysis of how affect and passions block or corrupt our thinking powers such that we neither experience ourselves as harmonious nor as rationally coherent and properly responsive to calls from our conscience. (More on this shortly.[4])

In Arendt's (Heidegger-inspired) framework, correspondingly, sexual love becomes determined by how one views the elements of what she calls "labor" (reproductive activity, which is sustaining of biological life and grounding of a private life); "work" (human life as enabled by our capacity for creating a human world by crafting useful objects, such as roads and houses and tools); "action" (human political life as enabled by our individual ability to use speech and act spontaneously to create a political society with a history); and "thought" (human theoretical life, including academic life as enabled by our ability to think and reason through abstract concepts).[5] Moreover, on her account, human evil centrally involves our inability to orient ourselves by a commitment to truthfulness and rational coherence in response to facts.[6] She relatedly and (in)famously proposes that some of the worst of evils are "banal,"[7] which she later explores through the concept of "thoughtlessness."[8] Like Kant, she thinks that human evil is characterized by incoherence and conflict internal to each person.

Arendt is one of the richest thinkers on the human tendency to evil, and one of her remarkable theoretical abilities is to identify blind spots and unintended consequences of assumptions or practices in the Western philosophical tradition. One such suggestion is that philosophy as a practice typically has ranked these various valuing activities (animalistically, biologically, socially, aesthetically, rationally) in a hierarchy, where philosophical contemplation reigns at the top. In contrast, activities involving labor (in her system) or our animality (in Kant's system) are typically found toward the bottom of the scheme. She proposes that this philosophical practice is interwoven with societal practices and structures, such as the activity of labor or our animality being the work of "natural slaves" or women, and as inherently private (in the home) rather than public in nature. In contrast, philosophy has typically identified the highest human ideal—the one closest to being as such or god—as philosophical search for eternal truths. The best human life, therefore, is thought to be philosophical contemplation of eternal principles constitutive of the universe (and not merely the earth), principles that our rational capacities are seen as giving us access to. One benefit of Arendt's analysis is that it enables us to see how racist, sexist, and heterosexist thoughts in classical philosophical texts often are internally related to a general disdain for animality. It can also help us see how members of these groups are often closely associated with their animality (so understood), and hence, like other animals, viewed as permissible to use or as needing to be controlled or managed. And it can help us see how these people are sometimes viewed as needing to be corrected or stopped, because they are allegedly even perverting animality. Moreover, since our animality in general has been viewed as a constitutive or primary component in private activities—the sphere of women and slaves—it follows that we should not give it priority in our lives if we can, and those who value or even prioritize engaging in these activities reveal their lower or perverted nature (or soul), namely one that seeks to function only in accordance with our "merely" biological, reproductive, or animalistic desires. Needless to say, on such conceptions, envisioning diverse, flourishing sexually loving human lives consistent with human dignity becomes impossible.

In *The Human Condition*, Arendt suggests that in modern life this traditional problem of philosophy has become a wider problem of society. She suggests that the problems related to a wrongly devalued animality moved from being primarily a philosophical problem—and possibly a problem for philosophers' lives—to becoming a general societal problem because modern technology and labor movements enabled the material conditions of human life to become less preoccupied by domestic life-sustaining activities. This also led to a blurring of distinctions between different kinds of valuing activities, and a lack of appreciation of the differences between the private and the public spheres, which in turn fostered overblown social senses of self. As contemporary social media shows so clearly, we have a serious temptation not to protect our own private lives and to let our sense of ourselves be determined by how others see us. As we will see in the final section, this analysis is helpful for shedding light on some of the challenges facing anyone who wants to realize a flourishing sexually loving life.

In the next section, in contrast, we put forward a Kantian account of human nature, and an interpretation of central features of good sex, that can overcome this problem in the philosophical tradition—including in much Kantian philosophical thought—and in modern human practice. More specifically, we use the writings of Kant to develop an account of the resources and challenges internal to us—as emotionally complex, particular human beings—that make sexual love a joyous and deeply meaningful, yet also difficult, life-long project to realize in good, sustainable ways. We argue that on this approach, good sexual love channels sexual desire that is grounded in our natural vital force in ways that are in line with our temperaments, are aesthetically and teleologically pleasing to us, and are pursued within moral boundaries set by our genuine commitment to truthfulness and to treating ourselves and others as having dignity. Hence, good sexual love non-accidentally involves being able to do things like form meaningful (even if not necessarily permanent nor even long-lasting) relationships of vulnerable and yet empowering intimacy with others, to appreciate and enjoy one's own embodiment in self- and other-regarding ways that are nourishing or conducive to flourishing, to become increasingly able to be genuinely curious, kind, open-minded, stabilizing, and playful around one's own and others' good sexual ways. This enables us, in section three, to focus on specific patterned challenges that face us given the modern historical societies—with their oppressive forces—we are inheriting.

Human Animality

In our view, good philosophical accounts of animality must accommodate the fact that many people experience certain distinctive and relatively constant fundamental drives in their sexually loving ways, both with respect to how they prefer their own body to be in these regards and to their preferences regarding others' sexually complementing bodies. That is to say, whether the focus is on sexual orientation or gender identity, a central component behind all of the letters LGBTQIA concerns how we experience ourselves as embodied, sexually loving beings. In the Kantian philosophical framework, some of this requirement can be captured by Kant's account of how sexual love necessarily involves our natural vital force as given direction by our animalistic sexual drive.

Second, on this Kantian account, our animalistic sexual drive is understood as fundamentally only reflexively, and not also reflectively, self-conscious.[9] This proposal opens up the possibility of a third option when it comes to sexual/gender identity and sexual orientation, between so-called (rationalistic or empirical) essence and (existential or constructivist) choice accounts.[10] Rather than viewing our sexually loving ways as something that is given at birth and then realizes itself through us more or less successfully (as many empiricist and rationalist essence accounts ultimately do) *or* as something we ultimately simply choose in more or less free ways (as many existentialist and (postmodern) constructivist accounts do), sexual love is viewed as typically having some phenomenological—subjective, reflexive, first-personal—tendencies or directions. There is, in other words, something first-personal to get right about these things, and getting it right can neither be reduced to an empirical (third personal) question, a (purely self-reflectively available) rational idea of soul types, or the fact that our rational reflective capacities enable us to choose (with more or less resistance) in response to our desires.

Moreover, a general suggestion here is that in order to capture sexually loving desire's typical motivational strength, as well as its somewhat unruly nature, categories like Kant's "natural vital force" and "animality" must do important philosophical work. These ideas enable us to capture ways in which there are reflexively (and not reflectively) self-conscious aspects to these parts of us. There are conscious, animalistic strivings of a self centrally involved in sexual love, but this doesn't mean that this striving is inherently reflective all the way down, although it is accompanied by reflective powers from the start. Rather, on this approach, realizing one's sexually loving capacity well involves learning to attend to how one feels a complexity of concerns with regard to one's own and other's sexual loving ways and how to develop, transform, and integrate these areas of ourselves into stable wholes that are consistent with treating oneself and others as having dignity. Hence, viewing our sexually loving ways as having certain, basic phenomenological tendencies is compatible with how mature and flourishing sexualities typically are characterized by a genuine openness to fluid experiences, new developments, and transformations. For example, it is not uncommon that people who both genuinely like sexual love and like living together in couples also enjoy occasional sexual experiences involving more people, such as those available in sex clubs.

On the Kantian account advanced here, the general story of our sexually loving selves goes something like this: we all have an embodied forcefulness (a natural vital force) that has three animalistic prongs to it, as they are enabled by the three relational categories of the understanding: self-preservation (substance), sex drive (causality), and affectionate community (community). These are conscious strivings of a self, and when things go well and these strivings are satisfied in good ways, we feel strong and harmonious.

To bring out the structure of this Kantian account, sketching an ideal of what good child development looks like, as well as some common challenges internal to this development, is useful. To start, on this account, a human baby drinks milk as long as the baby is hungry (as long as drinking is pleasant and the pain of hunger is going away) and as it is doing this, it feels strong and increasingly content until there is no more pleasure from the activity (and discomforts make an entrance). In Kantianese, this baby is striving to preserve its animality, and in so doing it feels its natural vital force as strength and this guides the activity (through pleasure and pain). This striving is, then, the baby's nascent ability to sustain itself—self-preservation—as a biological being. Similarly, a human baby is calmed and feels pleasure at being held close and being attended to affectionately; basic community (or sociality in this sense) is also a natural striving for us. Finally, a human baby is incredibly receptive to human touch, which is not only a way in which we seek basic community, but also reveals our innocent, nascent sex drive; a baby experiences it as pleasant to touch others and feels pleasure in light physical, playful touch. As with eating and affection, however, this is a nascent ability, the development of which is constitutive of growing and maturing. For example, as babies, after we have eaten ourselves full, been attended to in gently, sometimes playful ways, and then held in affectionate ways, we feel content, safe, and peaceful—and as this happens, we become ready to rest and often fall asleep. To express this point in Kantianese, we can say that in this case, our natural vital force is successfully attended to, engaged, and developed by our conscious strivings for self-preservation, sexuality (touch), and basic community. In addition to these natural, animalistic drives, Kant proposes that, from the start, we can see nascent human freedom present in the fact that only human babies scream: the scream is only possible to explain philosophically if we attribute to the baby the capacity to represent its inability to act as a frustration. A baby cannot do anything yet, which is experienced as extremely frustrating—and so we scream when we are born (A[11] 7: 268).[12] Also, notice that if all of these three animalistic drives are realized well together as a whole and the frustration regarding freedom is handled well by the caregivers, a baby is harmonious and strong; in Kantianese, its natural vital force is experienced as strong and calm. Correspondingly, caregivers' early aims centrally involve facilitating an environment in which a baby can strive in such unified, good ways.

The story about good human sexual love, in turn, becomes a story where one develops this starting point with the rest of Kant's account of human nature (humanity and personality, in addition to animality) as well as with other ideas from his aesthetic-teleological thought and practical philosophy.[13] For example, as a baby develops a social sense of self (as revealed in smiling at and with others and also starting to have emotions revealing a longing for attention, including as revealed in jealousy) and the ability to act (as evidenced in activities like reaching for objects, rolling around, crawling, walking, etc.), the baby is simultaneously developing several cognitive abilities. As we learn to navigate the world, we learn to associate objects with pleasures and pains, to master abstract concepts, to manage the power of choice (at first evidenced in the period of learning the power to say "no" when asked by caregivers if one wants various things), to experience aesthetic pleasures of the agreeable (what one likes) and of the beautiful and the sublime (to be aesthetically deeply pleased or struck), and to hold things together in teleological wholes (including as stimulated by various children's games involving combining). Moreover, as our baby selves are learning all of this, at some point, we are also able to relate to what we are doing not only through abstract concepts ("mummy," "milk," "play"), but also as an "I" (we switch from describing ourselves in the third person singular ("Hannah") to the first person singular) and we start to display and develop temperamental qualities. Some of us, for example, reveal ourselves as having more introverted or more extroverted emotional patterns, or, if we use Kant's categories, we reveal ourselves as having "sanguine" (joyous and spontaneous), choleric (dramatic and in need of much attention), melancholic (forcefully serious, thoughtful, and careful), or phlegmatic (calm and somewhat inactive or deeply reflective) patterns. In themselves, these temperamental patterns are neither good nor bad, but rather involve specific kinds of emotional dis-/advantages with regard to mastering feeling or activity.

As we are learning how to do all of this—and we'll get back to spelling this out regarding sexual love in a moment—we also experience the propensity to do bad things. On Kant's account, this characteristically involves three kinds of experiences: we have experiences where we know we shouldn't do something, but then we do it anyway ("frailty"); we realize that sometimes we do what is right, but that the reason why we do it is not praiseworthy or stable ("impurity"); and finally, we do bad things in the name of doing good things ("depravity"). Also, throughout all of this learning, transforming, and integration, we become increasingly able to act on practical reasons, meaning that we become increasingly able to use abstract concepts to set ends we want to pursue (act on "maxims") as well as to do or not do something just because it is right or wrong to do (act on a moral motivation or out of "duty"). The human being, in other Kantian words, becomes susceptible to "moral feeling" (the moral "ought") or develops a "personality" and a "conscience"; we become capable of moral responsibility. Living life as a growing and maturing little human being—or its caregivers—is, in other words, not easy, and complications are added by the fact that we live our first many years (before we are capable of even minimal moral responsibility) within the bubble of our given families with all their (often rationalized) immaturities and dysfunctions.

On this approach, it is therefore important to realize that we start this project in nascent ways much before we are capable of morality and we are extremely dependent on having caregivers around who provide us with an environment in which we can safely make this our own project—one that hence starts from a place of sexual innocence. In part, providing such an environment involves setting clear boundaries around each child in the home, boundaries that no one is permitted to intrude upon. The bare minimum in this regard is obviously for the caregivers not to cross these boundaries themselves by engaging their children in sexual activities or by permitting anyone else (adult or otherwise) to do so.[14] However, generally, a good home also involves intimate personal space for each where one can safely be with oneself and which everyone else regards as crucially important to stay out of. In addition, the challenge is for caregivers to provide safe support and non-judgmental guidance as their child tries to figure out how to love sexually in ways that are consistent with their self- and other-regarding sexual desires, their need to learn to preserve themself (to love sexually safely), and their need for affectionate love and care from others. Central here is therefore not only to learn how to feel sexual pleasure strongly, including how our own bodies work in these regards, as well as to learn to try and fail and try again, but also to manage not simply always to act on sexual desire or not to act in self- or other destructive ways. In addition, we have to learn how to develop, integrate, and transform our sexuality with our appreciation of aesthetic pleasure—feeling oneself and others as lovable and even beautiful and/or sublime—by developing attention to the ways in which others can both complement us in wholes of various aesthetic kinds and strike us as pleasing, wonderful, and/or sexually exciting to be around. And, as mentioned above, we have to learn to do this as the kind of persons we are. If we typically feel strongly and deeply over time (melancholic), for example, or instead tend to feel strongly and playfully in the moment (sanguine), then it's important to take this into account as we pursue sexual love. We have to figure out what we can handle emotionally and what is best for ourselves. For example, is it best for us to sexually love someone who is a little similar to or who is a little different from us in these regards, or does this difference not really matter? We have to learn to manage emotions of comparison, such as envy and jealousy, and we must learn how to reject or be rejected by others. What is ours to have is both something we want and the other wants us to have.

To illustrate this point with Kant's accounts of affect and passion,[15] Kant suggests that affect is prior to reason or involves barring or numbing reason, while passion involves deeply wanting something I cannot have—something that isn't mine to have—and not being able to treat it as such by obsessing about having it. So, if I find someone very sexually attractive, this can involve affect: I can find myself having a feeling that I don't want to explore conceptually (think about) but just go with it and see where it takes me. If the other person also wants this, then there is mutual affective attraction. This basic attraction is something the attracted parties can—individually and together—develop further by learning to understand the nature of the attraction, such as how to stimulate it further or turn it into a stabilizing and constructive force in their lives—by using associative, abstract conceptual, aesthetic-teleological means that are attuned to their temperaments and situations.

If the affect isn't mutual or if it stops being mutual, however, then the danger of passion develops. In fact, Kant argues that sexual love itself can develop into a damaging, obsessive passion is if it is not reciprocated (A 7: 266). Moreover, he argues that unrequited love, "[l]ike all passionate longings ... gnaw[s] and consume[s] the heart or, so to speak, bind[s] the vital force with shackles" (PMF[16] 15: 940) and "the capacity of the understanding is of little help against it; for the end of enchanted human being sees very well indeed the reasons against [their] favorite inclination, but ... [they feel] powerless to give them active emphasis" (MH[17] 2: 261).[18] For example, after a breakup, the danger is that not only affect—say, anger, disappointment, or sadness—but also that passion will develop because one is or both are unable to move on, and instead they obsess about what history has shown doesn't work well, such as by it not being transformable into something stabilizing. Hence passion is revealed in how one keeps prodding at something even when doing so doesn't make rational sense and entails that one gets trapped in the past and obsessed with what clearly cannot be. Importantly, such obsession isn't properly responsive to reasoning in that when one obsesses in these ways, one knows that doing so is bad. The reason why reflective cognitive powers are somewhat powerless with regard to passion is exactly because sexual love is, at its base, animalistic and only reflexively self-conscious. Knowing that there is nothing one can do or that the relationship was destructive in nature is sometimes not enough for us to be able to let go as the feelings involved are strong and somewhat irresponsive to practical reason (moral reflection).

On the flip side, strong sexual love is also unruly because of how it is so very energizing. It can give us subjective experiences of strong natural vitality as well as of being deeply attractive, important, and good. It can empower us, in other words, with aesthetically charged, complex subjective experiences of our animality, humanity, and moral feelings—individually and as a package. Moreover, developing sexual love well—including as it involves giving oneself to others in vulnerable, playful, and rightly trusting ways, and giving them affirmation and intimate attention—makes it easy to associate one's sense of self *only* with the other's presence, rather than holding onto it as also something one has in oneself. Losing such a relationship—whether through heartbreak or death—is consequently often experienced as ungrounding and/or as giving rise to destructive, passionate desires of obsession. Oftentimes, the best we are able to do for a while might just be to avoid doing bad things and to try to keep going on one's own in a different direction, since letting go fully or learning to live with the fact of how life will be from now on is emotional work that is partially unreflective and constitutive of our selves all the way down.

It follows from the above account that developing our raw sexual desires into more mature versions centrally involves three main activities of development, transformation, and integration—namely those involved in realizing it as integral to developing also the other parts of our animality (self-preservation and basic community) as well as our humanity (social sense of self and rational end-setting) and personality (moral responsibility)—and to do all of this in ways attuned to our type of temperament and aesthetic-teleological sentiments. All of these emotional structures are resources if we dare to learn to attend to, become familiar with, and understand them. In addition, developing our sexual desires constructively requires us to learn to manage our unruly temptation to act destructively toward ourselves and toward others (evil) and to support our partners in their efforts to do the same—and to hold ourselves and each other accountable when we fail. Moreover, in all these projects of healing, managing, integration, development, and transformation, we use combinations of associative, abstract conceptual, and aesthetic-teleological tools. If we do all of this well, our sexually loving way becomes constitutive of what Kant calls "the highest good," namely our individual unions between happiness and morality, where our commitment to morality—to treat others as having dignity—sets the framework within which we pursue happiness in the various situations in which we find ourselves.[19]

Another core Kantian proposal that we consider important is his suggestion that the most important general duty we have is to learn to be truthful. Kant's judgment is that most people never take on this challenge, and those who do typically don't make this commitment until in their thirties. Moreover, he suggests, if we look at those who really do try, becoming more reliably successful at truthfulness doesn't tend to occur until in their forties (A 7: 294). Why such emphasis on the importance of truthfulness? One reason, in this context, is simply that without truthfulness, it isn't possible to develop, transform, and integrate one's sexual loving aspects in associative, abstract conceptual, and aesthetic-teleological ways that are harmonious with one's natural forcefulness and temperament. After all, if one is not truthful, one will not learn to feel one's natural vital force strongly, nor how various kinds of engagements with oneself and others are consistent or inconsistent with harmony and stability in oneself, nor how to describe one's feelings accurately, nor to learn not to act just because the feelings are intense or fiery or deep, nor to be stably around and feel strong feelings without fear. Moreover, without truthfulness, one is much more likely to feel a lot of internal conflict—feelings going strongly in many conflicting directions—exactly because one is not able to ensure that what one is doing is deeply consistent with one's natural vital force, one's temperament, and one's aesthetic-teleological sensibilities. This, in turn, gives rise to the temptation to numb or suppress one's feelings or let them have fiery outlets of anger or sadness, which only can give rise to more dysfunction and destructive pathologies.

Another core reason why truthfulness is so important on this approach is that it is constitutive of learning how to develop, transform, and integrate all these emotional aspects of oneself in ways that are respectful of oneself and others as having one's own life to live—a life that we should treat as precious or as having dignity. More specifically, truthfulness is necessary to enable one's conscience to function properly: to hear the call of never treating oneself or others as mere means (to obey our practical reason's command of "perfect duties," as Kant would say) and to feel the obligation to learn to be kind, non-judgmental, and supportive of ourselves and each other in our efforts to feel and live better, more fully on our own and together (to obey our practical reason's command of "imperfect duties," as Kant would say). Living in this way is also only possible if one strives to develop a reliable ability to do what is right because it is right when this is called for (to develop "moral character," as Kant would say). Again, sometimes this only involves learning to let go or not to act even though one feels very aroused or sexually excited (so, not to act on affect); other times it is to learn to move on after somebody whose sexual love has been very important is no longer there, such as when sexually loved ones die or when the sexual love changes direction and attention (so as to not develop passion).

Overall, then, our aim regarding sexual love is to strive for a sustainable union between it (as part of our pursuit of happiness) and morality. Because of the emotional complexity of the project—including the different degrees of reflectivity and motivational strength of the various components and the fact that everybody else too has this complexity—realizing a good sexually loving life requires one to address and deal with much arrested development and dysfunction in oneself and one's sexually loving partners. And because sexual love naturally is also rather fiery[20] and unresponsive to reason, it can be tempting to give up on it and instead settle for trying to focus the energy on making the organic whole of one's animality sustainable through time by striving for predictably comfortable, pleasing sex with one partner and for affectionately loving this partner (and possible children) in self- and other-preserving ways. Saying that sex can also be more than this is not to say that there's anything wrong about predictably comfortable sex as such, nor to deny that these kinds of sex have a good and stabilizing role for many or in many circumstances. Rather, the point is simply that for many people sex can be more—including less predictable, more challenging, and more uncomfortable without thereby being bad—and that allowing and pursuing more when life allows can be a wonderful part of life.[21] It is typically not easy, though, given the societies we inherit with their histories of oppression and the related dehumanizing violences we can become subjected to.

Patterns of Dehumanization and the Possibility of Sexually Loving Lives

While our sexuality has a certain basic phenomenological direction to it, because our feelings are susceptible to associative thought, and because much oppression is expressed in rationalized ways, it is also the case that various social norms and institutions shape our sexualities in ways we are unaware of. The social world in which we live shapes our sense of which forms of sexual love are possible and which are deemed valuable and desirable for people like us. It informs which sexual experiences are open to us, which are encouraged or incentivized, which are forbidden (though possibly still desired), which feel frightening or unimaginable. Through an interrelated set of social rewards—ranging from realizing the possibility of being intelligible to oneself and one's community, to peer acceptance and support, to elevated social status, to being considered worthy of finding sexual connection or love—and social punishments—including harassment and assault, social ostracism, and outright group expulsion (or even just the mere threat of these things)—our sexualities are profoundly influenced by the families and communities in which we find ourselves. Many find themselves torn between the draw of having sex that strengthens their natural vital force and asserts and realizes who they actually are, on the one hand, and the limiting forces of social acceptance or intelligibility, on the other. But the human imagination is powerful, and our sexual desires tend to be unruly, and can refuse to bow to the powerful constraining forces of convention.

Arendt's account of how both philosophy and modern life undervalue our animality can make sense of some of the ways that oppressive forces cause distinctively sexual harms or injustices. For one, members of oppressed groups tend to be associated with their animality in ways that others are not. Simone de Beauvoir relatedly called out this tendency to depict women as "think[ing] with [their] glands," pointing out that "Man superbly ignores the fact that his anatomy also includes glands, such as the testicles, and that they secrete hormones. He thinks of his body as a direct and normal connection with the world, which he believes he apprehends objectively, whereas he regards the body of woman as a hindrance, a prison, weighed down by everything particular to it."[22] Women are still too often portrayed as beholden to their animality in ways men are not, held hostage by their irrational hormones, tied down to the animal functions of bearing and suckling children, incapable of transcending the immanence of mundane domestic realities in the ways men are. Women's unbridled animality is also portrayed as precariously unruly, as ironically both dangerously tempting to men and in need of male control, resulting in a variety of social and religious practices that valorize feminine modesty and make women responsible for men's lust. Similarly, Black people have been characterized as animalistic, both brutish and hypersexual, resulting in depictions of Black men as dangerous predators and Black women as "unrapeable" Jezebels. By associating these groups of people with what are ostensibly the least valuable aspects of human nature, these people are marked variously as not worthy of respect, as deserving harmful treatment, and as needing to be controlled both for their own good and for the good of society.

The above Kantian account of human nature, in conjunction with the Arendtian account of the distinctive pitfalls of our overblown modern social selves, can shed further light here, bringing into focus problems with the kinds of sex you might think we should be having but are not, and with the kinds of sex we are having instead. One way these social considerations can get in the way of the sex we should be having has to do with how the new identity categories that arose in modernity encourage in-group/out-group identifications that lead to division and conflict. These identifications result in patterned forms of social exclusion that have and will continue to prevent certain loving sexual relationships from even having the chance to begin. Racial and ethnic categories, for example, have been reified and interracial or inter-ethnic sexual relationships historically discouraged or outright forbidden. Anti-miscegenation laws that criminalized interracial marriage (and sometimes also sex between members of different racialized groups) originated in the late seventeenth century in North America and remained on the books well into the twentieth century. Nazi anti-miscegenation statues discriminated against Jewish, Roma, LGBT+, and Black people in Germany; laws in fascist Italy forbade marriages between Italians and Jews or Africans; even today, in some countries, marriages between, for example, Jewish, Christian, and Muslim people are culturally or even legally forbidden, restricted, or discouraged. Of course, sexually loving relationships can and do manage to exist across these arbitrary social lines, and the prospect of breaking a taboo can make boundary-crossing in such contexts sexually exciting. But doing so ethically requires, among other things, vigilant truthfulness and learning to feel and trust what is deeply good for oneself, insofar as such relationships inevitably involve coming up against prejudices held by oneself, one's partner(s), and society at large.

Another considerable temptation of bad sex arises when we unthinkingly adopt the identity categories foisted upon us by society. The risk of doing so is the risk of living a life of sexual bad faith, one that fails to stay true to ourselves. The existential importance of being able to live out one's sexuality in good ways is evident, at the extreme, in the higher suicide rates for those who are denied the possibility of doing so.[23] But for too many, a life of sexual bad faith will have us seek out actions or interactions we don't actually find sexually satisfying or fulfilling, but fall in line with what society says that "people like us" are supposed to want. Deeply entrenched heteronormative biases will thus discourage people from exploring queer desires, for example. Classist and racist biases will discourage people from exploring sexual relationships with people from different socio-economic and racialized backgrounds. Ableist biases discourage able-bodied people from exploring sexual relationships with people with disabilities and vice versa, and they discourage people with disabilities from pursuing any sexual interactions whatsoever. Transphobic biases discourage people from expressing their true gender identities, and they discourage cis people from pursuing sexual relationships with trans people.

What's worse, we impose these identity categories onto our children, denying them the opportunity to discover and explore their true sexual or gender identities. Compulsory heteronormativity, and its attendant homophobia and transphobia, denies many the opportunity to live authentic LGBTQIA+ lives. Racism and classism prevent them from forming sexual connections with others across racial, ethnic, or class lines. Regressive prudishness and loathing of animality (often religiously rationalized) give rise to repression and shame instead of acceptance and encouragement. Socio-economic class inequalities force many parents into low-paying jobs that leave them with little time or resources to direct toward protecting their children, much less promoting their sexual flourishing. Because racism is positively correlated with both socio-economic oppression and a higher danger of being sexually assaulted, this often puts POC families in a radically difficult situation in these regards.

Just as troubling as the sexual experiences and relationships we are discouraged or forbidden from pursuing are some of those we are encouraged to pursue in their stead, some of which are internally connected with dehumanizing violence and oppression. Stereotypes are often instrumental in this process, providing people with limiting social scripts that funnel them into personalities and behaviors that numb their *natural* vital force, and lead them to desire and pursue sexual interactions that are not merely unfulfilling but are immoral and hence numb their *moral* vital force. Gendered sexual stereotypes that characterize men as naturally driven by lust and women as naturally chaste, for example, lead to eroticized relationships of gendered dominance and submission that many feminists argue are directly responsible for the disproportionate rates of sexualized violence against women. Racialized sexual stereotypes such as that of the Latina spitfire and Latino machismo, or, again, of Black men as lustful brutes and Black women as unrapeable Jezebels, are internally connected with more sexual violence against women of color and unfair and ungrounded accusations of sexual violence imposed on men of color.

Another way that stereotypes influence the content of people's sexual desire and are internally connected with much dehumanizing violence and oppression occurs through the mechanism of mainstream pornography. Too many pornographic scripts fetishize the objectification, degradation, and humiliation of women at the hands of men. According to Catharine MacKinnon,

> Pornography sexualizes rape, batter[y], sexual harassment, prostitution, and child sexual abuse; it thereby celebrates, promotes, authorizes, and legitimizes them. More generally, it eroticizes the dominance and submission that is the [common] dynamic. It makes hierarchy sexy and calls that "the truth about sex" or just a mirror of reality. … What pornography does goes beyond its content: it eroticizes hierarchy, it sexualizes inequality.[24]

Mainstream pornography entrenches sexist stereotypes about what men's and women's sexual natures are supposed to be like: men are portrayed as sexually active, aggressive, and dominant, while women are sexually passive and submissive. Adherence to these narrow sexual stereotypes can encourage people to seek out sexual interactions that are unsatisfying at best, and demeaning or degrading at worst. When such pornography takes the place of a better informed, comprehensive sex education, it stunts people's sexual imaginations and can make them feel compelled to perform certain sexual roles not because doing so actually expresses their true sexual desires, but because they don't have a robust sense of what more authentic alternatives might be.

The concern here is not merely that such stereotypes deny people the space to explore their sexualities in authentic ways, however. The deeper concern has to do with how social and historic forces construct stereotypes that make it difficult or impossible to have morally permissible and emotionally healthy sex—sex that properly treats ourselves and others with respect and that for many can be an important component of a flourishing life.

Kant's thoughts on objectification can fruitfully illuminate failures of respect that arise when people are used as mere means in a sexual context—either when their sexual wishes or desires are ignored, or when they are treated as if what they experience in a sexual encounter does not matter. As explained above, because sexual desire is importantly only reflexive, not reflective, and because of the danger of affects and passion, there is a danger, according to Kant, that sexual desire results in such objectification. Because sexual desire and pleasure can bring us to subjective experiences of urge in such acute and powerful ways, they can drive out all reflection (other thoughts), including thoughts of the moral respect we're required to have for others. We can bring ourselves into a condition of being turned on that feels like our brains fritz in such a way that we can block our conscience's demand for us to respect someone and make sure that what we want is also what they want. "In that condition of mind," characterizes Martha Nussbaum, "one cannot manage to see the other person as anything but a tool of one's own interests, a set of bodily parts that are useful tools for one's pleasure, and the powerful urge to secure one's own sexual satisfaction will ensure that instrumentalization (and therefore denial of autonomy and of subjectivity) continue until the sexual act has reached its conclusion. At the same time, the keen interest both parties have in sexual satisfaction will lead them to permit themselves to be treated in this thinglike way by one another."[25] Kant thought this is a two-way street, that men and women do this symmetrically to each other. To desire someone sexually, then, can involve wanting to objectify them and wanting to let them objectify you. This is morally dangerous or problematic (depending on the interpretations), Kant thought, because it can lead us to both fail to respect other people and to fail to respect ourselves—and, so, can be damaging to all parties involved.[26]

Kant himself didn't realize this, but many such failures of respect are patterned in gendered ways. Contemporary feminists such as MacKinnon and Andrea Dworkin have taken up this Kantian account of what is wrong with sexual objectification but unlike Kant they do not believe that these dangers or problems are inherent to sexual desire itself. Instead, they lay the blame at the feet of how we are socialized erotically, into a paradigm shot through with gendered hierarchy and domination. The failures of respect resulting from sexual objectification are not symmetrical, according to these feminists. Rather, in ways reminiscent of Beauvoir, these thinkers argue that men are taught to experience sexual desire by dominating and using women, and women are taught to experience sexual desire by being dominated and turned into sex objects. Moreover, these gendered asymmetries in our erotic imaginations have real-world consequences. MacKinnon and Dworkin argue (somewhat controversially) that the eroticization of these gendered relationships of dominance and submission is directly causally responsible for women's disproportionate risk of sexualized violence at the hands of men. Rae Langton argues that these pornographic scripts create a social world in which women are objectified, depicted as existing "for sex," and where their attempts to refuse sex are portrayed as integral components of erotic scenes.[27] Even if a direct causal connection between these pornographic tropes and sexualized violence against women cannot be established, the morally problematic patterns are clear enough. Ours is a social world in which people are encouraged to internalize gendered identities and erotic scripts that make deep, flourishing sexual love impossible to experience for many and that make morally respectful interactions very difficult.

Concluding Remarks

In the societies in which we currently live, realizing sexual love well in less traditional ways—queer sex, for example, or kinky sex, or BDSM sex, or sex therapy, or sexual experiences enabled by sex experts—typically requires wisdom with regard to how to protect and realize one's sexuality in flourishing ways, as well as safe spaces in which desires can be explored. Such spaces can offer a respite from the oppressive forces of social convention precisely because these communities have had to struggle at the margins for social recognition and acceptance, and even the right to exist at all. Members of these communities are more likely to have done the work of reflecting deeply on their own desires, and are thus less likely to be susceptible to bad faith in important regards, because these communities are born of the need to learn to know one's authentic desires when acting on prominent social conceptions is impossible without being self-destructive. Such communities also tend to be profoundly incentivized to ensure that their members' desires are pursued in morally permissible ways, and thus tend to have highly evolved norms of sexual communication and consent—including, for example, practices such as safe words.[28] These spaces are also ones in which serious violence can occur, however, including by being tempting to people who have serious inabilities to own who they are and are thus inclined to engage in behavior that is destructive of both themself and others. Still, communities such as these, which typically exist at the margins of society, often have insights that would be helpful also if they were taken up by the mainstream.

Notes

1. Thanks to Lucy Allais, David Boonin, and Shelley Weinberg for their generous help along the way of writing this paper.
2. For more on Kant's philosophical mistakes and deep, personal discomfort around these issues as well as references to much of the related literature, see Helga Varden (2020). *Sex, Love, and Gender: A Kantian Theory.* Oxford: Oxford University Press.
3. By this, we simply mean that way in which we engage the world in part by viewing some things as constitutive of wholes. For example, the doors, stairways, windows, and walls are parts of—in the sense of constitutive of—a house.
4. Given our aims in this paper, engaging in all the related interpretive debates in the Kant community would take us in the wrong directions. For more on some of these debates, see Varden (2020), and for Kant's account of the predisposition to good and the propensity to evil in human nature, see his (1793/1999) *Religion with Boundaries of mere Reason: and Other Writings*, eds. A. Wood and R. di Giovanni (Cambridge University Press), starting with the sections "Concerning the Predisposition to Good in Human Nature" and "Concerning the Propensity to Evil in Human Nature" (6: 26–33).
5. See Arendt's (1958/2018) *The Human Condition. Chicago:* University of Chicago Press.
6. See Arendt's (1951/1973) *The Origins of Totalitarianism.* Harcourt, Brace, Jovanovich.
7. See Arendt's (1963/2006) *Eichmann in Jerusalem: A Report on the Banality of Evil.* Penguin Classics.
8. See Arendt's (2018) *Thinking without a Bannister: Essays in Understanding (1953–1975).* Schocken Books.
9. "Self-reflexive" consciousness means an awareness internal to any thought or action, whereas "self-reflective" consciousness means a first-personal *thinking about* what one is thinking or doing and that one is the one doing it. Self-reflective consciousness is a second-order awareness of what I am already self-reflexively conscious of.
10. Exploring some of the choice versus essence themes is a central topic in Varden (2020).
11. Immanuel Kant (2007). *Anthropology from a Pragmatic Point of View* in *Anthropology, History, and Education*, ed. by Robert B. Louden and Günter Zöller, transl. by Mary Gregor, Paul Guyer, Robert B. Louden, Holly Wilson, Allen W. Wood, Günter Zöller, and Arnulf Zweig. New York: Cambridge University Press.
12. We also cry, and we cry because we've been separated from the being we've been inside of, and so we long for unity again—something that is satisfied by being put on our caregivers' bodies and being swaddled. This, however, is consistent with the scream, and it is consistent with many other newborn animals being put together with others or being held close by their parents. The other animals don't scream when born, however, and that's what Kant wants us to be able to explain.

13. For more on all of these points and the other ones covered below, see Varden (2020). For many other Kantian ideas of emotionally healthy, morally responsible self-realization, see also Carol Hay (2013). *Kantianism, Liberalism, and Feminism: Resisting Oppression*. Palgrave Macmillan.
14. On this account, feeling such desires reveals that one has some rather deep, repressed sexual immaturities that one needs to take seriously and shed light on so that they can be matured in responsible ways—and because they will involve reflexive desires, the best way to do this is with the help of excellent therapists. For example, psychoanalysts are exactly trained in engaging our so-called subconscious desires.
15. Kant's discussions of affects and passions are spread throughout his works. A good place to start is his discussion in *Anthropology from a Pragmatic Point of View* in *Anthropology, History, and Education*.
16. Immanuel Kant (2007). "On the philosophers' medicine of the body" (PMB), in *Anthropology, History, and Education*.
17. Immanuel Kant (2007). "Essay on the maladies of the head" (MH), in *Anthropology, History, and Education*.
18. Note that we have changed the translation into the gender-neutral noun (they) here to make it more true to the original German.
19. Kant's idea of the highest good is found in many places in his practical works. For example, see "On the common saying: That may be correct in theory, but it is of no use in practice" (8: 279) and *Critique of Practical Reason* (5: 110–115). Both texts are printed in Kant, Immanuel (1996). *Practical Philosophy*, transl. and ed. by Mary J. Gregor. New York: Cambridge University Press.
20. For two classic, Kantian interpretations of sexual love's fiery nature, see Christine Korsgaard's proposal that it is inherently "possessive" (in her 1992 article "Creating the Kingdom of Ends: Reciprocity and Responsibility in Personal Relations," in ed. J. Tomberlin *Philosophical Perspectives, 6, Ethics*, pp. 305–332, Atascadero, CA: Ridgeview Publishing Company) and Barbara Herman's proposal that it is inherently "objectifying" (in her 1993 article "Could It Be Worth Thinking about Kant on Marriage?", in eds. Louise M. Antony and Charlotte E. Witt *A Mind of One's Own: Feminist Essays on Reason and Objectivity*, pp. 53–72, Cambridge: Westview Press).
21. For a wonderful Kantian interpretation of many of the functions and types of sex, see Jordan Pascoe (2012). "Hunger for you: Kant and Kinky Sex," in S. M. Kaye (ed.) *What Philosophy Can Tell You About Your Lover*, pp. 25–36. Open Court Press.
22. Simone de Beauvoir, *The Second Sex*, ed. and trans. H.M. Parshley (New York: Vintage, 1989 [1949]), xxi.
23. Ann P. Haas, Philip L. Rogers, and Jody L. Herman (2014). *Suicide Attempts among Transgender and Gender Non-Conforming Adults: Findings of the National Transgender Discrimination Survey*. New York and Los Angeles: American Foundation for Suicide Prevention and the Williams Institute (williamsinstitute.law.ucla.edu/wp-content/uploads/AFSP-Williams-Suicide-Report-Final.pdf).

24. Catharine MacKinnon, "Pornography, Civil Rights, and Speech," *Harvard Civil Rights-Civil Liberties Law* Review 20, no. 1 (1985): 462.
25. Nussbaum 266–267.
26. For more on some of these discussions, see (Hay 2013) and (Varden 2020).
27. Rae Langton, "Speech Acts and Unspeakable Acts," *Philosophy & Public Affairs* 22, no. 4 (1993): 293–330.
28. Quill Kukla, writing as Rebecca Kukla, "That's What She Said: The Language of Sexual Negotiation," *Ethics* 129, no. 1 (2018): 70–97.

CHAPTER 6

Sexual Jealousy and Sexual Infidelity

Natasha McKeever and Luke Brunning

Introduction

Few people have evaded the torment of some jealousy in their romantic life. Jealousy is painful and can be destructive. However, it is also widely believed that jealousy is a sign of love. Peter Goldie describes it as 'the price that has to be paid for there to be a certain sort of love from a certain sort of person',[1] and Jerome Neu argues that, 'at the centre of jealousy is the desire to be desired or for affection, the need to be loved'.[2]

In this chapter, we consider jealousy in the context of romantic love, questioning how we should best understand what jealousy is, and how we should respond to it. The structure of the chapter is as follows. In section one, we ask what jealousy is and suggest jealousy is best understood as an emotional response to the threatened loss of love or attention, to which one feels deserving, because of a rival. In section two, we consider the relationship between romantic love and jealousy, and the question of whether it can be consistent to feel jealousy toward someone we love. We argue that jealousy is not necessarily inconsistent with love, though it can make people behave in ways which are. Not all forms of romantic life involve love, however, and the general value of jealousy can be questioned, so in section three we evaluate claims made in support of jealousy's instrumental value, and argue that they have to be balanced against a range of potential harms. The risk of these harms arising needs to be

N. McKeever (✉)
University of Leeds, Leeds, UK
e-mail: n.mckeever@leeds.ac.uk

L. Brunning
University of Birmingham, Birmingham, UK
e-mail: l.brunning@bham.ac.uk

D. Boonin (ed.), *The Palgrave Handbook of Sexual Ethics*,
https://doi.org/10.1007/978-3-030-87786-6_6

taken seriously. In section four, we assess two potential ways of managing jealousy (which are not mutually exclusive)—firstly by adopting a policy of monogamy and secondly by engaging in emotional work. Neither of these methods is easy, and neither will solve jealousy altogether, but we argue that the second strategy should be taken more seriously.

What Is Jealousy?

Jealousy is a familiar experience, but its nature is more elusive. How does jealousy relate to other aspects of mental life? What are its foundations? Is it a valuable experience? In this section, we chart a course through the philosophical literature on jealousy. Our focus is jealousy in romantic life, but we will not presuppose that such jealousy differs radically from jealousy in other contexts. Sexual jealousy is one form of jealousy which occurs in romantic life—that is, jealousy experienced as a result of sexual activity. But jealousy in romantic life is not always about sex. For example, it might be about a partner's close emotional relationship with another.

When people describe being jealous, they are usually describing some form of episode. Most plausibly, they describe an emotional episode. Unlike a lingering mood, which has a diffuse object, if it has one at all, jealousy is usually brief with a sharp focus. Even when jealousy lasts longer, it does so as an episode, rather than being present all the time. Most people experience these episodes occasionally and, unless specified, they are our focus below. This is not to deny that people may be disposed to have pangs of jealousy or state correctly that they 'are a jealous person'.

Jealousy is messy. Philosophical attempts to understand this mess differ in how they describe jealousy's *components*, *structure*, and *content*. In turn, these attempts are situated within specific theories of emotion. Those theories differ in the emphasis they place on emotions' affective or cognitive nature. Most philosophers take a middle way between the view that emotions are just feelings, or just judgments, to suggest they involve a mixture of affect and cognitive content.[3] We follow suit here. Importantly, this mixture needn't be simple. Emotions involve changes to our physiology (pulse, sweat, muscle tension, pupil dilations) as well as distinct kinds of phenomenological tone (elation, foreboding). Cognitively they involve both beliefs and evaluations, about or of our environment ('*that's* a viper!' 'vipers are *dangerous*!'). Emotions also motivate and give behavior a typical shape. We may be drawn to the target of our emotion or we may be inclined to flee.

Different emotions can make us feel or behave in similar ways. The bodily changes associated with nervousness and excitement, for instance, are not that different, and both love and hate may see us fixate on someone. As a result, it is useful to distinguish emotions in terms of their intentional content: what they are *about*. This approach sets the stage for moral discussion of emotions, as they might be about, or refer to, moral notions (e.g., fairness, or respect). Yet we must not fixate on an emotion's content to the detriment of its other constituents, such as how it feels. Analyses of emotions in terms of their content must accord with their felt character. With this in mind, let's consider how jealousy makes people feel, before considering its content.

Imagine being jealous. A partner talks fondly of a new work colleague; a teacher's gaze sweeps over us as they pick on a classmate; a sibling is praised for yet another achievement. These experiences are often painful. But unlike other pains, such as sadness or disappointment, jealousy *agitates* and leaves us anxious and bewildered. We may feel on edge, fired up, ready for confrontation.

There is a fertility to jealousy, too, in eliciting other emotions. In some situations, jealousy is layered with anger or spite, and, if the rival is as hapless as we had hoped, schadenfreude. In other situations, jealousy is fringed with sadness, fear, or ennui. Jealousy can also be shadowed by envy and longing, which look outward toward what others have, or insecurity and self-loathing, which are more introverted. *The fact* that we are jealous rarely goes unnoticed either, so jealousy is often accompanied by meta-emotions like pride, shame, or humiliation, which have our agitated state as their object.

Jealousy also motivates in characteristic ways. Perhaps the most common desire associated with jealousy is that the rival (the work colleague, classmate, or sibling) is *set back*; they do not get attention, squander the limelight, or just go away. A second, arguably less common drive, is to protect or improve what is valued, to recapture attention, or to amplify love.[4] These desires lie behind our efforts to admonish or thwart others, or to make ourselves more visible. (All are common romantic comedy tropes.)

Jealousy is often discussed alongside fear of loss. But the two emotions yield different motivations. Although jealousy is similar to being startled, in that it prompts people to see what is going on, jealous people are less easily assuaged. They attend intently, focus on a rival, or seek out threats. This attention is sometimes visual, as in the intense persistent looking central to Alain Robbe-Grillet's novel *Jealousy*, but it can also be auditory; ears have prominence in many visual depictions of jealous people.[5] This attention lingers and can tip over into rumination and obsession. As such, jealousy is unlike fear. Settled fear of something can motivate action against it, but it more usually motivates us to withdraw, to turn away, or even to embrace denial.

Jealousy feels bad and motivates us in some characteristic ways, but what characterizes the emotion, and sets it apart from similar feelings like envy? One claim often made of jealousy is that it has a tripartite structure.[6] *We* are jealous of our *partner* over the attention they get from their *colleague*. This claim is used to distinguish jealousy from emotions like fear which are held to have a bipartite structure (*we* are scared of *spiders*). Although useful in drawing our attention to the potential importance of third parties when considering jealousy, it is debatable whether this claim about structure helps to distinguish jealousy from envy. Sara Protasi, for example, argues that envy should also be thought of as tripartite, to be distinguished from coveting;[7] *I* am envious of *the wealth* of *my sister*.

A more profitable strategy is to consider the *evaluative* aspect of jealousy. What are the assessments at the heart of the emotion? When we are jealous, we evaluate that *something which we value, is lost or changed, or under threat of loss or change, and this loss or change is due to a rival.*[8] And by changed, we mean changed for the worse. This is a vague starting point, but jealousy often feels vague. Right away, someone will ask: *what* is the central good we fear will be lost or changed? As we shall see, answers to this question in the form of a philosophical account of jealousy are open to counterexamples which may suggest a tidy account may not be forthcoming.

To begin with, we might think the good at stake in jealousy is another *person*; that we do not want to lose *them*. Stated thus, this view is mysterious: what is it to 'have' someone in the first place? The view also cannot account for cases where we feel jealous toward people we do not 'have'. In response, someone might suggest that *relationships* are actually at stake in jealousy, that we have to have a relationship with someone (a lover, parent, teacher) to feel jealous. But this also seems implausible. Much jealousy arises precisely because we want, but lack, a relationship with someone.[9] For example, I might feel jealous when a work colleague to whom I am attracted begins dating someone else.

So perhaps what is at stake in jealousy is the *love* of another person,[10] or their *affection* or *intimacy*?[11] These analyses capture cases where people care for each other but cannot enshrine that concern into a relationship, but they cannot account for situations where people are jealous *because* their love or affection is totally unrequited. We might love someone from afar and feel jealous when another person has better luck with them, for example, or feel jealous when our jaded or even ex-spouse turns elsewhere. This focus on love or intimacy can also make jealousy seem akin to disappointment, envy, or fear of loss where we are pained because we lack something we want. Not only would such a view struggle to give a distinct characterization of jealousy, it neglects how jealousy can be tumultuous and disorienting. Finally, although these analyses may accurately portray jealousy in romantic or family life, where love, affection, and intimacy are expected, those feelings are less central to other contexts where jealousy is common, like the office or classroom.

To accommodate these concerns, we can shift emphasis from love and affection to *attention*. Not just any attention, but that which shapes our sense of self. Leila Tov-Ruach argues that when we are jealous, we are gripped by the felt absence of someone's 'formative attentive regard'.[12] Her basic idea is that our sense of self is shaped by the interest and attention of other people over time, whether lovers, family members, friends, or mentors. We are jealous when that important kind of concern is threatened, and jealousy is disorienting precisely because it threatens our sense of self, rather than something trivial. This view overcomes some of the problems above, especially as it does not confine jealousy to one kind of context, and it can acknowledge variety in the kinds of formative regard that shape who we are. That said, it is vulnerable to the observation that we are often jealous in petty situations where our sense of self is not under threat.

Tov-Ruach's understanding of jealousy, and indeed any view which emphasizes love, intimacy, affection, or formative attention, is also prone to a final problem, namely, that we might experience *more* of these things from someone as a consequence of the fact that they are also attending to other people.[13] Parents with a new baby may be more attentive to their first born; a person in an open relationship may be more intimate with their spouse; a teacher who has to help stragglers may be more appreciative of a gifted student. In each case, that child, spouse, or student might still be jealous. Put simply: jealousy does not seem to track the 'amount' of love, affection, or attention we experience.

Let's take stock. The views above try to capture what is central to jealousy. In different ways, they emphasize our vulnerability in the face of other people and connect jealousy to fear of loss of love, affection, and attention: all aspects of interpersonal life that have a big impact on our sense of self, especially in romantic life. Three problems remain. First, these views can struggle to differentiate jealousy from envy or fear of loss. Second, people can be jealous when they have plenty of love and attention. And, finally, the role of the 'rival', or third party, takes a backseat in these views where perhaps it should not.[14]

To accommodate these worries, and foreground third parties, some theorists add a further condition to the analysis of jealousy. They suggest that we are pained not simply because something we care about is threatened or lost, but because that thing is, in some sense, not properly *ours* in the first place. Jealousy thus serves as a reminder that we have less control over that which we care about than we had previously thought. Moreover, this valued good is being threatened by a rival, which is usually, though not always, a person. Someone could, for example, be jealous of football if it seems like it is taking their partner away from them.[15] Viewed with these points in mind, jealousy is often seen as responding to *affront*, not mere absence. At stake is not just love, or the self, but also social concepts of legitimate expectations, entitlement, or desert.

As Peter Goldie puts it, for example, jealousy often arises when people feel their 'legitimate expectations' of a love relationship have been violated.[16] The notion of legitimacy, here, is similar to what other philosophers call *entitlement*. Aaron Ben-Ze'ev, for instance, thinks that 'at the very foundation of romantic jealousy lies the concept of entitlement' where the notion of entitlement is understood as 'our assumed right to exert control'.[17] However, a person can have legitimate expectations to things to which they are not entitled. We are not entitled to a birthday card from our partners, but we have a legitimate expectation they will get us one.

Perhaps desert is a more helpful concept here. Kristján Kristjánsson suggests that jealousy is an emotion that is sensitive to what people *deserve* (or what they think they deserve). We might think desert is a narrower, but stronger, concept than mere entitlement because people can be entitled to things that they do not deserve, and deserve things to which they are not entitled. For example, being married to someone might make us entitled to receive some of their wealth if we divorce, but having been a terrible spouse to them might mean that we deserve nothing. Conversely, being a kind and generous partner might mean that we deserve to be treated kindly and generously back by a partner, but we are not *entitled* to be treated as such. In practice though, Kristjánsson thinks that in many sexual and romantic contexts, jealousy is probably irrational because 'the sexually jealous person overlooks the fact that love is not a matter of will and no one deserves to be sexually attractive to another'.[18] Because it is often irrational, he thinks the question of whether such jealousy is *morally* justified does not emerge, unlike cases of jealousy in the family or toward a teacher.

Is Kristjánsson right here? Whether jealousy is rational, arational, or irrational will depend on how we are understanding what it would mean for jealousy to be rational. There are various different ways that we could approach this question; due to considerations of space we will try only one. Let us suppose that jealousy is rational if what it aims at is possible or plausible and irrational if it is impossible or implausible.

Kristjánsson is right that love is not a matter of will, but what jealousy aims at might be. Sometimes a partner might have a significant degree of control over their behavior which makes us jealous. In addition, there might be elements of control that we can exert in order to regain their attention. If our jealousy indicates that what we want is more of their attention, affection, or intimacy, then, by spurring us into action, jealousy can be a step on the way to our getting this. This might be particularly true if our partner has diverted attention to others because we have been neglecting them. To this end, jealousy can be rational. It can also be a useful reminder that what is threatened was never properly *ours* in the first place and must be worked for.

On the other hand, if what we want is for our partner to be attracted to us exclusively, then our jealousy will be rather hopeless. Such a desire is unlikely to be satisfied and, if it is, it might be only by chance. Our jealousy in this case might be thought of as irrational, since it is implausible and perhaps impossible that our partner will never be attracted to anyone else.

Therefore, some jealousy at least will be rational, and thus open to being assessed morally, and arguably even irrational jealousy can still be assessed morally, since we could be under a broader moral obligation to correct the irrational beliefs that underlie it. For example, a husband who, because of his patriarchal beliefs, becomes irrationally jealous whenever his wife talks to another man has a moral obligation to try and manage and correct those beliefs.

Jealousy and Romantic Love

Jealousy may be rational in some cases, but is it consistent with a loving attitude toward someone? Rachel Fredricks argues that the self-interested desires at the heart of jealousy are 'in tension with a caring attitude towards someone, where you value things for their sake'.[19] These jealous desires aim, in a purely self-interested way, at securing the thing we want as a jealous person, such as affection from a partner, or the exclusivity of their sexual attention. Thus,

Fredricks' view is that 'what is *paradigmatic* about jealousy from a moral perspective is that it undermines the moral value of relationships between jealous subjects and targets, and it does so in a way that is traceable to the particular kind of desire that is necessary for the jealousy'.[20]

It is, of course, true that jealousy is self-interested. The jealous person is preoccupied with securing whatever it is that they are jealous over, often at the cost of the well-being of others involved. On the face of it, jealousy thus seems inconsistent with love, insofar as love requires caring for the beloved for their own sake. If Jake loves Jemma, but Jemma wants to have sex with Paul, then, arguably, if Jake really cares about her, he should be happy for her to do so. As Harry Chalmers puts it: 'when we see our partner find joy in someone else, would it not make more sense for us to be happy for her? Would it not be truer to our love, truer to our good will, to share in her joy?'[21]

This seems like the right response for Jake to have if we accept a 'disinterested concern' view of love, such as Harry Frankfurt's, who argues that love 'consists most basically in a disinterested concern for the well-being or flourishing of the person who is loved'.[22] This means that in authentic love, the lover's only concern is for the interests of the beloved; the lover is not in love for self-interest. If we accepted a view of love like this, then Goldie would be wrong that jealousy often arises when people feel their 'legitimate expectations' of a love relationship have been violated[23] because there would be no such thing as legitimate expectations in a love relationship. To love in the expectation of getting something back would not really be love at all. And Kristjánsson would be wrong that jealousy could be a virtue because it shows sensitivity to what we justly deserve, since justice and desert would not come into love either.

However, there is also truth in the claim that romantic love *should* make us feel happy and good about ourselves, and that is one of the reasons why we want it in their lives.[24] Romantic relationships which make people unhappy are generally considered bad relationships and often the people within them are encouraged by third parties to leave the relationship. Indeed, romantic love without any self-interest might actually be undesirable or even seem to be not love at all, as illustrated in Monique Wonderly's example of Alejandro and Denise:

> Alejandro claims to love his fiancée Denise. His feelings for her are wholly selfless. In seeking to preserve and promote her well-being, he thinks nothing of himself or his own interests. In their interactions, he is concerned exclusively with how best to contribute to her welfare. He also sees her as special and unique, and cannot imagine valuing anyone else in the same way that he values her. This morning, Denise learned that she has been offered a wonderful opportunity to realize her life's dream of embarking on a six-year space mission. Unfortunately, during this time, she would be unreachable to all earth dwellers, including her beloved fiancé. When she relays the news to Alejandro, he is overjoyed and immediately offers to help her pack. When she expresses concern over being apart from him for such a long time, he cheerfully replies, "This is what's best for you, and that's really all that matters to me!"[25]

Wonderly writes that although there is 'something noble and pure' in Alejandro's response to Denise telling him he won't see her for six years, there is also something 'quite disturbing' about his attitude. This is because he does not seem to consider Denise to be 'important for him'; he does not see her 'as an essential source of enrichment in his life'.[26] Similarly, if Jemma had sex with Paul and then left Jake for Paul, and all Jake had to say is 'if that's what's best for you that's all that matters to me', Jemma might, fairly, question whether Jake really thought of her as important to his life.

In order to explain why romantic love has a self-interested element, Wonderly turns to attachment theory, first developed by psychologists John Bowlby and Mary Ainsworth in the mid-twentieth century. They looked at the way that infants develop attachment bonds with their primary caregivers, treating them as a 'secure base' and a 'safe haven'.[27] The need to have secure attachment bonds is a basic human need, which, if not satisfied, has a significant negative effect on functioning and flourishing. Psychologists have also argued that we develop attachment bonds with romantic partners and our partners fulfill the function of secure bases and safe havens for us in our adult life; we feel more confident when they are there, and we turn to them for comfort in times of distress.[28] Thus, one of the reasons romantic relationships are valuable to people is that they provide security via attachment bonds.

It is partly because we want romantic love to be a source of security and enrichment to our lives that Frankfurt does not view romantic relationships as 'very authentic or illuminating paradigms of love' as he construes it. He lists a number of other 'distracting elements'[29] of romantic love which render it inauthentic in his view: 'a hope to be loved in return or to acquire certain other goods that are distinct from the well-being of the beloved—for instance, companionship, emotional and material security, sexual gratification, prestige, or the like'.[30]

Frankfurt is right that people hope to be loved in return or to acquire other goods from romantic love, and this does make it self-interested, to an extent. He is also right that *if* love is disinterested concern, then romantic love is not paradigmatic of love. However, perhaps romantic love is just a different kind of love to that which interests Frankfurt; love rooted both in caring for the other and in an attachment bond to them. Or perhaps all love, even parental love, is like this.[31] On either approach, romantic love is valuable, in part, *because* it contributes to our self-interest. This does not make it selfish, or entail that caring is not an important feature of love. Romantic love is inherently reciprocal—even in cases of unrequited love the lover almost always *desires* for their beloved to love them back. The lover wants to be a source of security and enrichment to the beloved's life. Indeed, romantic love makes us feel good about ourselves in part because we know that we are contributing to the self-interest of our beloved.[32] Making the beloved happy makes the lover happy, and the beloved knowing that the lover is happy in turn makes them happy.

Incidentally, even on the disinterested concern view of love, we face a dilemma working out who should prevail between Jake and Jemma, since, if

Jemma loves Jake in a disinterested concern way, then her only goal will be to promote his well-being. Therefore, if having sex with Paul will make Jake feel jealous, if she loves him, she ought not to do it. However, if Jake loves her back in this way then he will want her to do it. Applied to reciprocal relationships, the notion of disinterested concern will often come unstuck in dilemmas such as this, because fulfilling reciprocal relationships require not only caring for each other, but also decisions over whose interests take priority, which will involve the application of principles such as fairness and reasonableness.

Self-interested desires do not, therefore, render romantic love inauthentic; they are just part of the distinct kind of love it is and part of a reciprocal and equal relationship. As Wonderly puts it, self-interestedness in romantic love does not 'spoil love, but rather adds to its shape and substance'.[33] In reciprocal romantic love, the lovers care about each other for both their own sake and the sake of the other. They *matter* to each other, and they want a *relationship* to continue and to be fulfilling for both of them. As Georgie Tsai notes, such desires mean that romantic love involves vulnerability: 'loving someone renders one susceptible to sorrow when the beloved is harmed or when the relationship is impaired'.[34] Furthermore, people can and often do treat each other badly in romantic relationships, and the moral value of such relationships lies not only in concern and caring for each other, but also in the ability to act fairly, and to negotiate and compromise so that all parties in the relationship get approximately the same degree of fulfillment from it.

Therefore, it *is* right that concepts such as legitimate expectations and desert come into romantic relationships, as they might come into all relationships. Put differently, La Rochefoucauld might have been correct to think that 'in jealousy there is more self-love than love',[35] only to overlook the fact that love is not *purely* other-interested. However, working out what one can expect from a relationship is not easy and this is where assessments of the reasonability of jealousy get particularly murky. If Jake is jealous of Jemma's friendship with Paul, does he *deserve* for her to stop being friends with Paul? Indeed, does he *deserve* for Jemma to not have sex with Paul because for her to do so would make him jealous? Or, conversely, does Jemma deserve to be able to do what she wants despite how Jake feels about it?

Jealousy may be compatible with a loving attitude toward someone, but not all jealousy arises in romantic relationships where people love each other, it can strike at the heart of more casual relationships, too. We should also ask whether there is anything else to be said for jealousy. Just because it may be compatible with loving attitudes does not mean it is the best way of relating to someone romantically. Does jealousy have further value in romantic life?

Jealousy's Value?

Is it good to be someone who can be jealous? Jealousy is said to have instrumental benefits. For instance, jealousy has expressive significance. In experiencing it, we may come to realize we care about someone,[36] and someone else may realize we care about them.[37] Jealousy's expressive power can also disturb any relationship between someone we care about and a rival.[38] As such, jealousy may help us defend our sense of self, insofar as that is bound up in the affection or attention of other people.[39] Jealousy is also said to prevent people from becoming indifferent to each other[40] and to strengthen existing relationships.[41] A specific way it could help relationships is by functioning as an erotic catalyst;[42] it may even be directly enjoyable as a masochistic pleasure.[43] More generally, jealous experiences may help people reflect on their relationship.[44] Undertones of jealousy may even make us better people, and better partners, as it purportedly encourages 'creativity and competitive achievement'.[45]

But jealousy is not straightforwardly beneficial. Many of these apparent benefits are attained only in a fragile way. To access some of them, for example, jealousy as a prompt to reflection, we have to already be the kind of people who can manage emotions well and situate them in the context of our mental life and circumstances. But if we are like this, we are less likely to need jealousy to learn about our feelings, or to show another person we care about them, in the first place.[46]

More worryingly, these benefits need to be balanced against a range of potential harms stemming from romantic jealousy. Rachel Fredricks documents a litany of these harms.[47] When we are jealous, we often go awry *epistemically*, with false beliefs about what we desire, the true extent of our entitlements, and rivals. This is especially true when jealous people ruminate on their predicament or tip over into more obsessive patterns of thought.

Morally speaking, jealousy eats away at the well-being of everyone involved. Suspicion, false accusation, snide behavior, and general acrimony all contribute to damage valued relationships and destabilize family life. Jealousy can also motivate violence. Indeed, it is the leading cause of the murder of partners and ex-partners, particularly female ones.[48] Jealousy wastes time and distracts people from what they care about. Jealousy can destabilize our sense of self; it generates painful meta-emotions of guilt and shame, can feed into anxiety or low self-esteem, and may be at odds with our self-image of ourselves as people in command of our feelings, or as people who want others to flourish.

Our jealousy may prevent us from respecting the autonomy of other people. The expressive force of jealousy, and the behaviors it motivates, pre-empts what is good for the other person or people involved, and we may fail to give *them* the appropriate scope to maintain boundaries for themselves. Jealousy is a backseat driver in the emotional life of other people, which is disrespectful to them.

These considerations lead us to doubt the value of jealousy as an emotional trait. Although it is compatible with loving attitudes, jealousy is risky to cultivate. A natural response to these risks would be to try and reduce or manage them. We now consider two strategies for doing this.

Responding to Jealousy

It is not in anyone's interests to simply let jealousy run riot. There are various ways to respond to jealousy. Here we consider the two which are probably the most common: (i) partners can adopt a policy of monogamy to minimize how much jealousy enters the relationship; or (ii) partners can learn to manage their jealousy, possibly by developing feelings of compersion (which we define below). These are not mutually exclusive.

Monogamy as a Response to Jealousy

Monogamy between people in marriages or romantic relationships is often recommended as a way to quell jealousy. Roger Scruton, for example, argues that erotic love is naturally monogamous and must be so because love is prone to jealousy and jealousy can be destructive; 'it is in the deepest human interest, therefore, that we form the habit of fidelity', a habit he describes as 'natural and normal' but also 'easily broken'.[49] On the flipside, views about the natural jealousy of humans are sometimes used to argue that polyamory is inherently unstable.[50]

Sexual and romantic fidelity involves sacrifice, however, which often goes unrecognized. People in monogamous relationships must give up the possibility of romantic love or sex with anybody else. Due to the restrictions placed on partners' powers to engage in sex or love, monogamy has been likened to private property[51] and has been argued to be morally unjustifiable.[52] Furthermore, as noted by Chalmers, monogamy does not get rid of jealousy, as it just helps us to avoid jealousy triggers, rather than dealing with the underlying fears. Chalmers argues that monogamy is a 'capitulation to jealousy' and actually perpetuates it, by encouraging a competitive view of love, and making us worry that our shortcomings and mistakes will lead our partner to wonder if they'd be happier with someone else.[53]

Thus, rather than being merely a response to our jealousy, it seems right that the norm of monogamy in turn reinforces and promotes jealousy, not only by encouraging people to think that it is reasonable to expect to have a monogamous romantic relationship, but also by encouraging other problematic beliefs, such as that if a partner is attracted to someone else then this implies that we are not 'enough' for them and that we ought to feel shamed by this. Jeffrie Murphy argues that the significance of the rival in jealousy and the shame we feel by being 'beaten out in a competition over a loved person', but not when that person dies[54] must be understood as, at least in part, 'based on a competitive account of human intimacy and love'.[55] This competitive account is intricately linked to the norm of monogamy—it is because our partner can only have *one* romantic partner that we are in constant competition with rivals to fill that position.

On the other hand, it is important to remember that even those who do not subscribe to a competitive account of love, or embrace the norm of monogamy, still experience jealousy.[56] While it seems right that the norm of monogamy reinforces jealousy, it is also likely such a dominant norm because most people have a predisposition toward jealousy or that jealousy is anchored deeply in our attachments to other people.[57] In addition, not everyone experiences monogamy as a sacrifice; some people find it fulfilling and are genuinely not interested in sexual and romantic relationships with others outside of their relationship. In those cases, monogamy might not be experienced as a response to jealousy, but simply as the most fulfilling sort of relationship. Having minimal opportunities to feel jealous would be an added bonus in such cases. However, there is a different approach to jealousy that can be taken which aims not at limiting the opportunities or actions of those involved but rather at managing the negative emotional response to those actions. Emotional management can be engaged in in conjunction with monogamy, but it is also a promising strategy for people who do not find monogamy fulfilling, or who do not find that monogamy stops them from feeling jealous.

Emotional Management as a Response to Jealousy

Whether it is reasonable to expect people to manage their jealousy will depend, at least in part, on the extent to which doing so is possible for them. It is probably true that humans have a natural propensity toward jealousy, and particularly toward sexual jealousy, based on biological and psychological factors related to our evolutionary history.[58] Some have argued that jealousy is useful from an evolutionary perspective as a way of protecting paternity by motivating people to retain a single partner,[59] or underlying a strategy to protect important caring relationships.[60] Buss argues that sexual jealousy is a so-called basic emotion, which is an emotion which evolved because it contributed to reproductive fitness.[61]

Nonetheless, this does not mean that we must simply accept jealousy as an inescapable feature of us and our romantic lives. We should not equate evolutionary value with moral value.[62] Racist traits, for example, are not morally valuable, even if in-group favoritism was evolutionarily useful. We have the capacity to manage and change our predispositions and natural impulses. Doing so is not easy, but if there is a compelling enough reason to do so, we ought to try.

What would managing jealousy involve? A first step might be to recognize that it is possible to do so. As Dossie Easton points out, 'we are taught that jealousy is the one intolerable emotion. We are taught to manage other emotions such as anger, aggression, or sadness. … But we are all taught emotionally that jealousy is some state where we would just explode.'[63] The second step is to try and interpret our emotions: 'anxiety, insecurity, fear, and other negative associations and memories are often mistaken for jealousy, and legitimate jealousy can be accompanied by other negative emotions. Moreover, jealousy is prompted by a range of characteristic reasons that can be easily confused with each other.'[64] A couple could work together to dissect their jealous experiences.

In doing so they might find, for example, that underlying it is low self-esteem and fear of abandonment, or lingering feelings of entitlement anchored in their experience of a patriarchal upbringing.

Further emotional work might involve reflecting upon our sense of entitlement, vulnerability, and ideals of romantic love, as well as gradually exposing ourselves to situations which ordinarily would make us feel jealous, and trying to focus on the flourishing of others rather than ruminating on threats to ourselves.[65]

The ultimate aim might be to feel compersion when a partner has sexual or romantic relationships with others. Compersion can be defined as feelings of joy and positivity toward a partner's romantic and sexual relationships with others and it is an emotional ideal for many polyamorous people.[66] Compersion is clearly much less destructive and harmful than jealousy, and the compersive partner shows that they care for their partner's well-being, thus retaining the benefit of jealousy that it can signal care or affection.[67]

However, despite compersion's clear benefits, managing jealousy is harder for some than others, and it will be extremely difficult for some people. As Kyle York points out, 'anxious attachment systems, poor mental or physical health, and low self-esteem are all predictors of higher jealousy' and not all jealousy operates at a psychological level which is responsive to rational reflection.[68] There are further factors to take into account such as whether the jealous person's partner is supportive of and reassuring toward them while they try to manage their jealousy, whether the jealous person has a support network outside of the relationship, and the time and resources they have to devote to their emotional work. Equality of opportunity and reciprocity should also factor. For example, in a case where a woman is at home looking after a baby, given the lack of opportunity she has for other sexual and romantic relationships, it could be unreasonable for her partner to ask her to engage in the emotional work necessary to develop a compersive attitude toward him having relationships with others.

Of course, though, monogamy is not free from emotional work either; it is just that the emotional work involved in monogamy is sanctioned by cultural and societal values and norms, so we are less likely to consider it to be work or to question whether it is reasonable or valuable. In the example above, if the couple are monogamous, they will still both have to engage in emotional work to maintain their relationship, which might involve managing other painful emotions, such as anger and frustration. Given that, for many people at least, monogamy does impose significant restrictions and sacrifice, and can be hard work, it ought not to be engaged in in an unreflective way, and there should be better understanding of and support for alternative ways of dealing with jealousy and developing compersion.

Conclusion

Jealousy is a complex emotion. We are jealous because we care about people, think we are deserving of their attention, and assume a competitive attitude toward romantic life. Often these attitudes are implicit, and not well articulated, but they are not always irrational. Jealousy is a self-interested emotion which can make it seem in tension with romantic love. But romantic love is seldom entirely selfless, and on reflection it would be strange if it was. Jealousy can therefore express the ordinary sense that to love someone is for them to be important to us, not just in themselves. People have tried to defend jealousy more substantively, attributing to it a range of instrumental benefits, but it is often harmful and, occasionally, extremely so. These harms must be taken seriously. Although we can be forgiven for a degree of self-interest in our romantic lives, this does not justify harm. These harms need to be mitigated directly. Embracing monogamy to do so fails to address the underlying issue, and for some, jealousy's potential to hurt is reason enough to try and get rid of the emotion entirely, or to replace it with compersion. Other people may think this is unrealistic. At the very least, we should try to manage jealousy in romantic life.

Notes

1. Goldie (2000: 240).
2. Neu (1980: 433).
3. For example, Ben-Ze'ev (2010). Exponents of the former view are influenced by William James; exponents of the latter, like Martha Nussbaum or Robert Solomon, are influenced by ancient Stoicism. For the purposes of this chapter we set aside questions about what *kind* of cognitive content features in emotions, that is, whether emotions are more like beliefs, or perceptions, or something else.
4. The fact that jealousy can motivate acts of shocking violence prompts our caution here.
5. Toohey (2014: Chapter 3).
6. For example, Farrell (1980). These claims are disputed. For instance, Kingsley Davis suggests jealousy is also oriented to a public, real or implied, and thus has a quadripartite form (c.f. Davis 1936: 395). For our purposes, these disputes matter only insofar as they encourage us to focus on typical features of jealousy, such as the role of a third party.
7. Protasi (2017).
8. Threat of loss seems more common than situations where one is jealous that a rival has secured some good we valued. Threat of change seems to cover cases where our relationship, affection, and so on continue (or improve) but is altered. Perhaps nonmonogamous cases are like this.
9. The possibility of jealousy in the absence of a relationship would also suggest that fidelity (Prinz 2004: 98–99) or exclusivity (Ben-Ze'ev 2010), understood as properties of *relationships*, is not always at stake in jealousy in a central way.
10. Neu (1980).

11. Solomon (2007).
12. Tov-Ruach (1980: 467).
13. Ben-Ze'ev (1990: 494).
14. Murphy (2002). The topic of the 'rival' is complex. We can distinguish between situations where other people are active competitors for the good we value, from those when they are not. In the latter, perhaps they have no interest in the person we care about at all. Similarly, jealousy can be elicited by imagined situations, mere suspicions, and perhaps even toward the dead (such as a partner's ex-spouse). One dispute concerning rivals is whether they are part of the object of jealousy, or not. Robert C. Roberts thinks that any of the lover, beloved, and rival can be the focus of the emotion (2003: 257); whereas others, like Justin D'Arms, think jealousy focuses more squarely on the person we care about: 'The jealous person's real locus of concern is the beloved—the person whose affection he is losing or fears losing—not his rival. [...] The envious person's locus of concerns is the rival. [...] Roughly for the jealous person the rival is fungible, and the beloved is not fungible. [...] Whereas in envy it is the other way around' (D'Arms 2009: 3–4).
15. For examples of these views, see Ben-Ze'ev (2010: 289–297); Farrell (1980); Goldie (2000: 225) and Roberts (2003).
16. Goldie (2000: 237).
17. Ben-Ze'ev (2010: 44).
18. Kristjánsson (2018: 105).
19. Fredericks (2013: 98).
20. Fredericks (2013: 100).
21. Chalmers (2018: 235).
22. Frankfurt (2004: 79).
23. Goldie (2000: 237).
24. Keller (2000: 163).
25. Wonderly (2017: 240).
26. Wonderly (2017: 240–241)
27. Wonderly (2016: 229).
28. Hazan and Shaver (1987, 1994); McCarthy (1999).
29. Frankfurt (2004: 43).
30. Frankfurt (2004: 83).
31. After all, attachment theory arose from studying parental bonds.
32. McKeever (2019: 214).
33. Wonderly (2017: 241).
34. Tsai (2016: 170).
35. La Rochefoucauld (2007: 91).
36. Neu (1980: 63).
37. White (1980).
38. Welpinghus (2017).
39. Wreen (1989: 651).
40. Ben-Ze'ev (2010: 324) and Toohey (2014: 190).
41. Kristjánsson (2002: 160). Kristjánsson later (2018) changed his mind on this after reading Fredericks (2013).
42. Perel (2017: 99–100).
43. Brunning (2020: 7–8).
44. Pines (1998: 199).

45. Toohey (2014: 221).
46. Brunning (2020: 15).
47. Fredericks (2013).
48. Buss (2013: 160).
49. Scruton (2001: 339).
50. Brunning (2018: 517).
51. McMurtry (1972).
52. Chalmers (2018).
53. Chalmers (2018: 237).
54. Murphy (2002: 148).
55. Murphy (2002: 147).
56. Mogilski et al. (2019).
57. Brunning (2020).
58. Buss (2000).
59. Buss (1994, 2000).
60. Chung and Harris (2018).
61. Buss (2013: 158).
62. Solomon (2007: 107–109).
63. Easton (2006: 646).
64. Brunning (2018: 522).
65. Brunning (2020: 236–237).
66. Brunning (2020: 226).
67. Brunning (2020: 239).
68. York (2019: 11).

References

Ben-Ze'ev, A. 1990. Envy and Jealousy. *Canadian Journal of Philosophy* 20: 487–516.

———. 2010. Jealousy and Romantic Love. In *Handbook of Jealousy*, ed. Sibyl Hart and Maria Legerstee, 40–54. Chichester: Wiley-Blackwell.

Brunning, L. 2018. The Distinctiveness of Polyamory. *Journal of Applied Philosophy* 34 (3): 513–531.

———. 2020. Compersion: An Alternative to Jealousy. *Journal of the American Philosophical Association* 6 (2): 225–245.

Buss, D. 1994. Individual differences in mating strategies. *Behavioral and Brain Sciences* 17 (3): 581–582.

Buss, D.M. 2000. *The Dangerous Passion: Why Jealousy Is as Essential as Love and Sex*. Free Press.

———. 2013. Sexual Jealousy. *Psychological Topics 22* (2): 155–182.

Chalmers, H. 2019. Is Monogamy Morally Permissible? *Journal of Value Inquiry* 53 (2): 225–241.

Chung, M., and C.R. Harris. 2018. Jealousy as a Specific Emotion: The Dynamic Functional Model. *Emotion Review* 10 (4): 272–287.

D'Arms, J. 2009. Envy. In *Stanford Encyclopedia of Philosophy*, ed. Edward N. Zalta. Spring. http://plato.stanford.edu/archives/spr2009/entries/envy/.

Davis, K. 1936. Jealousy and Sexual Property. *Social Forces 14* (3): 395–405.

Farrell, D.M. 1980. Jealousy. *The Philosophical Review 89* (4): 527–559.

Frankfurt, H. 2004. *The Reasons of Love*. Princeton University Press.

Fredericks, R. 2013. *Troubling Others and Tormenting Ourselves: The Nature and Moral Significance of Jealousy*. Doctoral diss., University of Washington.

Goldie, P. 2000. *The Emotions: A Philosophical Exploration*. Oxford University Press.

Hazan, C., and P. Shaver. 1987. Romantic Love Conceptualized as an Attachment Process. *Journal of Personality and Social Psychology* 52: 511–524.

Hazan, C., and P.R. Shaver. 1994. Attachment as an Organizational Framework for Research on Close Relationships. *Psychological Inquiry* 5 (1): 1–22.

Keller, S. 2000. How Do I Love Thee? Let Me Count the Properties. *American Philosophical Quarterly* 37 (2): 163–173.

Klesse, C. in conversation with Easton D. 2006. Expert Interview: The Trials and Tribulations of Being a 'Slut'-Ethical, Psychological, and Political Thoughts on Polyamory. *Sexualities* 9 (5): 643–650.

Kristjánsson, K. 2002. *Justifying Emotions: Pride and Jealousy*. Routledge.

———. 2018. *Virtuous Emotions*. Oxford University Press.

La Rochefoucauld, F. 2007[1671]. *Collected Maxims and Other Reflections*. Trans. E.H. Blackmore and A. M. Blackmore. Oxford University Press.

McCarthy, G. 1999. Attachment Style and Adult Love Relationships and Friendships: A Study of a Group of Women at Risk of Experiencing Relationship Difficulties. *British Journal of Medical Psychology* 72: 305–321.

McKeever, N. 2019. What Can We Learn about Romantic Love from Harry Frankfurt's Account of Love? *Journal of Ethic and Social Philosophy* 14 (3): 204–226.

McMurtry, J. 1972. Monogamy: A Critique. *The Monist* 56 (4): 587–599.

Mogilski, J.K., S.D. Reeve, S.C.A. Nicolas, et al. 2019. Jealousy, Consent, and Compersion Within Monogamous and Consensually Non-Monogamous Romantic Relationships. *Archives of Sexual Behaviour* 48: 1811–1828.

Murphy, J.G. 2002. Jealousy, Shame, and the Rival. *Philosophical Studies* 108 (1–2): 143–150.

Neu, J. 1980. Jealous Thoughts. In *Explaining Emotions*, ed. A. Rorty. Berkeley: University of California Press.

Perel, E. 2017. *The State of Affairs: Rethinking Infidelity*. Harper Collins.

Pines, A.M. 1998. *Romantic Jealousy: Causes, Symptoms, Cures*. New York: Routledge.

Prinz, J. 2004. *Gut Reactions: A Perceptual Theory of Emotion*. New York: Oxford University.

Protasi, S. 2017. 'I'm not envious, I'm just jealous!': On the Difference Between Envy and Jealousy. *Journal of the American Philosophical Association* 3 (3): 316–333.

Roberts, R. 2003. *Emotions: An Essay in Aid of Moral Psychology*. Cambridge University Press.

Scruton, R. 2001. *Sexual Desire*. Orion.

Solomon, R. 2007. *True to Our Feelings. What Our Emotions Are Really Telling Us*. Oxford: Oxford University Press.

Toohey, P. 2014. *Jealousy*. Yale University Press.

Tov-Ruach, L. 1980. Jealousy, Attention, and Loss. In *Explaining Emotions*, ed. Amélie O. Rorty, 465–488. Oakland: University of California Press.

Tsai, G. 2016. Vulnerability in Intimate Relationships. *The Southern Journal of Philosophy* 54 (Spindel Supplement): 166–182.

Welpinghus, A. 2017. Jealousy: a response to infidelity? On the nature and appropriateness conditions of jealousy. *Philosophical Explorations* 20 (3): 322–337.

White, G.L. 1980. Inducing Jealousy: A Power Perspective. *Personality and Social Psychology Bulletin* 6 (2): 222–227.

Wonderly, M. 2016. On Being Attached. *Philosophical Studies* 173 (1): 223–242.

———. 2017. Love and Attachment. *American Philosophical Quarterly* 54 (3): 235–250.

Wreen, M.J. 1989. Jealousy. *Noûs* 23: 635–652.

York, K. 2020. Why Monogamy is Morally Permissible: A Defense of Some Common Justifications for Monogamy. *Journal of Value Inquiry* 54 (4): 539–552.

CHAPTER 7

Sexual Use, Sexual Autonomy, and Adaptive Preferences: A Social Approach to Sexual Objectification

Patricia Marino

Sexual objectification is often understood to involve treating a person as a thing or using someone as a tool for your own purposes. This happens in a wide range of ways such as responding to a person as a body—or body part—rather than a full person; treating them primarily as a symbol of something, such as status; and being caught up in one's own emotional states and inattentive to their desires and wishes. Many cases involve ignoring a person's wants and desires to focus primarily on their body and on one's own gratification; this can happen in pornography, sex work, and what I call "commodified casual sex," in which sexual pleasure becomes a good to be exchanged—either in kind or for something else. From a social point of view, objectification is often understood as a central component in sexism and other forms of oppression: when women, racialized people, and people in other marginalized groups are objectified, hypersexualized, or fetishized, they are harmed. To say that an activity or attitude is sexually objectifying is often, therefore, to say that it is harmful and wrong.

But objectification can be puzzling. Much sex, and sexual attention, seems to potentially treat persons as sex objects; does this mean all sexuality is somehow suspect? As is often noted, ways of interacting that are potentially objectifying can frequently seem benign and even good, as when lovers are in thrall to the physical beauty and excitingness of their partners and thus momentarily

P. Marino (✉)
University of Waterloo, Waterloo, ON, Canada
e-mail: pmarino@uwaterloo.ca

D. Boonin (ed.), *The Palgrave Handbook of Sexual Ethics*,
https://doi.org/10.1007/978-3-030-87786-6_7

inattentive to their partner's mental states. Does this mean sexual objectification can be a good thing? Sometimes people choose to self-objectify—dressing and acting in ways that highlight their sexuality or encourage others to see them through a sexualized lens. A few years ago, Cameron Diaz was quoted as saying "I think every woman does want to be objectified. There's a little part of you at all times that hopes to be somewhat objectified, and I think it's healthy."[1] Maybe you've heard men say things like "What's wrong with sexually objectification? I'd love to be sexually objectified!" How should we understand the ways that people choose or want to be objectified?

Some analyses of the ethics of objectification focus on the nature of the relationship between the people in question, the reciprocity or mutuality of their pleasure, or the genuineness of their sexual desires. For example, emphasizing the multifacetedness of objectification, Martha Nussbaum suggests that what matters morally is that the activity is consensual and takes places in the context of a mutually respectful, caring, and equal relationship, one in which participants are treated as full human beings the rest of the time. In previous work, I've argued for a different view: that objectification is ethically benign when done in accordance with respect for autonomy. One can, in a sense, choose to be used, and as long as autonomy and consent are respected, there is nothing inherently wrong with treating persons as bodies, treating them as interchangeable with others, and so on. It would follow that as long as participation is freely chosen, pornography, sex work, and commodified casual sex are not morally problematic in virtue of objectifying persons.[2]

But what does it mean for objectification to be freely chosen? Obviously, coercion, deception, and force violate autonomy and prevent someone from making a genuinely free choice. But what about the more subtle social and cultural forces that lead us to the choices that we make? I've argued that the social background is crucial to ensuring autonomy in this domain. For example, insofar as a society is organized so that some people must allow themselves to be sexually objectified—because they are poor and have no viable alternative but to earn money from it or because they are valued primarily as sexual objects and thus cannot opt out—objectification can't be freely chosen and is therefore a problem. Part of what matters is therefore a kind of "social autonomy"—our ability to be who we want to be in the world and be seen as we want to be seen.

Pressures on choices from background context and threats to social autonomy have often been theorized through the idea of "adaptive preferences"—these have been understood in a wide range of ways, but the basic idea is that a preference is adaptive when it is shaped by social forces in some problematic way. In this chapter, I explore various ways that the ideas of social autonomy and adaptive preferences can be used to theorize choices regarding sexual objectification in social contexts. I argue that this framing can help us understand why objectification is complex with respect to sex, gender, and other factors. Generally, when women, or people who are racialized in various ways, or other people our society tends to categorize and stereotype, are valued only or primarily for the sexual pleasure they can give to others, their social autonomy is adversely affected. I argue that from a perspective focused on social autonomy, sexual objectification can be both a wonderful thing for particular individuals on particular occasions and also a widespread social harm.

Objectification and Autonomy

There are many ways to theorize sexual objectification.[3] Contemporary thinking in Anglophone culture has been much influenced by feminist scholarship beginning in the twentieth century emphasizing the connections between sexist social norms, sexual objectification, and discrimination. In many contemporary societies, women are valued more for their qualities as sex objects or for their appearance than for their accomplishments, intelligence, character, and so on. Furthermore, femininity is generally associated with passivity or submissiveness while masculinity is associated with action and dominance. Feminist scholars argue that being valued for attractiveness and submissiveness is linked to sexism and discrimination more generally: seeing women as sex objects leads people to dehumanize them, to ignore their individuality and personhood, and to feel permitted to discriminate against them. In her 1989 book *Toward a Feminist Theory of the State*, Catharine MacKinnon says that "All women live in sexual objectification the way fish live in water" (149). Part of the idea here is that not only are women surrounded by objectification, they also must be willing to be seen as sexual objects—and desirable ones—to get the things they need and want in life. A culture that relentlessly uses women's sexuality and bodies for various purposes is one in which constant objectification has a pernicious effect on women's lives.

But is objectification always a problem? Martha Nussbaum and others have argued that objectification can take on various aspects, can be part of the best sexual experiences, and is highly sensitive to context and tone.[4] Citing passages in D. H. Lawrence, Nussbaum shows how temporary absorption in one's lover's sexual and physical characteristics, even to the point of identifying a person with their genitals, is compatible with overall love and respect. When things are going well, we enjoy being the object of our partner's desire and lust; if our beloved temporarily fails to attend to our full humanity in sex, this is not a sign of ethical failing but rather of appropriate passion. Nussbaum distinguishes seven kinds of objectification and argues that they present different ethical aspects; objectification, she concludes, is not always bad.

There are several ways to try to distinguish desirable or benign forms of objectification from sinister ones. Nussbaum emphasizes the nature of the relationship of the participants: as long as the partners are in a respectful ongoing relationship, they are not using one another instrumentally when they temporarily lose sight of one another's full humanity in sex and objectify one another in the ways mentioned above. She and others also point to the importance of mutuality and reciprocity in sexual relations: to avoid wrongly using another person, both partners should give and receive sexual pleasure.

In previous work, I've argued for a different response that focuses on the autonomy of the choices in question. In this autonomy view, what matters is that consent to practices is genuine and autonomous, and this applies to objectifying practices such as pornography, sex work, and casual sex as well as any other kind of sex. One can choose to be objectified. There is a kind of instrumental use, common in sex, which is morally benign, and for which autonomous consent is what matters. That is instrumental use of a particularly weak sort: A objectifies B by treating B as a body, by treating B as interchangeable with others of B's type, by treating B as passive, by simply ignoring B's desires and wishes, and so on, all while respecting B's autonomous choice to participate. In this way of thinking about things, objectification is not a problem with pornography, sex work, or commodified casual sex as long as these are freely chosen; relationships, mutuality, and reciprocity are not required.[5]

In this framing, much depends on how we understand autonomy. At a basic level, respect for autonomy entails refraining from coercion, deception, and exploitation. This means that if you want to objectify someone sexually, you have some responsibility to know the circumstances of the choice and your role in those circumstances and to be honest about the reasons. If A asks B for oral sex in exchange for a place to stay for the night, and the streets outside aren't safe, the request is not in accordance with respect for sexual autonomy, because it's exploitative. If A claims to love B but is just using sex with B to get back at C, that is deceptive and not in accordance with respect for autonomy. If A sees B being attacked and demands sex in exchange for calling 911, that is coercive and doesn't respect B's autonomy.

The harder and more subtle cases involve the ways that social pressures and oppressive social structures affect our choices. The metaphor of the fish in water makes vivid the ways that choice is socially contextual. If a person chooses to consent to being objectified, or chooses to self-objectify, because that is part of the gender norm they have grown up with or because it is impossible to opt out of objectification, then the choice is socially constrained. This suggests that one element in theorizing freely chosen objectification might have to do with social factors and thus involve a kind of "social autonomy."

To motivate the idea that objectification has to do with social autonomy, consider this story from Dan Savage's 2005 book *The Commitment*—a memoir largely about his relationship with his husband and their decision to marry.[6] In this anecdote, Savage describes an encounter he had with a young man who was in the audience for one of his lectures. During the Q and A, this man explained that he had a strong desire for having a sexual experience of having cake "smashed in his face"—he had never had cake smashed in his face because no one was interested in having this experience with him. After he interpreted Savage's lighthearted complimentary joke—"I'd smash a cake in your face in a heartbeat"—as a serious offer, Savage agreed to do it, and they did: they went back to the young man's hotel, and the young man got into the bathtub, and Savage smashed a cake in his face.

In a sense, this young man is objectifying Dan Savage: he is ignoring Savage's particular desires and wishes and he is treating him as interchangeable with others, thus treating him as instrumentally useful and fungible. There is no real relationship, and there is no symmetry, reciprocity, or mutuality. And yet—and I am appealing to intuitions here—it does not seem the young man did anything wrong.

Looking at context helps us to see why this case is a non-problem and points us in a direction to understand what social autonomy is. Savage goes ahead with the encounter on his own terms—it isn't pushed on him or the result of social pressure. Savage is a white, American, able-bodied, good-looking man, with a range of sexual experience and a good job that is compatible with such behavior. The world doesn't treat people like him as worthy only of being sexual objects, and he isn't going to get fired for a sexual favor given out freely. Partly because of his situation, and partly because of the unusual nature of the request, people are not going to start denigrating him as "that whore who is willing to smash cake in your face." This is not to say that Dan Savage has global autonomy in other contexts. Gay people are still subject to serious abuse, harassment, and threats, and these obviously get in the way of them doing what they want to do. And gay culture can itself be objectifying. The point here is just that in these particular circumstances, Savage is enabled to act with a certain amount of self-determination.

One way of understanding this social autonomy has to do with the kinds of options he has in the situation and the costs of making various choices. In this case, whether he participates or not, there are few costs either way. This stands in sharp contrast to the situation that MacKinnon and her colleagues allude to when they point to the social omnipresence of certain kinds of objectification: if women live in sexual objectification the way fish live in water, then the costs of opting out, or refusing objectification, are high. A woman who refuses to dress sexily at work might be fired; women who complain about workplace sexual harassment can be vilified and treated as liars or crybabies. The social autonomy view thus presses us to ask not only whether someone consents to objectification but also why they are consenting, and the availability of options is one factor that affects the autonomy of the choices people make. In the next section, I consider how the lens of "adaptive preferences" can help explore this further.

Social Autonomy and Options: Adaptive Preferences and Conscious Bargaining

One commonly used framework for understanding socially constrained choices is that of "adaptive preferences." These are theorized in a range of ways but all have to do with the idea that an adaptive preference is a preference formed in social conditions that would not have been formed that way had those conditions been otherwise; often the intuition is that they are preferences that are not what the person would have otherwise wanted or that they are formed in response to social conditions that are constraining or unjust, so that the person would have been better off had conditions been otherwise. For example, a woman who lives in a society with sexist beauty norms may respond to the relevant social pressure by developing a desire, or preference, to conform to those norms.

There are a lot of different ideas regarding how to characterize "adaptive preferences," leading to definitions that aim to capture roughly the same intuitions but differ with respect to the precise phenomena that they pick out.[7] In the framing of political theorist Jon Elster, an adaptive preference is a preference shaped by a process that preempts a choice for what the person would have preferred, had it been available. That is, we would prefer A, but since A is not available, we not only choose B but come to prefer it to A, just as the fox came to prefer not to have the grapes in the fable of the sour grapes. A woman who lives in a society in which women cannot go to medical school, and who thus forms a desire to become a nurse and not a doctor, may have an adaptive preference if, given the option, she would have preferred to become a doctor. The preference is adaptive to the circumstances in Elster's sense.

Notably, Elster's characterization is neutral with respect to the quality or normative aspect of the choice in question: it applies to any non-available option whether that option is good or bad, harmful or helpful. By contrast, other theorists of adaptive preferences emphasize such normative aspects. Anita Superson, for example, characterizes "deformed desires" as those formed specifically in response to *unjust* social conditions—such as those found in a sexist and patriarchal society.[8] So a preference is "adaptive" only if it is adapting to a context of injustice. Serene Khader (2001) characterizes adaptive preferences as those "inconsistent with basic flourishing"—here, a preference is adaptive only if adapting it is bad for the person in some way.[9] These characterizations have a more normative aspect that Elster's: a person who develops a preference for apples instead of oranges because apples happen to be available and oranges are not could have an adaptive preference in Elster's framing while they typically would not in Superson's or Khader's—at least, assuming that the context of fruit availability is not a result of social injustice (for Superson) or that the choice for apples is not contrary to the chooser's well-being in some way (for Khader).

In this chapter, I will focus on the more normatively neutral framing suggested by Elster, though I discuss a comparison with other approaches briefly below and in the final section. As the examples so far suggest, many relevant cases are ones in which a person's options are constrained: they encounter a range of available options that are less desired than other, hypothetical, options that are not available. In these cases, choosing what would have been otherwise the most preferred option is only possible by incurring a substantive cost; achieving a desired end is only possible through means that are not desired. Again, imagine a woman who lives in a society in which conforming to gendered beauty norms is the only way to be taken seriously at work or to be seen as attractive, but who would otherwise prefer to dress differently. In such cases, even though the person has options, the social context constrains those options.

In contexts of social constraint, a person may engage in what I will call "conscious bargaining": knowingly making trade-offs by thinking through the various pros and cons of her various choices and considering what she will gain or lose in each case. For example, a woman may choose gendered beauty norms despite a clear perception of their costs because she has judged that the gains outweigh the costs.

However, in such cases one may instead adapt to the social constraint by forming a new positive preference for what would have been otherwise unappealing—a preference experienced as positively valenced toward the choices made. For example, if the option of rejecting beauty norms while also being seen as competent and attractive is not available, a woman may develop a felt preference in favor of conforming. As Marilyn Frye says in her analysis of sexism, people often become what they practice being: it may be that a woman who chooses conformity first because of societal pressures then comes to desire conformity itself, on grounds that it comes to feel right and familiar to her.[10] Extending Elster's specific formation, in this kind of case, where one can opt out but the costs of doing so are high, I propose that we see the developed preference as an adaptive preference. While strictly speaking a person in this kind of situation still has a choice, their choice is constrained by contextual availability, and their coming to prefer one of the previously unpreferred options on its own terms is adaptive in a similar way to the more straightforward cases in which a single preferred option is unavailable.

This means that a case in which a person experiences costs associated with an option can be either a case of conscious bargaining or a case of adaptive preferences. While these are sometimes conflated or brought together, in my view it is useful to distinguish them. The first has to do with knowingly making trade-offs and the second with the ways social contexts shape and inform our preferences themselves. In the case of gendered beauty norms, if there are costs associated with violating these norms, a person may weigh the relevant costs and decide what they want most overall; this would be conscious bargaining. But we may also say that the person would have preferred an option in which they could violate those norms and not experience the costs. If they have adapted to their context by developing a felt preference for conformity, this would be an adaptive preference.[11]

In a poignant forward to a recent collection on sex and women's desire, comedian and performer Margaret Cho lists the many reasons that made her say "yes" to sex even when she felt "no." For example: "I thought I was so ugly and fat that I should just take sex every time it was offered, because who knew when it would be offered again"; "I felt it was too much trouble to say no,"—that is, saying no led to hassle and arguments.[12]

The latter of these is an example of conscious bargaining. Cho knows she would prefer a world that gives her the option of saying "no" without the trouble of explanation or having an argument about it; she negotiates the available options as best she can. The former is ambiguous, because of the suggestion that she has judged her options incorrectly, but a person who chooses sex with someone they do not much like because they correctly judge themselves not to have the option for sex with someone they like more is also enacting conscious bargaining. Cases of adaptive preferences are different and would involve a felt preference for the chosen option.

One way to understand the relationship between adaptive preferences and conscious bargaining, on the one hand, and social autonomy, on the other, has to do with social context and the options available to a person. There are many ways of theorizing autonomy, and as Nomy Arpaly has emphasized, they are not mutually incompatible.[13] Different theories can address different kinds of autonomy: the kind of autonomy associated with internal self-governance is distinct from that associated with reflective endorsement, which in turn is distinct from that associated with worldly power. Various concepts of autonomy are likely relevant to choices regarding sexual objectification, but here I'll focus just on an idea that relates social autonomy to context and options.

The basic idea is that a high level of social autonomy is related to having a range of acceptable or preferred options and that a person whose options are all bad has a low level of social autonomy. This fits with the idea, sometimes seen in theories of positive liberty, that if the costs of not doing X are very high, then one can't choose X autonomously. Obviously, "have sex with me or I will kill you" does not give a person options; it just coerces them into obedience. The idea of social autonomy broadens this insight into a continuum. Several autonomy theorists who have concerned themselves with choices have endorsed versions of this idea. For example, in exploring the value of autonomy, Tom Hurka writes of his basic assumption that autonomy "involves choice … from a wide range of options."[14] In her social-relational account of autonomy, Marina Oshana writes: "I maintain that the self-governing individual must have access to an assortment of options."[15] Joseph Raz says that "autonomy requires the availability of an adequate range of options"[16]—a person who lives her whole life being chased around an island by a horrible and deadly beast may make many "choices"—about where to run, and so on—but we would hardly say she is autonomous.[17] Gerald Cohen makes an analogous point when he says that to "exercise a freedom" in choosing a hazardous job, a worker must have an "acceptable" or "reasonable" alternative.[18]

What makes an option "good" or "bad" in this context? A full answer to this question would be complex and would require analysis of the relationships among autonomy, well-being, and rationality—relationships that are variously understood and often contested. One important distinction is between frameworks in which the assessment of the option is from the point of view of the person making the choice and frameworks in which there is an objective or "third-person" perspective from which the quality of various options can be judged. Following Elster's value-neutral characterization of adaptive preferences, we could say that "good" or "bad" is from the point of view of the person making the choice and we could thus eschew a third-person evaluative perspective. By contrast, following accounts of adaptive preferences like those of Superson and Khader, we could say that there is a background theory of social justice or human flourishing that informs these evaluations. Again, I will focus here on the neutral, first-person evaluative perspective and on drawing out its implications; while a full discussion would be beyond the scope of this chapter, in the final section I offer a few considerations in favor of adopting this first-person evaluative approach in this context.

When it comes to conscious bargaining, the degree of social autonomy concerns the range of available options. Those who consent to being objectified because it is preferable to other bad options have a low level of social autonomy. For example, a person who must, against his or her wishes, dress sexily and put up with cat calls at work or risk losing a job has a low level of social autonomy; the options are all bad. But a person who enjoys the attention, has other good options, and has nothing to lose from opting out may choose to be objectified with a high level of social autonomy. This, I take it, helps to explain what is happening in the example of Dan Savage: he consents to being objectified, but since he chooses from a range of good options, his choice has a high degree of the relevant kind of autonomy.

Now consider adaptive preferences. Using the characterization described above, in a case of adaptive preference, a person comes to desire and prefer a given option because another option, which would have been preferred if it had been available or easy to access, is not available or comes with a cost. This is a counterfactual, which means that in many real-life situations we can never know whether something is an adaptive preference: how could we know what would have been preferred in other circumstances?

What we often do is conjecture, from our own point of view, that another person's preference—or our own—is adaptive. For example, we might reason that if options that seem good to us were available to others, others would come to prefer them. On the one hand, this framing can be understood as explaining some of our practices in contexts of oppression, in which a social reframing of a choice leads the chooser to see it as part of an oppressive system, rather than something they wanted for themselves. On the other hand, this framing can also be fraught, especially in cases where we risk making judgments about others whose lives and contexts we don't understand well. In a critique of Western women pronouncing on the practices of other women, Uma Narayan argues that imagining one's self as another risks oversimplification, stereotyping, and caricature. Narayan says that in this context, Western feminists invoke tropes of the "prisoner of patriarchy"—someone who has patriarchal oppression imposed on her "entirely against her will and consent" and the "dupe of patriarchy"—who imposes on herself the constraints of patriarchal oppression because she subscribes completely to its norms and practices.[19] Instead, women can often be seen as "bargaining with patriarchy." In the perspective being explicated here, this kind of example shows how determining whether to see another's preference as adaptive is complex.[20]

In a society where women are often evaluated on the basis of their appearance, a woman might find that to be taken seriously in whatever she is doing she must also seem sexually attractive. As we've seen, this might include self-objectification: dressing a certain way, with girly clothes and heels, wearing a certain hairstyle, developing a gentle and non-threatening demeanor. As above, a case like this can involve "conscious bargaining," in which a person makes a calculation about which option is best for them overall. But as we've seen as well, a case like this can also be an "adaptive preference." It presents an interconnection among options that the person cannot opt out of, namely, the option of dressing in comfortable clothes and sneakers and still being valued and taken seriously. And the costs of choosing to self-objectify or to opt out may both be high. So in these circumstances a woman may be developing an adaptive preference for self-objectification: she comes to feel that she wants and prefers it for itself. But it isn't really because that is who she is. Instead, it's like the fox and the sour grapes: she comes to prefer the better of the only remaining alternatives she has.

If social autonomy is related to options in the way I've described above, then if a preference is truly adaptive in this sense, then it is not fully autonomous in the sense of social autonomy. This is because by hypothesis, such a preference is for an option that a person cannot opt out of without incurring serious costs and an option that one would not prefer if other options were available. In one straightforward sense, then, adaptive preferences are not the person's own and are structured in undesirable ways by the person's surroundings.[21]

Even though we typically cannot know whether a preference is adaptive, there is a way to have reason for believing that a preference for X is not adaptive, and that is to see that relatively low-cost options exist that include not-X. Strictly speaking, not being able to opt out does not itself make a choice socially non-autonomous, but it puts us in a state of uncertainty whether a choice is socially autonomous. As we will see in the next sections, there are various ways that changing social context can affect preferences and our understandings of preferences.

Examples and Implications

In considering some examples and implications, consider first the novel *Story of O*, which features in the analyses of Dworkin and Nussbaum. In this mid-twentieth century novel, the main character, O, enters into a sexually submissive relationship with a "master," Sir Stephen—a relationship that gradually comes to dominate her entire life. From the first-person point of view, she describes this relationship and this life as thrilling and deeply satisfying. Andrea Dworkin presents it as a most sinister case of objectification, because O is "possessed," "an object," and virtually a slave to her master's desires. Nussbaum suggests that Dworkin sees O as "a paradigm of a pattern of relationship prevalent in our culture," but argues in response that simply seeing O as an object misses the different aspects of objectification that are in play. These include inertness, fungibility, and ownership, which she says are treated as inevitable consequences of denial of autonomy and instrumentalization. For Nussbaum, it is essential to see these as separable, and from her point of view the sinister aspect of O's objectification concerns not the fungibility, inertness, or even instrumentalization per se but rather the nature of O's relationship with Sir Stephen more generally: since the pattern of submission infuses their entire interaction, the relationship lacks the kind of mutuality, equality, and full respect that would make objectifying actions and moments ethically acceptable.

As I've discussed in my previous work, from the autonomy point of view, the ethical question concerns not the ways in which O is obedient but rather whether she chooses the relationship autonomously. In one sense, the novel presents the answer as "yes," since O has many good options that do not involve this relationship at all: she is a successful photographer with an apartment, friends, and so on, and she is explicitly invited to leave should she wish to. Given that from her point of view she is choosing from among good options, and given that she has good options that involve opting out entirely, she has, in one sense, a high degree of sexual autonomy and thus we have prima facie reason to think that this is not an ethically troubling instance of objectification.

However, cases like this are highly contextual, and here the framework of adaptive preferences can allow for a more subtle reading. It is possible that a preference like O's for being a sexual object in general is somehow adaptive. Insofar as a woman lives in a society that makes it difficult or impossible for women to opt out of sexual objecthood, her choice may not be autonomous. If women are generally valued for their status as sex objects, then O's choosing one way to be a sex object rather than another may be adaptive in the sense that she cannot opt out of sexual objecthood at all, or because the costs of doing so are high.

Generally speaking, when a person cannot opt out, they cannot make a choice. This inability to opt out is particularly troubling when it comes to sexual objectification because of the particular way that being treated as a sexual object—not by a given individual but by one's cultural and social world—can take over a person's entire life in such particularly pernicious ways. People who are always treated only or primarily as sexual objects do not have any choice in the matter and are robbed of important elements of a good life in modern society, because they aren't taken seriously as citizens, workers, and friends.

When women, or people of certain races, or other people our society tends to categorize and stereotype, are valued only or primarily for the sexual pleasure they can give to others, they are harmed in this way. This is one reason that men whistling at women in the street can be a sinister and serious form of objectification: it is a reminder of the ways in which a women's role in society is still thought of as that of being a provider of sexual pleasure, in a context in which it is impossible to opt out. For many men, being whistled at in a similar way does not present a relevantly similar situation and thus the ethical aspects are very different.

The autonomy view helps us understand why these asymmetries exist and focuses on the contexts of acts of objectification as well as the content of acts of objectification. Because women are frequently objectified in ways that go beyond their control, to rank women's looks on a 10-point scale in conversation can be very bad ethically, while seemingly more objectifying acts aimed at men—such as paying a man for a sexual service—can be fine. As is often pointed out, the former reinforces a problematic status quo while the latter generally does not. The autonomy view explains in a simple and straightforward way why there is such an asymmetry and why the ethical aspects are related to the status quo: in adding to the conditions that make opting out impossible, the first contributes to an erosion of the person's autonomy and thus objectifies them in a pernicious sense. The second does not.

In a different example, the framework of social autonomy helps explain why there is a difference between women's appearance in soft-core pornography aimed at men and their appearance at Pride parades—even though on the face of it, they may involve similar kinds of flaunting and showing of bodies. Both activities can involve being ogled, seen through a limited sexualizing lens, and so on. But from the point of view of social autonomy, the pornography both reflects and reinforces the same kind of problematic existing social patterns that we saw in the "Objectification and Autonomy" section in the metaphor of the fish in water—that women are treated as valuable for their sexual attractiveness so much more than men are. Women are objectified in ways that make it difficult to opt out. Pride events, by contrast, seek to create space in which the status quo is challenged, in which women are not just sexual objects and men are not just sexual subjects.

Of course, gay culture also has patterns of objectification, and people subject to oppression, stereotyping, and marginalization can experience social patterns constraining choices in a range of ways, especially when it comes to sexuality and objectification. As is often pointed out, racialized people are often "hypersexualized"—perceived in ways that place their sexuality at the center of their identity. Robin Zheng (2016) discusses the ways that the hypersexualization of Asian-American women causes them to be seen through a sexualizing, one-dimensional lens, valued for a set of perceived characteristics related to stereotypes such as being submissive. Niels Teunis (2007) analyzes patterns of sexual interaction in a community of gay men and finds that, often, white men want Black men as sexual partners only when the Black men are willing to play a particular, and racially stereotyping, role—being dominant, in control, and on top. Teunis says that this fixation on a particular trait is wrongly objectifying.

Relatedly, in an analysis of sexual racism, Nathaniel Adams Tobias Coleman points out that when their partners demand that Black men play specific roles, Black men are at a disadvantage with respect to having the kind of sex they want: to engage in the activities they prefer, on their own terms, they may have to do more of what the other person wants them to do.[22] Trans people are often objectified and hypersexualized, partly through being fetishized by people with sexualized stereotypes of trans bodies.[23] People with disabilities are sometimes sexually fetishized, seen as sexually interesting in virtue of their disability.[24] Since people experience oppression along multiple axes, a fuller analysis of the social conditions that impact social autonomy would require taking an intersectional approach that would analyze people's experiences with interlocking systems—examining, for example, how trans people of color experience objectification.[25]

Objectification as Inherently Wrong?

From the point of view developed here, there is nothing inherently wrong with sexual objectification as objectification: as long as other ethical violations are avoided and consent is respected, the normative questions concern the individual's evaluative perspective, the reasons for their choices, and the ways that the social and political background context influences those choices. This perspective might be contrasted with one in which sexual objectification is viewed as inherently bad—in which, regardless of social context, no autonomous and healthy person would reasonably choose it.

There are, I believe, two related but distinct ways to frame the issue. One would be to say that while autonomy is central, it is not choice autonomy that matters; it is, instead, a richer, internal form of autonomy—a form that would show immediately that one cannot autonomously choose certain forms of sexual objectification. For example, from the perspective of a perfectionist idea of autonomy such as that of Marina Oshana, we might be led to the idea that sexual objectification—or at least some kinds of it—is always a problem. This would involve rejecting the framing in which it is from the point of view of the person choosing that the options can be said to be good or bad.

A second frame would be to say that while a focus on social autonomy and options is on the right track, the relevant notion of adaptive preference is not the one I have been appealing to here. In the framing I have been using here, conscious bargaining is distinct from adaptive preference and the latter involves normative evaluation of options from the point of view of the person choosing. As we've seen above, alternative framings of adaptive preferences are more normative: in Superson's notion of "deformed desire," desires formed in response to unjust social conditions, and in Khader's theory, in which adaptive preferences are those "inconsistent with basic flourishing." These alternative framings aim to capture what is relevant about socially constrained preferences through characterizations that are conceptually distinct and thus pick out different phenomena. These alternative framings might entail that a choice to be objectified or self-objectify is always problematic in some sense—either because it is a response to an injustice in the background context or because there is something about objectification that harms or hinders one's well-being.[26]

I cannot fully explain my disagreements with these contrary perspectives in such a short chapter, and my main aim here has been to elaborate the conclusions that follow from the first-person evaluative framing. But I will suggest at least two considerations in favor of my perspective. The first has to do with why we might think objectification inherently bad, rather than contextually bad. One possibility may be a felt intuition that objectification is always a problematic choice in some way, that it cannot be consistent with well-being. As a possible expression of this felt intuition, consider Nancy Bauer's discussion of young women's choice to give and not receive sexual pleasure. She writes of her mystification that her female students want to give sexual pleasure to men and to spend their weekend evenings giving "unreciprocated blow jobs to drunken frat boys."[27] She says the students tell her it gives them a feeling of power, and that they enjoy it, but that they are unable to explain why they enjoy it or why it matters so little to them to get pleasure in return. Bauer conjectures that her students have moved from "pretending to enjoy" to "actually reveling in" their own "self-effacement."[28] The implication seems to be that if these women are choosing to consent to objectification, they must be acting in ways that are not in their best interests—they are choosing contrary to their well-being because of sexist social pressure.

The matter is complex, but it's worth noting that we are used to encountering discussions of objectification in which the objectified persons are women in contexts of societal sexism: this may obscure the mechanism by which the cases become problematic. In support of this, consider that being excited or pleased by the sexual excitement or pleasure of another person (as Bauer's students are) may not seem odd in other contexts, even when it is unreciprocated. Furthermore, in the absence of sexist social pressures, people still choose to be objectified. I've already mentioned the example of Dan Savage, who I take to be choosing to be objectified in an unproblematic way. When some straight white men talk of their desire to be sexually objectified, it is possible to understand them as expressing an intelligible and unsurprising wish to be the object of sexual desire—to be found attractive, arousing, and so on, and perhaps to be found so because of their bodies and physicality, as passive, or as givers and not receivers of sexual pleasure. It is because they are typically treated as full persons in our society that this choice for them makes sense and does not seem strange or ethically troubling. In this same direction, some male submissives enjoy being dominated in very one-sided ways. Some gay men self-objectify in filming themselves for others to enjoy watching: one-sided, unreciprocated. Some gay men are willing to pay to give, and not receive, oral sex.[29] These examples show that even if there is something problematic in the unequal division of sexual labor and pleasure Bauer describes, it need not be traced to the fact that choosing to be objectified is somehow contrary to well-being.

Again, these matters are complex. But I'll mention one further reason to resist a third-person objective theory about good, bad, worth wanting, worth rejecting, in the domain of sexual objectification. The history of attempts to say what is good and bad in sex includes some very bad missteps, such as the classification of homosexuality as a disease or disorder, the Victorian idea that women do not or should not enjoy sex, and the presentation of vaginal orgasms as preferable to those arising from stimulation of the clitoris. We are, perhaps, not good at judging or predicting what is good for others in the sexual realm. This might support a framing in line with the one developed here, in which the evaluations of choices of sexual objectification are understood from the point of view of the person choosing.

Conclusion

I've put forward here a view in which sexual objectification is understood through the lens of social autonomy and adaptive preferences. On this view, the ethics of sexual objectification is a matter of autonomous consent, where one kind of autonomy that is relevant is "social autonomy"—our ability to be who we want to be in the world and to be seen as we want to be seen. One threat to social autonomy is when our choices are socially constrained, which can be understood through the lens of adaptive preferences and conscious bargaining.

One possibility emerging from this framework is that in certain social and political settings it becomes difficult or impossible to autonomously consent to being objectified because you can't opt out: you can't choose not to be objectified. I've argued that the idea that autonomy has a social dimension helps us understand why objectification is complex with respect to sex, gender, and other factors. Generally, when women, or people who are racialized in various ways, or other people our society tends to categorize and stereotype, are valued only or primarily for the sexual pleasure they can give to others, their social autonomy is adversely affected. Among other things, this helps us make sense of various complexities including the idea that objectification can be a wonderful thing and can also be harmful and destructive.[30]

Notes

1. Huffington Post, "Cameron Diaz: 'Every Woman Wants to Be Objectified,'" *Huffington Post*. Available at www.huffingtonpost.com/2012/11/20/cameron-diaz-every-woman-wants-to-be-objectified_n_2164965.html (accessed December 30, 2020).
2. Patricia Marino, "The Ethics of Sexual Objectification: Autonomy and Consent," *Inquiry* 51 (2008), 345–364. To say that there is nothing inherently wrong with objectification in certain contexts leaves open the possibility that some kinds of objectifying practices have problematic effects and thus leaves open the possibility that pornography, sex work, and commodified casual sex are wrong for other reasons. See, for example, Jeff Gauthier, "Prostitution, Sexual Autonomy, and Sex Discrimination," *Hypatia* vol. 26, no. 1 (Winter, 2011).
3. In addition to work discussed below, see Sandra Lee Bartky, *Femininity and Domination: Studies in the Phenomenology of Oppression* (New York: Routledge, 1990), Susan Bordo, *Unbearable Weight* (Berkeley, CA: University of California Press, 1993), and Ann Cahill, *Overcoming Objectification: A Carnal Ethics* (Routledge, 2012).

4. Martha Nussbaum, "Objectification," *Philosophy and Public Affairs*, 24(4) (1995), 249–291.
5. Marino, "The Ethics of Sexual Objectification."
6. Dan Savage, *The Commitment: Love, Sex, Marriage, and My Family* (New York: Plume, 2005).
7. Jon Elster, *Sour Grapes: Studies in the Subversion of Rationality* (Cambridge: Cambridge University Press, 1983). In addition to texts cited below, see also Amartya Sen, *On Ethics and Economics* (Blackwell, 1988); Martha Nussbaum, *Women and Human Development: The Capabilities Approach* (Cambridge 2001); Marilyn Friedman, Autonomy, Gender, and Politics (Clarendon, 2003); Deepa Narayan, "Conceptual Frameworks and Methodological Challenges," in *Measuring Empowerment: Cross-Disciplinary Perspectives* (World Bank Publications, 2005); Uma Narayan, "Minds of Their Own: Choices, Autonomy, Cultural Practices and Other Women," in *A Mind of One's Own. Feminist Essays on Reason and Objectivity*, L. Antony and C. Witt (eds.) (Westview Press, 2002), 418–432. See discussion in Superson, Anita, "Feminist Moral Psychology", *The Stanford Encyclopedia of Philosophy* (Summer 2020 Edition), Edward N. Zalta (ed.), URL = <https://plato.stanford.edu/archives/sum2020/entries/feminism-moralpsych/>.
8. Anita Superson, "Deformed Desires and Deformed Desire Tests," *Hypatia* 20(4) (2005): 109–126.
9. Serena Khader, Adaptive Preferences and Women's Empowerment (Oxford University Press, 2001) and Khader, Serene J., "Must Theorising about Adaptive Preferences Deny Women's Agency?" *Journal of Applied Philosophy* 29, no. 4 (2012): 302–317.
10. Marilyn Frye, *The Politics of Reality* (Freedom, CA: Crossing Press, 1983).
11. In "Minds of Their Own: Choices, Autonomy, Cultural Practices and Other Women," Uma Narayan develops the idea that a person may encounter a "bundle of elements," some of which they want and others of which they do not, where they are powerless to undo the bundle (p. 422). In such contexts, Narayan emphasizes the possibility of bargaining and the dangers of assuming that others are "duped" or forced by patriarchy rather than knowingly making trade-offs and "bargaining" with patriarchy. See further discussion of Narayan below.
12. Margaret Cho, Introduction to *Yes Means Yes: Visions of Female Sexual Power and a World Without Rape.*
13. Nomy Arpaly, *Unprincipled Virtue: An Inquiry into Moral Agency* (Oxford University Press, 2002).
14. Thomas Hurka, "Why Value Autonomy?" *Social Theory and Practice*, 13 (1987) (3), 361–382, p. 362.
15. Marina Oshana, *Personal Autonomy in Society* (Ashgate, 2006), p. 84. Oshana goes on to deny that "the adequacy of options can only be decided by the individual whose options they are." This marks a divergence from my own approach; I discuss this contrast below and in "Objectification as Inherently Wrong?" section.
16. Joseph Raz, *Morality of Freedom* (Oxford University Press, 1988), p. 410. The full passage there says, "while autonomy requires the availability of an adequate range of options, it does not require the presence of any particular option among them." See also page 373.
17. Raz, Morality of Freedom, pp. 373–374.

18. Gerald. A. Cohen, *History, Labour, and Freedom: Themes from Marx* (Clarendon Press, 1988), p. 282.
19. Narayan, "Minds of Their Own," pp. 418–419.
20. My framing of adaptive preferences may seem to echo Narayan's characterization of a "dupe." But in its emphasis on the first-person evaluative point of view, my account shares with hers two implications: (1) that there are deep epistemological challenges involved in judging that someone's preferences are adaptive and (2) that we should hesitate to assume that we know what would be best for someone else.
21. As Elster himself has it, when a person comes to prefer an option because he or she hasn't other good alternatives, or comes to desire something because of unjust or coercive social pressures, the resulting "choice" is not fully autonomous or free. Of course, if social autonomy is related to options in a different way this wouldn't follow. Khader argues that adaptive preferences should not be seen as autonomy failures, and this may seem to flatly contradict what I have said here, but when we examine the details, the tensions are less obvious. First, Khader directly argues that adaptive preferences do not undercut a person's autonomy as conceived as various kinds of "procedural" autonomy or "substantive" autonomy. My specific view about linking options to autonomy directly makes no claim about adaptive preferences undercutting these forms of autonomy as opposed to social autonomy. Khader also says that we should not use the term "non-autonomous" for adaptive preferences. But she supports this point by saying the terminology suggests that people who make choices inconsistent with flourishing because of their circumstances do so in a way that is less consciously aware than other choosers. In my own conceptualization of these matters, both adaptive preferences and conscious bargaining can be ways of making choices for bad options. The deeper tension has to do with whether we restrict ourselves to a first-person evaluative point of view as I am doing here; I discuss this briefly in "Objectification as Inherently Wrong?" section.
22. Nathanial Adam Tobias Coleman, "What? What? In the (Black) Butt," *The Newsletter on Philosophy and Lesbian, Gay, Bisexual, and Transgender Issues*, 11 (1) (2011): 12–15.
23. Mirella J. Flores, Laurel B. Watson, Luke R. Allen, Mudiwa Ford, Christine R. Serpe, Ping Ying Choo, and Michelle Farrell. 2018. "Transgender People of Color's Experiences of Sexual Objectification: Locating Sexual Objectification within a Matrix of Domination.: *Journal of Counseling Psychology* 65 (3): 308.
24. See discussion in Elizabeth Emens, "Intimate Discrimination: The State's Role in the Accidents of Sex and Love," *Harvard. Law. Review* 122 (2008): 1307–1402; see especially pages 1343–1344.
25. See, for example, Flores, Watson, Allen, Ford, Serpe, Choo, and Farrell, "Transgender People of Color's Experiences of Sexual Objectification."
26. Superson and Khader do not say this; rather I am drawing out my own thoughts on how their framing of adaptive preferences might lead to conclusions about objectification.
27. Nancy Bauer, "Pornutopia," *n+1* 5 (Winter 2007): 63–73, p. 72.
28. Bauer, "Pornutopia," p. 73.
29. See, for example, Kerwin Kaye, "Male Prostitution in the Twentieth-Century: Pseudohomosexuals, Hoodlum Homosexuals, and Exploited Teens," *Journal of Homosexuality* 46(1/2) (2003): 1–77.

30. This paper has been a work in progress for some time and I have benefitted from many conversations with colleagues and friends. I would particularly like to thank Alysha Kassam for her feedback on previous drafts. I'm also grateful to David Boonin, Jonathan Dewald, Raff Donelson, Brian Earp, and Helga Varden. My thanks also to the audiences at the McMaster University Philosophy Department and at the Northwestern University Queertopia 6.0 conference for their questions and comments.

CHAPTER 8

Masturbation and the Problem of Irrational and Immoral Sexual Activity

Michael Tooley

I shall be defending two main theses in this chapter. The first is that aside from sex that aims at reproduction, most human sexual activity is deeply problematic in two ways: it is irrational and immoral. As regards my second thesis, I shall borrow a phrase from a deeply pessimistic poem, *Dover Beach*, by Matthew Arnold, the ending of which is as follows:

Ah, love, let us be true
To one another! for the world, which seems
To lie before us like a land of dreams,
So various, so beautiful, so new,
Hath really neither joy, nor love, nor light,
Nor certitude, nor peace, nor help for pain;
And we are here as on a darkling plain
Swept with confused alarms of struggle and flight,
Where ignorant armies clash by night.

My second thesis is then that when it comes to thinking about sex, we too are on a darkling plain where ignorant armies clash by night, and if we are to move on to a rational and morally acceptable approach to sex, we need to clearly recognize, repudiate, and, hopefully, ultimately vanquish those armies.

Who are the ignorant armies? First of all, there are those who believe that, provided that one practices so-called safe sex, the enjoyment of sexual pleasure

M. Tooley (✉)
University of Colorado, Boulder, CO, USA
e-mail: michael.tooley@colorado.edu

D. Boonin (ed.), *The Palgrave Handbook of Sexual Ethics*,
https://doi.org/10.1007/978-3-030-87786-6_8

is both morally unproblematic and, given the intensity of sexual pleasure, something very much to be pursued. Second, there is the 'no safe sex' contingent. Third, and overlapping greatly with the second group, there are those who were indoctrinated by their parents and religious institutions with a view of sex associated with most of what are called the 'great religions' of the world, according to which the enjoyment of sexual pleasure is morally wrong except with a person of the opposite sex to whom one is married.

These three views are not the only unsound approaches to human sexual behavior, but given an understanding of why the views just mentioned are unsound, it would not be especially challenging to see why virtually every view that one has ever encountered concerning when the pursuit of sexual pleasure is both rational and morally acceptable is philosophically untenable.

Why Most Sexual Behavior Is Irrational

Human irrationality when it comes to sex is not unfamiliar. Thus, as Robin Williams is said to have quipped, "God gave men both a brain and a penis, and only enough blood to run one at a time." The relevant irrationality here, however, is not that involved in the pursuit of sex, but in sex itself.

The defense of my first thesis will be as follows. First, I shall sketch the negative outcomes to which most sexual behavior may lead. Then I consider the rejoinder that one can greatly reduce the chances of suffering those negative outcomes by practicing 'safe sex', with the result that the pleasure of sex outweighs possible negative consequences. My response will be that one can enjoy at least as much pleasure, and arguably more, without significantly exposing either oneself or one's partner to the risks that most sexual behavior involves, and thus that sexual behavior that does expose oneself and others to such risks is both irrational and morally wrong.

Possible Negative Consequences of Sexual Behavior

Sexual behavior can have two types of unwelcome consequences: first, diseases where an infectious agent is passed from one person to another; second, pregnancy, when that is not one's goal.

Sexually Transmitted Diseases

To have a vivid appreciation of why most sexual behavior is irrational and morally problematic, one needs to know of the dangers involved. The following summaries are based on the Center for Disease Control and Prevention (CDC) website as of November 2020.[1]

The most common sexually transmissible diseases (STDs) are as follows: chlamydia, genital herpes, gonorrhea, hepatitis A and B, HIV/AIDS, human papillomavirus (HPV), and trichomoniasis. Of these, **chlamydia** is a relatively mild disease and curable, but if not treated, it can have serious effects for women, such as pelvic inflammatory disease, which can involve long-term pelvic pain, infertility, and an extremely dangerous and potentiality deadly ectopic pregnancy—a condition where a fertilized egg develops outside the uterus. If left to grow, it may damage nearby organs and cause life-threatening loss of blood. To prevent complications, treatment is needed. In the early stages, medications may be sufficient; later stages require surgery.

Genital herpes is a very common STD, with more than one out of every six people between the ages of 19 and 49 having it. It can give rise to blisters and painful sores, although most people with herpes are unaware that they are carriers of the disease. Herpes can also increase the risk of contracting HIV, and a woman may pass herpes onto her baby at birth, potentially causing a fatal infection. There is no cure for herpes, but medications can shorten outbreaks as they occur.

Gonorrhea, an infection of the genitals, rectum, and throat, is now very common among young people aged 15–24. Men may have no symptoms and women only mild ones. If not treated, however, it can have serious health consequences, especially for women, where it can cause pelvic inflammatory disease. Moreover, a pregnant woman can pass it on to her baby, resulting in serious health problems for the child. Finally, untreated gonorrhea can increase one's chances both of contracting and of transmitting HIV.

Is gonorrhea curable? At present, medication can eliminate the infection, but it will not repair damage that has already occurred. In addition, drug-resistant strains of gonorrhea are emerging.

Hepatitis comes in three varieties, A, B, and C, the first two of which can be transmitted sexually. Hepatitis A and C are curable, while B is not, but there is a vaccine against it. All three forms may result in serious liver disease.

HIV/AIDS: Acquired human immunodeficiency disease is the most serious sexually transmitted disease. The HIV virus attacks and destroys a specific type of immune system cell in the body, making it harder for the body to fight off other infections, so that even minor infections like colds can be severe. AIDS is incurable, but strict adherence to anti-retroviral therapy can dramatically slow the disease's progress, prevent secondary infections and complications, and prolong life. Anti-retroviral therapy, however, is very expensive—estimated as between $14,000 and $20,000 a year—and there can also be significant side effects.

Human papillomavirus (HPV): Many types of HPV infection have been identified, at least 40 of which can infect the genital area, and most sexually active people become infected at least once in their lifetimes. HPV infections are, however, generally self-limited and also asymptomatic and unrecognized. Two types, however, can cause cancers in the throat and genital areas, while two others cause genital warts and recurrent respiratory papillomatosis—a rare medical condition involving benign tumors along the aerodigestive tract.

There is a vaccine that prevents cancer caused by HPV, which the CDC recommends be given to 11–12-year-olds. Unfortunately, there are strong anti-vaccination movements, especially in the United States, with many parents also unwilling to accept that their children will probably become sexually active teenagers. Consequently, many people may wind up not protected against the serious dangers of HPV.

Syphilis is a very dangerous disease with four stages. Primary-stage syphilis typically involves sores at the site of the original infection, while symptoms of secondary-stage syphilis include skin rash, swollen lymph nodes, and fever. In both of these stages, the symptoms can be quite mild and thus go unnoticed. Next is the latency stage, where there are no symptoms, and which is followed, between 10 and 30 years after the original infection, by the tertiary stage, which is extremely serious, typically involving damage to many organ systems, including the heart, blood vessels, the brain, and nervous system—damage that can result in death.

Syphilis can be cured by the right antibiotics, but they do not repair any damage already done, and here it is important to note that while symptoms in the primary and second stages are mild, even in those stages, without treatment, syphilis can do serious damage to one's brain and nervous system (neurosyphilis), possibly resulting in paralysis or dementia, and to one's vision (ocular syphilis), possibly resulting in blindness.

Trichomoniasis involves a protozoan parasite, and it is a very common sexually transmitted disease. Symptoms include itching and burning sensations and discharge from the penis or vagina. The symptoms are generally mild, so most people with the parasite do not know that they are infected. Trichomoniasis can, however, increase the likelihood both of contracting and of transmitting other sexually transmitted diseases, including HIV/AIDS, and women who are pregnant may give birth prematurely and so have lower birth weight babies. Lastly, if one is aware that one has the disease, it can be cured with a single dose of a prescription antibiotic.

Finally, there is **pelvic inflammatory disease**, which differs from all of the preceding STDs in that while the infectious agent is transmitted from one person to another—from a man to a woman via vaginal intercourse—the serious consequences of the infection are ones from which only women suffer. The symptoms of pelvic inflammatory disease (PID) are quite varied and include the following: pain in the pelvis, abdomen, lower back, or vagina, which can occur during intercourse or urination, and vaginal discharge, chills, fatigue, fever, nausea, and vomiting. In addition, pelvic inflammatory disease may also cause tubal infertility and ectopic pregnancy.

This concludes my summary of sexually transmitted diseases that one is potentially exposed to through sexual behavior—a summary that I believe provides strong *prima facie* grounds for concluding that it is very much in one's interest to minimize one's chances of contracting such diseases if that can be done without any counterbalancing loss.

A Second Possible Negative Consequence of Sexual Behavior: Pregnancy

Not all sexually active women engage in vaginal intercourse, but the proportion seems quite high. Thus, a study carried out by Melissa Bisson and Timothy Levine on university students in 'friends with benefits' relationships found that

56% engaged in all forms of sexual activity, while 22.7% engaged in intercourse only[2], meaning that over three-quarters of such people engaged in vaginal intercourse. It seems, then, that a high proportion of women expose themselves to *some*, perhaps significant, risk of pregnancy.

Given the extreme irresponsibility of American men as regards children fathered outside of marriage or a stable relationship, and a legal system that refuses to deal appropriately with such behavior, the result is that 21% of American children live with a single mother whereas only 4% live with a single father.[3] The consequences for women are very bad indeed, especially since many are young and will probably have to curtail their education, thereby being doomed to low-paying jobs while struggling with the extremely difficult task of raising a child on their own.

One can, of course, choose to have an abortion, and since the *Roe v. Wade* Supreme Court decision—which currently is in great danger of being overturned by the Supreme Court—women often do so. Many women, however, unaware that all anti-abortion arguments are open to *decisive* philosophical objections, mistakenly believe that abortion is seriously wrong and thus are unlikely to choose an abortion. The fathers of such children, however, almost always quickly hit the road in search of other women to have sex with—and quite possibly impregnate—with the result that the woman, unless she comes from a well-off background, is in all probability doomed to a very difficult life.

Responses to the Negative Consequences of Sexual Behavior

I have set out a *prima facie* argument for the view that most human sexual behavior not aimed at reproduction is irrational in the sense of being contrary to the individual's self-interest, first because of the risks of sexually transmitted diseases, some of them very unwelcome indeed, and second, in the case of women, because of the risk of an unwanted pregnancy. Let us now consider, then, possible responses to that *prima facie* case.

'Safe Sex' and the Avoidance of Sexually Transmitted Diseases

The 'safe sex' response can be put as follows.

(1) The use of condoms greatly reduces the probability of contracting a sexually transmitted disease.
(2) Many things that one does involve risks. If one drives to a movie theater, rather than watching a video at home, one risks dying in a traffic accident. Or consider skiing or rock climbing, both of which involve the risk of death or serious injury.
(3) It is not irrational to engage in such activities, since the harm is sufficiently unlikely that the danger is outweighed by the pleasure provided by those activities.
(4) When one considers how intense sexual pleasure is, surely that outweighs the greatly reduced risk of STDs involved if one uses condoms.

At least four things are wrong with this argument. First, does the use of condoms greatly reduce the probability of contracting an STD? Looking online, one might think that it does, since Planned Parenthood, a very responsible source, says that "Condoms are great at preventing both pregnancy and STDs," and then goes on to say, "If you use condoms perfectly every single time you have sex, they're 98% effective at preventing pregnancy." Many less responsible websites then cite this 98% figure. But Planned Parenthood then goes on to say, "But people aren't perfect, so in real life condoms are about 85% effective—that means about 15 out of 100 people who use condoms as their only birth control method will get pregnant each year."[4]

This latter figure, moreover, agrees reasonably well with what the CDC says about the effectiveness of condoms for preventing *pregnancy*, since it says that the probability that a woman will become pregnant in a year where male condoms are the only birth control method is 18%[5], so the probability that one will not become pregnant is therefore only 82%.

What does this 82% figure mean, however, with regard to the probability of contracting a sexually transmissible disease from a partner who has one? Well, women are fertile only for 6 days out of every 28, whereas viruses will receive a warm welcome on any day of the month. The upshot is that, given a partner with a sexually transmissible disease, the probability that one will not contract it in a year of active sex using condoms is 82% raised to the power of (28/6), giving a probability that one will *not* contract an STD from an infected partner with whom one has regular sexual intercourse for a year of just under 40%.

A second shortcoming in the above argument is that it fails to consider the fact that the dangers are greater for women than for men and in three ways. First, and most obviously, only women can become pregnant. Second, and as noted above, some sexual diseases have potentially worse consequences for women than for men. Finally, it is "twice as easy for men to infect women with any STD as for women to infect men."[6]

The third shortcoming is that there are great differences between men and women with regard to the pleasure resulting from vaginal intercourse. This is discussed in the first chapter of Elizabeth Lloyd's outstanding book, *The Case of the Female Orgasm*, where Lloyd sets out and critically comments on the findings of 32 studies.[7] Here, however, it will have to suffice to say that Lloyd's conclusions are fully in accord with those of the "National Health and Social Life Survey"[8] that found that while 75% of men always reach orgasm during sexual intercourse, only 29% of women do so. Hence, even if the risk were worth it for men, it might very well not be so for women.

Sex that Is Completely Safe

Finally, the fourth, and the most important thing wrong with the 'safe sex' argument is that it does not consider crucial alternatives: it is as if the only alternatives were 'safe sex', unsafe sex, and no sex. Once this is noticed, my point will be an obvious one, but it is one that is made surprisingly infrequently—so much so that I recall vividly how taken aback I was—but totally delighted—when in 1993 I saw the point being made, and vigorously, by an advice columnist, Ann Landers.

The column was entitled "Safe Sex—*Teens need to get the message*," and a reader had sent Landers a letter that the reader thought "could save thousands of lives," where the concluding sentence of the letter was "To quote Dr. Robert C. Noble, a University of Kentucky infectious disease expert, 'We should stop kidding ourselves. There is no safe sex. If the condom breaks, you may die.'"[9]

In teaching an "Introduction to Ethics" course over many years, I would hand out the first part of the column, without Landers's response, and ask students what they thought her response would be. If memory serves, no student ever guessed right. Landers's response[10] was as follows:

> **Dear Readers:** *Powerful piece, isn't it? Well, now I am going to stick my neck out by suggesting a far more realistic solution than abstinence.*
>
> *The sex drive is the strongest human drive after hunger. It is nature's way of perpetuating the human race. Males reach their sexual peak as early as 17. There must be an outlet. I am recommending self-gratification or mutual masturbation, whatever it takes to release the sexual energy. This is a sane and safe alternative to intercourse, not only for teenagers, but for older men and women who have lost their partners.*
>
> *I do not want to hear from clergymen telling me it's a sin. The sin is making people feel guilty about responding to this fundamental human drive. I love my readers, and my mission is to be of service. This could be the most useful column I have written since I started 38 years ago.*

It is well past time, then, to hear no more of the 'no safe sex' nonsense. Masturbation is completely safe as regards sexual diseases and unwanted pregnancies, with the only danger being the slight chance of repetitive stress injury to one's wrist if one is especially enthusiastic.

But Isn't Masturbation a Poor Substitute for the 'Real Thing'?

How do the advocates of 'safe sex' respond to the point just made? The answer is that they virtually never do, the reasons for which are very unclear. Surely they are not guilty, dictionary definitions notwithstanding, of thinking of masturbation as 'self-abuse'—rather than as self-gratification—and it is hard to believe that many are in the grip of the Catholic view that there is a morally perfect deity who was justified in killing a man named 'Onan' for unnatural sex in spilling his seed upon the ground.[11]

In any case, two responses deserve to be considered, the first of which is that, simply from the point of view of sexual pleasure, intercourse is better than sexual self-satisfaction. So is this the case?

It is certainly not so in the case of women where, in contrast to the 29% who always obtain sexual orgasm via intercourse, Shere Hite found that 95% of women who masturbate "could orgasm easily and readily, whenever they wanted."[12] Moreover, the percentage is even higher in the case of women who have, for example, what Betty Dodson referred to as "the Cadillac of vibrators"—a Magic Wand (invented by Hitachi, who then later, rather than continuing to take pride in their enormous contribution to human happiness, distanced themselves from it).[13]

In the case of men, online searches yielded no information about the percentage of men who *can* always achieve orgasm via masturbation, but surely it is higher than the 75% who always achieve orgasm via intercourse. So, for men, the difference is less dramatic, but for them too, masturbation is a surer route to orgasm than intercourse. Moreover, men can augment the sexual pleasure from masturbation, both by adding sexual toys, produced by companies such as Tenga, and by techniques such as edging, described on many Internet websites, such as one by The Promescent Team, "How To Masturbate For Men: 15 Tips."[14]

Another 'Safe Sex' Response: There's Much More to Sex than Orgasm

My argument is also open, however, to the following rejoinder: while orgasm is certainly a climactic part of sex, sex with another person involves much more than what is present when one masturbates. First of all, one's partner can use his or her hands to stimulate any part of your body, so that one experiences a much greater variety of sensations. Second, the other person's arousal contributes to one's own sexual arousal and vice versa. Third, there is the satisfaction of contributing to another person's happiness. Finally, there's just the fun of doing something together with another person, as with non-sexual activities. All of these are absent in solitary sex.

Almost Completely Safe Sex

The points just made are fair enough, but they fail to rebut the claim that engaging in 'safe sex' is deeply irrational. The reason is that while masturbation is the only type of sexual activity that is completely safe, there is also mutual handjob sex—aka 'mutual masturbation'—which with a little care and thought, such as not using saliva or vaginal secretions for lubrication, and washing one's hands at appropriate points, is almost completely safe.

Consequently, one can have two-person sex that almost completely eliminates the dangers of 'safe sex,' and that is not inferior to the latter *in any way*. Hence it is irrational to engage in 'safe sex.'

The Superiority of Mutual Handjob Sex: ESO

The irrationality, however, is deeper than indicated so far, since mutual handjob sex is not only far superior to 'safe sex' as regards both safety and the risk of unwanted pregnancy, but also not inferior in any way to 'safe sex' involving, for example, sexual intercourse: mutual handjob sex is also superior in that it enables people to have sexual experiences that are both much longer and much more exciting.

This should not be surprising, since the penis and vagina are surely primitive instruments for sexually pleasuring another person in comparison with the human hand, given both the latter's strength and its extremely fine motor control, the result of eons of evolutionary history.

For a full account of the sexual superiority of mutual handjob sex, the reader will need to consult *ESO—How You and Your Lover Can Give Each Other Hours of Extended Sexual Orgasm*, by Alan and Donna Brauer. Here, however, is a brief quote:

> ESO is not merely extended foreplay. It is not simply "multiple orgasm." Nor is it some ancient system of subtle spiritual discipline. *At its ultimate level, for women, ESO is deep, continuous orgasm of ever-increasing arousal lasting thirty minutes to an hour or more. ... For men, ESO at its ultimate level is first-stage orgasm, that momentary peak of intense pleasure just before a man feels he is going to ejaculate, extended in time for thirty minutes to an hour or more.*[15]

It may well be, of course, that ESO is not to everyone's taste—or at least not all of the time!—and that one prefers, sometimes or always, a sprint to the finish line. No matter. The point is still that given the greater speed, and variation in speed, at which hands can move, and the greater pressure, variation in pressure, twisting action, and so on, that are possible in a ten-finger exercise—perhaps topped off, if one is a man, and so inclined, with a bit of post-climax stimulation—the aforementioned traditional instruments for pleasuring another person cannot compete with hands-on sex.

Summing Up: The Irrationality of 'Safe Sex'

This completes the argument for the conclusion that 'safe sex'—where that involves vaginal, anal, or oral intercourse, using a condom—is seriously contrary to one's own self-interest and thus irrational. My main points have been, first, that 'safe sex' is far from safe, exposing one, as it does, to a serious chance of contracting a sexually transmitted disease; second, that there is a completely safe alternative that is actually superior as regards orgasmic pleasure, namely, masturbation; third, that when one wants to experience the things that two-person sex offers, and which masturbation does not, there is mutual handjob sex, which, while not completely safe, is almost nearly so; and finally, mutual handjob sex opens the door to ESO, the most exciting of all sexual experiences.

Why Most Sexual Behavior Is Immoral

A friend has been visiting, you've been playing pool, and your friend, who is about to set off for home, says, "Barb, we haven't had anything to drink tonight, and I've heard that your margaritas are margaritas to die for! Could I try a couple before I set off for home?" You reply that the road down the mountain from your place is steep and winding, and quite dangerous at night, to which your friend replies, "I know that, but life would be boring without some exciting challenges."

Your friend's request is both voluntary and fully informed, but surely you should refuse it, on the grounds that it is seriously irrational, that is, highly incompatible with her own self-interest. Accordingly, one has the following the argument for the serious immorality of 'safe sex':

(1) It is *prima facie* seriously wrong to *enable* someone to engage in behavior that is very irrational in the sense of being *seriously* contrary to that person's self-interest.
(2) Engaging in 'safe sex' with someone enables that person to act in a way *seriously* contrary to his or her self-interest.

Therefore

(3) It is *prima facie* seriously wrong to have 'safe sex' with someone.

Irrationality and Immorality: Men Versus Women

That completes the defense of my two theses. A final important point, however, has so far been left unsaid, namely, that in the case of heterosexual lovers, the charges of irrationality and immorality do not fall equally upon the man and the woman. As regards the charge of irrational behavior, it is heavier in the case of the woman, because of (a) the existence of a sexually transmitted disease from which only women can suffer—pelvic inflammatory disease, (b) the greater chance of a woman's contracting an STD from a man than vice versa, and (c) the possibility of pregnancy. On the other hand, as regards the immorality of engaging in 'safe sex', here the greater fault lies with the man, since he is enabling the woman to expose herself to greater risks than those to which he is exposed.

The Changes that Are Needed

Focusing on Sexual Behavior that Is Completely Safe

The crucial first step needed to reduce the very harmful sexual behavior that people engage in is to focus on sexual behavior that is completely safe—masturbation—and on sexual behavior that is almost completely safe—mutual handjob sex. In the case of ordinary people, the failure to do this is understandable, given the very negative way in which masturbation has traditionally been viewed. What is much less understandable is that philosophers, including those working in applied ethics, have been almost totally silent on this matter, so that if one were to peruse a hundred anthologies used in introduction to ethics courses, it is unlikely one would find a single article on this topic.

Masturbation: A Historical Perspective
In thinking about masturbation, I believe it is important to be aware of the views that dominated Western society for centuries, since an awareness of those views may help one to free oneself of any remnants of earlier negative attitudes toward masturbation.

Philosophical Arguments

Philosophers have an uninspiring track record as regards discussing the moral status of masturbation, both because the topic has only rarely been addressed and because, when it has been, the arguments offered have not been such as many present-day philosophers would embrace. This will become clear if we consider, first of all, the arguments of the two historical figures whose views on masturbation are probably best known—namely, St. Thomas Aquinas, the most famous Catholic philosopher, and Immanuel Kant—followed by the arguments of some present-day writers.

Aquinas's discussion of masturbation is set out in his *Summa Theologica*, II–II, Question 154, where, after characterizing "unnatural vice" in Article 11 as involving sexual actions "from which generation cannot follow," he then asks, in Article 12, "Whether the unnatural vice is the greatest sin among the species of lust?," and his answer is that it is, on the following grounds:

> As stated above (Articles 6 and 9) wherever there occurs a special kind of deformity whereby the venereal act is rendered unbecoming, there is a determinate species of lust. This may occur in two ways: First, through being contrary to right reason, and this is common to all lustful vices; secondly, because, in addition, it is contrary to the natural order of the venereal act as becoming to the human race: and this is called "the unnatural vice."

The argument, thus, is that while all lustful vices are "contrary to right reason," any unnatural sexual vice—which includes masturbation—is also contrary to nature, because it is contrary to "the natural order of the venereal act," and, thereby, is "the greatest sin among the species of lust."

Aquinas's argument seems clearly fallacious as it stands, since he has offered no reason for thinking, for example, that there are no sexual vices that, though not unnatural, are morally worse than the unnatural vice of masturbation because they are *much more seriously* contrary to right reason.

What is crucial in the present context, however, is not that Aquinas's argument is fallacious but simply that he is defending a view that entails that unnatural vice, including masturbation, is a greater sin than, for example, sexual intercourse between unmarried adults, even though the former is harmless, whereas the latter may have very unfortunate effects indeed, especially for the woman.

Immanuel Kant's discussion of the moral status of masturbation is found in *The Doctrine of Virtue*—which is Part II of *The Metaphysics of Morals*. There, in *Article II. On Carnal Self-Defilement*, Kant says:

> That such an unnatural use (and so misuse) of one's sexual power is a violation of *duty to oneself* and, indeed, one which is contrary to morality in the highest degree occurs to everyone immediately, along with the thought of it, and stirs up an aversion from this thought to such an extent that we consider it indecent even to call this vice by its proper name.[16]

This claim—with its 'highest degree' and 'to everyone' contentions—is a strong claim indeed, and it is very hard to believe that it was true in Kant's day. But setting that aside, Kant realizes that something more is needed by way of argument, so he proceeds as follows:

> But it is not so easy to produce the rational proof that the unnatural, and even the merely purposeless use of one's sexual power is a violation of duty to oneself (and indeed, as far as the first is concerned, a violation in the highest degree).—The *ground of proof*, of course, is that man surrenders his personality (throwing it away) by using himself merely as a means to satisfy his animal instincts. But this does not explain the high degree of the violation of humanity in one's own person involved in unnatural vice, which by its form (the attitude embodied in it) seems to surpass even self-murder in its viciousness. It consists, then, in this: that the man who defiantly casts off the burden of life is at least not making a feeble surrender to animal impulse in throwing himself away; self-murder requires courage, and in this attitude there is always room for reverence for humanity in one's own person. But unnatural vice, which is complete abandonment of oneself to animal inclination, makes man not only an object of enjoyment but, still further, an unnatural thing, *i.e.* a *loathsome* object, and so deprives him of all reverence for himself.[17]

There is much that might be said about this argument, such as that there are many impulses that we share with animals—to eat, to drink, to move about, and so on—that do not seem morally problematic in themselves, so why should this be so as regards the desire for sexual pleasure? But perhaps further reflection on Kant's argument can be left to the reader.

Some present-day philosophers have also offered arguments for the immorality of masturbation. They tend to be Catholic writers, who are thus committed to the view that masturbation is not only wrong but seriously so, since, as is stated in the *Catechism of the Catholic Church*, "Both the Magisterium of the Church, in the course of a constant tradition, and the moral sense of the faithful have been in no doubt and have firmly maintained that masturbation is an intrinsically and gravely disordered action," (para 2352) and as such, if done with "*full knowledge* and *complete consent*" (para 1859) is a mortal sin—which places one in danger of undergoing eternal torment in Hell. It is not surprising, then, that the Catholic writers Robert George and Patrick Lee, in their book *Body-Self Dualism in Ethics and Politics* hold that masturbation is morally wrong:

> We also argue that nonmarital sexual acts such as masturbation, fornication, and sodomy cannot actualize any basic human good and therefore involve instrumentalizing one's (and possibly others') body (or bodies) for the sake of a mere experience (without the reality) of unity or for the sake of self-fulfillment.[18]

Three comments. First of all, as Michael Zimmerman and Ben Bradley point out in their *Stanford Encyclopedia of Philosophy* article "Intrinsic vs. Extrinsic Value," Plato in his dialogue *Protagoras* (353e) "concludes that pleasure is in fact good as such and pain bad, regardless of what their consequences may on occasion be,"[19] and they then go on to say, "Many philosophers have followed Plato's lead in declaring pleasure intrinsically good and pain intrinsically bad." So George and Lee are rejecting a very widely accepted philosophical view, and they offer no plausible argument for doing so.

Second, as Steven Macedo,[20] Andrew Koppelman[21], and Gary Chartier[22] have argued, the above argument is open to a *reductio ad absurdum* objection. Thus, if masturbation is wrong because one is producing a valuable experience unaccompanied by some sort of unity or self-fulfillment, what about immersing one's body in a hot bath simply to enjoy the sensation of warmth, or drinking a nutrition-free diet soft drink, simply to enjoy the taste? If these things—and many similar cases that will readily occur to the reader—are morally wrong by George and Lee's lights, as they are, then I think that few would agree with their argument.

George and Lee themselves, however, are unmoved by this type of objection, since they do not consider such implications absurd, and David Boonin[23] has argued that, because of this, familiar *reductio ad absurdum* objections are less than completely satisfying. Boonin then focuses, however, on three types of cases that, he argues, ground stronger objections. These are, first, cases where one uses one's body as a means of preventing painful experiences[24]; second, cases where one uses one mind to create better physical states[25]; and finally, cases where one uses one's body to protect someone else's body, and in each of these types of cases Boonin argues, convincingly, that the basic principle that George and others employ leads to conclusions that they themselves cannot accept.

A very different type of argument for the view that masturbation is morally wrong has been offered by another recent philosopher—Roger Scruton—in his book *Sexual Desire: A Moral Philosophy of the Erotic*:

> Sexual purity does not forbid desire: it simply ensures the status of desire as an interpersonal feeling. The child who learns 'dirty habits' detaches his sex from himself, sets it outside of himself as something curious and alien. His fascinated enslavement to the body is also a withering of desire, a scattering of erotic energy and a loss of union with the other.[26]

Scruton's argument, unlike those of Aquinas, Kant, and George and Lee, seems to rest upon an empirical claim, to wit, that the child (or older person?) who masturbates will become enslaved to his body in such a way that he has no interest in relations with other children or people. Scruton offers no evidence at all for this claim.

A second argument that Scruton offers involves an even more exciting empirical claim: "The harmless wanker with the video-machine can at any moment turn into the desperate rapist with a gun."[27] Again, no support is offered for this extraordinary claim.

This concludes my brief survey of arguments that philosophers have offered for the immorality of masturbation. I believe both that they are as good as any that one can find and that they are very weak indeed.

Medical Views

Given the extremely strong view concerning the immorality of masturbation embraced and propagated by the Catholic Church, it is not surprising that ordinary people in Western European society came to have negative views of masturbation and thought of it as a dirty habit that must be strongly discouraged in children. It is surprising, however, to learn that the medical profession sometimes was guilty of providing support for such a view, via claims that had no scientific basis at all.

A dramatic illustration of this is given by Wayland Young in his book *Eros Denied—Sex in Western Society*. Young describes the case of William Acton, the author of the book *The Functions and Disorders of the Reproductive Organs*, which "went through six editions between 1857 and 1875."[28] In that book, as Young points out,[29] Acton offers the following translation[30] of a passage from a three-volume medical work by a French author, Claude-François Lallemand:

> However young the children may be, they become thin, pale, and irritable, and their features assume a haggard appearance. We notice the sunken eye, the long, cadaverous-looking countenance, the downcast look which seems to arise from a consciousnessConsciousness in the boy that his habits are suspected, and, at a later period, from the ascertained fact that his virility is lost. ... Habitual masturbators have a dank, moist, cold hand, very characteristic of vital exhaustion; their sleep is short, and most complete marasmus comes on; [wasting of the body: S.O.E.D.] they may gradually waste away if the vile passion is not got the better of, nervous exhaustion sets in, such as spasmodic contraction, or partial or entire convulsive movements, together with epilepsy, eclampsy, and a species of paralysis accompanied with contraction of the limbs.[31,32]

It is hard to believe that such views could be put forward by outstanding authors of medical textbooks; it illustrates in a very dramatic way how negative the attitudes toward masturbation were at that time.

Intellectuals Who Have Led the Way

Where does one turn to find writers who have realized the importance of focusing upon masturbation if one is to arrive at a sound ethical view of human sexuality? The answer is that two figures have towered above the field, both of them women, and both of whom focused especially on female sexuality. First of all, there was Shere Hite (1942–2020), who in 1976 published her groundbreaking book *The Hite Report: A Nationwide Study of Female Sexuality*, followed, in 1981, by *The Hite Report on Men and Male Sexuality*. The reaction to her writings was, as Katherine Seelye points out,[33] "a tidal wave of anger and resentment against her," including death threats, as many men could not accept the view that women needed neither sexual intercourse nor men to achieve sexual satisfaction. As a result, Hite renounced her US citizenship in 1995 and went to live in Europe.

The other great contributor was Betty Dodson (1929–2020), who, while a struggling artist, published *Liberating Masturbation: A Meditation on Selflove* in 1978, followed in 1983 by *Selflove and Orgasm*, which was then reprinted, in a slightly different form, as *Sex for One—The Joy of Selfloving*. The reception of her early books convinced her that her career should be that of a sex educator, where she conducted hands on seminars and invented the "Betty Dodson Method" of masturbation, involving the aforementioned "Cadillac of vibrators"—the Hitachi Magic Wand.

The following passage will give one a sense of Betty Dodson, and her writing:

> My retirement fantasy, if it ever got aired on prime time, could change the image of old age. There are thirteen of us old folks living together in a commune. Every full moon we gather in front of the TV set to watch a new pornographic video of politically incorrect sex. After brewing a strong pot of tea, we plug in our vibrators and settle down for an evening of orgasms. The rocking chairs creak, the vibrators hum, and occasionally one of us smiles and nods yes after a particularly good one.[34]

A Sound Approach to the Sexual Education of Our Children

While sexually transmitted diseases affect individuals of all ages, they take a particularly heavy toll on young people. The CDC, for example, in discussing sexual activity among adolescents and young adults, estimates that "youth ages 15–24 make up just over one quarter of the sexually active population, but account for half of the 20 million new sexually transmitted infections that occur in the United States each year."[35] A sound approach to sex education is therefore crucial, and given that masturbation is completely safe, full information about masturbation should clearly be part of sex education.

In 1993, Joycelyn Elders was appointed Surgeon-General of the United States by President Bill Clinton. While attending a United Nations conference on AIDS in 1994, "Dr. Elders was asked by Dr. Rob Clark, a psychologist who was attending the AIDS conference, about the prospects for 'a more explicit discussion and promotion of masturbation' as a means to limit the spread of the virus," and in reply she said, "I think that is something that is a part of human sexuality, and it's a part of something that should perhaps be taught."[36] The result was that Clinton demanded her resignation.

Elder's statement, in view of the "perhaps," was very mild; moreover, in a telephone interview, "Dr. Elders told The Associated Press that she had intended to relate that masturbation is a natural part of human sexuality—not that schoolchildren should be taught how to masturbate."[37]

Some groups spoke out against the forced resignation of Elders, but Americans in general did not protest, thereby showing how widespread irrational views of masturbation were.

What would a sound sex education look like? First of all, adolescents should be taught that masturbation is completely safe and a source of great pleasure.

Second, they should receive information about how to masturbate and about ways of increasing the pleasure that one experiences. In the case of girls, they should certainly learn of the Magic Wand, while boys should learn, for example, about edging, so that even if masturbation tends to remain, for them, a one-shot affair, it can at least be more of a marathon rather than just a quick sprint.

Third, children should be intellectually inoculated against two of the ignorant armies. Thus, on the one hand, they should receive detailed information both about the dangers of 'safe sex' and about the advantages, especially if one is a woman, of masturbation, which is completely safe, and of mutual masturbation/handjob sex, which is almost completely safe. Then, as regards those who hold that all sex outside of marriage is wrong, and who attempt to indoctrinate children with the view that they should abstain from all sex, children should be exposed to the arguments such people advance, including religiously based arguments, and should learn why those arguments are unsound.

Fourth, girls are especially at risk in engaging in unsafe sexual activities, so they should learn of the findings of Shere Hite concerning female sexuality.

Finally, there should be no opting out of sex education: here, as elsewhere, parents should clearly not be allowed to do things that put the health and possibly even the lives of their children at risk.

Talking Openly and Freely About Sexual Behavior

Although we live in a hookup society, with sexting, and apps such as *Tinder*—so that our society is in some ways very unpuritanical—few people indeed talk about sexual behavior in the way one talks about most other things. How often does it happen, for example, that one phones home to wish Mom or Dad a happy birthday, getting no answer the first time one calls, and then, when one calls later, and says, "I called earlier, but you must have been out," Mom or Dad says, "Birthdays are one of our favorite times for some very hot sex, so we turn off our phones to avoid unwelcome interruptions"? Or suppose you're a philosophy student, and you call a friend to ask her if she would like to drop by on Friday to discuss Hume's argument for inductive skepticism. How likely is it that she might say, "Friday won't work, since after a week of thinking intensely about philosophical problems, I really need to unwind, so Fridays I watch some porn and spend the evening masturbating. But Saturday would work well."

When Sally Meets Harry: Interviewing Prospective Lovers

It may be a while before—if ever—people in general talk freely about sex, but that is no reason not to do so oneself, since the advantages are great. Suppose, for example, that Sally and Harry are taking some philosophy classes together and have had several interesting conversations about philosophical arguments and problems. At some point, Harry asks Sally if she would be interested in going on a date. How might Sally respond?

If Sally is happy to talk freely about sex, she need not waste her time going on a date. Instead, she can find out about Harry's views on sex by a few short, direct questions: "Do you masturbate?," "How do you feel about masturbation?," "How do you think masturbation compares with intercourse as a source of sexual pleasure for a woman?," "Do you think that vaginal, or oral, or anal intercourse is preferable to mutual handjobs?"

If Harry fails the interview, he and Sally can happily remain philosophical companions, while if he passes, then when Sally introduces Harry to Mom and Dad, she can set their minds at rest by telling them that she and Harry spend a good deal of time together, mainly talking philosophy and developing their ESO skills.

Rational Consent as What Matters

If there were time machines, and one could travel back to the 1950s, people who thought that all premarital sex was morally wrong might be heard to remark, "Conscience is soluble in alcohol." Alcohol is now a focus of concern in a different way, with many holding that if you have consumed quite a bit of alcohol, you are no longer capable of consenting to sex.

Is this plausible? Imagine that you are so completely drunk that you are lying on the floor, unable to get up, with things in the room spinning around your head at something approaching the speed of light. Suppose, further, that you are a heterosexual and that someone now invites you to enjoy some homosexual fun. Would you find it impossible to say no? Highly implausible, surely?

If so, then alcohol does not have the effect, unless you are unconscious, that you cannot consent. What *is* true is that alcohol may impair *rational* consent—consent in accord with your own best interests. But it is not only alcohol that may impair rational consent: sexual arousal not only may, but is quite likely to, do so. That means, in turn, that a rather common view about sex and consent, namely, that consent is required at every step, for every new type of sexual activity, is unsatisfactory, since as the level of sexual excitement grows, one may no longer be capable of rational consent.

The correct solution to the problem of consent, accordingly, is that there should be an agreement, before either person is sexually aroused, concerning what sexual activities are on the table, unless forbidden at some later point, and what sexual activities are not to be engaged in, *where the latter cannot be revised as sex proceeds.* The danger of consent that is impaired by sexual arousal is thereby eliminated.

The Last of the Ignorant Armies: Religion and Sexual Morality

The final ignorant army to be defeated consists of people who are adherents of virtually all of the 'great religions' of the world and who accept the teachings of their own religion concerning sexual morality.

Here I shall focus on the dominant religion of Europe and North and South America—Christianity—along with its forerunner, Judaism. To start with the latter, the teachings of the Jewish Bible, also known as the Christian Old Testament, concerning sex and morality, are contained mainly in Leviticus. There one finds, for example, the passage much beloved by Christian fundamentalists, namely, Leviticus 20:13–4, which says that if a man lies with a man as with a woman, it is an abomination, and both should be put to death. Rather less beloved, and less commonly quoted, however, are the nearby Leviticus 20:10, 20:11, and 20:14–15, which say, respectively, that adulterers should be executed, that if a mother and her son have intercourse, they should both be put to death, and that if a man has intercourse with a woman and her daughter, all three should be burnt to death.

There are also interesting passages in Deuteronomy, including one that would have considerable application today, namely Deuteronomy 22:20–21, which says that a woman found on her wedding night not to have been a virgin should be taken to the door of her father's house and stoned to death.

Christians, of course, may claim that these vigorous and forthright moral teachings were set aside with the coming of Jesus and Christianity. So what did Jesus say as regards sex and sexual morality? For one thing, according to Matthew 15:19–20, he said that fornication—sexual intercourse outside of marriage—"defiled" a person, while according to Matthew 5:28–30, he also said that "everyone who looks at a woman lustfully has already committed adultery with her in his heart," and then immediately went on to say, "If your right eye causes you to sin, pluck it out and throw it away," while if, instead, your right hand was your undoing, his injunction was to "cut it off and throw it away; it is better that you lose one of your members than that your whole body go into hell." Finally, in a later passage—Matthew 19:10–12—when his disciples asked him whether it might be better not to marry, Jesus said "there are eunuchs who have been so from birth, and there are eunuchs who have been made eunuchs by men, and there are eunuchs who have made themselves eunuchs for the sake of the kingdom of heaven. He who is able to receive this, let him receive it."

Such views are surely incompatible not only with the approach to sex education defended above but with any approaches other than those whose aim is to indoctrinate defenseless children with the view that anything less than complete abstinence from *all* sexual activity until marriage is morally wrong.

One can, of course, think that a certain religion is basically right without accepting every view advanced by that religion. One might think, for example, that the Catholic religion is the religion that is closest to the truth, while rejecting its teaching on birth control or abortion on the grounds that those beliefs, rather than being 'revealed truths', are views grounded on unsound philosophical arguments. However, thinking that Jesus is God in the flesh, while rejecting his views on sex as morally mistaken, is hardly such a case. Any jurisdiction that contains too many devout Christians, therefore, is likely to find it difficult incorporating a sound system of sex education into its public school system.

Philosophy: The Slumbering Giant

An overwhelming proportion of parents indoctrinate their children, and most philosophers, strangely, say very little about this, even though it is clear that such indoctrination almost always results in people who are intellectually crippled for life and who are unable to critically examine the religious views with which they were raised—beliefs, moreover, that not only can, but generally do, have highly undesirable effects on society at large by preventing rational and morally defensible legislation in areas such as abortion, recreational drugs, euthanasia, prostitution, and the sexual education of children.

How can this indoctrination be counteracted, so that people can learn to think critically? Since primary and secondary schools would face a public outcry if they made any serious and effective attempt to equip students with a disposition to think critically, it falls upon universities to act, and within universities, philosophers are by far best equipped to enable students to think critically about the beliefs that were instilled in them at an early age. So how are philosophers doing in this regard?

First of all, within philosophy, one of the areas most directly relevant is philosophy of religion. How are philosophers doing in this area? Judging from philosophy of religion textbooks, it would seem that they are doing very poorly, since virtually no philosophy of religion textbook contains a section that focuses on objections to Judaism or Christianity. Why is this? The answer becomes clear if one considers the results of a 2009 PhilPapers survey, which found that while, among philosophers in general, less than 15% either accepted or were inclined to theism, in the case of those specializing in philosophy of religion, slightly more than 72% either accepted or were inclined to theism.[38]

The failure of philosophers to encourage critical examination of religious beliefs is not, moreover, confined to philosophy of religion. Thus, while introduction to philosophy courses often have a section on arguments for and against the existence of God, the kinds of arguments that carry the most weight with ordinary believers—namely, arguments from supposed miracles and arguments from religious experiences—are virtually never topics in introduction to philosophy textbooks. Or consider critical thinking textbooks. How many contain material focusing on critical thinking about religious beliefs? Or peruse introduction to ethics textbooks. The indoctrination of children by parents is arguably one of the world's great evils, but this again is a topic that is virtually never present.

The moral seems clear: when it comes to encouraging students to think critically about the religious beliefs in which they have been indoctrinated, philosophers are failing badly.

Summing Up

I have argued that most sexual behavior that is not aimed at reproduction is both irrational and immoral. I have also argued, however, that this does not mean that sacrifices are called for, since anyone can easily jettison problematic sexual behavior in favor of sex that is at least as satisfying, in every respect—and arguably more so—and that is not problematic in any way. But while any individual can easily make that change, what is needed is for people in general to do so, and for that to occur, one needs a very different form of sex education, and one to which, in the United States at present, there would be extremely strong religiously based objections. Still, religious beliefs have declined quite dramatically in some countries—and especially in Europe, where a 2010 Eurobarometer study found that 49% do not believe that God exists[39]—and they have also begun to do so in the United States, where between 2007 and 2019 the percentage of Christians fell from 78% to 65%, while the percentage with no religious affiliation rose from 16% to 26%.[40] So perhaps there is hope that we can escape from the darkling plain where ignorant armies clash by night and arrive at a widely shared approach to sexual behavior that is neither irrational nor morally problematic.[41]

Notes

1. https://www.cdc.gov/std/default.htm.
2. Bisson and Levine 2007, p. 68.
3. Livingston 2018, p. 1.
4. Planned Parenthood, p. 1.
5. "Effectiveness of Family Planning Methods," https://www.cdc.gov/reproductivehealth/UnintendedPregnancy/PDF/Contraceptive_methods_508.pdf.
6. "National Health and Social Life Survey, Biology 203, "National Health and Social Life Survey, Biology 203," https://www.siue.edu/~cbwilso/203nhsls99.htm, p. 2.
7. Lloyd 2005, pp. 1–43.

8. Laumann et al. 1992 and 2008.
9. Landers 1993, p. 1.
10. Ibid., p. 1.
11. Genesis: 38: 9–10.
12. Hite 1976, p. 59.
13. "Hitachi Magic Wand," https://en.wikipedia.org/wiki/Hitachi_Magic_Wand
14. https://www.promescent.com/blogs/learn/how-to-masturbate
15. Brauer and Brauer 1983, p. 3.
16. Kant 1797; 1964, p. 88.
17. Kant 1797; 1964, p. 88–9.
18. George and Lee 2009, p. 3.
19. Zimmerman and Bradley 2019, p. 2.
20. Macedo 1995, p. 282.
21. Koppelman 1997, p. 88.
22. Chartier, Gary, 1999, p. 115.
23. Boonin 2013, pp. 236–8.
24. Ibid., pp. 238–42.
25. Ibid., pp. 242–5.
26. Scruton 1986, 340–1.
27. Ibid., p. 386.
28. Young 1964, 205.
29. Ibid., p. 208
30. Acton 1862, p. 8.
31. This is a translation by Acton of a passage from Claude-François Lallemand's three-volume work *Des pertes séminales involontaires*, France, Béchet Jeune et Labé, 1836–42, volume i, p. 462.
32. For a fuller account of medical views of masturbation in Victorian times, see Diane Mason's book *The Secret Vice: Masturbation in Victorian Fiction and Medical Culture*
33. Seelye, Katherine Q., *New York Times*, September 11, 2020.
34. Dodson 1987, p. 4.
35. https://www.cdc.gov/std/life-stages-populations/adolescents-youngadults.htm
36. Jehl 1994, p. 3.
37. Ibid.
38. Bourget, David, and Chalmers, David, 2009, PhilPapers Survey. https://philpapers.org/surveys/results.pl
39. European Commission 2010, p. 204.
40. Pew Research Center—Religion and Public Life 2019, p. 1.
41. I am indebted to David Boonin, Rob Lovering, and Graham Oddie for very helpful comments on an earlier draft of this paper.

References

Acton, W. 1862. *The Functions and Disorders of the Reproductive Organs in Childhood, Youth, Adult Age, and Advanced Life, Considered in Their Physiological, Social, and Moral Relations*. United Kingdom: Churchill. Available as a Google book: https://

www.google.com/books/edition/The_Functions_and_Disorders_of_the_Repro/cVX7PaDkzGYC?hl=en

Aquinas, St. Thomas, (1265-74). *Summa Theologica*, II-II, Question 154, Article 12: Whether the unnatural vice is the greatest among the species of lust. Second and Revised Edition, 1920. Literally translated by Fathers of the English Dominican Province.

Bisson, Melissa A., and Timothy R. Levine. 2007. Negotiating a friends with benefits relationship. *Archives of Sexual Behavior* 38 (1): 66–73.

Boonin, David. 2013. Reductio ad Absurdum Objections and the Dis-Integration Argument against Merely Instrumental Sex. *Journal of Social Philosophy* 44 (3): 233–249.

Brauer, Alan P., and Donna Brauer. 1983. *ESO—How You and Your Lover Can Give Each Other Hours of Extended Sexual Orgasm*. New York: Warner Brothers.

CDC, "Adolescents and Young Adults." https://www.cdc.gov/std/life-stages-populations/adolescents-youngadults.htm

CDC, "Effectiveness of Family Planning Methods." https://www.cdc.gov/reproductivehealth/UnintendedPregnancy/PDF/Contraceptive_methods_508.pdf

CDC, "Reproductive Health." https://www.cdc.gov/reproductivehealth/unintendedpregnancy/pdf/contraceptive_methods_508.pdf

CDC, "Sexually Transmitted Diseases." https://www.cdc.gov/std/default.htm

Chartier, Gary. 2001. Natural Law, Same-Sex Marriage, and the Politics of Virtue. *UCLA Law Review* 48: 1593–1632.

Department of Nursing, University of Southern California, America's Sex Education: How We Are Failing Our Students, https://nursing.usc.edu/blog/americas-sex-education/

Dodson, Betty. 1974. *Liberating Masturbation: A Meditation on Self Love*. New York: Self-published.

———. 1983. *Selflove and Orgasm*. New York: Self published.

———. 1987. *Sex for One—The Joy of Selfloving*. New York: Crown Publishers, Inc.

European Commission. 2010. *Special Eurobarometer 341—Biotechnology Report*. https://ec.europa.eu/commfrontoffice/publicopinion/archives/ebs/ebs_341_en.pdf

George, Robert, and Patrick Lee. 2009. *Body-Self Dualism in Contemporary Ethics and Politics*. Cambridge: Cambridge University Press.

Hite, Shere. 1976. *The Hite Report: A Nationwide Study of Female Sexuality*. New York: Macmillan.

———. 1981. *The Hite Report on Men and Male Sexuality. New York*: Alfred A. Knopf.

Jehl, Douglas. 1994, December 10. "Surgeon General Forced to Resign by White House," New York: *The New York Times*.

Kant, Immanuel. 1797. *The Doctrine of Virtue*—Part II of *The Metaphysics of Morals* Translated by Mary J. Gregor, New York: Harper & Row, 1964.

Koppelman, Andrew. 1997. Is Marriage Inherently Heterosexual? *The American Journal of Jurisprudence* 42 (1): 51–95.

Landers, Ann. 1993. "Safe Sex—*Teens need to get the message*," Creators Syndicate.

Laumann, Edward O., Gagnon, John H., Michael, Robert T., and Michaels, Stuart. National Health and Social Life Survey, 1992: [United States]. Inter-university Consortium for Political and Social Research [distributor], 2008-04-17. https://doi.org/10.3886/ICPSR06647.v2

Libreria Editrice Vaticana. 2019. *Catechism of the Catholic Church*, Second Edition, United States Conference of Bishops. Available online: https://www.usccb.org/sites/default/files/flipbooks/catechism/

Livingston, Gretchen, "About One-Third of U.S. Children Are Living with an Unmarried Parent," Pew Research Center, April 27, 2018. https://www.pewresearch.org/fact-tank/2018/04/27/about-one-third-of-u-s-children-are-living-with-an-unmarried-parent/

Lloyd, Elisabeth A. 2005. *The Case of the Female Orgasm—Bias in the Science of Evolution*. Cambridge, MA: Harvard University Press.

Macedo, Steven. 1995. Homosexuality and the Conservative Mind. *Georgetown Law Review* 84 (2): 261–300.

Mason, Diane. 2008. *The Secret Vice: Masturbation in Victorian Fiction and Medical Culture*. Manchester and New York: Manchester University Press.

Pew Research Center—Religion & Public Life. 2019. "In U.S., Decline of Christianity Continues at Rapid Pace" https://www.pewforum.org/2019/10/17/in-u-s-decline-of-christianity-continues-at-rapid-pace/

Planned Parenthood, "How effective are condoms?" https://www.plannedparenthood.org/learn/birth-control/condom/how-effective-are-condoms

Scruton, Roger. 1986. *Sexual Desire: A Moral Philosophy of the Erotic*. New York: Free Press.

Seelye, Katharine Q. 2020. "Shere Hite, Who Challenged Myths of Female Sexuality, Dies at 77," *New York Times*, September 11, 2020. https://www.nytimes.com/2020/09/11/books/shere-hite-dead.html

The Promescent Team, "How To Masturbate For Men: 15 Tips (Ultimate Guide)" https://www.promescent.com/blogs/learn/how-to-masturbate

Wikipedia, "Hitachi Magic Wand". https://en.wikipedia.org/wiki/Hitachi_Magic_Wand

Young, Wayland. 1964. *Eros Denied—Sex in Western Society*. New York: Grove Press.

Zimmerman, Michael J., and Bradley, Ben. 2019. "Intrinsic vs. Extrinsic Value, https://plato.stanford.edu/entries/value-intrinsic-extrinsic/

PART II

Sex and Traditional Values

CHAPTER 9

Virgin Versus Chad: On Enforced Monogamy as a Solution to the Incel Problem

Dan Demetriou

Introduction

In April 2018, 25-year-old Alek Minassian mowed down more than two dozen Toronto pedestrians with his van, killing ten. A self-described "incel" (an internet portmanteau of "involuntary celibate"), Minassian was inspired by Elliot Rodger, whose own mass killing was, according to the latter's manifesto, similarly motivated by a deep resentment of female rejection.[1] "The Incel Rebellion has already begun!" Minassian posted on Facebook just before his killing spree. "We will overthrow all the Chads and Stacys! All hail the Supreme Gentleman Elliot Rodger!"[2]

Soon after, University of Toronto psychologist and public intellectual Jordan Peterson was asked by *New York Times* journalist Nellie Bowles for his thoughts about the tragedy.

> "He was angry at God because women were rejecting him," Mr. Peterson says of the Toronto killer. "The cure for that is enforced monogamy. That's actually why monogamy emerges." Mr. Peterson does not pause when he says this. Enforced monogamy is, to him, simply a rational solution. Otherwise women will all only go for the most high-status men, he explains, and that couldn't make either gender happy in the end. "Half the men fail," he says, meaning that they don't procreate. "And no one cares about the men who fail."
>
> I laugh, because it is absurd.

D. Demetriou (✉)
University of Minnesota, Morris, Morris, MN, USA
e-mail: ddemetri@umn.edu

D. Boonin (ed.), *The Palgrave Handbook of Sexual Ethics*,
https://doi.org/10.1007/978-3-030-87786-6_9

> "You're laughing about them," he says, giving me a disappointed look. "That's because you're female." [...] In situations where there is too much mate choice, "a small percentage of the guys have hyper-access to women, and so they don't form relationships with women," he said. "And the women hate that."[3]

Peterson's endorsement of "enforced monogamy" drew the ire of the commentariat, many of whom interpreted the phrase as a *Handmaid's Tale*-type scenario or associated it with the crudest forms of sexual coercion. Common among these critiques was the claim that Peterson was assuming some sort of right to sex on behalf of incels.[4] Peterson replied in subsequent interviews and tweets that "enforced monogamy" is a well-established term in anthropology and related fields (when I restrict the search to 2017 and earlier, Google Scholar reports 550 publications using that phrase and about 103 using "normative monogamy," which means the same thing). Peterson noted in a blog post about the controversy that he doesn't base his case for enforced monogamy on a right to sex or spouses, but rather on the pragmatic social benefits of monogamy.

> Men get frustrated when they are not competitive in the sexual marketplace (note: the fact that they DO get frustrated does not mean that they SHOULD get frustrated. Pointing out the existence of something is not the same as justifying its existence). Frustrated men tend to become dangerous, particularly if they are young. The dangerousness of frustrated young men (even if that frustration stems from their own incompetence) has to be regulated socially. The manifold social conventions tilting most societies toward monogamy constitute such regulation.
>
> That's all.
>
> No recommendation of police-state assignation of woman to man (or, for that matter, man to woman).
>
> No arbitrary dealing out of damsels to incels.
>
> Nothing scandalous (all innuendo and suggestive editing to the contrary).
>
> Just the plain, bare, common-sense facts: socially-enforced monogamous conventions decrease male violence. In addition (and not trivially) they also help provide mothers with comparatively reliable male partners, and increase the probability that stable, father-intact homes will exist for children.[5]

In this chapter, I will construct an argument premised on the "incel problem" and concluding with the remedy of "enforcing monogamy," which I will henceforth refer to as the "Enforced Monogamy Inference" or EMI. I cast the issue in terms of an "inference" rather than a mere call for "enforced monogamy" in order to focus our attention on the fact that we're talking about a proposed solution to a certain problem—namely, high rates of involuntarily celibate men—as opposed to a general call for enforced monogamy or a monogamist position motivated by another rationale, such as some religious scruple against extra-marital sex. If EMI were the brainchild of Peterson alone, discussion of it would not be warranted here. But EMI is it not idiosyncratic to

Peterson: it, in whole or in part, has a great deal of currency in the manosphere, and it might be beneficial for students and scholars of sex ethics to think through it. I stress this is an *acquaintance* with the inference from the "incel problem" to enforced monogamy: in a discussion of this size, it is impossible to advance a thoroughgoing analysis of its soundness. Readers are welcome to fill gaps in my reasoning and/or pursue lines of thought only gestured at here.

Note that Peterson is concerned not just about unhappily-alone young men, but also the larger social implications of the incel phenomenon (since inceldom is an identity, I will try to reserve "incel" for men who self-identify as incels and "involuntarily celibate" for the much larger class of involuntarily celibate men). Advocates of EMI-type thinking tend to be social pragmatists who feel our mating patterns not only profoundly affect, but are also beholden to, social well-being. Thus I begin by advancing a moral premise that advocates of EMI are likely to sympathize with, which sees mating patterns as morally constrained by social or civilizational sustainability. I then summarize some well-established social scientific research suggesting that polygamy, as compared to monogamy, is indeed significantly disadvantageous to civilizational advance. After that I pivot away from how EMI applies to polygamous cultures to how it applies to our own. This raises a conceptual problem for EMI: is it about *relieving intrasexual competition* on behalf of involuntary celibates by restricting the availability of the best-positioned men, as one might gather from the monogamy versus polygamy framing? Or is it about *decreasing the percentage* of involuntarily celibate men, which is a much broader project? I explore the former as it might apply to a sexually unregulated but socially monogamous mating market. Advocates of EMI can and do draw from the sexual economics literature to argue that liberal sexual mores, in conjunction with certain facts about innate sex differences and reproductive technology, result in a degraded form of polygamy, thereby encumbering modern socially monogamous societies with some of polygamy's downsides. In the final section I step back and conclude that "enforced monogamy," in the sense of *normative anti-polygamy*, is probably justified for non-polygamous societies if we accept civilizational sustainability as a baseline. On the other hand, "enforced monogamy," in the sense of imposing *stricter sexual and marital monogamous norms*, is not yet warranted, since less intrusive interventions—interventions Peterson himself is exemplary of undertaking—are likely to be more effective.

Social and Civilizational Sustainability as a Necessary Condition for Sex Ethics

Environmentally conscious people generally agree that we must strive to live sustainably. Of course, accepting environmental sustainability as a baseline is perfectly compatible with acknowledging a defeasible right to high-consumption lifestyles: if you like "rolling coal" in big trucks as opposed to buzzing about in tiny hybrids, or if you prefer to spend your leisure time crisscrossing the globe instead of gardening, that is your business. Even so, we generally think the *prima facie* right to such freedoms might be outweighed if the costs of exercising these freedoms are externalized to others and to future generations. Although in some cases we should call for an absolute prohibition on a particular high-consumption industry, product, or practice in the name of sustainability, we will typically call merely for an internalization of the real costs of making these things sustainable, which may or may not price them out of the market—calls to remove beef subsidies are one example.[6]

Granted, which practices are sustainable is a difficult question, since the set of sustainable practices grows, shrinks, or shifts based on a host of changing social, psychological, and technological factors, each of whose relevance to environmental outcomes is also disputed. And granted, what even counts as a "sustainable" solution conceptually is also controversial. (Is a *presently* unsustainable practice "sustainable" if it is predictably sustainable in the long term? How *anthropocentric* may a "sustainable" outcome be? Is a high-consumption practice "unsustainable" if it is *counterbalanceable* by other measures?) But even in the absence of consensus on their answers, we usually hold that we must try to make our societies environmentally sustainable. Most of us would add that, even in the absence of laws or regulations enforcing sustainable behavior—and perhaps even more urgently in the absence of such regulations—we have a significant *prima facie* obligation to create and adhere to norms that do a good job of incentivizing people to consume more responsibly, to invent green technologies, to create businesses that make what are likely to be sustainable practices profitable, convenient, and fun. In sum, it isn't very controversial to say that environmentally conscious people should want to raise the costs of environmentally unsustainable lifestyles and lower the costs of environmentally sustainable lifestyles.

More controversial than environmental sustainability is a commitment to *social* or *civilizational* sustainability. Or maybe not: for although conservatives are more likely to worry about social and civilizational sustainability using those terms, it is not uncommon for people across the political spectrum in any polity to be concerned about high crime rates, violent factionalism, unhappy citizens, bad parenting, or shrinking birthrates. These concerns we should label "social" rather than "civilizational," since tribal elders in the remotest villages fret over such matters at least as much as our leaders do. By "civilizational sustainability" I refer to a civilization's need for massive public and private investment in things such as reliable physical infrastructure, science and technology hubs, cultural/arts institutions, and commons such as publicly funded schools, parks, hospitals, and edifying media. Extraordinary private and public investments in education, creative productivity, and public health are justifiable in no small part because we need to develop human capital if we hope to maintain, let alone advance, civilization: all those tech complexes, factories, labs, and studio spaces are worthless without healthy, talented, and motivated people to put them to use. (The link between civilizational and environmental sustainability isn't merely analogical, since the climate and energy crises themselves will be *solved* only by an advanced, creative civilization that has done a good job of developing and harnessing human capital.[7]) So our lifestyles should be not only environmentally sustainable, but socially and civilizationally sustainable as well.

Importantly for our purposes, a society's mating patterns[8] are as critical as anything else to its prospects for social and civilizational survival. For instance, no reasonable person who cares about Japan can shrug off statistics showing that over 40% of Japanese people from 18 to 34 report never having sex or that half of Japanese marriages are sexless.[9] The negative consequences of a bad strategy with regard to sex, reproduction, and marriage are less immediate, perhaps, than those of a bad strategy with regard to agriculture or war, but no less disastrous in the long run. Thus, if we have good reasons to think some mating pattern is a threat to social or civilizational sustainability, then we should call for norms which, if not prohibiting that sexual lifestyle, at least make it costly. Prominent evolutionary psychologist Geoffrey Miller is commendable in this regard for acknowledging the above points even though he demurs on the question of enforced monogamy as it intersects polyamory.[10] "Sex-positive activists often argue that sexual relationships are matters of individual choice, and nobody else's business," he writes.[11] "Yet," Miller goes on,

> sexual relationships can impose good and bad side-effects ("positive and negative externalities") on children, communities, economies, civilizations, and future generations. Mating markets matter. Sexual ethics matter. Reproductive choices matter. Families matter. That's why we evolved instincts to stick our noses into other people's sex lives, and why human sexuality has often been the most controversial domain of human politics and religion.[12]

We may, after examining the evidence, come to different conclusions about the threats posed by some proposed or actual mating pattern, but we mustn't dismiss sustainability objections *ex ante* because we assume there can be no connection between mating patterns and social and civilizational sustainability.

Is Monogamy Civilizationally Adaptive, as Compared to Polygamy?[13]

Nellie Bowles laughed off as "absurd" Peterson's appeal to hierarchies, high rates of involuntary celibacy, violence, and reproductive desperation in justifying enforced monogamy and its effect of, in an uncharitable manner of speaking, "redistributing" women to less desirable men.[14] I suppose it may sound absurd to someone unfamiliar with the literature, much as the claim that cow flatulence could contribute to climate change sounds absurd to those who never bothered to research the question. But in fact, Peterson's worries, if not his prescription, are perfectly familiar in the fields concerned with the societal impact of marriage patterns. In this section I summarize the strongest case for worrying about any mating pattern that (by design or not) distributes women to a significantly smaller fraction of men: that of evolutionary biologist Joe Henrich and his colleagues. Henrich's 2010 *amicus* brief advising against legalizing polygamy in British Columbia[15] served as the basis for his "Puzzle of Monogamous Marriage" (2012),[16] a *locus classicus* of cultural evolution theory he co-authored with evolutionary biologists Robert Boyd and Peter Richerson. The social and civilizational advantages of "normative monogamy" (Henrich et al.'s favored term for enforced monogamy) also figure prominently in Henrich's *The WEIRDest People in the World: How the West Became Psychologically Peculiar and Particularly Prosperous* (2020), which seeks to explain the global dominance of northwest European cultures in the modern world.[17]

As Henrich et al. explain, the stability and effects of marriage patterns, which are socially enforced, cannot be understood without reference to underlying mating strategies, which are biological.[18] Humans are a significantly dimorphic species—not just physically but also psychologically and behaviorally. As it concerns mating, the relevant behavioral dimorphism concerns, on the female side, strong observed sexual preferences for males of high status and/or genetic quality. For short-term pairings or extra-pair couplings, genetic quality (as assessed by looks and charm) figures relatively highly for women, while for long-term pairings qualities such as wealth, power, protectiveness, and commitment come to the fore.[19]

These innate mating preferences, some of which are not unique to humans, are largely explained by the different levels of investment required by males and females to pass on their genes.[20] Begetting children is far more costly to females, not only energetically but also in terms of opportunity costs for additional children, since women have a low cap on how many children they can have, even in the best of reproductive circumstances. This means evolution favored choosiness in females more than it did for males, since bad sexual decisions are more costly for females. It also means females are more attached to their children, since they have fewer chances than males do to pass on their genes. These considerations, when added to the fact that humans mature at a very slow rate, go a long way toward explaining the female bias in favor of mates possessing the genetic, material, and psychological means to help them produce and raise offspring who will not only survive to maturity but also be well-positioned to pass on *their* genes because of *their* attractiveness on the mating market.[21]

The upside for females is that most will reproduce: historically, the strong majority of women had children, and more than that husbands. Not so for men, whose chances of having a spouse or children depend a great deal more on the culture they find themselves in. The mainstream account seems to be that, although most men probably had mates in small-scale foraging groups, the rise of large-scale agricultural societies resulted in bigger socio-economic gaps that in turn led to larger harems for elites and correspondingly fewer females for males at the bottom of society. Researchers suggest that it is not uncommon for around 40% or more of men to go without a mate in a polygamous society.[22]

Although Peterson's concern for low-ranking men was received with great hostility, it is noteworthy that feminist philosophers often object to polygamy on the grounds that polygamy harms underprivileged women.[23] Yet there is reason to think that, in material terms at least, it is *high*-status women, and not low-status ones, who are the losers under polygamy.[24] Whether women and girls are given away by their families or choose partners themselves, men of means are highly sought after and (quite unlike married women of means) these men are usually receptive to additional fertile-age spouses. Thus in such cultures, high-status men get high-status women as senior wives. Senior wives retain special prerogatives throughout their lifetimes but nonetheless cede some clout as their husbands acquire junior wives. In contrast, securing high-status husbands is an important way up the social hierarchy for young and attractive lower-status females.[25]

On the male side of the ledger, polygamy can benefit successful men not only in terms of sex and children but also materially: in traditional West African societies, for example, more wives mean more workers for a man's plot of land. All these advantages redound to additional status and influence.[26] (Once, in northern Ethiopia, I was informed by locals of a marriage taking place between a 50-something-year-old Afar man and a 14-year-old girl—his sixth or seventh wife, some having died. This patriarch had almost 60 children and was able to procure a highly sought-after administrative sinecure simply because he had so many sons and grandsons that he posed a local security risk if not mollified by the government.) Although winners in this game, polygamous fathers are worse than monogamous ones on average, given that the resources they would otherwise spend on parental investment are spent courting new wives. A number of studies show significantly higher mortality rates for children in polygamous families in Mormon and African households, for instance, even after controlling for relevant variables.[27]

Biologically speaking, low-status men are the worst off of all four sex/status demographics, since they alone are at serious risk of being deprived of any genetic legacy. As one would suspect, unattached males are far more prone to engage in self-destructive behaviors, such as gambling and drug use. Not all of these self-destructive behaviors are the *cause* of their being single, either: research suggests that men forsake much of their self-destructive behavior when they find mates—especially wives—since the domesticating influence of wives for at-risk men sees them working harder and spending less time hanging out with friends, drinking, using drugs, or wandering about in seedy areas looking for sex.[28]

Preventing the loss of human, social, and civilizational capital of dissolute lives should be a concern for all of us, but it is those unattached men who do *not* silently succumb to despair who pose the greater threat to our societies. Evolution-based analyses predict,[29] and we find, that polygamous cultures (and even polygamous subcultures within mixed societies) are consistently more violently criminal and unstable. Henrich, Boyd, and Richardson summarize research concluding that, controlling for a variety of relevant variables, marriage reduces the probability of a man's criminality by 35%–50%. In their original work based on datasets drawn from crime reporting across 157 countries and controlling for a variety of variables such as GDP, inequality, and democracy, they find that the pool of unmarried men is a significant predictor of rape and murder. Various natural experiments that resulted in spikes of males relative to females (such as China's one-child policy or massive influxes of male immigrants) also precede spikes in crime that, after many variables are controlled for, point to usual effects of desperate males who can improve their chances of passing on their genes only by engaging in risky, criminal, and/or revolutionary behavior.[30]

Polygamy results in a scramble for unattached women, since even husbands continue to look for new wives. This drives down the age of marriage in such societies and (since established males are favored over less-established males) widens the age gap between husbands and wives—often to the tune of 20 years. Child marriage is not at all uncommon in polygynous societies, and in various such cultures girls may be promised before being born. Moreover, the mere fact that a wife has no monopoly on her husband's reproduction or sexual satisfaction further lowers her bargaining power in the relationship. All these factors contribute to the domineering hyper-patriarchal family dynamics often seen in polygynous societies.[31] Polygamy, for instance, encourages parents to aggressively interfere with their daughter's choices. Because of the pressing demand for young women, bride prices are common, which incentivize parents to direct their daughters in more or less forceful ways to wealthy suitors (this money is often used to pay the bride price a brother needs to pay *his* fiancé's family).[32] Cultures practicing female genital cutting are almost always polygamous, and it is commonly said in these cultures that cutting is intended to "reduce the sexual demands on her husband, allowing him to have several wives" and lower the odds that a young, sexually underserved young wife will stray when her polygamous husband is tending herds, raiding cattle, or otherwise occupied—a cheaper recourse than having one's personal harem guarded by eunuchs.[33]

For their part, Henrich and his collaborators theorize that monogamy was a cultural adaptation that allowed its adoptees to outcompete their polygamous rivals. The monogamous culture that has come down to us began in the Greek city-states of deep antiquity, where it served to reduce male intrasexual competition and build social cohesion. [34] Yes, high-caste Greek men of the archaic and classical era had a variety of sexual outlets, from barbarian prostitutes and slaves to young male mentees (consorting with Greek women other than one's wife could mean death, however): for males, ancient Greeks and Romans advocated only for social monogamy. But prohibitions against additional wives or even concubines gave any male citizen a real chance of marriage, unlocking an untapped store of domestic energy from low-ranking men who, for the first time, had biological "skin in the game" for the survival and success of their polities. Monogamy clearly proved advantageous for Greek and, later, Roman societies, who played a large if not definitive role in shaping Jewish and Christian disapproval of polygamy. The spread of Christian monogamy across polygamous Europe encouraged equality, higher social trust, and higher levels of human capital, all of which played a crucial role in European civilizational dominance. Colonialism spread monogamous norms around the world.[35]

In concluding this section, I note that many of the social and civilizational downsides of polygamy are predictable even if that mating system doesn't come in its "traditional" form. Indeed, many of the behaviors commonly associated with patriarchal and religious elements of traditional polygamy are probably *effects of* polygamy, not some independent force of patriarchy or religion. Given our evolved preferences, allowing unattached *and* attached men to pursue unattached women cannot but lead to a higher percentage of unattached males, more patriarchal control of women to protect them from this scramble, less parenting investment by males lucky enough to be successful at procuring mates, less civic buy-in from males shut out by the competition, and more internal tension between the reproductive haves and have-nots. As philosophers Cheshire Calhoun, Polycarp Ikuenobe, Andrew March, and others have argued, liberal norms would seem to permit polygamy.[36] We have raised above an argument by analogy to environmental sustainability to the effect that the *prima facie* right people have, even on liberalism, to engage in consensual sexual and marital arrangements may be outweighed if those choices are civilizationally unsustainable. Whether polygamy's liberal advocates have appreciated the dangers of polygamy in that regard is not clear. My sense is that philosophers usually disbelieve, ignore, or downplay the innate mating preferences that, when given any option of polygamy, foster the negative consequences described above.

EMI as Applied to Our Unregulated Sexual Market

The rate of sexlessness among young American men is unprecedented in modern times and sloping sharply upward. According to a recent major report, from 2008 to 2018, the share of 18–24-year-old men having no sex in the previous year rose from 12% to 31%, and for men between 25 and 34 that number went from 7% to 14%.[37] Similar trends are found internationally.[38] Insofar as these figures represent men being outcompeted in the mating market, it is tempting to consider enforced monogamy as a regulatory intervention—we have called this the "enforced monogamy inference" or EMI.

That said, the *sole* explanation for these trends cannot be out-competition by elite males. Young women are having less sex, too, although the dropoff for females is not as dramatic and is more pronounced later in their lives.[39] Factors commonly cited as causal of this more general "sex recession" include troubling increases in housing costs, student loan debt, obesity rates, anti-depressant use, porn consumption, and cell-phone addiction.[40] But how on earth would enforcing monogamy encourage young people to get off their cell phones or help them out of their parents' basements? We are thus forced to say a bit more about what the EMI concerns. Let's disambiguate between two interpretations. A *narrow* interpretation of EMI says that we should impose monogamous norms or policies specifically to reduce the competition involuntarily celibate men face from better-positioned (attractive, high status, rich, etc.) rivals. A *wide* interpretation, on the other hand, says that monogamous norms would help involuntarily celibate men secure mates for a variety of reasons.

Wide EMI is not fixated on direct male intrasexual competition. We can reduce the number of involuntary celibates if, say, we removed some barrier that's keeping men and women from getting together. (For instance, in Japan, the same percentage of men and women are going without sex, so sexlessness there is not obviously a matter of direct intrasexual competition.) Granted, conceptually, it seems inapt to say that (say) student loan forgiveness, which might help many reticent men (and women) feel more optimistic about starting a relationship and thereby reduce the number of involuntary celibates, would be "enforcing monogamy." But other measures that an advocate of wide EMI might favor can be reasonably thought of as enforcing monogamy. For instance, while social scientists debate its significance, among incels, at least, porn addiction is widely blamed for their situation.[41] (One "coomer" meme (the "coomer" is a compulsive masturbator with red, watery eyes) shows him staring at his screen while his possible-reality "trad" girlfriend disintegrates in the background.) Chastity norms also reasonably fall under "enforced monogamy," interpreted in the wide sense. Underappreciated by researchers and usually moralistically dismissed by journalists who study incels is a widespread disgust for promiscuous women in this demographic. However offensively they may express the sentiment,[42] it is as important to know if many men would rather be single than be with a woman with a high partner count as it is to know if women would rather be single than be with a man who is poor. Efforts to curb pornography addiction or female promiscuity, whether justified or not, could count as monogamy enforcement. (Obviously, a partisan of wide EMI might reject any number of initiatives that *conceptually* fall under enforced monogamy, just as an environmentalist might reject any number of ill-advised sustainability proposals.)

For his part, Peterson is reasonably read as meaning EMI in the narrower sense of blaming direct intrasexual competition for the lopsided distributions of female attention. In an unquoted passage of his blog post on the matter, Peterson adverts to Henrich's research on polygamy; and even in Bowles' quotes he seems to have intrasexual competition in mind, as when he says that, in situations with "too much mate choice," "a small percentage of the guys have hyper-access to women, and so they don't form relationships with women."[43] This suggests Peterson thinks an unregulated mating market reproduces some of the same problems associated with polygamy. Does he have a point? In what remains of this section, I will develop the case for this "narrow" interpretation of EMI, as it might apply to the mating patterns of most developed countries today.

We begin, as some manosphere theorists of the male predicament do,[44] with "Sexual Economics Theory" (SET), pioneered by psychologists Roy Baumeister, Kathleen Vohs, and colleagues.[45] According to SET, a number of factors of the modern world conspire to split the traditional mating market in two: one for sex and one for long-term mates/marriage. We need not repeat the material already discussed about how parental investment helped shape male and female mating strategies. But one prediction of parental investment theory ignored above but relevant to SET is that, although both women and men desire sex and long-term relationships, women will be more interested in long-term relationships, and men will be more interested in casual sex. This means women have a bargaining advantage when it comes to sex, and men have a bargaining advantage with regard to long-term relationships.[46] SET predicts, accurately enough it seems, that in earlier societies, men on the mating market were primarily motivated by exclusive sexual access to young women, and young women were primarily motivated by the benefits of marriage to high-quality men—benefits that include not only quality genes but also provisioning and protection for themselves and their children.[47] As we have seen, whether the society is polygamous or monogamous matters a great deal to how these preferences are *directed*, but the transition from polygamy to monogamy didn't change the ultimately transactional nature of mating: women "trade" sexual exclusivity for commitment and resources. The search for exclusive sexual access incentivized men—in the decade or so where their intelligence, creativity, and physical prowess are at their peak—to build homesteads, develop their talents, and elevate themselves socially. Likewise women—even in monogamous societies where women had relatively more safety and autonomy and thus married at a later age—still married in their prime reproductive years. However, the advent of effective birth control marked a monumental shift in human relations by detaching sex and reproduction. The significance of that development for SET is that birth control "bifurcated" the mating market into one for recreational sex and another for marriage. As harmless as that might sound, its consequences might be quite dire.

First, the separation of sex from reproduction allows young people to pursue relatively riskless sex. For SET theorists, this means that women wishing to "lower" the price of sexual access could do so without much risk to themselves—indeed, offering sex at a lower price gives a young woman more access to male attention than she otherwise might have. Any vestiges of "slut shaming," without being buttressed by a real fear of pregnancy, are insufficient to keep a significant number of women from lowering the commitment costs of mating with them. This serves to break the "cartel" that women (and their parents) maintained on female sexual access. For although men usually *respect* women who show sexual restraint, and would prefer one for a spouse, most men will have a hard time committing to one sexually restrained woman when many other women are offering sex without commitment. This causes a race to the bottom that leaves the many young women who are *not* interested in a string of meaningless liaisons and who would *like* to find a lifelong mate with whom to start a family forced to choose between having sex earlier, more often, and almost certainly with more men than they'd prefer to or being celibate and alone.[48]

The predictable results of cheap sex and commitment-shy men are more out-of-wedlock births, lower rates of marriage, and later age of first marriage, all of which have happened (although of course other factors have played a large contributory role too, such as an increasing interest among women in getting educated and establishing careers). Less obvious and more controversial is that the mere cheapness of sex might be serving as, as Baumeister and Vohs put it, an "opiate of the (male) masses."[49] How so? Well, they and other sexual economics theorists take it as a theorem that men "will do whatever is required in order to obtain sex, and perhaps not a great deal more."[50] And *that* matters for the rest of us since, if they are correct, the contributions of ambitious young men are, for reasons ultimately biological, key to civilizational advance. Baumeister and Vohs do not mince words:

> [Women] are more risk averse, resulting in fewer entrepreneurs and inventions. (Baumeister 2010 noted an appalling gender imbalance in new patents; nobody is seriously suggesting that the U.S. Patent office systematically discriminates against women, but women simply do not apply for patents in anything close to the rate that men do.) Women are less interested in science and technology fields. They create less wealth (for themselves and others). [...] Female sociality is focused heavily on one-to-one relationships, whereas male sociality extends to larger groups networks of shallower relationships (e.g., Baumeister and Sommer 1997; Baumeister 2010). Crudely put, women hardly ever create large organizations or social systems. That fact can explain most of the history of gender relations, in which the gender near equality of prehistorical societies was gradually replaced by progressive inequality—not because men banded together to oppress women, but because cultural progress arose from the men's sphere with its large networks of shallow relationships, while the women's sphere remained stagnant because its social structure emphasized intense one-to-one relationships to the near exclusion of all else (see Baumeister 2010). All over the world and throughout history (and prehistory), the contribution of large groups of women to cultural progress has been vanishingly small.[51]

Whether Baumeister, Vohs, and their fellow travelers overstate this case or not is an empirical matter. What's not empirical is the moral relevance of this fact, if it is a fact, when conjoined with a moral outlook saying that we should raise the cost of cheap sex if it discourages human capital formation in civilizational-threatening ways.

So for SET the danger is *easy* sex—men, its theorists say, rarely exceed what is demanded of them to procure sex. But sex is *still* not all that easy for some men, as the phenomenon of inceldom demonstrates. Could we have fallen into a mating pattern that looks like social monogamy when it comes to marriage, but a degraded form of polygamy as far as sex is concerned?

Because of female selectivity, sex without risk results in a windfall of uncommitted sex for men with the best short-term mating value. It is imperative to stress *short-term* mating value, because this observation lay at the core of Minassian's resentment of "Chads." Sexual liberation splits apart not only the mating market but also the features that made one a desirable mate. Harkening back to points made above, when selecting for short-term mates, where provisioning/protecting and commitment aren't at issue, women will tend favor genetic quality as advertised by good looks and display. This means that the male beneficiaries of young women's sexual favors will now tend to be those who in previous generations would be considered peacocks, bad boys, or otherwise poor marriage material. Independent lines of research have suggested that men may actually be born predisposed to pursue either long-term "dad" (high sexual fidelity and parental investment) or short-term "cad" (more promiscuous and less parental investment) strategies, as evidenced by correlations between markers for testosterone (such as digit ratio), sexual precociousness, and "dark triad" (narcissism, Machiavellianism, psychopathy) traits.[52] Thus the big winners in the *sex* (not marriage) market tend to be effective short-term strategists.

It is commonly claimed in manosphere forums that the distribution of female attention follows the Pareto Principle, which applied in this context would translate to 20% of the men getting 80% of female interest. Sociologist Mark Regnerus reports in his book *Cheap Sex* that, based on *Relationships in America* surveys, 20% of the heterosexual American men between 25 and 50 are having 70% of the sexual partnerships and 10% are having about half of the sexual partnerships. (Although similar stats are seen on the female side of things, a heterosexual woman's having lots of sexual partners reveals more about her sexual choices than her value on the mating market, for reasons discussed above.) Regnerus is quick to point out a few fallacious takeaways from this fact. First, it doesn't mean that the most sexually successful men are having sex with 70% of the sexually active *women*; rather, they are merely involved in 70% of the sexual pairings, and most of these are likely to be with the fraction of most promiscuous women. Second, it doesn't mean that these men have actual harems: overlaps in their (brief) relationships are fairly rare and short-lived, so a woman who has a relationship with one of these skillful short-term strategists is, in principle, still available for average guys.[53] That said, Regnerus perceptively notes that, "if among the remaining 80 percent of men fidelity and a modest sexual history [for potential mates] are key values, then yes—the 20/70 phenomenon may well be dragging down marriage rates in the peak years of fertility (twenties to early thirties)."[54] If indeed long-term strategists find women with higher partner counts less attractive as potential mates, then successful short-term strategists are not so much outcompeting them on the marriage market as sabotaging them.

We mustn't confuse the *most successful short-term strategists* with the *most desirable men* (even on the uncommitted mating market): many of the former simply have low standards and hustle more for sex. More threatening for average men is what we see in the virtual marketplace of online dating, which to a greater extent favors what we might term a sexually attractive "Chad" or at least a type of Chad. Infamously, women rate 80% of men's Tinder profiles as "below average,"[55] while on the dating app Hinge, half of female interest is directed to 15% of men, causing one data scientist to quip, "while the mythological ship-launcher, Helen, was a woman, it turns out that it's actually the most attractive men who soak up the most disproportionate amount of their gender's affection."[56] When we apply the Gini coefficient (a measure of inequality often appealed to in economic arguments for wealth redistribution) to female interest for males on dating apps, it is equivalent to that of the most unequal countries in the world, such as Haiti, whereas the Gini coefficient for male interest in females is equivalent to the economies of Western Europe.[57] It remains unclear how much of that interest translates into sex, however. It is also unclear whether these apps shape the way women shop for partners: it would appear that these apps favor handsome men who know how to market themselves visually and over chat functions, while disadvantaging (say) charismatic or laconic men, who might be very attractive in person. Anecdata suggests that the overwhelmingly negative feedback men experience on these apps is demoralizing huge swaths of quite eligible men.

Where does this leave the rest of the young males who are naturally more sexually restrained, less physically attractive, and maybe not particularly charismatic, but who in former generations might impress marriage-minded young women (and especially their parents) by creating businesses, developing their economically productive talents, and building homesteads? These males will have a much rougher time of it in a bifurcated mating market, where women can have risk-free sex without commitment. It is no exaggeration to say that how to respond to this dilemma may be the fundamental divide in manosphere thought. One response is to work harder at becoming a cad or a Chad, and these are main themes of the "pick-up artist" (PUA) and "Redpill" movements. Another response is simply to drop out of the mating market altogether, as represented by the MGTOW ("men going their own way") and incel movements, whose partisans hold that the costs of pursuing, and even winning, a woman in today's mating market are not worth it.

In any event, we might reasonably conclude that either male demographic at the mating extremes—those who get sex too easily and those for whom it is (or perceived to be) unattainable—is to be minimized if we are concerned for what's good for society and civilizational advance. The former have relatively less incentive than women do to settle down, and often continue to pursue reproductively prime-age women (low 20s), replicating some of the pressure for young women we see in polygamous cultures. In small numbers they are sustainable; but if the only way to win a young woman's affections are, or are felt to be, to be a peacocking "alpha" or dark triad quasi-psychopathic narcissist, then we will see far too many men diverting their efforts from long days at the office to long workouts at the gym, opting to spend their money on clothes, cars, and nightclubs instead of on down payments for homes. On the other extreme, discouraged young involuntary celibates also fail to contribute their talents to the civilizational commonweal. Of course, even many self-described "incels" work hard and find positive outlets in productive hobbies. But the talents and energies of too many are frittered away in escapist pursuits such as political message boards (where they are often radicalized), video games, pornography, and occasional prostitutes. Insofar as they still contemplate a relationship, they fantasize about removing themselves to some traditional society or impoverished country where they think they'll be more attractive. In short, they act in some ways like reproductively unsuccessful men in polygamous cultures do, who often despair, revolt, or abandon everything for a shot in a foreign land.

To sum up this section, it has been unclear up to this point how "enforced monogamy" is to be interpreted. Is it supposed to ease the competition involuntary celibates face from other men, who are dominating female attention? Or is it about increasing sexual restraint in a variety of ways (e.g., fostering norms against pornography and promiscuity), which in turn will result in a more egalitarian mating pattern and a smaller percentage of involuntarily celibate men? Although incels themselves acknowledge the role of factors other than direct intrasexual competition (such as porn overuse and a strong aversion to sexually experienced women) in causing their woes, Peterson seemed to have the former, narrower interpretation in mind when he prescribed enforced monogamy. And indeed, although not the whole story by any means, a lopsided distribution of female attention may be a significant stressor for involuntarily celibates, given the bifurcated mating market and the unequal distribution of attractive traits for short-term mates—traits, advocates stress, whose rewarding disincentivizes male contributions to civilization.

Evaluating the EMI

If we accept the premise that society has the right to impose informal and even formal sanctions on socially and civilizationally unsustainable mating patterns, and if we accept the thrust of Henrich et al.'s case for monogamy's being an important civilizational adaptation, then social and legal barriers to polygamy seem warranted, at least for monogamous societies. If we apply the standards of environmental sustainability to civilizational sustainability, I think the only reasonable conclusion is that polygamy is to civilizations what the cattle industry or gas-fueled car engines are to the environment—unsustainable. So if polygamy isn't one of our problems, let's not let it become one.

That said, and again in parallel with environmental sustainability, it is often practically (and almost always theoretically) *possible* to render an "unsustainable" practice "sustainable" by making changes elsewhere. If a country's economy is rather inflexibly based on (say) coal for whatever reason, it *may* maintain that coal-based economy sustainably if (say) it captures the greenhouse gasses produced by its factories, meaning that sustainability advocates will have a harder time justifying anti-coal policies for *that* country relative to other countries who don't rely on coal as much or at all. Similarly, it is less problematic to reinforce existing anti-polygamy norms than to reform the sexual norms of a sexually liberated, post-contraceptive society. Censuring or even banning polygamy presents minimal costs on socially monogamous societies. The same cannot be said about enforcing monogamy on sexually unregulated but socially monogamous societies, where stricter norms of sexual monogamy would be highly disruptive. Although civilizational sustainability may demand reform of our mating patterns nonetheless, if there are less intrusive ways to effectively address an unsustainably high percentage of involuntary celibates (or unsustainably low birthrate or any other mating-related civilizational threat), those should be pursued first.

A presumption in favor of values such as liberty and autonomy would entail that, when it comes to unsustainable consumption and all else being equal, we should prefer social sanction to laws, prefer nudges[58] to social sanction, and prefer education and persuasion to nudges. If that presumption is correct, then whether "enforced monogamy" is manifested in the form of laws, social sanction, or even nudges to reduce mating inequality, we should first try to educate ourselves out of the problem. Some of that education involves disseminating the (obviously still-controversial) message that highly unequal mate distributions, whether in the form of marriages or hookups, should be a matter of grave concern for all of us. But the most important educational intervention would be practical. Just as we may try to address a racial or sexual employment gap not merely through demanding hiring quotas but also (and perhaps preferably *first*) through efforts to better prepare underrepresented groups to be competitive in the job market, our parents, educators, coaches, and counselors should be making a more deliberate effort to instruct young women on how to be savvier about their mating choices (given their long-term interests), and to train young men on how to be more attractive mates and better marriage material.

The proposal is hardly radical: preparing young people for romance and marriage used to be the central goal of church functions, community dances, and finishing schools. Granted, even if the traditional cultural scripts taught in these venues were to become attractive once again, the challenges of modernity (some of which were touched on above) make them almost impossible for young people to actually follow. Nonetheless, we mustn't throw the baby out with the bathwater. *If* traditionalist messaging is incorrect about what makes us more attractive mates and partners and/or demands the impossible of young people, then we should change the content, not eliminate the training altogether. Sexual assault training and sex-education courses do this to an extent; but safe sex, free of coercion, is a very low bar indeed.[59] Happily, we have some excellent examples of thoughtful, scientifically informed, and popular courses at universities designed to educate students about human nature, the virtues of dating with discernment, and courtship dynamics.[60] Indeed, Peterson's *own* transition from popular intellectual to guru is largely attributable to the way his lectures on sex differences, personality, and responsibility inspired and equipped legions of young men to be less resentful, awkward, and self-sabotaging. So yes, the increased competition posed by cads and Chads who, on traditional monogamy, would have been avoided or quickly married off is probably an additional challenge facing today's young men looking for love. Nonetheless, it seems fair to say that intrasexual competition is not the involuntary celibate's biggest obstacle to sex and marriage and thus that enforced monogamy probably shouldn't be first on our list of remedies for their predicament.

Notes

1. Dan Bilefsky and Ian Austen, "Toronto Van Attack Suspect Expressed Anger at Women," *New York Times*, April 24, 2018, https://www.nytimes.com/2018/04/24/world/canada/toronto-van-rampage.html.
2. Mack Lamoureux, "A Brief History of 'Incel,' the Misogynistic Group Allegedly Cited By Toronto Van Attacker," *Vice*, April 24, 2018, https://www.vice.com/en/article/pax9kz/a-brief-history-of-incel-the-misogynistic-group-allegedly-cited-by-toronto-van-attacker.
3. Nellie Bowles, "Jordan Peterson, Custodian of the Patriarchy," *New York Times*, May 18, 2018, https://www.nytimes.com/2018/05/18/style/jordan-peterson-12-rules-for-life.html.
4. See, for example, David Futrelle, who writes

   > You can't solve the problem of aggrieved male entitlement by engineering some weird and at least semi-coercive program of "enforced monogamy" built on the assumption that men inherently deserve some sort of access to women's bodies—regardless of what the women inhabiting these bodies want. You can't rid the world of violence born of aggrieved male entitlement with a "solution" that reinforces that sense of entitlement.

   "The Dark Corners of the Internet that Spawned Ideas Like 'Enforced Monogamy,'" *Vice*, May 25, 2018, https://www.vice.com/en/article/d3k3ex/jordan-peterson-enforced-monogamy-incels.
5. Jordan Peterson, "On the New York Times and "Enforced Monogamy," *Jordan B Peterson* blog, https://www.jordanbpeterson.com/media/on-the-new-york-times-and-enforced-monogamy/.
6. Christina Sewell, "Removing the Meat Subsidy: Our Cognitive Dissonance Around Animal Agriculture," *Journal of International Affairs*, February 11, 2020, https://jia.sipa.columbia.edu/removing-meat-subsidy-our-cognitive-dissonance-around-animal-agriculture.
7. In addition, and too-often ignored, is the fact that asteroids are inevitable and grave environmental threats, and cannot be identified and diverted by anything other than an advanced civilization—one quite a bit more advanced than our own. David J. Eicher, "Why the Asteroid Threat Should Be Taken Seriously," *Astronomy*, n.d., https://astronomy.com/bonus/asteroidday.
8. I'll use "mating pattern" as a catchall term to cover not only "marriage patterns" (discussed in the next section) but also patterns of sexual behavior, along with the concomitant norms for each.
9. Justin McCurry, "Record Numbers of Couples Living in Sexless Marriages in Japan, Says Report," *The Guardian*, February 14, 2017, https://www.the-

guardian.com/world/2017/feb/14/record-numbers-of-couples-living-in-sexless-marriages-in-japan-says-report.

10. Because of space constraints, I will not discuss polyamory here. Suffice it to say that I believe that, although polyamory-acceptance may indeed lead us down a slippery slope to polygamy acceptance and legalization for reasons of philosophical or legal consistency, and although I think (for reasons discussed just below) that polygamy should remain illegal and socially unacceptable in societies that are currently monogamous, polyamory itself is orthogonal to polygamy, and anti-polygamy legislation does not (and should not) infringe on polyamory.
11. Miller is not being uncharitable here: one book he assigns in his groundbreaking "Polyamory and Open Sexuality" course on the topic, Mark Michaels and Patricia Johnson's *Designer Relationships* (Jersey City: Cleis Press, 2015), advocates for complete social neutrality between monogamy, polygamy, polyamory, or any other mating arrangement in order to allow people to choose what works best for them, and is dismissive of any social critiques in the very brief passages where social impacts are considered at all.
12. Geoffrey Miller, "Polyamory Is Growing—And We Need To Get Serious About It," *Quillette*, October 29, 2019, https://quillette.com/2019/10/29/polyamory-is-growing-and-we-need-to-get-serious-about-it/.
13. Because of considerations of space, I consider only polygyny here and ignore the extremely rare polyandrous form of polygamy.
14. Cf. the controversy generated by Robin Hanson's post "Comparing Income & Sex Redistribution," *Overcoming Bias* blog, June 26, 2018, https://www.overcomingbias.com/2018/06/comparing-income-sex-redistribution.html.
15. Joseph Henrich, "Polygyny in Cross-Cultural Perspective: Theory and Implications," affidavit S-097767 submitted to the Supreme Court of British Columbia, Canada, July 15, 2010. Available: https://stoppolygamyincanada.files.wordpress.com/2010/12/joseph-henrichs-research-paper.pdf.
16. Joseph Henrich, Robert Boyd, and Peter Richerson, "The Puzzle of Monogamous Marriage," *Philosophical Transactions of the Royal Society* B 367 (2012): 657–669.
17. Joseph Henrich, *The WEIRDest People in the World: How the West Became Psychologically Peculiar and Particularly Prosperous* (New York: Farrar, Straus, and Giroux, 2020).
18. Henrich, "Polygyny in Cross-Cultural Perspective," 26.
19. David Buss, "Sex Differences in Human Mate Preferences: Evolutionary Hypotheses Tested in 37 Cultures" *Behavioral and Brain Sciences* 12, no. 1 (1989): 1–14.
20. Robert Trivers, "Parental Investment and Sexual Selection," in *Sexual Selection and the Descent of Man, 1871–1971*, ed. B. Campbell (Chicago: Aldine, 1972), 52–95.
21. See for updated research on mating strategies pt. 3 of David Buss, *Evolutionary Psychology* (New York: Routledge, 2019).
22. Henrich, "Polygyny in Cross-Cultural Perspective," 40.
23. See, for example, Emily Crookston, "Love and (Polygamous) Marriage? A Liberal Case Against Polygamy," *Journal of Moral Philosophy* 12 no. 3 (2014): 267–289, esp. 283ff.
24. In fairness, Crookston's (see *supra*) concern is a moral injury, not a material one.

25. William Jankowiak, Monika Sudakov, and Benjamin Wilreker, "Co-Wife Conflict and Co-operation," *Ethnology* 44, no. 1 (2005): 81–98.
26. William Tucker, *Marriage and Civilization* (Washington DC: Regnery, 2014), ch. 8. That said, I know a Cameroonian noble woman who procured additional wives while her high-status husband was on a lengthy trip abroad, as she needed more help in their fields. When he returned, she informed him about his new wives, and it is a running joke in that household that these junior wives called her "husband."
27. Henrich, Boyd, and Richerson, "Puzzle of Monogamous Marriage," 665–666.
28. Henrich, Boyd, and Richerson, "Puzzle of Monogamous Marriage," 661–663.
29. Martin Daly and Margo Wilson, *Homicide* (New York: De Gruyter, 1988).
30. Henrich, Boyd, and Richerson, "Puzzle of Monogamous Marriage," 661–663.
31. "We find," say two researchers on the question,

> statistically significant relationships between polygyny and an entire downstream suite of negative consequences for men, women, children, and the nation-state, including the following outcomes: discrepancy between law and practice concerning women's equality, birth rate, rates of primary and secondary education for male and female children, difference between males and females in HIV infection, age of marriage, maternal mortality, life expectancy, sex trafficking, female genital mutilation, domestic violence, inequity in the treatment of males and females before the law, defense expenditures, and political rights and civil liberties.

Rose McDermott & Jonathan Cowden, "Polygyny and Violence Against Women," *Emory Law Journal* 64, no. 6 (2015): 1767–1814. Available at: https://scholarlycommons.law.emory.edu/elj/vol64/iss6/4.

32. Gary Becker, *A Treatise on the Family* (Cambridge: Harvard University Press, 1981), 56ff. See also Robert Guest, "The Link Between Polygamy and War," *The Economist*, December 19, 2017, https://www.economist.com/christmas-specials/2017/12/19/the-link-between-polygamy-and-war.
33. Anika Rahman and Nahid Toubia, *Female Genital Mutilation: A Guide to Laws and Policies Worldwide* (United Kingdom: Bloomsbury Academic, 2000), 5–6.
34. Walter Scheidel, "A Peculiar Institution? Greco–Roman Monogamy in Global Context," *The History of the Family*, 14 no. 3 (2009): 280–291.
35. Henrich, "Polygyny in Cross-Cultural Perspective," 17–19; Henrich, Boyd, and Richerson, "Puzzle of Monogamous Marriage," 666–667.
36. Cheshire Calhoun, "Who's Afraid of Polygamous Marriage? Lessons for Same-Sex Marriage Advocacy from the History of Polygamy," *San Diego Law Review* 42 (2005): 1023–1042; Polycarp Ikuenobe, "The Monogamous Conception of Romantic Love and Western Critiques of Polygamy in African Traditions," *Philosophical Papers* 47, no. 3 (2018): 373–401; Andrew March, "Is There a Right to Polygamy? Marriage, Equality and Subsidizing Families in Liberal Public Justification," *Journal of Moral Philosophy* 8 (2011): 246–272.
37. Peter Ueda, Catherine Mercer, Cyrus Ghaznavi, and Debby Herbenick, "Trends in Frequency of Sexual Activity and Number of Sexual Partners Among Adults Aged 18 to 44 Years in the US, 2000–2018, *JAMA* 3(6) (June 12, 2020): 1–15.
38. "Demographics of Inceldom," *Incelswiki*, accessed November 2, 2020, https://incels.wiki/w/Demographics_of_inceldom#Young_incels_in_the_U.S.

39. Christopher Ingraham, "The Share of Americans Not Having Sex has Reached a Record High," *Washington Post*, March 29, 2019, https://www.washingtonpost.com/business/2019/03/29/share-americans-not-having-sex-has-reached-record-high/.
40. Kate Julian, "Why Are Young People Having So Little Sex?", *The Atlantic*, December 2018, https://www.theatlantic.com/magazine/archive/2018/12/the-sex-recession/573949/.
41. Although more sexually active men also consume more pornography, there is reason to believe that, for many men, porn consumption serves as an adequate-enough substitute for sex such that they are disincentivized from seeking out relationships. For more on this debates, see Jean Twenge, "Possible Reasons US Adults Are Not Having Sex as Much as They Used To," *JAMA* 3(6) (June 12, 2020), https://jamanetwork.com/journals/jamanetworkopen/fullarticle/2767063; Philip Zimbardo, *Man Interrupted* (Newburyport, MA: Red Wheel, 2016).
42. "In the elaborate incel taxonomy of participants in the sexual marketplace, I am a Becky, devoting my attentions to a Chad. I'm probably a "roastie," too—another term they use for women with sexual experience, denoting labia that have turned into roast beef from overuse." Jia Tolentino, "Rage of the Incels," *New Yorker*, May 15, 2018, https://www.newyorker.com/culture/cultural-comment/the-rage-of-the-incels.
43. Bowles, "Jordan Peterson, Custodian of the Patriarchy."
44. "Roy Baumeister," *Incelwiki*, https://incels.wiki/w/Roy_Baumeister; see also an online discussion by psychologist Todd Grande on the incel-sexual economics connection, "Sexual Economics Theory vs. Feminist Theory: MGTOW, INCEL, & Science," June 17, 2019, https://youtu.be/M1R_pRxj9Zg.
45. Roy Baumeister and Kathleen Vohs, "Sexual Economics: Sex as Female Resource for Social Exchange in Heterosexual Interactions," *Personality and Social Psychology Review* 8, no. 4 (November 2004): 339–63.
46. Roy Baumeister, Kathleen Catanese, and Kathleen Vohs, "Is There a Gender Difference in Strength of Sex Drive? Theoretical Views, Conceptual Distinctions, and a Review of Relevant Evidence," *Personality and Social Psychology Review* 5, no. 3 (August 2001): 242–73.
47. Roy Baumeister, Tania Reynolds, Bo Winegard, Kathleen Vohs, "Competing for Love: Applying Sexual Economics Theory to Mating Contests," *Journal of Economic Psychology* 63 (2017): 230–241.
48. Baumeister and Vohs, "Sexual Economics," 350ff; Mark Regnerus, *Cheap Sex: The Transformation of Men, Marriage, and Monogamy* (New York: Oxford University Press, 2017); Steven Rhoads, "Hookup Culture: The High Costs of a Low 'Price' for Sex," *Society* 49 (2012): 515–519.
49. Roy Baumeister and Kathleen Vohs, "Sexual Economics, Culture, Men, and Modern Sexual Trends," *Society* 49 (2012): 520–524, 523.
50. Baumeister and Vohs, "Sexual Economics, Culture, Men, and Modern Sexual Trends," 521.
51. Baumeister and Vohs, "Sexual Economics, Culture, Men, and Modern Sexual Trends," 522ff. For more discussion of male overrepresentation at the poles of accomplishment, see Roy Baumeister, *Is There Anything Good About Men? How Cultures Flourish By Exploiting Men* (New York: Oxford University Press, 2010).

52. Kristina Durante, et al. "Ovulation Leads Women to Perceive Sexy Cads as Good Dads," *Journal of Personality and Social Psychology* 103 no. 2 (2012): 292–305. For layman summaries, see Mary Duenwald, "For a Good Time, Well, Don't Call Dad," *New York Times*, Dec. 2, 2003, https://www.nytimes.com/2003/12/02/science/for-a-good-time-well-don-t-call-dad.html; Lawrence Josephs, "How to Pick a Daddy Type," *Psychology Today*, Apr 18, 2020, https://www.psychologytoday.com/us/blog/between-the-sheets/202004/how-pick-daddy-type.
53. Regnerus, *Cheap Sex*, 85–88.
54. Regnerus, *Cheap Sex*, 87.
55. Anonymous, "Tinder Experiments II," *Medium*, March 24, 2015, https://medium.com/@worstonlinedater/tinder-experiments-ii-guys-unless-you-are-really-hot-you-are-probably-better-off-not-wasting-your-2ddf370a6e9a.
56. Aviv Goldgeier, "What's The Biggest Challenge Men Face On Dating Apps?: A Q&A With Aviv Goldgeier, Junior Growth Engineer," *Hingeirl*, n.d., https://hingeirl.com/hinge-reports/whats-the-biggest-challenge-men-face-on-dating-apps-a-qa-with-aviv-goldgeier-junior-growth-engineer/.
57. Bradford Tuckfield, "Attraction Inequality and the Dating Economy," *Quillette*, March 12, 2019, https://quillette.com/2019/03/12/attraction-inequality-and-the-dating-economy/.
58. Richard Thaler and Cass R. Sunstein, *Nudge: Improving Decisions About Health, Wealth, and Happiness* (New York: Penguin Books, 2009).
59. See Rhoads, "Hookup Culture" for a longer critique. Indeed, it could appear at times that sex and relationship education is almost designed to undermine the psychological wellbeing and reproductive futures of those it's meant to serve.
60. See psychologist Alexandra Solomon's "Marriage 101" course at Northwestern, or Boston College's Kerry Cronin's recorded talk "Hanging Out and Hooking Up," February 9, 2015, https://youtu.be/60K2-LEDPyg.

CHAPTER 10

The Ethics of Cohabitation

Christopher Kaczor

By "cohabitation" I mean living together in the same legal residence with a sexual partner who is not a spouse. Cohabitation might appear to be a lifestyle that has no significant ethical implications. But, as one study puts it, "Living together before marriage may seem like a harmless or even a progressive family trend until one takes a careful look at the evidence."[1] So, in this chapter, I intend to provide empirical evidence that calls into question the prudential wisdom and moral acceptability of cohabitation, especially for young people. I will presuppose the truth of the ethical premise that it is prima facie morally wrong to do things that substantially risk lowering human well-being (the endangerment principle). What makes drunk driving wrong? Well, it isn't that every time someone drives drunk people die, get injured, or damage property. Drunk driving is wrong because it substantially risks lowering human well-being. To determine whether an activity substantially risks lowering human well-being, we have to look at the empirical evidence, in this case the empirical evidence about cohabitation and marriage.

Some people have what I'll call a "low" view of marriage. They think that marriage is just a piece of paper and a meaningless ceremony. What really matters is love, commitment, faithfulness to your partner, and all these things can be had just as easily without marriage. These people might consider cohabitation as an alternative to marriage. Other people have what I'll call a "high" view of marriage. They think that marriage is incredibly important, and they do not want to get divorced. They really want to have a good marriage. Such people might consider cohabitation as a good preparation for marriage. Let's

C. Kaczor (✉)
Loyola Marymount University, Los Angeles, CA, USA
e-mail: ckaczor@lmu.edu

D. Boonin (ed.), *The Palgrave Handbook of Sexual Ethics*,
https://doi.org/10.1007/978-3-030-87786-6_10

consider cohabitation both as an alternative to marriage and as a preparation for marriage.

Cohabitation as an Alternative to Marriage

Some people think that marriage doesn't matter. What is marriage other than a meaningless ceremony, a piece of paper, and an exchange of rings? Some people who do not care about marriage do care about having a loving relationship. Glenn Stanton describes this view:

> To become legally married would weigh down a beautiful and unencumbered love with the ugly legalities of a business contract. Such baggage would only serve to choke out the lifeblood of the relationship. Does love require a legal document to legitimize itself? Without such entanglements, two people could remain together because they desire to be together, not because some legal entity says they must. Cohabitation would allow the warmth of love, not cold legalities, to regulate the temperature of domestic relationships.[2]

The idea here is that if you want a lasting relationship, if you want your partner not to cheat on you, if you want a peaceful union, then these things will most likely be found in cohabitation as an alternative to marriage.

The scholarly research comparing couples who cohabit to married couples reveals significant differences on average between the two groups.[3] Cohabiting couples are much more likely to break up than married couples. This likelihood of breaking up is particularly when children are involved, "Only one out of three children born to cohabiting parents remains in a stable family through age 12, in contrast to nearly three out of four children born to married parents."[4] Popenoe and Whitehead point out, "Children need and should have parents who are committed to staying together over the long term. Cohabiting parents break up at a much higher rate than married parents and the effects of breakup can be devastating and often long lasting."[5] Cohabiting couples are more likely to experience domestic violence than married couples. Cohabiting couples are more likely to be unfaithful to each other sexually than married couples. Popenoe and Whitehead point out, "Cohabiting couples report lower levels of happiness, lower levels of sexual exclusivity and sexual satisfaction, and poorer relationships with their parents."[6] Cohabiting couples are more likely to argue and fight than married couples. Popenoe and Whitehead note:

> Annual rates of depression among cohabiting couples are more than three times what they are among married couples. And women in cohabiting relationships are more likely than married women to suffer physical and sexual abuse. Some research has shown that aggression is at least twice as common among cohabitors as it is among married partners. Two studies, one in Canada and the other in the United States, found that women in cohabiting relationships are about nine times more likely to be killed by their partner than are women in marital relationships.[7]

Popenoe and Whitehead point out that, "children living in cohabiting unions with stepfathers or mother's boyfriends are at higher risk of sexual abuse and physical violence, including lethal violence, than are children living with married biological parents."[8] They continue, "In general, the evidence suggests that the most unsafe of all family environments for children is that in which the mother is living with someone other than the child's biological father. This is the environment for the majority of children in cohabiting couple households."[9] As W. Bradford Wilcox points out, "a recent federal report on child abuse found that children in cohabiting stepfamilies were 98 percent more likely to be physically abused, 130 percent more likely to be sexually abused, and 64 percent more likely to be emotionally abused, compared with children in married step-families."[10] So, if someone cared about having a lasting relationship, not getting cheated on, fighting and arguing less, and providing a more safe environment for children, all these things would give them a strong reason to prefer to be part of a married couple than part of a cohabiting couple.

Why do cohabiting couples have more infidelity, more trouble, and more breaking up on average than married couples? The differences may have something to do with the power of public commitment. It is common in our culture to require those taking up an important position to make public oaths to discharge their duties faithfully. The Marine entering the armed forces, the witness before testifying in court, and the President of the United States when being sworn into office all take oaths. Marriage involves taking an oath, vow, or promise in public before witnesses to act in certain ways toward the husband or wife. In one common version of the marriage vow, both spouses promise, "I take you to be my lawful spouse, forsaking all others. I promise to love you in good times and in bad, for richer or poorer, in sickness and in health, until death do us part." Taking marriage vows does not guarantee that either party will carry out what they have promised. But Rhoades and Stanley note, "making a clear, deliberate decision to commit to one option and reject alternative options strengthens a person's tendency to follow through on the commitment. Wedding ceremonies ritualize the foundation of commitment."[11] The public nature of marriage is important here. To say, "I'll love you forever" in private is one thing. But to make a vow, before some public authority, like a judge, priest, or rabbi, and normally in front of friends and family, that you'll love someone as your spouse so long as you both shall live is to involve oneself in a more formal, official, and definitive way. As Rhoades and Stanley point out, "commitment is strengthened when it is publicly declared because individuals strive to maintain consistency between what they say and what they do."[12] As the book *Influence: The Psychology of Persuasion* by Robert B. Cialdini points out, when people make a free, public, and active commitment to some cause, they are much more likely to follow through than if a free, public, and active commitment has not been made.[13] Marriage is a free, public, and active commitment. If one or both of the parties is forced to go through a marriage ceremony, they are not, in fact, properly married but may get an annulment. If an exchange of vows takes place in private with no witnesses and no public authority, it is not a true marriage. If either party does not perform the act of commitment, let's say the man or the woman refuses to say the marriage vows, then there is no marriage. Marriage on average produces better results for couples in part because people who publicly vow to love one another have an extra incentive to love one another brought about by the public oath they have taken. According to cognitive dissonance theory, people don't want to view themselves as inconsistent, and they don't want to be viewed by others as inconsistent, so vows serve as an incentive to act in accordance with what has been publicly promised. Couples without marital vows lack this extra incentive to make the relationship work. They also receive less support from extended family for making the relationship work.[14]

Cohabitation as a Preparation for Marriage

Most young people today wish to get married someday. Indeed, "a huge majority of our nation's young people still express the strong desire to be in a long-term monogamous marriage."[15] As Galena K. Rhoades and Scott M. Stanley put it:

> And once we find "the one," do the choices we make and experiences we have together as a couple before and on the big day influence our ability to have a successful marriage? These questions are important, of course, because about 80 percent of today's young adults report that marriage is an important part of their life plans. (Hymowitz et al. 2013)[16]

College graduates in particular are much more likely to marry and to want to marry than high school dropouts.

What we do leading up to marriage makes a significant difference for our chances of success in marriage. As Rhoades and Stanley put it, "Our first major conclusion challenges what we'll call the Vegas Fallacy—the idea that what happens in Vegas stays in Vegas. Actually, what people do before marriage appears to matter. Specifically, how they conduct their romantic lives before they tie the knot is linked to their odds of having happy marriages."[17] It is a mistake to think that whatever I do in college and prior to marriage doesn't really matter for my future happiness. The Vegas Fallacy is obvious in other areas of life. It is counterproductive to seek a sound financial future in our 30s and 40s by racking up huge credit card debt in our 20s. It is counterproductive for having good health in middle age to eat only McDonalds and never exercise for the decades before. What we do in our 20s matters for who we are in our 30s and 40s.

If you live with your partner before marriage, it is sometimes thought, you'll be better able to evaluate whether marriage to this person is a good idea. If a good marriage is really important to you, if you do not want to get divorced, then cohabitation may be a good way to arrive at these goals. Some people view cohabitation as a testing period before deciding about marriage as a good way to avoid an unhappy marriage and divorce.

The scientific evidence indicates cohabitation is, in fact, linked to higher rates of divorce and more marital problems. Whether this link is a matter of causation or mere correlation, we'll consider later. A 2019 Stanford University study in the *Journal of Marriage and Family* found that "Married couples who did not cohabit before marriage had more stable unions after 5 years of marriage."[18] The researchers found, "The marital survival curves showed that married couples who did not cohabit before marriage had more stable unions after 5 years of marriage consistently across the three marriage cohort categories."[19] Cohabitation, according to these researchers, provides short-term benefits but long-term costs. Indeed, one study found that couples who cohabit are 80% more likely to divorce.[20]

Even if a cohabiting couple does not get divorced after marrying, the research indicates they are likely to have a less successful marriage in comparison to married couples who did not cohabit before getting married. Indeed, the quality of the relationships of couples that cohabit prior to marriage and couples that do not is significantly different on average. One study summarized the evidence in the following way, "Compared to those who did not cohabit premaritally, those who did cohabit have lower marital satisfaction (Stanley et al., 2004), more negative observed communication (Cohan & Kleinbaum, 2002), more conflict, (Thomson & Colella, 1992), more physical violence (Brownridge & Halli, 2000; Stanley et al., 2004), and higher rates of infidelity for wives (Forste & Tanfer, 1996)."[21] Those who do not cohabit have a higher likelihood of higher marital satisfaction, better communication, less conflict, less domestic abuse, and less cheating.

Moreover, the research indicates that the longer the couple cohabits prior to marriage, the more likely the couple is to divorce.[22] Brad Pitt and Angelina Jolie, for example, lived together for nine years and then divorced after just two years of marriage. It may seem counterintuitive, but the empirical evidence indicates that what happened to Brad Pitt and Angelina Jolie, a long period of cohabitation followed by marriage and then divorce, is not surprising.

Particularly likely to divorce are those who engage in "multiple cohabitation," for example, living with one person from 22 to 24, then another person from 26 to 27, and then finally marrying a third person after cohabiting from 29 to 31. The Stanford study found that divorce odds are "4.66 times higher for women who had 2+ cohabitations before marrying their first husband."[23] Why would multiple cohabitation be especially linked to divorce? "A reason for this could be that the experience of dissolving one cohabiting relationship generates a greater willingness to dissolve later relationships. People's tolerance for unhappiness is diminished, and they will scrap a marriage that might otherwise be salvaged."[24] Popenoe and Whitehead point out, "Contrary to popular wisdom, you do not learn to have better relationships from multiple failed cohabiting relationships. In fact, multiple cohabiting is a strong predictor of the failure of future relationships."[25] Since most cohabiting couples do not get married, someone who is evaluating potential marital partners by means of cohabitation will likely engage in multiple cohabitation, thereby putting themselves in the category of those four times more likely to get divorced.

Cohabitation and Divorce: Cause or Mere Association?

Given this empirical evidence, students of social science will ask about what explains the relationship between cohabitation and these negative consequences for men, women, and children. Does cohabitation itself *cause* an increased likelihood of divorce? Or is it the case that the kind of person who *selects* cohabitation is also the kind of person who is more likely to divorce regardless of whether they cohabit before getting married? I suspect that people wearing leather jackets are more likely to die in motorcycle accidents. But there is no reason to hold that wearing leather jackets causes motorcycle accidents. Rather, the kind of person who chooses to ride a motorcycle is also the kind of person who is more likely to wear a leather jacket. If the fashion sense of the Hells Angels were to radically change, you might find a link between wearing a bow tie and dying in a motorcycle accident. Moreover, in some situations, there is bidirectional causality between two things. A smiling face invites a deepening of friendship, and a deepening friendship gives rise to a smiling face.

If cohabitation is merely associated with divorce, if the kinds of people who cohabit are also the kinds of people who divorce, this is important information to consider in making decisions. If you don't want to get divorced, it is reasonable not to marry someone who is more likely to get a divorce. If you don't want to be cheated on, it is reasonable to avoid the kinds of people who are more likely to cheat. If you want a long-lasting relationship, it makes sense to seek those who are more likely to have a lasting relationship. So, if the link between cohabitation and cheating, breaking up, and divorcing is that the kind of person who cohabits is the same kind of person who cheats, breaks up, and divorces, then a wise person takes this information into account when choosing a partner for a long-term relationship.

On the other hand, some scholars think that cohabitation is not simply linked with higher rates of divorce but actually causes higher rates of divorce. Although selection plays a role in the link between cohabitation and marital problems, one study pointed out that, "selectivity does not account for all of the cohabitation effect. For example, Cohan and Kleinbaum (2002) and Kamp Dush et al. (2003) carefully controlled for a number of possible selection variables and continued to find that those who had cohabited premaritally were are higher risk for distress and divorce (also see Amato & Rogers, 1999; Stanley, Amato, Johnson, & Markman, 2006)."[26] In other words, "even when this 'selection effect' is carefully controlled statistically, a negative effect of cohabitation on later marriage stability still remains."[27] In the words of one study, "And no positive contribution of cohabitation to marriage has been ever been found."[28]

Indeed, much scholarly debate is focused not on *whether* cohabitation causes increased likelihood of divorce but on *why* cohabitation causes increased likelihood of divorce. The following explanations have been proposed to explain why cohabitation increases divorce rates.

The first explanation of why cohabitation causes divorce focuses on establishing habits that hinder successful marriages. Habits arise from our actions, and our actions are linked to our identities. To be a husband or a wife is a social identity that comes with social expectations in accordance with these roles. When a couple is not married, they do not have the social identity of spouses. They have not publicly and solemnly promised to love each other in good times and bad, in sickness and in health, until death do them part. So, although they live together, they live out their identities as people who are not yet married. Popenoe and Whitehead point out, "One study found, for example, that 'living with a romantic partner prior to marriage was associated with more negative and less positive problem solving support and behavior during marriage.' A reason for this, the authors suggest, is that because long-term commitment is less certain in cohabitation, 'there may be less motivation for cohabiting partners to develop their conflict resolution and support skills.'"[29] Popenoe and Whitehead also note, "cohabitants tend not to be as committed as married couples in their dedication to the continuation of the relationship and reluctance to terminate it, and they are more oriented toward their own personal autonomy. It is reasonable to speculate, based on these studies, that once this low-commitment, high-autonomy pattern of relating is learned, it becomes hard to unlearn."[30] The more we are habituated to living without commitment and as unmarried persons, the more difficult it becomes to take on the new role of a committed spouse. Popenoe and Whitehead put the point as follows, "The longer you live together with a partner, the more likely it is that the low-commitment ethic of cohabitation will take hold, the opposite of what is required for a successful marriage."[31] This explanation is consistent with the research showing that the longer a couple cohabits prior to marriage, the higher their likelihood of divorce.

A second explanation of why cohabitation causes divorce focuses on cohabitation changing attitudes to marriage. Popenoe and Whitehead write:

> The results of several studies suggest that cohabitation may change partners' attitudes toward the institution of marriage, contributing to either making marriage less likely, or if marriage takes place, less successful. A 1997 longitudinal study conducted by demographers at Pennsylvania State University concluded, for example, "cohabitation increased young people's acceptance of divorce, but other independent living experiences did not." And "the more months of exposure to cohabitation that young people experienced, the less enthusiastic they were toward marriage and childbearing."[32]

The idea is that cohabitation changes attitudes toward marital commitment by making a couple less committed to faithful love. Rosenfeld and Roesler suggest that, “cohabitation teaches people that romantic relationships are impermanent and disposable.”[33] If cohabitation moves couples from thinking of marriage as a bond of unconditional love to thinking of marriage as conditional, temporary, and uncommitted coupling, then couples may act in accordance with their changed attitude. Rosenfeld and Roesler note, “as couples advance from cohabitation to marriage, they advance with their relationship commitment already eroded by the casual and informal nature of cohabitation.”[34] Couples with attitudes to low commitment are more likely to divorce.

A third explanation of why cohabitation causes divorce focuses on shifting attitudes of the potential spouses for one another. In *Sex in the City*, Samantha said, “Honey, before you buy a car, you have to test drive it.” This analogy is revealing. How do we treat things that we buy as opposed to things that we test drive in a rental? Dan Gilbert explains, “Committed owners attend to a car’s virtues and overlook its flaws, thus cooking the facts to produce a banquet of satisfaction, but the [rental test driver, potential] buyer for whom escape is still possible (and whose defenses are not yet triggered) is likely to evaluate the new car more critically, paying special attention to its imperfections as [he or she] tries to decide whether to keep it.”[35] How then could this apply to cohabitation? Committed spouses attend to their spouse’s virtues and overlook their flaws, thus cooking the facts to produce a banquet of satisfaction. By contrast, cohabiters are likely to evaluate the potential spouses more critically, paying special attention to imperfections as they try to decide whether to marry them. The longer this critical attitude continues, the more deeply ingrained it will become. As one woman put it, “I felt like I was on this multiyear, never-ending audition to be his wife.”[36]

Why does cohabitation cause divorce? A fourth explanation is sometimes called “sliding, not deciding.” Galena K. Rhoades, Scott M. Stanley, and Howard J. Markman describe this explanation in the following way:

> the average couple would likely find it harder to break up if they were cohabiting than if they were dating and not cohabiting, all other things being equal, because cohabitation involves a higher level of constraints. For example, cohabitation may involve increased financial commitments (e.g., a lease), increased difficulty to move on (e.g., moving out and finding another place to live), and increased social pressure to stay together (e.g., friends and family beginning to expect more of the relationship, including, in many cases, marriage). Thus, cohabitation could increase the likelihood of marriage, even among couples who are at higher risk for divorce or marital distress.[37]

After living together, particularly after many years, the ties that bind the couple multiply. “Cohabiting couples buy furniture together, adopt pets, sign leases, and get used to a routine of living in a certain place together, all constraints that may keep people in a relationship even when they are not sure they want to stay. Some may therefore slide into a marriage that they would have otherwise avoided.”[38] Given the shared furniture they bought, their dog, a cell phone plan, and a baby, the couple then slides into marriage thinking, "well, we might as well get married." But the man or the woman would have never settled for this partner or a relationship of this quality had they not been so constrained by the mounting cost of breaking up. Unfortunately, in many cases, the quality of the partner or the relationship ends up in the end not being good enough to sustain the marriage. Galena K. Rhoades, Scott M. Stanley, and Howard J. Markman note:

> couples who slide into cohabitation may not only be risking the increase of constraints prior to the full development of dedication, they may be building a relationship on a shakier foundation to begin with merely by the absence of a clear, focused commitment at a critical phase of relationship development. A stronger commitment, based on dedication, may more fully support motivation to follow-through during the inevitable tougher times that many, if not most, relationships will experience.[39]

Almost all marital relationships have times of trouble. When someone makes a free commitment to love his or her spouse, that person is more likely to carry out this marital commitment.[40] By contrast, the cohabiting couple may experience less freedom given the mounting costs of breaking up. The choice of cohabitors is already somewhat constrained, since breaking up involves for them not just breaking up, but questions about who gets the furniture, what to do about the cell phone plan, the dog, and perhaps even child custody issues. By contrast, couples who don't cohabit make an explicit decision to marry and live together. "Why would making a decision matter? Making decisions consciously is related to a stronger commitment to follow through on those choices."[41] What we vow to do, we have an extra incentive to do, lest we see ourselves as inconsistent and become viewed by others as untrustworthy.

How then can we evaluate a potential husband or wife if not by cohabitation? Rhoades and Stanley note, "cohabiting may not be a good test of a relationship because constraints (e.g., merging finances, sharing friends, adopting a pet) may make it more difficult to end the relationship when you realize a partner has failed the test. Other ways to test a relationship, without increasing constraints, may be more beneficial: plan a trip together, meet each other's parents, observe your boyfriend/girlfriend in many different settings, or seek other people's opinions."[42] If the research on cohabitation is sound, then cohabitation does not appear to be a good way of evaluating a potential spouse. Indeed, as Popenoe and Whitehead point out, "There is no evidence that if you decide to cohabit before marriage you will have a stronger marriage than those who don't live together."[43]

The "sliding, not deciding" explanation also explains why the longer a couple cohabits, the more likely they are to get divorced. Such couples have built up a life together over years, and so they eventually slide into marriage. But after sliding into marriage, the quality of the relationship is not high enough to make the marriage endure.

Does Cohabitation Affect Men and Women Equally?

In some respects, cohabitation affects men and women equally. After all, if a cohabiting couple later marries and then divorces, both the man and the woman are divorced. But in other respects, cohabitation results in significant disparities between the sexes.

It is important to begin by noting differences on average between men and women with respect to what cohabitation means. In his essay "Men and Women Often Expect Different Things When They Move In Together," W. Bradford Wilcox points out:

> the majority of cohabiting young men do not endorse the maximum indicator of relationship permanence: 52 percent of cohabiting men between ages 18 and 26 are not "almost certain" that their relationship is permanent. Moreover, a large minority (41 percent) of men report that they are not "completely committed" to their live-in girlfriends. By contrast, only 39 percent of cohabiting women in the same age group are not "almost certain" their relationship will go the distance, and only 26 percent say they are not "completely committed."[44]

On the other hand, "Women tend to see it as a step toward eventual marriage, while men regard it more as a sexual opportunity without the ties of long-term commitment."[45] Such differences between men and women may partially explain the higher levels of conflict for cohabitors in comparison to spouses. Scott M. Stanley, Galena Kline Rhoades, and Howard J. Markman argue that cohabitation places women at greater risk:

> women may have more to lose by sliding through relationship transitions. Women get pregnant, not men; women disproportionately do the work of raising those children when the relationship ends. Women are also more likely than men to be trapped in the most dangerous types of domestically violent relationships. Further, another line of research suggests that men's attitudes about sacrificing for their female partners are far less associated with levels of commitment to the future than those of women.[46]

These considerations suggest an asymmetry when a young man and a young woman cohabit that subjects her to greater risk for adverse outcomes than him. But there is another gender asymmetry worth exploring.

I know a couple whom I'll call Jessica and Bill. They both wish to get married someday, and they moved in together when they were both 25 years old. After seven years of living together, they broke up. Since most cohabiting couples end up not getting married, the case of Jessica and Bill is typical. "After 5 to 7 years, 39% of all cohabiting couples have broken their relationship, 40% have married (although the marriage might not have lasted), and only 21% are still cohabiting."[47] After breaking up, Jessica and Bill begin the process of looking for someone else to marry.

When looking for a marriage partner, men and women both highly value friendliness and intelligence. But men and women on average differ in terms of how they value other characteristics in a potential marriage partner. A 2020 study in the *Journal of Psychological Science* surveyed men and women in 45 countries. They found, "Support for universal sex differences in preferences remains robust. Men, more than women, prefer attractive, young mates, and women, more than men, prefer older mates with financial prospects."[48] In their book, *Premarital Sex in America*, Mark Regnerus and Jeremy Uecker note, "For women, age is a debit in sexual economics theory, gradually drawing down physical attractiveness and perceived fertility. For men, age is a credit, heightening their access to resources and making them more attractive to women. It's also a more reliable predictor of maturity. Young women are generally more interested in settling into a stable and secure relationship than men of the same age are, and women often presume that older men have already acquired some relationship skills."[49] A 2019 study found, "On average, women show stronger preferences for mates with good earning capacity than men do, while men show stronger preferences for physically attractive mates than women do."[50]

Do these differences in mate preferences result from male-dominated societies where women cannot earn well? Do these differences in mate preferences decrease in countries with greater gender equality like Norway, Sweden, and Denmark? You might ask, as did a 2019 study, "Are sex differences in preferences for physical attractiveness and good earning capacity in potential mates smaller in countries with greater gender equality?" The scholars looking at the evidence came to the following conclusion:

> Although women preferred mates with good earning capacity more than men did and men preferred physically attractive mates more than women did, we found little evidence that these sex differences were smaller in countries with greater gender equality [like Norway, Sweden, and Denmark]. ... Collectively, these results provide little support for the social roles account of sex differences in mate preferences.[51]

In other words, even in Nordic countries with greater gender equality, women still indicated stronger preferences for marriage partners with good earning capacity than men, and men still desired physically attractive mates more than women do.

If evolution has shaped sexual desires in human beings, these findings are unsurprising. Evolution accounts for why men more than women would desire youth and beauty in their marriage partners. As David Buss points out in his book *The Evolution of Desire: Strategies of Human Mating*, the fertility of women reaches its peak in their 20s. Men who desired young women would be more likely to pass on their genes than men who preferred women in their 50s. Since men can reproduce even in old age, seeking a youthful spouse is not as much priority in terms of female reproductive strategies. On the other hand, women who mated with men who could support them and their children were more likely to raise children to the age when the children could reproduce. Evolutionary psychologists point out that evolutionary pressures, shaping the human species over countless generations, have produced differences between male and female coupling strategies.[52]

Let's return to the case of Jessica and Bill. After seven years of living together, they break up and each starts looking for a new partner to marry. In the course of their years living together, Bill gained more of what most women value in a marriage partner. At 25, he was a graduate student making $25,000 a year. At 32, he was a tenure track professor making $78,000 a year. Bill is now more mature, has a better job, and is better able to provide for a family. He has more of what most women are looking for in a marriage partner. But Jessica has not advanced her position in the marriage market in terms of what men especially desire in a mate. At 32, she has less youth and beauty than she did at 25. So, relative to where they began to cohabit, Bill has secured an advantage in the marriage market, and Jessica suffers a disadvantage in the marriage market. Cohabitation, in other words, works to the systematic disadvantage of women and the advantage of men in the marriage market.

Conclusion

Given the evidence referred to in this chapter, cohabitation endangers the well-being of those who engage in it. Cohabitation is not in reality a good alternative for marriage. Cohabiting couples, on average, have more infidelity, violence, and breaking up than married couples. Cohabitation as a preparation for marriage also fails to deliver on the promise of lower rates of divorce and more satisfying marriages. Living together before marriage does not in fact lower a person's likelihood of divorce but rather increases it. Couples that cohabit before marriage report lower relationship satisfaction, more infidelity, and more negative styles of communication. Moreover, cohabitation is particularly detrimental to women looking to marry who are put at a disadvantage in comparison to men should the relationship not end in marriage (the most common result) and both parties seek newer partners to marry.

How does this relate to morality? It is prima facie wrong to substantially risk lowering human well-being (the endangerment principle). If the research I've cited in this article is correct, cohabitation (both as an alternative to marriage and as a preparation for marriage) substantially risks human well-being. Cohabitation as an alternative to marriage is a form of living more likely to result in relationship dissolution, physical violence, drug abuse, and abuse of children. Cohabitation as a preparation for marriage is more likely to lead to divorce, negative styles of taking, infidelity in the wife, lack of commitment in the husband, and a lower-quality marital relationship. Moreover, this form of cohabitation is particularly detrimental to women who typically lose in the course of cohabitation more of what most men particularly care about in a potential marriage partner, leaving women at a disadvantage (in comparison to their male cohabitors) in the marriage market. If the empirical evidence cited in this chapter and the endangerment principle are correct, cohabitation is prima facie wrong.

Notes

1. David Popenoe and Barbara Defoe Whitehead, *Should We Live Together? What Young Adults Need to Know About Cohabitation Before Marriage*, 2nd Ed. (The National Marriage Project: Rutgers, The State University of New Jersey, 2002) p.6. The empirical research on this topic has focused on opposite sex couples, so my remarks will be limited to cohabitation of opposite sex couples.
2. Glenn Stanton, *Why Marriage Matters: Reasons to Believe in Marriage in Postmodern Society* (Colorado Springs, CO: NavPress, 1997), 56.
3. Popenoe and Whitehead, *Should We Live Together?*
4. As cited in, Scott Stanley, "Cohabitation is Pervasive" June 20, 2018. https://ifstudies.org/blog/cohabitation-is-pervasive.
5. Popenoe and Whitehead, *Should We Live Together?* p.17.
6. Popenoe and Whitehead, *Should We Live Together?* p.8.
7. Popenoe and Whitehead, *Should We Live Together?* p.9, internal notes omitted.
8. Popenoe and Whitehead, *Should We Live Together?* p.4.
9. Popenoe and Whitehead, *Should We Live Together?* p.10.
10. W. Bradford Wilcox, "Why the Ring Matters," *The New York Times*, June 30, 2011.
11. Rhoades and Stanley, *Before "I Do"* p.13.
12. Rhoades and Stanley, *Before "I Do"* p.14.
13. Robert B. Cialdini, *Influence: The Psychology of Persuasion* (New York: HarperBusiness, 2006), 92.
14. Popenoe and Whitehead, *Should We Live Together?* p.9.
15. Popenoe and Whitehead, *Should We Live Together?* p.18.
16. Galena K. Rhoades and Scott M. Stanley, *Before "I Do" What Do Premarital Experiences Have to Do with Marital Quality Among Today's Young Adults?* (The National Marriage Project, University of Virginia, 2014) p.4.
17. Galena K. Rhoades and Scott M. Stanley, *Before "I Do" What Do Premarital Experiences Have to Do with Marital Quality Among Today's Young Adults?* (The National Marriage Project, University of Virginia, 2014) p.5.
18. Michael J. Rosenfeld and Katharina Roesler, "Cohabitation Experience and Cohabitation's Association with Marital Dissolution." *Journal of Marriage and Family* 81.1 (2019): 42–58, at p.50.
19. Ibid.
20. Glenn Stanton, *Why Marriage Matters: Reasons to Believe in Marriage in Postmodern Society* (Colorado Springs, Colo.: NavPress, 1997), p.59.
21. Galena K. Rhoades, Scott M. Stanley, and Howard J. Markman. "Working with Cohabitation in Relationship Education and Therapy." *Journal of Couple & Relationship Therapy* 8.2 (2009): 95–112.
22. Popenoe and Whitehead, *Should We Live Together?*, p.15.
23. Rosenfeld and Roesler, "Cohabitation Experience and Cohabitation's Association with Marital Dissolution." p.54.
24. Popenoe and Whitehead, *Should We Live Together?* p.7.

25. Popenoe and Whitehead, *Should We Live Together?* p.4.
26. Galena K. Rhoades, Scott M. Stanley, and Howard J. Markman. "Working with Cohabitation in Relationship Education and Therapy." *Journal of Couple & Relationship Therapy* 8.2 (2009): 95–112, at 96.
27. Popenoe and Whitehead, *Should We Live Together?* p.5.
28. Popenoe and Whitehead, *Should We Live Together?* p.7.
29. Popenoe and Whitehead, *Should We Live Together?* p.7.
30. Popenoe and Whitehead, *Should We Live Together?* p.7.
31. Popenoe and Whitehead, *Should We Live Together?* p.4.
32. Popenoe and Whitehead, *Should We Live Together?* p.5.
33. Rosenfeld and Roesler, "Cohabitation Experience and Cohabitation's Association with Marital Dissolution." P.43.
34. Rosenfeld and Roesler, "Cohabitation Experience and Cohabitation's Association with Marital Dissolution." P.43.
35. Daniel Gilbert, *Stumbling on Happiness* (New York: Vintage, 2007), 204.
36. Meg Jay, "The Downside of Cohabiting before Marriage," *New York Times*, April 15, 2012, SR4.
37. Galena K. Rhoades, Scott M. Stanley, and Howard J. Markman. "Working with Cohabitation in Relationship Education and Therapy." *Journal of Couple & Relationship Therapy* 8.2 (2009): 95–112, at 99.
38. Rhoades and Stanley, *Before "I Do"* p.10.
39. Galena K. Rhoades, Scott M. Stanley, and Howard J. Markman. "Working with Cohabitation in Relationship Education and Therapy." *Journal of Couple & Relationship Therapy* 8.2 (2009): 95–112, at p.101.
40. Robert B. Cialdini, *Influence: The Psychology of Persuasion* (New York: HarperBusiness, 2006), 92.
41. Rhoades and Stanley, *Before "I Do"* p.10.
42. Rhoades and Stanley, *Before "I Do"* p.11.
43. Popenoe and Whitehead, *Should We Live Together?* p.4.
44. W. Bradford Wilcox, "Men and Women Often Expect Different Things When They Move In Together" *The Atlantic* July 8, 2013.
45. Popenoe and Whitehead, *Should We Live Together?* p.17.
46. Stanley, Scott M., Galena Kline Rhoades, and Howard J. Markman. "Sliding Versus Deciding: Inertia and the Premarital Cohabitation Effect." *Family Relations* 55.4 (2006): 499–509, at p.508.
47. Popenoe and Whitehead, Should We Live Together?, 2.
48. K. V Walter, et al. "Sex Differences in Mate Preferences across 45 Countries: A Large-Scale Replication." *Psychological Science* 31.4 (2020): 408–423, at 408.
49. Mark Regnerus and Jeremy Uecker, *Premarital Sex in America: How Young Americans Meet, Mate, and Think About Marrying* (New York: Oxford University Press, 2011), p.78.
50. Lingshan Zhang, et al. "Are Sex Differences in Preferences for Physical Attractiveness and Good Earning Capacity in Potential Mates Smaller in Countries with Greater Gender Equality?" *Evolutionary Psychology* 17.2 (2019): 1474704919852921.
51. Lingshan Zhang, et al. "Are Sex Differences in Preferences for Physical Attractiveness and Good Earning Capacity in Potential Mates Smaller in Countries with Greater Gender Equality?" *Evolutionary Psychology* 17.2 (2019): 1474704919852921.
52. David M. Buss and David P. Schmitt. "Mate Preferences and Their Behavioral Manifestations." *Annual Review of Psychology* 70.1 (2019): 77–110.

CHAPTER 11

Why Is Sexual Assault Special?: Transactional Sex and Sacred Intuitions

Francis Joseph Beckwith

Sexual assault and sexual harassment are grievous wrongs that deserve more severe punishments than their non-sexual counterparts. At the same time, for those of us who live in North America or Western Europe, we inhabit cultures that have gone to great lengths in demystifying and disenchanting sex, treating it as just another recreational activity. This has given us everything from college hook-up culture, sexting, social network "dating" services like Grindr, and an endless supply of free online hardcore pornography, accessible to every man, woman, and child with a computer, smart phone, or tablet. In this chapter, I am going to argue that our intuitions about the seriousness of sexual assault and sexual harassment are in tension with contemporary political and cultural trends that imply that sexual activity is just one of many purely transactional endeavors. When I say that an activity is merely transactional I am saying that its moral impermissibility (or whether it should be legally proscribed) depends entirely on whether the parties in the agreement choose freely, have ownership of the goods they are exchanging (e.g., their bodies, financial instruments, etc.), and are satisfied by the results. Take, for example, the viaticals market,[1] which arose in the late 1980s and early 1990s at the beginning of the AIDS epidemic before the vast improvement of treatment for the disease. (In those days, an AIDS diagnosis was a death sentence.) An investor pays an AIDS sufferer (or someone else with a terminal illness) a lump sum of cash and promises to pay the premiums of his life insurance policy in exchange for him designating the investor as the policy's beneficiary. Imagine a terminal patient is told

F. J. Beckwith (✉)
Baylor University, Waco, TX, USA
e-mail: Francis_Beckwith@baylor.edu

D. Boonin (ed.), *The Palgrave Handbook of Sexual Ethics*,
https://doi.org/10.1007/978-3-030-87786-6_11

that he has one year to live but needs about $45,000 to purchase pharmaceuticals and medical treatment so that he can live comfortably for his remaining days but does not have the cash on hand. The investor gives the patient the $45,000, pays his premiums, and collects the $100,000 benefit upon the patient's death. Of course, if the patient happens to recover and lives another 30 or 40 years, the investor will not collect on his investment. So, for the investor to make the most of his investment he has to hope for the patient's imminent demise. If you think, as some people do, that the viaticals market is deeply immoral, then the fact that both parties (patient and investor) are acting freely, have ownership over the goods they are exchanging, and are satisfied by the results is not sufficient to determine whether the activity is morally permissible. In other words, you do not think that the activity is *merely transactional*. In the same way, if you think that the moral permissibility of sexual relations depends exclusively on whether it is a free exchange of goods that satisfies both parties, then you think that sex is merely transactional. But in that case, as I will argue, this view is in tension with our intuitions about the serious wrongness of sexual assault and sexual harassment.

Harvey Weinstein and the Rise of the MeToo Movement

In October 2017, several accusations of sexual harassment were leveled against movie mogul Harvey Weinstein. According to a report in the BBC, "among the accusations are that he forced women to massage him and watch him naked. He also promised to help advance their careers in return for sexual favours."[2] Over the following weeks not only did many other women come forward to level similar accusations against Weinstein, a variety of other celebrities and high profile individuals were credibly accused of sexual misconduct including sexual assault and using their power and influence to procure sexual favors from underlings.[3] As more and more victims told their stories, the MeToo movement was born and became an international phenomenon. The social media tsunami that followed was so enormous that by October 2018 the hashtag #MeToo had been tweeted over 19 million times![4]

There is little doubt that the growth, scope, and power of the MeToo movement is the result of once-silent victims who now feel liberated to tell their stories in public, as well as millions of others who, after hearing those stories, experience visceral moral revulsion at the thought of powerful predators sexually assaulting or extracting sexual favors from unwilling or pressured victims with little or no power. Although this moral reaction seems so obviously justified to virtually everyone (including this writer), given the contemporary cultural trends mentioned above—trends that seem to teach that sex is mere recreation—it seems natural to raise the following questions: Why is *sexual* assault special? Why believe that *sexual* assault and transactional *sexual* favors are worse than their non-sexual counterparts?[5] To better understand the importance of these questions, and why trying to answer them highlights the tensions between sexually libertine cultural trends and our moral revulsion to sexual misconduct, consider the following three sets of fictional scenarios.

Weinstein's Quid Pro Quo Chef

Imagine that Weinstein is auditioning a young out-of-work, though talented, actress named Janet to play a supporting role in his upcoming film, *The Philosopher*. She realizes that if she lands the role, it could easily become her big break. Weinstein realizes that as well. So, he makes his offer: "I really think you're perfect for this role. However, there are a few other, more established, actresses who would probably be a better draw at the box office. So, I can only offer you $250,000 for the part, which is considerably less than what I can offer someone else. But I can give you more—let's say $400,000—if you agree to become my personal chef for a year. You can come to my home and prepare dinner for me and my family once a week for 52 weeks. I've heard from your references that you are an outstanding chef." Janet replies, "Did you make the same offer to the other finalists for this role?" Weinstein says, "Yes, three declined, and the fourth said, 'yes,' but since I heard that you were a better chef I wanted to wait for your answer." Janet agrees to the offer.

Is there anything morally wrong with this arrangement? It is certainly possible to say that using studio money to pay for personal culinary work is a violation of business ethics (unless it is contractually permissible or Weinstein owns the studio), but very few would say they were morally revolted by the agreement as they would if Weinstein had made this offer to Janet: "I really think you're perfect for this role. However, there are a few other, more established, actresses who would probably be a better draw at the box office. So, I can only offer you $250,000 for the part, which is considerably less than what I can offer someone else. But I can give you more—let's say $400,000—if you agree to become my personal lover for a year. You can come to my home and have sex with me once a week for 52 weeks. I've heard from your former boyfriends that you are really good in bed." In the first case, Janet is asked to make dinner once a week for a year in exchange for a movie role, but in the second case she is asked to make love once a week for a year in exchange for a movie role.

What precisely is the difference between these cases? What is it about the second case that elicits revulsion that is absent from the first case? It can't be, as some scholars may argue,[6] that in the second case Weinstein is using his power relationship while in the first he isn't. He is using his power relationship in both. In fact, he could use his power relationship with Janet to indirectly procure sex in a third case that would not elicit the same revulsion we sense when we first hear the second case. In this third case Weinstein makes the same request as he does in the first, except with this caveat: "Here's the ultimate reason why I want to procure your personal services as a chef: my wife loves well-prepared food so much that she will not have sex with me after she consumes a mediocre meal. But I need to have sex at least once a week. So, if you are as good a chef as your former boyfriends claim, the meals you prepare will serve as an instrumental cause for my sexual satisfaction."[7] No doubt this caveat will sound creepy to most people. But that would have more to do with the imprudence of Weinstein revealing a motive that he should have kept private than it would have to do with the morality of the motive itself. It would be like an amorous husband blurting out to the flower shop lady that her design for the arrangement he bought for his wife is "certain to get him laid." To be sure, such an utterance is imprudent, crude, and awkward, but the motive revealed in the utterance in and of itself is not morally revolting.

Weinstein and His Goon Squad

Weinstein was not only credibly accused of proposing quid pro quo sexual arrangements but also of committing sexual assault including rape.[8] The Federal Bureau of Investigation (FBI) defines rape as "penetration, no matter how slight, of the vagina or anus with any body part or object, or oral penetration by a sex organ of another person, without the consent of the victim."[9] There are, of course, other types of assault besides rape. Suppose, for example, Weinstein and his security detail, like mafia goons, physically intimidate (in a non-sexual way) one of Weinstein's prospective actresses, suggesting by their actions that she would suffer physical harm if she did not agree to accept a bit part in *The Philosopher.* Under the common law, this would count as an assault. According to the FBI, an assault is "an attempt with force or violence to do a corporal injury to another; may consist of any act tending to such corporal injury, accompanied with such circumstances as denotes at the time an intention, coupled with present ability, of using actual violence against the person."[10] Now suppose that Weinstein and his goon squad corner the actress Cate Blanchett in an alley, threaten her, and rough her up a little to get her to agree to play the leading female role in *The Philosopher*. The "roughing up" consists of non-consensual pushing and grabbing of Blanchett's arms and a slap in the face. Although she suffers no physical injury, Blanchett has been assaulted. But it would not be a *sexual* assault. Now imagine instead that the "roughing up" consists exclusively of the non-consensual touching of Blanchett's breasts and genitalia with the hands and genitalia of Weinstein and members of his goon squad. This would indeed be a *sexual* assault.

Although both the non-sexual and sexual assaults are immoral and a violation of Blanchett's rights, it would seem to most everyone that the latter is far worse. In fact, this widely held intuition tracks with how the criminal law punishes non-sexual and sexual assaults. Under Texas law, for example, sexual assault is "a felony of the second degree,"[11] meaning that if an individual is convicted of the crime he or she "shall be punished by imprisonment ... for any term of not more than 20 years or less than 2 years" and in addition "may be punished by a fine not to exceed $10,000."[12] But if the sexual assault results in bodily harm or involves kidnapping or impairing the victim, it becomes an aggravated assault, which is a "felony of the first degree."[13] This is punishable "by imprisonment ... for life or for any term of not more than 99 years or less than 5 years," and in addition the assailant "may be punished by a fine not to exceed $10,000."[14] On the other hand, ordinary non-sexual assault—as long as it does not cause "bodily injury to another"[15]—is virtually always "a Class C misdemeanor,"[16] which "shall be punished by a fine not to exceed $500" with no jail time.[17]

Now imagine a third assault on Blanchett that is exactly like the second except instead of penetrating and fondling her genitalia, Weinstein and his goons lick their index fingers and place them in Blanchett's ears, giving her about a dozen "wet willies."[18] Just as in the second case, the assailants employ their appendages to fondle and penetrate one of the victim's orifices. But they have not engaged in *sexual* assault. Most everyone's intuitions suggest—in line with the criminal law—that being wet-willed is an assault and that Weinstein and his goons should be punished for their crime, but no one believes that they should spend up to 20 years, let alone 99 years, in prison for committing the assault. So, the difference between decades in prison and a mere fine depends entirely on where and with what the victim is touched.

Weinstein the Promise Breaker

Imagine that the second Weinstein scenario comes to pass, that Janet agrees to sexual relations with Weinstein once a week for 52 weeks in exchange for the movie role and a higher salary. According to the MeToo Movement this quid pro quo arrangement is immoral because it is not really consensual due to the power imbalance between movie mogul and ambitious actress. But there's more to this contrived sexual affair than Janet and Weinstein. There is at least one other person that has a moral stake in this arrangement: Weinstein's wife. Suppose that about two months into the affair Weinstein's wife grows suspicious that her husband is committing adultery. She begins to notice that every Friday afternoon between 2 and 4 pm he needs to "go to the office and get some work done," even though when she calls the office Weinstein's colleagues cannot seem to find him. After rummaging through his email and texts, she finds evidence of his agreement with Janet and then confronts him: "You bastard! You're having an affair, aren't you?" Weinstein confesses and asks for his wife's forgiveness. Not able to forgive him, she files for divorce. He violated his promise, made on their wedding day, that he would remain true to her and "forsake all others." His betrayal is just too great for her to overcome.

Now imagine that the third Weinstein scenario comes to pass, that Janet agrees to prepare dinner for Weinstein and his family once a week for 52 weeks in exchange for the movie role and a higher salary so that Weinstein can increase the frequency of sex with his wife. But Weinstein's wife is not entirely pleased with the arrangement. She had told her husband to seriously consider hiring her sister's personal chef, Michael. Although Weinstein promised his wife that he would interview Michael, he did not follow through with the promise.

When she asks him to explain what happened with Michael, Weinstein replies, "I talked with Michael last week about cooking for us. Unfortunately, he is right now overbooked and could not possibly include us in his schedule." Because Weinstein believes that Janet is a much better chef than Michael, and thus hiring her improves his chances of having sex with his wife more frequently, he intentionally does not contact Michael.

Suppose that after about two months of Janet cooking for the Weinstein family, Weinstein's wife grows suspicious that her husband broke his promise. She begins to notice that every time she brings up Michael, Weinstein gets fidgety and tells her that he needs to "go to the office and get some work done," even when all the studio's projects are going smoothly. After rummaging through his email and texts, she finds evidence that he never spoke to Michael. She then confronts him: "You bastard! You never spoke to Michael, did you?" She reminds him that on the day that they were married they made a solemn promise, in addition to their ordinary vows, that they would never lie to each other. Overwhelmed with guilt, Weinstein confesses and asks for his wife's forgiveness. Not able to forgive him, she files for divorce. For he not only violated the solemn promise they made on their wedding day, he violated his promise, made only two months prior, that he would interview Michael for the chef's position. His betrayal is just too great for her to overcome.

In both cases, Weinstein broke his promise to his wife. In the first he broke his promise to "forsake all others," to be faithful to her in all matters concerning sexual intimacy. In the second he broke two promises: that he would never lie to her and that he would interview Michael for the position of family chef. Yet, virtually no one would judge these cases of promise breaking as having equal gravity. There just seems to be something sacred about the sexual compact that is not present in either the promise not to lie or the promise to interview Michael. (When I use the term "sacred" I do not necessarily mean it in the religious sense. My understanding of the term is best captured by the definition offered by social psychologists Jesse Graham and Jonathan Haidt.[19]) Adultery seems to most people as a morally justified grounds for divorce.[20] On the other hand, although most would agree that it is generally immoral to break a promise never to lie, divorce seems to be a disproportionate punishment for the offense, especially if that's the first promise the offender ever broke![21]

What we have seen in these three sets of scenarios—Weinstein's Quid Pro Quo Chef, Weinstein and His Goon Squad, and Weinstein the Promise Breaker—is that when sexual activity is treated as a quid pro quo commodity, is integral to an assault, or is procured in violation of a marital promise, we typically issue a much harsher moral judgment than we would if it were a parallel case involving a non-sexual activity. This is reflected most especially in our criminal statutes and institutional rules that more severely punish sexual assaults and sexual harassment than other types of assaults and harassment. This, I believe, is a good thing. But this relies on a view of sexual activity—that it is sacred and not merely transactional—that runs contrary to prevailing cultural trends.

Sexual Activity as Transactional: What the Culture Is Telling Us

Pervading most of popular culture, and a great deal of academic culture, is the view that sex is at root purely transactional, and thus one may engage in it solely for pleasure or as a means to a variety of ends that citizens may choose in a liberal polity that should be neutral as to what constitutes the good life. In most popular sitcoms, movies, and songs, characters and performers often treat sex as if it were mere recreation, though occasionally engaged in by persons who actually love each other. But love—or anything remotely resembling a permanent commitment—is rarely if ever offered as a necessary condition for sexual relations. For this reason, under a transactional view, casual sex (or promiscuity) is not inherently wrong. Take, for example, these lyrics from the huge 2020 hit song by Cardi B. and Megan Thee Stallion, "WAP:"

> I said, certified freak. Seven days a week. Wet a-- p---y. Make that pull-out game weak, woo (Ah).
>
> Yeah, yeah, yeah, yeah. Yeah, you f---ing with some wet a-- p---y. Bring a bucket and a mop for this wet a-- p---y. Give me everything you got for this wet a-- p---y ....
>
> Talk your sh--, bite your lip. Ask for a car while you ride that di-- (while you ride that di--). You really ain't never gotta f--- him for a thang. He already made his mind up 'fore he came. Now get your boots and your coat. For this wet a-- p---y. He bought a phone just for pictures. Of this wet a-- p---y. Pay my tuition just to kiss me. On this wet a-- p---y. Now make it rain if you wanna. See some wet a-- p---y.[22]

According to *Billboard*, in its first week "the song drew 93 million U.S. streams," which was "the most for any track ever in its first week of release."[23] A reviewer in the *Los Angeles Times* praises the song as laying "out an astonishing array of boasts and desires from two female rappers proud to follow in the sex-positive footsteps of Lil Kim, Khia, Foxy Brown and Trina."[24] After noting that a Republican congressional candidate, James P. Bradley, was critical of the performance, and acknowledging that "not everyone was won over by song,"[25] the writer quotes a response from a Twitter user who points "out that Rick Ross and 2 Chainz ... rapped explicitly about sex without causing anywhere near the hubbub that Cardi and Megan did."[26] Attributing these differing judgments to "the gendered expectations we have of artists," the implication is that female performers who celebrate transactional sex like their male counterparts have advanced the cause of women's equality: "In an age when the leader of Bradley's party [President Donald Trump] brags about grabbing women by their private parts, women rapping about their parts carries a political weight that men rapping about theirs doesn't."[27] So, apparently, it is a source of female empowerment for women artists to extol transactional sex because of the goods and commodities they may acquire from men as a result of engaging in it. It turns out then that sexual quid pro quos are not always wrong if they're practiced for a good cause, such as deconstructing gendered expectations.

In the academic world, the transactional view of sex is often presumed to be the correct one. The rules that govern sexual conduct between students at most universities and colleges—with the exception of those with a conservative religious mission—require nothing more than consent between the participants. Title IX bureaucracies that were created by colleges and universities to prosecute assailants and protect victims of sexual harassment, sexual assault, and so on, assume the transactional view, most always peppered with a few restrictions for the purpose of avoiding the sorts of conflicts of interest that may occur in non-sexual business relationships. Take, for example, this passage from the University of Texas' policy on consensual relationships:

> The University recognizes that consenting individuals associated with the University should be free to enter into personal relationships of their choice. At the same time, such relationships must not put at risk the fundamental interest of every member of the University community to participate in University activities free from conflicts of interests, favoritism, and/or exploitation. Romantic relationships between certain categories of individuals affiliated with the University risks [*sic*] undermining the essential educational purpose of the University and can disrupt the workplace and learning environment.[28]

This is why a growing number of universities and colleges prohibit sex between a faculty member and a student unless they are married to each other.[29] When such cases do arise, the faculty member's employment is sometimes terminated. This is not because the institution believes there is anything intrinsically wrong about casual non-marital sex, but rather, it is because of the imbalanced power relationship between faculty member and student.[30] If, let's say, a married 60-year-old male professor at the University of Texas were to hook up for a one-night stand with a 25-year-old Texas A & M female graduate student he meets at an academic conference, the University of Texas would have no grounds for his dismissal. Under this policy—in which the goal is to eliminate "conflicts of interests, favoritism, and/or exploitation"—having consensual sex with one's student is morally equivalent to a department chair hiring her husband's restaurant to cater the department's receptions or the graduate director telling his doctoral assistant that he must mow his lawn once a month as one of his fellowship duties.[31]

Perhaps the most influential philosophical defense of what I call the transactional view can be found in the work of Alan H. Goldman.[32] He argues that once one eliminates the external goals that agents impute to their sexual activity—for example, love, reproduction, communication, and interpersonal awareness—it is at bottom "a desire for contact with another person's body and for the pleasure which such contact produces; sexual activity is activity which tends to fulfill such desire of the agent."[33] Just as eating and exercise can be enjoyed morally apart from the goals of nutrition and fitness, sexual activity can be enjoyed morally apart from the variety of ends that agents may attribute to it.[34] For this reason, "there is no morality special to sex: no act is wrong simply because of its sexual nature. Sexual morality consists in moral considerations that are relevant elsewhere as well being applied to sexual activity or relations. This is because the proper concept of sexual activity is morally neutral."[35] But that doesn't mean, argues Goldman, that there are not moral rules proper to sex, such as prohibitions against rape, since rape by its nature requires sexual activity. However, "the immorality of rape derives from its being an extreme violation of a person's body, of the right not to be humiliated, and of the general moral prohibition against using other persons against their wills, not from the fact that it is a sexual act."[36]

Goldman also mentions the immorality of having sex with subordinates, minors, and so on, arguing that "such relations are … exploitative and in a slightly less obvious way coercive."[37] So, for Goldman, as with The University of Texas (and I suspect Cardi B. and Megan Thee Stallion), what would make sexual conduct immoral is what would make any act immoral in a community committed to neutrality on virtually all matters of personal conduct: "Primarily that forced on others without their rational consent."[38] Yet, as we saw in the three sets of scenarios, rational consent by itself cannot do most of the heavy lifting. There still seems to be something irreducibly sacred and special about sex that cannot be captured by our liberal intuitions about choice and consent. We just know that enduring unwelcomed wet willies is not the same as enduring unwelcomed touching of one's private parts, even if the latter were momentary and the former lasted for several minutes.

Recall the scenario of Weinstein the Promise Breaker. There I made the observation that most everyone has the intuition that breaking a marital promise to be sexually faithful to one's spouse is more morally wrong than breaking a solemn promise (made as part of one's wedding day) to never lie to your spouse, and *much more* morally wrong than breaking a promise to your spouse (made two months earlier) to interview the chef of your spouse's sister. I came up with this illustration to respond to Goldman's claim that "the desire for sexual exclusivity is not … irrational or wrong, and where such an understanding or implicit promise exists, breaking it is wrong precisely as the breaking of a serious promise."[39] But how does one know that a promise is serious?

Suppose one answers: if the Weinsteins *believe* that their marital vows and solemn promises are serious, then one has to conclude that *for the* Weinsteins their promises of marital fidelity and to never lie to each other are indeed serious. But this appeal to the Weinsteins' own subjectivity just won't do, since it falls prey to all the problems highlighted in the three sets of scenarios. After all, if Weinstein and his wife had added another promise to their wedding day— "We promise to attend at least two New York Yankees home games every year together" or "We promise to watch all the seasons of The Sopranos twice in our first year of marriage"—it is doubtful that anyone would think that it is on par with their other promises to marital fidelity and to never lie to each other, even if the Weinsteins audaciously labeled this new promise as "solemn." It seems, then, that there is something inherently sacred about marital promises to sexual exclusivity and not lying to each other that is absent from a promise to attend Yankees games or watch an HBO mafia show, with sexual exclusivity ranked higher than not lying *ceteris paribus*.

Conclusion

If the best way to ground our moral intuitions about the evils of sexual assault and sexual harassment is to believe that our sexual parts and powers are somehow sacred (or special), then a *purely* transactional view of sexual activity is likely wrong. On the other hand, because this implies that *some aspects* of traditional sexual morality are likely correct, then the sexual libertinism embedded in many contemporary cultural trends is at cross-purposes with cultivating in our souls a moral revulsion to sexual misconduct.

Notes

1. I take this illustration from Michael J. Sandel's account of the viaticals market in his *What Money Can't Buy: The Moral Limits of Markets* (New York: Farrar, Strauss and Giroux, 2012), 136–141.
2. "Harvey Weinstein timeline: How the scandal unfolded," BBC News (29 May 2020), https://www.bbc.co.uk/news/entertainment-arts-41594672
3. Dan Corey, "A growing list of men accused of sexual misconduct since Weinstein," NBC News (8 November 2017), https://www.nbcnews.com/storyline/sexual-misconduct/weinstein-here-s-growing-list-men-accused-sexual-misconduct-n816546
4. Dalvin Brown., "19 million tweets later: A look at #MeToo a year after the hashtag went viral," *USA Today* (13 October 2018), https://www.usatoday.com/story/news/2018/10/13/metoo-impact-hashtag-made-online/1633570002/
5. While working on this chapter, I discovered that another author, David Benatar, had broached similar questions in his article, "Two Views of Sexual Ethics: Promiscuity, Pedophilia, and Rape," *Public Affairs Quarterly* 16.3 (July 2002): 191–201.
6. Fiona Wilson and Paul Thompson, "Sexual Harassment As An Exercise of Power," *Gender, Work, and Organization* 8.1 (January 2001): 61–83
7. Special thanks to Richard Eva (Baylor PhD candidate) with whom I worked through some of these examples in an extended conversation in my office.
8. In February 2020 Weinstein was convicted of rape and sexual assault by a New York court. See "Harvey Weinstein jailed for 23 years in rape trial," BBC News (12 March 2020), https://www.bbc.co.uk/news/world-us-canada-51840532
9. "An Updated Definition of Rape," The U.S. Department of Justice Archives (6 January 2012), https://www.justice.gov/archives/opa/blog/updated-definition-rape

10. "1610. Assault—18 U.S.C. 351(E)," The U.S. Department of Justice Archives (17 January 2020), https://www.justice.gov/archives/jm/criminal-resource-manual-1610-assault-18-usc-351e
11. Texas Penal Code, Sec. 22.011 (f), https://statutes.capitol.texas.gov/Docs/PE/htm/PE.22.htm.
12. Texas Penal Code, Sec. 12.33 (a)(b), https://statutes.capitol.texas.gov/Docs/PE/htm/PE.12.htm
13. Texas Penal Code, Sec. 22.021 (e), https://statutes.capitol.texas.gov/Docs/PE/htm/PE.22.htm.
14. Texas Penal Code, Sec. 12.32 (a)(b), https://statutes.capitol.texas.gov/Docs/PE/htm/PE.12.htm
15. Texas Penal Code, Sec. 22.01 (a)(1) and (b-3), https://statutes.capitol.texas.gov/Docs/PE/htm/PE.22.htm
16. Texas Penal Code, Sec. 22.01 (c), https://statutes.capitol.texas.gov/Docs/PE/htm/PE.22.htm
17. Texas Penal Code, Sec. 12.23, https://statutes.capitol.texas.gov/Docs/PE/htm/PE.12.htm
18. I adapted my use of the "wet willy" from a similar example used by Benjamin H. Arbour in his article, "Sex and Sexual Assault in the #MeToo Era," *Think* 55, vol. 19 (Summer 2020): 39.
19. "*Sacredness* refers to the human tendency to invest people, places, times, and ideas with importance far beyond the utility they possess. Trade-offs or compromises involving what is sacralized are resisted or refused. In prototypical cases, these investments tie individuals to larger groups with shared identities and ennobling projects, and so trade-offs or compromises are felt to be acts of betrayal, even in nonprototypical cases in which no group is implicated." (Jesse Graham and Jonathan Haidt, "Sacred Values and Evil Adversaries: A Moral Foundations Approach," *The Social Psychology of Morality: Exploring the Causes of Good and Evil*, eds, Mario Mikulincer and Phillip R. Shaver [New York: APA Books, 2012], 14 [italics in original]).
20. Because I am Catholic, my own views on divorce align with those of the Catholic Church. See *Catechism of the Catholic Church: Revised in Accordance With the Official Latin Text Promulgated by Pope John Paul II*, 2nd ed. (Washington, DC: United States Conference of Catholic Bishops, 2000), 2380–2391.
21. Technically, when Weinstein lied about having interviewed Michael, he had only broken one promise, since Weinstein could have immediately fired Janet and interviewed Michael and thus could have still kept the promise to interview him.
22. Cardi B. and Megan Thee Stallion, vocalists, "WAP," 2020, by Cardi B. (lyricist), as found in Heran Mamo, "Here Are the Lyrics to Cardi B's 'WAP,' Feat. Megan Thee Stallion," *Billboard Magazine* (28 September 2020), https://www.billboard.com/index.php/articles/news/lyrics/9456405/cardi-b-wap-lyrics-megan-thee-stallion Letters in obscene words replaced with dashes by *Billboard*.
23. Gary Trust, "Cardi B & Megan Thee Stallion's 'WAP' Debuts at No. 1 on Billboard Hot 100 With Record First-Week Streams," *Billboard Magazine* (17 August 2020), https://www.billboard.com/articles/business/chart-beat/9435213/cardi-bs-wap-debuts-no-1-hot-100/
24. Mikael Wood, "Review: Cardi B and Megan Thee Stallion's 'WAP' is a savage, nasty, sex-positive triumph," *Los Angeles Times* (7 August 2020), https://www.

latimes.com/entertainment-arts/music/story/2020-08-07/cardi-b-megan-thee-stallion-wap-review

25. Wood, "Review."
26. Wood, "Review."
27. Wood, "Review."
28. The University of Texas at Austin, "Consensual Relationships" (3 January 2017), https://policies.utexas.edu/policies/consensual-relationships
29. Colleen Flaherty, "Relationship Restrictions: Academe sees a new wave of faculty-student bans in the era of Me Too," *Inside Higher Education* (24 May 2018), https://www.insidehighered.com/news/2018/05/24/academe-sees-new-wave-faculty-student-relationship-restrictions-era-me-too
30. See, for example, Jana Allen, "'It cannot be full consent': OU drama alumna reports sexual relationship with professor who resigns amid Title IX inquiry," *OUDaily* (8 December 2019; updated 29 January 2020), http://www.oudaily.com/news/it-cannot-be-full-consent-ou-drama-alumna-reports-sexual-relationship-with-professor-who-resigns/article_255fb1e4-19f3-11ea-a679-737cc8055218.html
31. I once had a Baylor colleague—no longer employed at the university—who actually required his grad assistants to mow his lawn. He was eventually busted.
32. Alan H. Goldman, "Plain Sex," *Philosophy & Public Affairs* 6.3 (Spring, 1977): 267–287; Alan H. Goldman, "Why Sexual Morality Doesn't Exist: We must get rid of ideas of what sex should be," *The Institute for Arts & Ideas News* 71 (12 February 2019), https://iai.tv/articles/why-sexual-morality-does-not-exist-auid-1212
33. Goldman, "Plain Sex," 268.
34. Goldman, "Plain Sex," 268, 272.
35. Goldman, "Why Sexual Morality Doesn't Exist."
36. Goldman, "Plain Sex," 281.
37. Goldman, "Why Sexual Morality Doesn't Exist."
38. Goldman, "Why Sexual Morality Doesn't Exist."
39. Goldman, "Why Sexual Morality Doesn't Exist."

CHAPTER 12

Deception and Sexual Harassment

Jessica Flanigan

New Zealand decriminalized sex work in 2003. This means that sex work is treated like similar professions, such as massage therapy, performance art, marriage counseling, or health work. Some sex workers in New Zealand work in legal brothels. In 2014, the owner of one legal brothel in New Zealand was sued for sexual harassment. Aaron Montgomery was accused of sexually harassing a 22-year-old sex worker who was his employee. Montgomery told her that "weekends were his play time" and that "he could do what he liked with the girls who worked for him." New Zealand's human rights tribunal awarded the worker $25,000 (NZ) in damages for the emotional harm of her former employer's harassment. The tribunal wrote,

> Context is everything. Even in a brothel, language with a sexual dimension can be used inappropriately in suggestive, oppressive, or abusive circumstances…Sex workers are as much entitled to protection from sexual harassment as those working in other occupations. The fact that a person is a sex worker is not a license for sexual harassment—especially by the manager or employer at the brothel.[1]

Implicit in the tribunal's decision is the claim that language with a sexual dimension in a workplace is oppressive or abusive if workers did not consent to it as part of their job—sex workers may consent to sexually explicit advances from clients but not their bosses. This is the context of brothel workers in New Zealand, which is why Montgomery's comment in the workplace was wrong.

In this essay, I develop a further defense of the New Zealand human rights tribunal's reasoning in this case. My central claim is that sexual harassment

J. Flanigan (✉)
University of Richmond, Richmond, VA, USA
e-mail: jflaniga@richmond.edu

D. Boonin (ed.), *The Palgrave Handbook of Sexual Ethics*,
https://doi.org/10.1007/978-3-030-87786-6_12

violates workers' rights when they did not consent to work in an environment that exposed them to harassment. Sexual harassment violates rights because it is a form of deception. If workers are informed that their workplace will involve sexual harassment beforehand, then they cannot claim that their rights against sexual harassment were violated. I develop this view in the first section "Sexual Harassment as Deception."

I contrast this view of sexual harassment with three other views of the wrongfulness of sexual harassment in the second, third, and fourth sections "Sexual Harassment as Harm," "Sexual Harassment as Coercion," and "Sexual Harassment as Discrimination." The first contrasting view is that sexual harassment violates a person's rights because it causes emotional distress. Second, I consider the view that sexual harassment violates a person's rights because it is coercive. Third, I consider the view that sexual harassment violates a person's rights because it is a form of discrimination or unequal treatment. These alternative views of sexual harassment implicitly assume an implausible theory of rights or they do not faithfully describe paradigm cases of sexual harassment.

In the fifth section, "Objections to Sexual Harassment as Deception," I consider and reply to four objections to the view that sexual harassment is a form of wrongful deception. The first objection is that conceiving of sexual harassment as a form of deception fails to fully account for the moral reasons against it. Second, one may object that this view of sexual harassment offers inadequate protections for women because most labor agreements are underspecified. The third objection is that conceiving of sexual harassment as deception burdens or blames the victims of harassment. The fourth objection is that this view of sexual harassment potentially legitimates the practice.

Sexual Harassment as Deception

Begin with some paradigm cases of sexual harassment.[2]

- *Verbal Harassment*: Jane turns work discussions to sexual topics and asks her fellow employees about their sexual experiences and preferences. These discussions make Jenny uncomfortable.
- *Sexualized Work Environment*: Kyle and Katie share an office. Kyle looks at pornography on his computer. He also frequently stares at her while she is working.
- *Unnecessary Touching*: Lisa and Lenny work together. Lisa occasionally approaches Lenny and rubs his shoulders. She also hugs him when they meet in professional contexts and she always stands very close to him.
- *Quid Pro Quo*: Martin strongly insinuated to Max that he would write a negative performance review for them if they did not agree to have sex with an important client.

In these cases, Jane, Kyle, Lisa, and Martin created a hostile work environment for their coworkers. Jenny, Katie, and Lenny did not consent to work in conditions where their peers or supervisors frequently discussed their sex lives, watched porn at work, or rubbed their shoulders. Max presumably did not know that having sex with Martin's clients would be considered a relevant consideration in their performance review. Though people use the term sexual harassment to describe a range of cases, these examples broadly represent the kinds of behaviors that the term refers to.

Yet we can imagine similar cases where employees do not face a hostile work environment, even though the environment includes similar behavior. For example, imagine that Jane and Jenny are performers at a strip club. In this context, Jane may reasonably assume that discussing her preferences and experiences with her coworkers is not inappropriate. Similarly, while Martin's *quid pro quo* was inappropriate if Max was a nurse or gardener, it would not be inappropriate, in contexts where sex work is legal, if Max's job included the provision of sexual services. In this context, if Max were a sex worker, it would be appropriate for someone to write a negative performance review of Max for failing to provide sexual services to clients.

Or, imagine that Kyle and Katie are content monitors for a social media site or that they work in a law firm that handles digital copyright. In this context, it would not be inappropriate for Kyle to look at pornography at work. If Katie were an airplane mechanic, it would be necessary for Kyle to stare at her while she is working to prevent errors. It's harder to see how it would ever be reasonable for coworkers to rub a coworker's shoulders, but in part this is because of the contingent, conventional norms about personal space that have evolved in existing workplaces. Family members who treat each other this way at Thanksgiving are not accused of creating a hostile holiday environment. And other kinds of workplace touching are still considered acceptable, such as shaking hands, high fives, slaps on the back, and, in some professional sports, a friendly pat on the butt.[3]

These examples support the claim that sexual harassment is wrong only because workers do not consent to work in a sexually explicit environment. If workers did consent to work in a sexually explicit environment, for example, if they were employed as sex workers, then the behaviors that constitute sexual harassment in most workplaces would not be objectionable in their workplaces. More generally, the context matters. In some workplaces, employees are very likely to be exposed to sexually explicit content or physical content, and employees generally know this when they decide to accept a job. Yet in the paradigm cases of sexual harassment, workers did not know, and could not reasonably foresee, that their jobs would involve exposure to sexually explicit material or physical touching so they could not consent to the terms and conditions of labor agreements that included these behaviors.

In this way, the harm or rights violation of sexual harassment is similar to the rights violation of other cases of rape by deception or fraud. Tom Dougherty develops the claim that a person's sexual autonomy can be violated if they are deceived into having sex. Dougherty's argument begins with the premise that it is seriously wrong to have sex with someone who does not give morally valid consent. Deception makes it impossible for a person to give morally valid consent, since a deceived person cannot know whether she is consenting to sex under conditions that include a potential deal breaker for sexual consent. In these cases, deal breakers are things that, if disclosed, would have changed a person's decision to consent. Therefore, it is seriously morally wrong to deceive someone into sex.[4] The same is true for employment. Just as a person is unable to consent to a sexual relationship when a potential partner concealed or misled their partner about potential deal breakers, she is also unable to consent to an employment relationship when a potential employer concealed or mislead her about a potential deal breaker for employment.

In any case where the workplace provision of sexual services or exposure to sexual harassment would be a deal breaker for a potential employee, and the expectation of sex work was not disclosed, sexual harassment violates employees' enforceable rights.[5] In these cases, the rights violation is due to the fact that employers engaged in deceptive reticence by failing to disclose information about a job that most people would reasonably deem relevant. Deceptive reticence is wrong for the same reasons that explicit deception is wrong. By engaging in deceptive reticence and failing to disclose that a job included the expectation of sex work, an employer prevents potential employees from being the authors of their own lives. Instead, he treats potential employees as a means to his own ends, which is disrespectful. For this reason, deceptive reticence is inconsistent with norms of mutual respect, because a respectful relationship would be one where people communicated honestly and transparently, rather than strategically.

Sexual harassment is also harmful because of the opportunity costs associated with taking a job. Perhaps some victims of sexual harassment would consent to be exposed to sexually explicit content or touching at work, but only if the work paid them additional wages as compensation. Sexual content and touching don't have to be a deal breaker for taking a job in order for an employee who did not consent to sexual content and touching to have a valid claim. For example, a waitress may be willing to work in a sexually charged workplace that encourages their staff to flirt with customers and tolerate some forms of physical contact, but only for higher wages. So if sexualized working conditions were not disclosed at the point of employment, she would have been unable to weigh her options and make an employment decision that reflected her judgment of the value of additional wages in exchange for tolerating physical contact with customers.

Summing up, the deception-based explanation for the wrongfulness of sexual harassment goes like this:

(1) Employers should disclose all reasonably foreseeable potential workplace-related deal breakers at the point of employment.
(2) Sexual harassment is a workplace-related deal breaker that is not typically disclosed at the point of employment.

(3) If a workplace-related deal breaker is not disclosed at the point of employment, then employees have a valid claim against it.
(4) Employees typically have a valid claim against sexual harassment.

On this view, most cases of sexual harassment are wrong in practice, but this argument does not rule out that workplaces that appear to tolerate sexual harassment could in principle be permissible if the presence of sexual harassment was disclosed before the point of hire.

In many cases, whether something is disclosed and whether it should be viewed as a potential deal breaker will depend on the broader context. Employers don't have a duty to specify every specific task involved in a job, but they should give potential employees an opportunity to learn about aspects of the job that may make them consider a different job. For example, if an employer at a retail store will require workers to use on-call scheduling to do some light housekeeping, then she should disclose these requirements to potential employees because many retail workers would plausibly prefer a job with regular hours and a job that doesn't require cleaning. But it would not be reasonable for employees to expect that they will only be required to fold product and ring up customers, because other tasks such as merchandising, inventory, and unpacking stock are all standard requirements of all retail jobs. A factory manager should disclose if a job requires workers to occasionally lift heavy objects or if there aren't bathroom breaks for workers on the line. The manager needn't disclose that the job will require workers to show up regularly in order to keep their jobs.

For anything related to sex though, employers should disclose whether workers will be exposed to sexual harassment or touching at work, even if it seems clear, because the moral risks of failing to disclose outweigh the risks of disclosure. After all, sexual content and touching are widely considered a deal breaker for many workers, and even for workers who may voluntarily work in jobs that involve some sexual content, their comfort levels and boundaries likely rule out other kinds of sexual content at work. For this reason, legal institutions should generally require that labor agreements do not include expectations that employees provide sexual services or work in a sexually explicit environment unless those expectations are specifically disclosed at the beginning of the employment relationship or explicitly introduced at a point of renegotiation. This means that laws that prohibit sexual harassment should be the default in how courts interpret labor contracts but that employees and employers should also be permitted to mutually opt out of being subject to these laws.

The same reasons that support an "affirmative consent" model of sexual consent in relationships are reasons in favor of an affirmative consent approach to consent for labor agreements. Because most people view sex as an intimate and personal decision, employers should adopt a wide presumption against assuming that workers consent to sexual content at work unless their consent is clear and explicit. Or, as the New Zealand tribunal wrote, context is everything, so it's important that workers and employers are very clear about the context of a labor agreement and that they continue to update and refine it as the relationship progresses.

The sexual harassment as deception model has several advantages. First, it can explain why sexual harassment is typically wrong, on the grounds that it is generally wrong to deceive people, especially when it comes to intimate and personal decisions where people have deal breakers which inform what they would consent to. Second, the deception-based model is compatible with the view that sexual harassment is wrong because it is disrespectful (since deception is disrespectful to people more generally) and also that it is wrong because it is bad for people's well-being (since deception prevents employees from taking jobs that would be a better fit or negotiating for higher wages). Third, the sexual harassment as deception model doesn't rule out the possibility that some kinds of morally permissible employment relationships, such as sex work, may involve sexual content, touching, or the provision of sexual services.

Sexual Harassment as Harm

As an alternative to sexual harassment as deception, there are at least three competing explanations of the wrongfulness of sexual harassment that merit further consideration. These are the view that sexual harassment is wrong because it is harmful to workers, that sexual harassment is wrong because it is coercive, and that sexual harassment is wrong because it is discriminatory against women. In this and the following two sections I argue against these competing explanations of the wrong of sexual harassment and I show that the deception-based explanation has many of the same advantages as these competing explanations while avoiding their drawbacks. I begin with the harm-based explanation.

The harm-based explanation for the wrong of sexual harassment goes like this:

(1) Employers should not initiate or tolerate harmful social norms or practices within a workplace.
(2) Sexual harassment is a harmful social norm or practice.
(3) Employers should not initiate or tolerate sexual harassment.

On this view, sexual harassment is wrong because it harms employees. For example, victims of sexual harassment may find the experience traumatic. They may feel depressed, powerless, resentful, or disrespected. Victims of sexual harassment may also view it as harmful because it prevents them from doing their jobs and advancing on their own merits. Sexual harassment can be distracting and degrading at work. This view appeals to a comparative account of harm. It says, relative to a workplace that is free of sexual harassment, workers are worse off in a workplace that initiates or tolerates it.

For example, several discussions of sexual harassment focus on the fact that sexual harassment is correlated with subsequent injury, illness, or poor mental health.[6] In some cases, a single instance of sexual harassment may not be harmful, but a broader workplace environment of harassment can be harmful on balance. Harm-based accounts of sexual harassment can accommodate this intuition, maintaining, as Samantha Brennan argues, that harassment may consist in a series of unintentional harms that create micro-inequalities, which together create a harmful work environment for women.[7]

There are three challenges for the harm-based account of the wrongfulness of sexual harassment. The first problem is that, when assessing whether sexual harassment is harmful, the harm-based account may defer too much or too little to workers' own subjective experiences of harassment, failing to yield a consistent standard for wrongful harassment. Relatedly, the harm-based account potentially rules out the possibility that sexual harassment could be wrong even if the victim never learned about it or thought they were made worse off. Second, the harm-based account cannot explain why a labor agreement that includes sexual harassment but replaces a greater harm is not a benefit. It is not always clear that the relevant comparison class is workplace that is free of sexual harassment rather than unemployment or a workplace that is worse in other ways. Third, in some cases sexual harassment may involve denying people opportunities or failing to benefit them, but it is not typically harmful to withhold benefits from employees in other cases. These challenges all arise for the same reason—the harm-based account of the wrongfulness of sexual harassment inherits many of the problems associated with any harm-based account of wrongfulness.

Consider first the difficulty with defining the wrong of sexual harassment in terms of a specific theory of well-being. On many accounts of harm, an act is harmful if it makes a person worse off in some way or if it exposes her to some state that is objectively bad in a non-comparative sense. But defining sexual harassment as harmful in these ways rules out the possibility that sexual harassment could be wrong even if the target of the harassment and everyone else in the workplace enjoys it and even if no one is injured, degraded, or exposed to pain as a result. The broader point is that theories of harm often appeal to some account of well-being, but there are few accounts of well-being that would rule out what are nevertheless clear cases of morally objectionable sexual harassment.

Imagine, for example, a workplace that has one female employee, Tanya, and six male employees. When Tanya was hired, she didn't know that her male coworkers would call her names like "sweetheart" and "sexy" and flirt with her in a sincere way, but it turns out that she actually really appreciates it and finds the dynamic empowering.[8] Maybe Tanya's coworkers' comments actually improve Tanya's subjective well-being and her objective interests are not undermined by her coworkers' behavior either. Perhaps Tanya finds this kind of sexually charged workplace empowering and affirming and that no one in the workplace views Tanya as a subordinate and no one talks about her in a way that is objectively degrading, whatever that could mean.

On a harm-based account of sexual harassment, Tanya's male coworkers got morally lucky. It would be false, on this account, that their sexually charged comments harmed Tanya or her coworkers because no one in this situation was made objectively or subjectively worse off, either by any single act of sexual harassment or by the broader culture of harassment.[9] So, if sexual harassment is wrong because it subjectively or objectively makes people worse off, it wouldn't be wrong in this case. But this behavior is widely considered to be wrongful sexual harassment even if it turns out that everyone in the office appreciates it.

Another problem for locating the wrong of sexual harassment in its harmfulness is that some people may be made worse off by encountering sexually explicit content at work without being victims of sexual harassment. For example, imagine that Tilly works at an art gallery where she occasionally is asked to evaluate and select artists' work for display. One day, Tilly evaluates some paintings by Savannah, which include a series of abstract flowers/nudes that look very sexually explicit. Tilly views Savannah's art as obscene, and the experience of working in the presence of Savannah's pictures makes her feel uncomfortable and a little degraded. She suspects that Savannah wants viewers to experience her art in this way. If a harm-based explanation of the wrong of sexual harassment appeals to Tilly's subjective well-being or her objective interest in avoiding degradation, then it would designate Tilly's workplace as one where she was harmed by exposure to sexually explicit content. But Tilly should have reasonably expected that she may encounter some sexually explicit art in her work as a gallerist, just as movie theater employees should reasonably expect to see R-rated content as part of their jobs. If viewing sexually explicit art was a deal breaker for Tilly, she should have disclosed it when she took the job. These examples suggest that the harm-based explanation is potentially overly exclusive because it cannot explain why Tanya's workplace is objectionable; it is also potentially under inclusive because it deems workplaces that are free of sexual harassment like Tilly's as objectionable.

A second problem with the sexual harassment as harm account is that most accounts of harm define harm relative to a comparison class.[10] That is, an act is harmful if it makes someone worse off than they were or otherwise would have been. But sexual harassment doesn't always work in this way. Imagine that Pete was about to be fired from his job for good reason—he dropped the ball on an important project and cost the company tens of thousands of dollars. Pete knows this and he is desperate to keep his job. His boss, Peggy, tells Pete that she will give him a second chance if he also goes out with her sister for a few dates. In this case, it's difficult to know the relevant counterfactual when determining whether Peggy's offer makes Pete worse off. Peggy's offer makes Pete worse off compared to keeping his job without dating Peggy's sister, but better off compared to losing his job.

The third problem is similar. Some cases of sexual harassment involve omissions. For example, imagine that Sally was denied the opportunity to join an important committee after turning down the boss's advances. Or imagine that Jo didn't get a promotion because their supervisors thought that they were too sexy for a management position. In these cases, what looks like sexual harassment doesn't make Sally or Jo worse off, but they missed out on benefits due to previously unstated expectations that they would date the boss or dress conservatively to get ahead.

In light of these kinds of problems with well-being and interest-based accounts of harm, or counterfactual comparative accounts of harm, some philosophers have suggested that the best theory of harm isn't about whether an act makes someone worse off. Rather, an act is a wrongful harm if it violates a person's rights.[11] I turn to this theory in the next section.

Sexual Harassment as Coercion

The harm-based explanations for the wrongfulness of sexual harassment either cannot account for seemingly wrongful sexual harassment that is not bad for anyone's interests or well-being, or they cannot make sense of seemingly harmful actions which do not make a person worse off because they are imperceptible, preventing a greater harm, or because they consist in mere omissions. A harm-based account that defines harms in terms of a theory of rights can avoid these general problems with theories of harm. The coercion-based explanation takes this form. On this view, sexual harassment is wrong because it coerces people or violates their enforceable rights. The coercion-based explanation for the wrongfulness of sexual harassment goes like this:

(1) Employers should not tolerate or engage in workplace coercion or rights violations.
(2) Sexual harassment is a form of workplace coercion or a rights violation.
(3) Employers should not tolerate sexual harassment.

Many accounts of sexual harassment characterize the wrong of sexual harassment in this way. On this account, coercion is defined as a threat of a rights violation. So if people have rights to their bodily integrity or sexual autonomy, then sexual harassment is wrong because either it constitutes a threat to a worker's sexual autonomy or it directly violates their bodily rights or sexual autonomy.

Before describing the sexual harassment as coercion view in more detail, we need a clearer definition of coercion. People use the word coercion in different ways. Some use the word in a non-moralized way. On this view, an act is coercive if it involves force or threats of force, and then it is a separate question whether the coercion is wrongful.[12] But when people advance the sexual harassment as coercion account, they use the word "coercion" in a moralized way, like the word "harm." On these views, coercion is presumptively wrong. It may be justified in some cases, but in the absence of overriding or excusing reasons, coercion involves some kind of a rights violation.[13] For example, coercion may be wrong because it impairs their ability to consent.[14] Or another way to put this—coercion wrongly prevents a person from exercising power over the proceedings of her own life.[15] This moralized account of coercion explains why coercion is presumptively wrong, because it presumptively involves a rights violation.[16]

With this account of coercion in hand, consider some versions of the sexual harassment as coercion account. Some accounts of coercion also appeal to a comparative account of harm. For example, Vrinda Dalmiya writes that "what is happening in harassment is not a barter but coercion." On her view, harassment is coercive because it infringes on the freedom of the victim. It infringes on the victim's freedom because she typically would not have chosen to be in the situation that harassment puts her in and because harassment makes her worse off.[17] Similarly, Elizabeth Anderson writes that "the core wrong in the paradigm case of sexual harassment is not discrimination, but sexual coercion; not a group injury, but an individual one."[18] James Rocha elaborates on the specific individual injury that sexual harassment causes when he argues that,

> In one difficult sexual harassment case, someone is offered a promotion in exchange for sex, but there is, due to the arrangement of the case, no implied threat or repercussion for refusal. I argue this case counts as coercive since the offer-making attempts to recast the agent's self-image simply by making the offer: the harasser attempts to define his employee as a sexualized object in the workplace to whom it is acceptable to make this kind of offer.[19]

Others argue that sexual harassment violates people's rights to have control over their self-presentation at work.[20]

Some of these accounts of sexual harassment as coercion appeal to the related concept of exploitation. For example, some argue that sexual harassment may involve a rights violation if people have rights against suffering from material deprivation, and workplace sexual harassment that implicitly carries the threat of losing a job is often effectively threating workers with material deprivation. On these views, sexual harassment is exploitative, because harassers take unfair advantage of economic vulnerability or precarity in the labor market. I am considering these exploitation-based accounts of sexual harassment alongside the coercion-based accounts because these accounts also locate the wrongfulness of sexual harassment in the fact that it involves a rights violation or a threat of a rights violation.

The sexual harassment as coercion view has three problems. First, sexual harassment as coercion accounts appeal to a theory of rights that would limit a range of other sexual communication that is seemingly unobjectionable. Second this account would seemingly deem all workplaces as wrongfully coercive, not just those where sexual harassment occurs, so it fails to yield a specific judgment of what in particular makes sexual harassment wrong. And third, if this account of the wrongfulness of sexual harassment does distinguish work that involves sexual harassment from work that is bad in other ways, then this account would potentially rule out the possibility that sex work can be permissible. But sex work can be permissible.

Consider first the theory of rights that is implicit in the coercion-based explanation for the wrongfulness of sexual harassment. On this view, in order for sexual harassment to be the kind of offense that officials ought to prohibit, it must include a rights violation or a threat of a rights violation. For example, say people have rights against interference. One may argue that sexual harassment is wrong because it interferes with workers by constraining their options at work. But as Japa Pallikkathayil writes, in order to determine whether it is impermissible for an employer to constrain an employee's options through sexual harassment, for example, by saying something like "I will fire you if and only if you refuse to have sex with me," one must first determine whether the employer was entitled to fire the employee or if he had some institutional obligations to continue her employment. In some cases, an employer may have contractual duties that prohibit firing his employee, in which case this kind of sexual harassment is wrong not because it is wrong to ask for sex but because it is wrong for him to threaten to violate his duty. Or, if the employer had a duty to fire the employee, similarly, making an announcement that gives her the option to retain her job for sex would amount to violating a duty or threatening to violate a duty. Either way, the wrongfulness of this kind of sexual harassment would not derive from making continued employment contingent on sex, but rather from making continued employment contingent when the employer had a duty to either fire or not fire his employee.

Say an employer were permitted but not required to fire an employee, but not for bad reasons. Proponents of the coercion-based explanation may argue that there is something special about workplace propositions when they include the provision of sexual services. Even if an employer were permitted but not required to fire an employee, for example, offering to trade a job for sex may violate the employee's rights in a way that other changes to a job description would not. On this view, sexual harassment could be coercive not because it includes a threat of wrongful termination (or employment) but because it is a labor agreement that requires the provision of sexual services.

For this view of coercion to explain the coerciveness of sexual harassment though, it must assume that people have rights against being put in circumstances where keeping their jobs is contingent on providing sexual services. But this would seemingly rule out the possibility that sex work could ever be non-coercive, since sex work is often a job that involves trading sex for money. Some critics of the sex industry press this claim. They argue that sex work is always a rights violation. But if people have rights to choose their sexual partners and the conditions under which they have sex, then they should also have rights to choose sexual partners who pay them. Many sex workers report that the decision to sell sex is voluntary and that their rights are not violated by clients who pay them for sex. So the claim that sex work is always coercive is also at odds with sex workers' testimony about their own experiences in the industry.

Another version of the coercion-based explanation is that sexual harassment is coercive because the targets of harassment are often unable to exit a workplace without losing the income that they need to meet their basic needs. So on this view, the rights violation included in a threat like "sleep with me or you're fired" is the threat of material deprivation. Yet this characterization of the coerciveness of sexual harassment at work potentially characterizes a very wide range of other occupations and features of work as objectionable as well. After all, many people only accept an unpleasant or degrading workplace environment because the job meets their basic needs. It may well be a rights violation that workers are required to work to meet their needs, but if work is coercive in this way, such a view does not establish that sexual harassment is especially coercive compared to other jobs.

Some people accept the conclusion that the coercion-based arguments against sexual harassment are more general arguments against work. For example, if most workplaces have at-will employment policies then most workplaces are potentially coercive in that workers always labor under the threat of being fired and are therefore forced to tolerate significant mistreatment. For example, Lea S. VanderVelde argues that workplace bullying and non-sexual harassment are also coercive and that it is a mistake to view sexual harassment as distinct from these other practices.[21] Elizabeth Anderson argues similarly that "the coercive and hierarchical structure of work invites many kinds of humiliations and abuses," and so even if officials address sexual harassment they would still offer workers inadequate protections from abuse more generally.

For those who accept this account of work, one might think that sexual harassment is still especially wrongful, even if all work is in some sense coercive. The deception account can also account for the intuitive wrongfulness of workplace bullying and humiliation, to the extent that these deal beakers were not disclosed at the outset. Yet unlike the coercion-based account, the deception-based account avoids the conclusion that work is objectionable whenever workers are employed to meet their basic needs. And for those who reject the claim that all work is prima facie objectionable, the sexual harassment as deception view can explain the wrong of sexual harassment but doesn't rule out all work. Because most work has informed consent, most work does not violate workers' rights in the way that sexual harassment often does.

Sexual Harassment as Discrimination

Many discussions of sexual harassment focus on the fact that sexual harassment typically involves men's domination of women. Some people view sexual harassment as a symptom of a broader problem, namely, discrimination against women in the workplace. And this view is reflected at several points in the law where anti-discrimination legislation includes provisions that are intended to protect employees from sexual harassment. The discrimination-based explanation for the wrongfulness of sexual harassment goes like this:

1. Employers should not tolerate a workplace that disadvantages some workers on the basis of their sex or gender identity, either in principle or in practice.

2. Sexual harassment disadvantages some workers on the basis of their sex or gender identity, in principle and in practice.
3. Employers should not tolerate sexual harassment.

Many people are proponents of the sexual harassment as discrimination account.[22] In the United States, the discrimination account has framed policies against sexual harassment. Legally, sexual harassment by an employer is prohibited on the grounds that it is illegal employment discrimination on the basis of sex or sexuality. In the 1970s, civil rights lawyers and activists developed affirmative action agreements and guidelines that included anti-sexual-harassment clauses.[23] When one of these lawyers, Eleanor Holmes Norton, became the chair of the US Equal Employment Opportunity Commission (EEOC), she included an anti-sexual-harassment clause in the commission's employment guidelines.[24] In 1986, the Supreme Court then cited these guidelines in a unanimous ruling that sexual harassment violates the part of the Civil Rights Act that prohibits employment discrimination on the basis of sex.[25]

Since the 1986 ruling, many legal scholars have affirmed the view that sexual harassment is a form of employment discrimination or a failure to provide equal protection to women in the workplace.[26] This is grounded in MacKinnon's claim that the wrongfulness of sexual harassment is closely tied to patterns of gender-based and sexuality-based oppression and violence.[27] Another version of this claim links sexual harassment to broader forms of structural injustice against people who are socially or economically vulnerable.[28] A more general version of this argument states that sexual harassment is wrong because it does not reflect an egalitarian disposition or because it undermines egalitarian relations between people in a workplace.[29]

These egalitarian explanations for the wrongfulness of sexual harassment, including the dominant discrimination-based view of sexual harassment, face three challenges. First, the first premise of this argument appeals to the idea that discrimination on the basis of sex and gender is always wrong at work. This premise is too broad though, because sometimes sex and gender are relevant to a job description. Discrimination is only objectionable when it targets traits that are "normatively extraneous" to a person's role.[30] For example, if a counseling service is hiring a new social worker to help their clients, they may have good reason to choose a man who can work with male clients, in which case sex would not be "normatively extraneous" to the job, nor would it be discriminatory for managers to treat a male employee differently, for example, by assigning him mostly male clients who request a male counselor.[31] Often, a person's sex or sexuality is relevant to a job description, so the mere fact of discrimination on such grounds is not sufficient to designate behavior in a workplace wrong.[32]

If the discrimination-based argument against sexual harassment allows for sex-based discrimination when it is relevant to a job, then the deception-based account can better capture what is typically wrong with sexual harassment, namely, that workers did not know that the provision of sexual services was required for the job. If the discrimination-based argument does not allow for forms of job-specific sex discrimination, then the account is also overly restrictive because it rules out the possibility of permissible sex work.

The second premise of this argument is also questionable because it characterizes sexual harassment as disadvantageous to the targets of harassment. This may be generally true, as an empirical matter, but it fails to account for the wrongfulness of sexual harassment that takes the form of sexual favoritism. To the extent that discrimination-based accounts of sexual harassment require harassment to be a disadvantage, such accounts rule out as wrongful cases where individuals or groups in an industry could benefit from exposure to sexual harassment but still have a claim against it.

Third, as a conceptual matter, sexual harassment need not entail differential treatment on the basis of sex or sexuality. Here the discrimination-based account encounters some of the same philosophical and practical challenges as other accounts in that it fails to explain why paradigmatic cases of workplace sexual harassment are wrong (e.g., clear examples of sexual harassment by people of the same sex and sexuality). And like other accounts, this account also fails to explain why sexual harassment at work is typically distinctively bad as a form of discrimination, in contrast to other forms of workplace discrimination.

Objections to Sexual Harassment as Deception

The first objection to the deception-based explanation for sexual harassment is that disclosing that workers may encounter sexual harassment or sexually explicit content at work does not make sexual harassment permissible.[33] I grant that sexual harassment could still be bad even if it is disclosed. It may be inefficient, degrading, and unnecessary to expose a worker to sexual harassment. It may also be inefficient, degrading, and unnecessary to require workers to wear uniforms that display colorful pins, or to require workers to clean the floors with hand tools rather than machines, or to search workers' bags as they leave their retail jobs. In general, managers should strive to create a workplace that is as efficient as possible, that only restricts workers' choices and control over their environments when necessary, and which doesn't expose them to a degrading atmosphere. Nevertheless, these obligations shouldn't be legally enforceable, as is the obligation to refrain from creating a workplace that exposes workers to sexual harassment in cases where the possibility of harassment was disclosed.

A related version of this objection is that disclosing that workers may encounter sexual harassment or sexually explicit content at work could be bad because it would deter women from working in certain industries, which would undermine gender parity in those industries.[34] Here again, I agree that it could be bad if women were deterred from working in an industry because the possibility of sexual harassment was disclosed. It would be even worse if the possibility were not exposed and women devoted time and resources to joining an industry that involved undisclosed expectations of women's sexual labor. However, it is important to distinguish again whether the badness of sexual harassment is merely criticizable or if it is the kind of thing that ought to be illegal. The fact that women, when informed about what a job entails, may not choose it is not a sufficient reason to mandate changes to the job. Say that women chose professions that had flexible hours, so disclosing long and inflexible hours in a subfield of an industry deterred women from joining that subfield. This fact alone would not justify a law that required the industry to offer more flexible hours.

A second objection to the deception-based explanation for sexual harassment is that contracts are underspecified at the point of employment, so a mere disclosure requirement would fail to address legitimate concerns about sexual harassment. This is true, but it overstates the ambiguity in most contracts. In general, contracts are created against a legal baseline. For example, they may specify that any contractual disputes will be settled through mediation when possible, or they may rely on the courts for interpretation. Different enforcement agencies will uphold different legal baselines.[35] Against those baselines, workers have the option to opt into additional contractual protections at the point of contracting. Workers also have the right to opt out of whatever features of a contract constitute the baseline legal assumptions. This means that public officials can interpret legal agreements as not including the expectation of sexual labor unless it is specified. In a way, the case for disclosure requirements for the risk of sexual harassment already assumes that labor agreements do not implicitly include any kind of disclosure of a risk of sexual harassment.

A related objection to the deception-based explanation for sexual harassment is that employers may disclose some risk of exposure to sexual harassment in workplaces in order to avoid liability.[36] If so, then the disclosure-based approach that I have advanced may, paradoxically, make the problem of sexual harassment worse by implicitly characterizing sexual harassment as acceptable or licensing sexual harassment in workplaces. This is an empirical possibility, but it is not a principled objection to the view. If workers and their employees did negotiate labor agreements that included disclosure about the potential for sexual harassment, then sexual harassment would not violate employees' rights in these cases. There are other reasons to think that firms may have good reasons to avoid disclosing these risks, since it could impair their ability to attract good employees and open them up to public shaming. For this reason, I think that the prediction that conceiving of sexual harassment as wrongful deception is unlikely to open the floodgates to harassment at work. In any case, a system where the risk of sexual harassment was disclosed as a risk associated with most jobs is still preferable to the current system, where workers are not notified that they may encounter sexual harassment.

Another concern about the deception-based explanation for sexual harassment is that it is in some ways a form of victim blaming, since disclosure requirements, if effective, would place the onus on women to reject contracts that expose them to sexual harassment rather than requiring employers to prevent it in their workplaces.[37] As mentioned above, a legal baseline for contracts needn't include the presumption that workers will be exposed to sexual harassment, so the onus would remain with employers to either disclose or prevent sexual harassment. More generally, one may object that legitimating the practice of sexual harassment or changing baseline social norms against promiscuity and sexual explicitness is burdensome to everyone who would prefer a more sexually conservative environment.[38] But to say people are wronged by the normalization of sexual harassment is to say that they are entitled to a particularly conservative social environment. But people are not entitled to control patterns of voluntary behavior in their neighborhoods simply on the grounds that object to the ways that this behavior would change social norms.[39] Mormons, for example, may find it frustrating that so much of social life happens in coffee shops and bars, but that would not entitle them to enforce laws that prohibited or discouraged coffee and alcohol consumption.

Conclusion

I have argued that public officials may enforce a baseline assumption in labor agreements that workers will, as a default, be protected from sexual harassment in the workplace unless the risk or reality of sexual harassment is disclosed at the point of employment. An opt-in approach to sexually explicit workplaces therefore requires employers to go out of their way to limit their liability for sexual harassment in the workplace and perhaps to pay higher wages to workers in exchange for this limited liability. Such an approach has three advantages over alternatives I considered here. First, conceiving of sexual harassment as deception respects people's autonomy, including their economic and sexual freedom to work in the sex industry. Second, this approach prompts people to be transparent about the boundaries of sexually explicit content and relationships at work. Third, this approach does not impose an unwarranted degree of sexual moralism on industries that have different norms and values.

According to the view I defended, behavior that people currently view as sexual harassment is not in principle wrong; it is only wrong in the context of existing labor agreements where people did not consent to a sexually charged workplace. To some, this implication of my argument will seem so intuitively implausible that they may consider rejecting the view simply on the grounds that it is extensionally inadequate. But as scholars of sexual ethics have noted before me, people's intuitive views of sexual ethics are notoriously unreliable.[40] In the past, public officials, and the public, were far too permissive of sexual assault and harassment and far too prohibitive with respect to interracial and gay sexual relationships as well as premarital sex.[41] The current approach to sexual ethics would strike people who lived a few hundred years ago as extraordinarily implausible and revisionary, and we have little reason to believe that public opinion has perfectly converged on the correct intuitions about sex and harassment in the time since.

As with previous views of sexual ethics, the current approach to sexual harassment is overly permissive and overly restrictive. It is overly permissive because it fails to effectively prevent sexual harassment in contexts where people did not consent to work in a sexually charged environment but are nevertheless expected to perform sexual services, permit unwanted touching, or participate in sexually explicit conversations at work. Yet the current approach is too restrictive because workers are denied the opportunity to voluntarily and consensually agree to contract for the provision of sexual services. These restrictions are paternalistic and counterproductive, and they fail to address the underlying reason that sexual harassment is wrong in the first place.

It is an open empirical question whether conceiving of the wrong of sexual harassment as a failure of disclosure, rather than as a harm, act of coercion, or discriminatory behavior, would reduce the behavior that people currently consider sexual harassment in workplaces. But this is the wrong question to ask. Public officials and employers should not aim to eliminate all sexual harassment from all workplaces; they should aim to eliminate all forms of non-consensual sexual harassment while leaving room for legally permissible forms of sex work. Sexual harassment should not be treated as distinct from other risky or unpleasant aspects of workplaces. And I am skeptical that policies that rule out labor agreements that allow sexual harassment will eliminate sexual harassment. Rather, such policies will perpetuate sexist policies that treat female sexuality as sacrosanct while paternalistically limiting women's workplace autonomy.

Notes

1. Michelle Duff, "Sex Worker Gets $25,000 over Harassment," *Stuff NZ*, March 1, 2014, https://www.stuff.co.nz/business/industries/9777879/Sex-worker-gets-25-000-over-harassment.
2. U.S. Department of the Interior Office of Civil Rights, "Sexual Harassment" (U.S. Department of the Interior, July 1, 2015), https://www.doi.gov/pmb/eeo/Sexual-Harassment.
3. Zeynep Yenisey, "Why Athletes Celebrate Victory With a Slap on the Butt," *Maxim*, January 27, 2016, https://www.maxim.com/maxim-man/athletes-and-butt-slapping-2016-1.
4. Tom Dougherty, "Sex, Lies, and Consent," *Ethics* 123, no. 4 (2013): 717–44, https://doi.org/10.1086/670249. pg. 720.
5. In this way, the deception explanation also captures the idea that judgments of sexual harassment are partly dependent on what an employee could reasonably expect to encounter at work. This idea is implicit in the legal standard of sexual harassment. For a review and critical assessment of this approach see: Matthew W. Green Jr, "What's So Reasonable about Reasonableness-Rejecting a Case Law-Centered Approach to Title VII's Reasonable Belief Doctrine," *U. Kan. L. Rev.* 62 (2013): 759.
6. Kathleen M. Rospenda et al., "Is Workplace Harassment Hazardous to Your Health?," *Journal of Business and Psychology* 20, no. 1 (September 1, 2005): 95–110, https://doi.org/10.1007/s10869-005-6992-y; Sharyn Lenhart,

"Physical and Mental Health Aspects of Sexual Harassment," in *Sexual Harassment in the Workplace and Academia: Psychiatric Issues*, Clinical Practice Series, No. 38 (Arlington, VA, US: American Psychiatric Association, 1996), 21–38; Dr Erica van Roosmalen PhD, Susan A. McDaniel PhD, and FRSC, "Sexual Harassment in Academia: A Hazard to Women's Health," *Women & Health* 28, no. 2 (February 5, 1999): 33–54, https://doi.org/10.1300/J013v28n02_03.

7. Samantha Brennan, "Rethinking the Moral Significance of Micro-Inequities," *Women in Philosophy*, 2013, 180–196.
8. Peter Fleming, "Sexuality, Power and Resistance in the Workplace," *Organization Studies* 28, no. 2 (February 1, 2007): 239–56, https://doi.org/10.1177/0170840606068307.
9. Depending on the theory of wellbeing behind one's sexual harassment as harm account, it may even fail to rule out cases where people did suffer disadvantages due to sexual harassment, but never experienced it in that way. For example, in a society where sexual harassment is normalized and accepted a victim may not realize that she missed out on opportunities due to others' treatment of her. If imperceptible or unexperienced harms are not harmful, then these instances of sexual harassment also may not be wrong on the sexual harassment as harm account.
10. Ben Bradley, "Doing Away with Harm," *Philosophy and Phenomenological Research* 85, no. 2 (2012): 390–412.
11. Warren Quinn, Philippa Foot, and James Woodward, for example, seem to endorse this kind of a theory of harm.
12. Scott Anderson, "Of Theories of Coercion, Two Axes, and the Importance of the Coercer," *Journal of Moral Philosophy* 5, no. 3 (2008): 394–422.
13. Alan Wertheimer, *Coercion* (Princeton: Princeton University Press, 2014).
14. Japa Pallikkathayil, "The Possibility of Choice: Three Accounts of the Problem with Coercion," *Philosopher's Imprint* 11, no. 16 (November 4, 2011), http://hdl.handle.net/2027/spo.3521354.0011.016.
15. Christine M. Korsgaard, *Creating the Kingdom of Ends* (Cambridge University Press, 1996).
16. It also explains why people can be less responsible for what they do when they are coerced, because coercion in some way prevents people from being the authors of their own lives. Pallikkathayil, "The Possibility of Choice."
17. Vrinda Dalmiya, "Why Is Sexual Harassment Wrong?," *Journal of Social Philosophy* 30, no. 1 (1999): 46–64.
18. Elizabeth Anderson, "Recent Thinking about Sexual Harassment: A Review Essay," *Philosophy & Public Affairs* 34, no. 3 (2006): 284–311.
19. James Rocha, "The Sexual Harassment Coercive Offer," *Journal of Applied Philosophy* 28, no. 2 (2011): 203–16, https://doi.org/10.1111/j.1468-5930.2010.00508.x.
20. Drucilia Cornell, *The Imaginary Domain* (New York: Routledge, 1995).
21. Lea S. VanderVelde, "Coercion in At-Will Termination of Employment and Sexual Harassment," 2003.
22. Catherine MacKinnon is perhaps the most widely cited proponent of this position. Catharine A. MacKinnon, *Sexual Harassment of Working Women: A Case of Sex Discrimination*, 19 (Yale University Press, 1979).

23. Raina Lipsitz, "Sexual Harassment Law Was Shaped by the Battles of Black Women," October 20, 2017, https://www.thenation.com/article/archive/sexual-harassment-law-was-shaped-by-the-battles-of-black-women/.
24. Lipsitz.
25. Meritor Savings Bank, FSB v. Vinson, 477 US, No. 84-1979 (Supreme Court 1986).
26. Joanna L. Grossman, "Moving Forward Looking Back: A Retrospective on Sexual Harassment Law," *Boston University Law Review* 95 (2015): 1029.
27. MacKinnon, *Sexual Harassment of Working Women*. Satz makes a similar point about markets in sexual labor, in arguing that these markets are objectionable because they stem from and contribute to the perception that women are not equal to men. Debra Satz, "Markets in Women's Sexual Labor," *Ethics* 106, no. 1 (October 1995): 63–85, https://doi.org/10.1086/293778.
28. For an overview of these arguments, and see Margaret Crouch, *Thinking About Sexual Harassment: A Guide for the Perplexed* (Oxford: Oxford University Press USA, 2000).
29. Elizabeth Anderson, for example, argues against workplace domination including sexual harassment on relational egalitarian grounds.
30. Sophia Moreau, "What Is Discrimination?," *Philosophy & Public Affairs* 38, no. 2 (2010): 143–79, https://doi.org/10.1111/j.1088-4963.2010.01181.x.
31. For an example of this case see Benedict Carey, "Need Therapy? A Good Man Is Hard to Find," *The New York Times*, May 21, 2011, sec. Health, https://www.nytimes.com/2011/05/22/health/22therapists.html.
32. And even when employment discrimination is objectionable, such as when employers do treat some employees differently on the basis of normatively extraneous traits, it is not clear that the victims of employment discrimination should necessarily have an enforceable claim against their employer, though they do under existing US law. Matt Zwolinski, "Why Not Regulate Private Discrimination?," SSRN Scholarly Paper (Rochester, NY: Social Science Research Network, April 11, 2007), https://papers.ssrn.com/abstract=978937.
33. This objection may stem from a more general skepticism that mere disclosure is sufficient to secure valid consent. I am taking this objection to assume that there is some level of disclosure that would be sufficient to secure valid consent, but that there may be some occupational conditions that, even if voluntary, are still impermissible. For a discussion of whether disclosure is enough to secure valid consent, see Tom Dougherty, "Informed Consent, Disclosure, and Understanding," *Philosophy & Public Affairs* 48, no. 2 (2020): 119–50, https://doi.org/10.1111/papa.12164.
34. This is a concern about the software industry, for example. For a further discussion of sexual harassment in Silicon Valley see: Sheelah Kolhatkar, "The Tech Industry's Gender-Discrimination Problem," *The New Yorker*, November 13, 2017, https://www.newyorker.com/magazine/2017/11/20/the-tech-industrys-gender-discrimination-problem.
35. I discuss the states' role in providing the background conditions for contracts in more detail elsewhere. Jessica Flanigan, "Rethinking Freedom of Contract," *Philosophical Studies* 174, no. 2 (2017): 443–463.
36. Margaret Jane Radin, "Boilerplate Today: The Rise of Modularity and the Waning of Consent," *Mich. L. Rev.* 104 (2005): 1223; Margaret Jane Radin,

Boilerplate: The Fine Print, Vanishing Rights, and the Rule of Law (Princeton University Press, 2012).

37. In this way, recharacterizing sexual harassment as an avoidable feature of a workplace may be viewed as analogous to characterizations of assault that focus on whether the victim could have avoided being assaulted. However, this analogy would be inapt because even if assault is in some sense avoidable, it is not consensual. In contrast, on my view, workers can consent to work in an environment that includes behaviors that would otherwise constitute sexual harassment.
38. Scott A. Anderson, "Prostitution and Sexual Autonomy: Making Sense of the Prohibition of Prostitution," *Ethics* 112, no. 4 (July 1, 2002): 748–80, https://doi.org/10.1086/339672.
39. This is not to deny that social norms are a legitimate subject of justice, only to deny that public officials may coercively intervene to shape these norms. For a broader discussion of social norms and justice, see Aaron James, "Power in Social Organization as the Subject of Justice," *Pacific Philosophical Quarterly* 86, no. 1 (2005): 25–49, https://doi.org/10.1111/j.1468-0114.2005.00213.x.
40. Martha C. Nussbaum, *Hiding from Humanity: Disgust, Shame, and the Law* (Princeton University Press, 2009).
41. Dougherty, "Sex, Lies, and Consent."

CHAPTER 13

Homosexuality, Bestiality, and Necrophilia

David Benatar

A "Queer" Argument

In some parts of the world, homosexuals are now enjoying much more social acceptance than they did even until very recently. Within a relatively short period, homosexual sex has been decriminalized, same-sex marriage has been recognized, and other forms of discrimination on the basis of sexual orientation have been prohibited. This includes the freedom to serve in armed forces without hiding one's sexual orientation.

These advances are not universal. There are places where homosexual men can still be put to death for engaging in sexual activity with other men.[1] In other places, where there has been progress, it has been uneven—some advances have been made but others have not. Even where much progress has been made there continue to be people who see at least some of the progress as regress or who, while accepting of the progress already made, do not think that any further social acceptance is warranted.

Those who are critical of homosexuality offer a wide range of arguments for their view. Others have responded to these arguments. My aim here is not to rehearse these debates in any detail, although there will be brief reference to some of them. Instead, I propose to focus on one argument that opponents of homosexuality sometimes raise and that defenders of homosexuality quickly dismiss as ridiculous. I both agree and disagree that it is ridiculous. That may sound like a queer view to have—in the original sense of "queer," namely strange. However, it is not quite as strange as it may at first seem. My view is

D. Benatar (✉)
University of Cape Town, Cape Town, South Africa
e-mail: philosophy@uct.ac.za

D. Boonin (ed.), *The Palgrave Handbook of Sexual Ethics*,
https://doi.org/10.1007/978-3-030-87786-6_13

that it is an argument that should be taken much less seriously by many of those who propose it but much more seriously by many of those who reject it.

The argument is sometimes advanced against homosexuality itself and sometimes against allowing or recognizing same-sex marriage. In its former incarnation[2] it can be presented as follows (even though it is rarely if ever presented this explicitly by those who advance it):

1. If there is nothing morally wrong with homosexuality, then there is also nothing morally wrong with bestiality and necrophilia.
2. There is something morally wrong with bestiality and necrophilia.
3. Therefore, there is something morally wrong with homosexuality.[3]

When directed against same-sex marriage,[4] the argument takes something like the following form:

1. If two people of the same sex should be allowed to marry, then a human being should be able to marry a non-human animal.
2. We should not allow inter-species marriage.
3. Therefore, we should not allow same-sex marriage.

The first of these versions may be the more basic, but either way it is the version more clearly focused on sexual ethics. An argument directed against homosexual marriage is not obviously an argument against homosexual *sex*, unless one thinks that sex is permissible only within the confines of marriage. For this reason, I shall focus on the first version. However, I shall consider the second version too because it seeks to withhold access to marriage from some people based on their sexual preferences or orientation, and that too is an ethical issue pertaining to sex.

Responding to the First Version

One possible response to the first version of the argument is to deny the second premise. Perhaps there is nothing inherently morally wrong with either bestiality or necrophilia. There may be *instances* of such actions which are wrong. These would include bestial acts that harm animals. They might also include necrophiliac acts with the corpses of those who, while they were still alive, did not agree to such use of their remains. On at least some views, they could also include those acts of bestiality and necrophilia that pose a health risk to the person engaging in these activities. However, bestiality need not always harm animals, necrophilia can be engaged in with the prior consent of the person whose corpse is involved, and there can be safe sex with animals and corpses.

Thus, there is a question whether bestiality and necrophilia are always wrong, as is typically assumed. I shall set aside that question for the moment in order to focus on the first premise. One reason for this is that many of those who defend homosexuality do not want to defend bestiality and necrophilia too. Indeed, that is the rhetorical force of the argument. If approval of homosexuality were said to commit one to approving of apple pie, the argument would be devoid of any force. It is because so many defenders of homosexuality do not accept that bestiality and necrophilia are acceptable that we should shift our focus to the first premise. (However, as we shall see, the discussion of the first premise will soon draw us back to the second premise.)

One response to the first premise is that there is an important difference between homosexuality, on the one hand, and bestiality and necrophilia, on the other. If there is such a difference, then we can reject the first premise. In other words, we cannot infer the acceptability of bestiality and necrophilia from the acceptability of homosexuality.

Perhaps the most obvious criterion for separating homosexuality from bestiality and necrophilia is that homosexual sex can be the expression of mutual romantic love between two people, just as heterosexual sex can. By contrast, if you have sex with your dog or your goat, you might be expressing love toward him or her, but your and the animal's understanding of the activity is, so far as we know, so different that there can be no mutual expression of love via the sex. No mutuality whatsoever is possible when you have a sex with a corpse. Not only can there be no mutual love expressed through the sex, there cannot even be mutual pleasure.

This way of differentiating homosexuality from bestiality and necrophilia implicitly appeals to a particular view of sexual ethics. According to this "significance view," for sex to be morally permissible it must be the expression of mutual romantic love. This condition, according to the significance view, is necessary but not sufficient. In other words, sex cannot be morally permissible unless it is an expression of mutual romantic love, but being the expression of such love is not enough to make it morally permissible. Other conditions might also need to be met.

Much, but certainly not all, homosexual sex is the expression of mutual romantic love and thus meets the requirement imposed by the significance view. The same is true of much, but again not all, heterosexual sex. Given this, those who accept the significance view of sex can differentiate *some* heterosexual activity from all bestiality and necrophilia. However, they will have to disapprove of some homosexual—and some heterosexual—sex, namely, casual sex, which is not the expression of mutual romantic love.[5]

Some defenders of homosexuality may be willing to embrace that implication. In other words, they may be willing to accept that sexual activity, whether heterosexual or homosexual, between people who love one another is morally acceptable, but that casual sex, whether heterosexual or homosexual, is not. However, others will resist the idea that casual sex is morally impermissible (or even merely morally flawed). The way to permit such sex is to adopt the "casual view" instead of the "significance view" of sex.[6]

According to the casual view, sex is just like any other potentially pleasurable activity. It is subject to the usual moral constraints but is no more morally laden than any other activity. Thus, just as there is nothing wrong with enjoying a film or playing tennis with mere acquaintances or even strangers, so there is nothing wrong with enjoying sex with people one barely knows or does not know at all. If having a meal with a different stranger each night is not impermissible culinary promiscuity, then having sex with a different person each night is not impermissible sexual promiscuity.

The casual view's implications for promiscuity will be congenial to those who think that there is nothing wrong with promiscuity. However, the casual view also implies that there is nothing inherently wrong with bestiality and necrophilia. As noted earlier, the usual moral constraints apply—including not to harm animals and perhaps[7] not to utilize dead bodies in ways to which the person whose corpse it is did not give prior consent, when they were alive. Beyond those sorts of constraints, however, everything is permissible. Just as one may (non-sexually) pet and play with animals, one may sexually pleasure animals or derive sexual pleasure from them. Similarly, on the casual view there is no unavoidable bar to deriving sexual gratification from a corpse. If it is morally permissible to put the dead on display (e.g., at a wake) and to stroke their heads in a last "goodbye," then on the casual view of sex, it should sometimes also be permissible to engage in sexual activity with them.

Thus, while the significance view can distinguish the permissibility of *some* homosexual activity from the impermissibility of *all* bestiality and necrophilia, according to the casual view *all* of these practices are permissible under some circumstances. In other words, while the significance view rejects the first premise of the argument we are considering, the casual view accepts the first premise but rejects the second. The upshot of either view is that there is nothing wrong with homosexual activity. Whether one accepts the significance view or instead the casual view matters only to the permissibility or otherwise of bestiality and necrophilia.

This has some interesting implications. Many—but not all—of those who oppose homosexuality are sexual conservatives who accept the significance view. The foregoing shows that they should be much less enamored of the argument that purports to connect homosexuality with bestiality and necrophilia. Those who accept the significance view are precisely the people who have the theoretical resources to oppose bestiality and necrophilia while permitting homosexual sex in the same circumstances in which they would permit heterosexual sex.

At the same time, many—but again not all—of those who defend homosexuality are sexual liberals who accept the casual view of sex. They cannot, on pain of contradiction, dismiss the equivalence of homosexuality with bestiality and necrophilia by appealing to the significance view. They are faced with the choice of either abandoning the casual view or accepting its permissive implications for bestiality and necrophilia. The latter is perfectly coherent,[8] but since many defenders of homosexuality are not willing to take that step, they have a problem.

The options are presented in Table 13.1.

Table 13.1 Two views of sexual ethics and their implications

	Homosexuality	*Bestiality*	*Necrophilia*
Significance view	Permissible	Impermissible	Impermissible
Casual view	Permissible	Permissible	Permissible

Advocates of the casual view cannot appeal to a range of other arguments to rule out bestiality and necrophilia (while allowing homosexuality). For example, they cannot claim that these practices are disgusting, or non-procreative, or unnatural, or abnormal. These are all appalling arguments. Why they are so appalling need not be spelled out here because the very same arguments can be applied, albeit no more successfully, to homosexuality. Many heterosexuals find homosexual sex disgusting and regard it as unnatural. Homosexual sex is not procreative, and it is abnormal, at least in the statistical sense (that a minority of the population engage in homosexual sex). The point is that defenders of homosexuality cannot selectively adopt these poor arguments to tarnish bestiality and necrophilia if they (correctly) reject them when they are advanced against homosexuality.

Is there any other way in which those who accept the casual view can distinguish homosexuality from bestiality and necrophilia? Some might want to appeal to consent, arguing that it is consent, rather than the significance view's condition, that is required in order for sex to be permissible. That would render consensual casual and homosexual sex permissible. We have already seen that such a consent criterion would not rule out all necrophilia. More specifically, it would not rule out necrophilia where the person, while they were still alive, gave prior consent to this use of their corpse.

It cannot be argued that the consent must be given immediately before the sexual activity begins and then continue through to its end. In other words, it cannot be argued that advanced consent is insufficient to render later sex with the corpse impermissible. If this were the case then the (advance) consensual use of corpses for medical education, medical, forensic and related experimentation, and for organ donation, would all be wrong. The idea of advance consent—that is, consent given in advance of death, or even of incompetence during life—is widely established and recognized.

Advocates of the significance view can argue that sex is different. They have the resources for arguing that because each party to the sexual activity must be expressing romantic love for the other party through the act, it is not possible for sex with corpses to be morally permissible.[9] Advocates of the casual view, by contrast, are unable to say this. The essential characteristic of their view is that sex is not different. They think it is just like everything else.

According to the casual view, could the consent condition fare any better at ruling out bestiality, given that animals (presumably) cannot give informed consent to sex with a human? They might be capable of a lower level of agreement, sometimes called "assent" rather than "consent," but they do not seem capable of giving full consent. They cannot give it in the moment, but they also cannot give it well in advance, as a person can before they die. Is this a way to distinguish homosexual sex (to which the parties can consent) from bestiality?

Advocates of the casual view must answer in the negative. This is because they cannot explain why the consent of animals is needed for non-harmful sex with them. When beings, such as non-human animals, are unable to consent because they lack competence to make decisions, others have to decide on behalf of those beings. There are certainly constraints on how those decisions may be made. Such constraints typically include a duty not to harm. However, if we are considering instances of bestiality that do not harm the animal, then this constraint is not violated. The mere absence of consent cannot be sufficient to render bestiality problematic. Consent is always absent in our dealings with animals. If you may rub your dog's back without his or her consent, why, on the casual view, may you not rub her or his genitals, perhaps with yours?

In this particular example, the animal might derive some pleasure, in which case it may actually be benefited by the sexual contact. However, in making decisions for incompetent beings, we do not always have to choose in ways that benefit them. Some decisions we reasonably and permissibly make for incompetent beings are ones that neither benefit nor (seriously) harm them. For example, a parent might do no wrong in authorizing a blood draw from his or her child for research rather than therapeutic purposes. It is thus not clear why consent is unnecessary only in cases in which the animal is actually benefited by the sexual event.[10]

Responding to the Second Version

The focus in this chapter has been on the first version of the argument that attempts to plot a slippery slope from accepting homosexuality to accepting bestiality and necrophilia. This version of the argument focuses on the permissibility of *sex* with people of the same sex, with animals, and with corpses.

What about the second version of the argument, which alleges a slippery slope from allowing gay marriage to allowing inter-species marriage? Although this version has some similarity to the first, there are also some crucial differences. Most important is that it seems possible to separate same-sex from different-species marriage without having to appeal to the significance view of sex (although one could develop an argument that included such an appeal[11]).

One need merely appeal to a particular and quite plausible understanding of what marriage is in order to distinguish same-sex and different-species marriages. According to this view marriage is a (particular kind of) temporally extended contractual relationship. Of course, people can have enduring relationships with animals, as many people do, typically non-sexually, with their companion animals. However, relationships between humans and animals cannot be *contractual*, at least not in the explicit way that marriage is. There can be an implicit agreement between humans and animals, as happens, for example, when a human agrees to care for a dog and the dog stays with the human even if it has the opportunity to leave. However, a marriage is typically a more explicit kind of agreement. Because non-human animals cannot understand the institution of marriage, they cannot enter into an explicit marital contract with a human.

One way to highlight this is to consider the *termination* of a marital contract—a divorce. A dog can leave a human—*separate*, if you will—but a dog cannot *divorce* a human. When I say "cannot," I do not mean that the law or societal conventions do not allow it. Instead, I mean that a dog lacks the capacity to terminate a relationship in the kind of way that a divorce involves. If that is the case, then, given that divorce and marriage require the same kind of contractual capacity, a dog also cannot enter into a marital relationship.

This is not to say that there cannot be some other understanding of what marriage is, such that a dog could marry a human. If, for example, one understood marriage in such a way that only an implicit agreement or even merely an enduring relationship were required, then a dog could marry a human. However, those who want to distinguish same-sex from different-species marriages are not obliged to accept that less demanding (and arguably less plausible) understanding of marriage. Thus, they have a mechanism for differentiating same-sex from different-species marriage.

We should distinguish this appeal to what marriage is, from a different idea of marriage to which conservative opponents of gay marriage often appeal. They insist that marriage is, by definition, a union of a man and a woman, and not the union of two men or two women. They think that whether we should recognize gay marriage can be resolved definitionally.

The problem with this appeal to an understanding of marriage is that it is arbitrary in a way in which appealing to marriage as a specific kind of explicit contractual relationship is not. It stipulates the sex of those who have historically been allowed to marry rather than appealing to a sex- and species-neutral capacity[12] that quite plausibly can be thought to be essential to marriage. It just so happens that, so far as we know, no animals have this capacity. If there are some animals who have the capacity, they and we are unable to communicate at a sufficiently sophisticated level to be able to enter into explicit contracts, including marital contracts, with one another. Nor is it the case that *all* humans have the requisite capacity to enter into a marriage contract. A human who is profoundly cognitively impaired and who cannot, as a result, understand what a marriage is lacks the capacity to get married, whether to somebody of the same or different sex.[13]

Conclusion

Homosexuality and gay marriage continue to elicit opposition. That this opposition is inappropriate, does not mean that we should ignore the arguments of those who continue to condemn it or restrict its social acceptance. Perhaps engaging their arguments will not lead them to change their minds, but ignoring their arguments is even less likely to bring about such a change. Short of a

Damascene moment, those critical of homosexuality and gay marriage are not going to change their minds unless we take their arguments seriously and engage them.

However, as my argument in this chapter shows, it is not only *opponents* of homosexuality who might be enlightened by such an exchange. Reflecting on the arguments of homosexuality's critics can shed helpful and critical light on the views of many who have no objection to homosexuality. It is not that their views on homosexuality should change. Instead, by reflecting on why their opponents' argument is mistaken, they may need to adjust some of their views about other sexual practices.[14]

Notes

1. At the time of writing, these places include, but are not limited to, Brunei, Iran, Somalia, and Sudan. "Death penalty for homosexuality," *Wikipedia*, https://en.wikipedia.org/wiki/Death_penalty_for_homosexuality (Accessed 25 July 2019).
2. For example, this argument was advanced in an amicus brief by the states of Alabama, South Carolina, and Utah to the United States Supreme Court in the case of *Lawrence* v *Texas*. The authors of the brief were writing in support of a Texas anti-sodomy law. There are also many jurisdictions that include prohibitions on bestiality and sodomy in the same sections of the penal code, thereby suggesting some kind of equivalence. See Lucas Paoli Itaborahy and Jingshu Zhu, *State-Sponsored Homophobia: A world survey of laws: Criminalisation, protection and recognition of same-sex love*, Eighth Edition, International Lesbian Gay Bisexual Trans and Intersex Association, May 2013.
3. Another way to understand the argument is as a *reductio ad absurdum*. In that case, we could take something like the following form:

    1. Assume that there is nothing wrong with homosexuality.
    2. If there is nothing wrong with homosexuality, then there is nothing wrong with bestiality and necrophilia.
    3. But there is something wrong with bestiality and necrophilia.
    4. Therefore, the original assumption must be incorrect and there is something wrong with homosexuality.

4. Bill O'Reilly is among those who has advanced this argument. Andrew Walzer, "O'Reilly's Ark: Gay marriage could lead to goat, duck, dolphin, and turtle marriage," *Media Matters for America*, 12 May 2009, https://www.mediamatters.org/research/2009/05/12/oreillys-ark-gay-marriage-could-lead-to-goat-du/150069 (Accessed 25 July 2019).
5. There is a possible view intermediate between the significance view and the casual view. On this weakened significance view, a necessary condition for sex to be morally permissible is that it be the expression of at least mutual affection or friendship, which is a lower bar than mutual romantic love. This view would render more homosexual and heterosexual sex permissible and would rule out both bestiality and necrophilia. However, it too would not permit truly casual

sex. (There are costs to accepting this view over the significance view. For example, unlike the significance view, it may not be able to rule out all pedophilia. See David Benatar, "Two views of sexual ethics: Promiscuity, Pedophilia, and Rape," *Public Affairs Quarterly*, Vol. 16, No. 3, July 2002, pp. 191–201).

6. I differentiated these two views in "Two views of sexual ethics: Promiscuity, Pedophilia, and Rape," *Public Affairs Quarterly*, Vol. 16, No. 3, July 2002, pp. 191–201. In that paper, I did not spell out the implications of these two views for either bestiality or necrophilia.
7. I say "perhaps" only to acknowledge the view of some that there are no post-mortem interests and that we cannot either harm or wrong the dead.
8. This is the view of Peter Singer, for example. See Peter Singer, "Heavy Petting," *Prospect*, 20 April 2001, https://www.prospectmagazine.co.uk/magazine/heavypetting (Accessed 23 July 2019).
9. The claim here is not that those espousing the significance view *must* reject necrophilia. Instead, the claim is that they *can*. They would have to reject necrophilia if the expression of mutual romantic love must be synchronic. However, if the expression of mutual romantic love were permitted to be diachronic, then some necrophilia could be permissible according to the significance view.
10. We are sometimes even justified in seriously harming others (without their consent), whether or not they are competent. Sometimes, for example, we are justified in taking self-defensive action that may harm an innocent bystander. However, I leave aside such cases because sex that is harmful to an animal is very unlikely to be among such instances.
11. For example, those who combine the significance view of sex with the view that sex is integral to an ideal marriage might argue that only marriages which could meet the significance condition ought to be recognized. This would preclude inter-species marriages but would permit many homosexual ones.
12. By "species-neutral capacity," I obviously do not mean a capacity that members of all species have. Instead, it is a capacity that makes no reference to species. It may still be the case that only members of one (or some) species have a species-neutral capacity.
13. Understanding marriage as a temporally extended contractual relationship also explains why the second version of the argument would be even more ridiculous if it were to allege a slippery slope from gay marriage to marriage by a living person to a corpse. There cannot be a (two-way) relationship with a corpse. Moreover, unless corpses are embalmed, they do not last for very long, thereby precluding the possibility of a temporally extended marriage.
14. I am grateful to David Boonin for his comments on this chapter.

CHAPTER 14

The Immorality of Premarital Sexual Abstinence

Alastair Norcross

Introduction

In this chapter, I'm going to argue that the moral seriousness of marital promises, whether that derives from the nature of the promises themselves or from the effects of breaking them, supports the conclusion that premarital sexual abstinence is seriously immoral. That is, that it is seriously immoral to marry someone without first engaging in extensive sexual activity with them (and probably also at least a period of cohabitation).

Consider the following cases:

Coffee Invite: You are sitting in your office at 10:45 when a friend calls and invites you to coffee at 11. You accept. Your friend asks you to promise to be there, because she will be leaving her office now to walk to the coffee shop, and her cellphone isn't working, so you won't be able to notify her if you change your mind, and the last time you said you'd come for coffee, you didn't show up. You recall that you have an important meeting at 11 either today or tomorrow, but you can't remember which. If the meeting is today, you will go to the meeting instead of coffee with your friend. You can easily check, but you don't, and simply make the promise.

Friend in Need: A friend phones from a deserted gas station with only a payphone (one of the few that still works) and a coffee machine. She has lost her cellphone and has enough change left either to make two more phone calls or to buy some coffee. The night is cold, and she needs a ride. Your cellphone is about to lose its charge. You say you will give her a ride. She asks you to promise, because she will use her remaining change for coffee, if she is sure you will come;

A. Norcross (✉)
University of Colorado Boulder, Boulder, CO, USA
e-mail: alastair.norcross@colorado.edu

D. Boonin (ed.), *The Palgrave Handbook of Sexual Ethics*,
https://doi.org/10.1007/978-3-030-87786-6_14

otherwise, she will try one or two other friends. You know that there is a 50 percent chance that your brother has borrowed your car. You can check whether this has happened, by walking over to your window and looking out. However, you don't do that but instead simply make the promise. Your friend spends the rest of her change on coffee.

Doughnuts: There is a plate of doughnuts in front of you. You know that there is a 50 percent chance they are all jelly-filled and a 50 percent chance they are all apple-filled. You love both kinds. Your friend loves jelly-filled, but is allergic to apple-filled, and would become quite ill if she were to take even one bite of one. She asks you to pass her a doughnut, if you can promise it is jelly-filled. You can easily check to see whether the doughnuts are jelly-filled or apple-filled, but you don't. You simply promise it is jelly-filled and pass her a doughnut.

Making the promise in these situations is clearly morally problematic, increasingly so with each case. The more there is at stake in each promise, and thus the more serious the promise, the worse it is. Furthermore, it seems that your behavior would be bad in each case, even if things turn out OK. Even if it turns out that your important meeting was tomorrow, and not today, you shouldn't promise to meet your friend for coffee without first checking. Even if your brother hasn't borrowed your car, and you succeed in giving your friend a ride from the cold gas station, you shouldn't have promised without first checking. And especially in **Doughnuts**, your behavior is morally outrageous, even if the doughnuts turn out to be jelly-filled, and your friend suffers no ill effects. Once again, you should obviously check first, before promising your friend that the doughnut you are passing is jelly-filled. Notice that the checking process here would have been quite pleasurable, given how much you enjoy both kinds of doughnut.

So, what explains the badness of your behavior in these situations? There are a few obvious, interconnected reasons. (a) Breaking promises is bad. Making a promise that you know you have a significant chance of breaking involves running a significant risk. The more significant the promise is to the person you make it to, the worse it is to break it. (b) Making a promise involves knowingly getting the promisee to believe something, usually that you will perform the promised action, or simply that the thing you promise is true is really true. If you know that there is a significant chance that you will not perform the promised act, or that the thing you promise to be true isn't true, you do something that you know has a significant risk of deceiving the promisee. The worse the consequences of the deception (especially for the person who would be deceived), the worse it is to risk those consequences. (c) In many cases, there are bad consequences to breaking a promise, and often there are worse consequences to making and breaking a promise than to not making the promise in the first place. The worse the consequences (especially for the promisee), the worse it is to risk those consequences by making the promise.

The Relevance of This to Premarital Abstinence

What does all this have to do with premarital abstinence? Consider this part of a traditional wedding ceremony.

> **The Declaration of Consent**
> *The Celebrant says to the woman*
> N., will you have this man to be your husband; to live together in the covenant of marriage? Will you love him, comfort him, honor and keep him, in sickness and in health; and, forsaking all others, be faithful to him as long as you both shall live?
> *The Woman answers*
> I will.
> *The Celebrant says to the man*
> N., will you have this woman to be your wife; to live together in the covenant of marriage? Will you love her, comfort her, honor and keep her, in sickness and in health; and, forsaking all others, be faithful to her as long as you both shall live?
> *The Man answers*
> I will.
> *The Celebrant then addresses the congregation, saying*
> Will all of you witnessing these promises do all in your power to uphold these two persons in their marriage?
> *People* We will.

This is from the *Book of Common Prayer*. Many wedding ceremonies take different forms, often with no religious elements at all, but most of them include an exchange of vows, in which both parties make commitments to each other, very similar to the traditional ones above. Notice that the celebrant refers to the commitments expressed by the spouses as "promises." You clearly don't have to utter the words "I promise" in order to make a promise. Pretty much every wedding ceremony involves the spouses making promises to each other. Certainly every one I have attended or officiated at. Notice also the seriousness and scope of the promises. Again, pretty much every wedding involves similarly serious promises. You are not promising your spouse merely that you will meet them for coffee or give them a late-night ride home. You are promising to *love*, *comfort*, *honor*, and *keep* them, in *sickness* and in *health*, *exclusively*, *for the rest of your lives*! For most people, the promises they make at their wedding(s) are the most serious and consequential promises they will ever make. Breaking your wedding vows may very well lead to much worse consequences for your spouse and for you than giving your allergic friend an apple-filled doughnut (which will make them quite sick, but only temporarily, and not fatally). But, as is well-known, for any randomly selected promising partner exchanging vows at a wedding, there is a high chance that they will break those vows. Although the divorce rate has been dropping recently (as has the marriage rate), there is still somewhere between a 40 percent and 50 percent likelihood of divorce for each marriage. Clearly, divorce results in breaking marriage vows (the causal relationship often goes both ways). But many marriages that *don't* end in divorce also involve broken vows. It is not uncommon for couples to remain married, while at least one person ceases (either temporarily or permanently) to love, comfort, honor, remain faithful to, or engage in any number of other promised behaviors toward the other. Such breaking of vows often has severely negative effects on the well-being of some or all involved in a marriage, including any relevant children. The negative effects on children of either divorce or the kind of marital strife caused by failure to keep marriage vows can be both severe and long-lasting. This information is not esoteric. Pretty much everyone who gets married is aware of the high rate of divorce and other forms of breaking of marriage vows and of the potential negative consequences thereof.

Presumably, with the boringly predictable exceptions of many people in the fields of entertainment and politics, spouses don't make these serious marital promises while at the same time believing that they will break them, or even that they are likely to break them. Most people making marital promises believe that they can, and indeed will, keep them. But are they justified in these beliefs? In the examples with which I began this chapter, your moral culpability stems from the fact that you knew there was an easy way to check whether you would, in fact, keep the promise you were making, but you didn't check. Even if you strongly believed that you would keep your promise, you shouldn't have believed that and certainly shouldn't have made the promise without checking. Is there anything similar in the case of marital promises? Many different factors affect whether couples divorce or whether one or the other breaks their marital vows. Some of these factors are extremely difficult to predict (such as whether the couple will encounter medical and/or financial struggles). But others are amenable to investigation prior to marriage. Sexual and personal compatibility are strongly relevant to the probability that a couple will honor their marriage vows. To take an extreme example, if you find someone both physically and personally repugnant, you almost certainly shouldn't promise to love, honor, cherish, and remain faithful to them for the rest of your life.[1] The mere fact that you don't find your proposed life partner physically and personally repugnant, though, is pretty weak evidence of the kind of sexual and personal compatibility required for a good chance of not breaking your marital vows. Fortunately, it is possible to get significantly more, and better, evidence. Plenty of sexual activity, certainly of the kind you expect to engage in once married, and extensive close personal contact (perhaps including a period of cohabitation) will provide both partners with significant evidence concerning sexual and personal compatibility. Such evidence is unlikely to be enough to guarantee that you will not break your marriage vows, but it's a darn sight better than not having it at all.

It appears, then, that there are powerful moral reasons not to make the kinds of promises involved in most marriage vows without first engaging in quite extensive sexual activity and close personal contact. Otherwise, you risk making a promise that you will not keep, and risk causing, and enduring, the many serious harms that may result from broken and/or unhappy marriages.

The inescapable conclusion is that, in order to honor and properly engage in any version of marriage that involves the kinds of promises common to anything like the traditional wedding ceremony, you must engage in extensive premarital sex and close personal contact.[2] In other words, premarital sexual abstinence is seriously immoral.

Well Duh!

This all might, and indeed should, seem like moral common sense. "Well duh!, Captain Obvious," you say, only a moral monster would jump headlong into the kinds of serious, lifelong commitments involved in most marriages, without first getting plenty of evidence that they'll be able to keep their promise and won't instead be causing extensive suffering, especially when it's possible, and even sometimes enjoyable, to get such evidence. But my extensive research has turned up the existence of many organizations opposed to these methods of gathering evidence and thus at odds with the moral seriousness of most marriage vows. Many, perhaps most (maybe even all) versions of Christianity (one of the most popular religions in the world, certainly in the top ten) frown on premarital sex. For example, Roman Catholicism, whose priests, cardinals, popes, and so on are themselves forbidden from marrying, both encourages most of its adherents to participate in the institution of marriage and forbids them from gathering evidence of compatibility by engaging in premarital sex. This is, of course, not the only thing that the Catholic church (and many other churches) forbids. But there are also organizations, whose sole (or at least principal) purpose seems to be to oppose the kind of gathering of evidence that one will honor one's marriage vows that I am here showing to be morally required. For example, in 1993, Southern Baptists (a denomination of Christianity particularly popular and influential in the United States) formed a group with the name "True Love Waits" (TLW), to promote sexual abstinence outside of marriage (which entails premarital abstinence) for teenagers and college students. TLW focuses its disturbingly successful recruitment activity on teenagers and college students but has preyed on victims as young as 11 (e.g., the popular singer Joe Jonas). The impressionable youth ensnared in the TLW web recite the following pledge: "Believing that true love waits, I make a commitment to God, myself, my family, my friends, my future mate and my future children to be sexually abstinent from this day until the day I enter a biblical marriage relationship." Many of them also wear a "purity ring." My argument against premarital abstinence is based in moral common sense, focusing, as it does, on the strong moral reasons not to make serious promises without first gathering extensive evidence that you will keep them and not to risk serious harms to others and to yourself. But, despite the commonsensical nature of my argument, its conclusion is widely rejected by large and influential groups. Hence the necessity of my stating the bleeding obvious.

What I Am Not Saying

I'm not arguing that you should have sex with everyone you are attracted to. My argument is perfectly compatible with the view that sex should be reserved for serious relationships. In fact, you could have sex with only one person your whole life and still not fall foul of my argument. If you only ever have sex with the person who ends up as your spouse, you will be fine, so long as you didn't wait until marriage before having sex with that person. You will also be fine (by my argument, at least), if you only ever have sex with one person and don't end up marrying them (either because you stay together but never marry or because you break up and don't have sex with anyone else). My argument is about the immorality of marrying someone you haven't previously had (plenty of) sex with. It is not about the morality of having various amounts of sex (a little, a moderate amount, lots, truly Wilt Chamberlainesque quantities, etc.) with various numbers of people.

I'm not making any claims about whether you should or should not get married at all. However, since the evidence you get by having extensive sexual contact (and other forms of close personal contact) with a proposed partner, while better than not doing those things, still leaves a substantial risk of breaking your marital promises, you might think that the argument also counts against marrying at all. Whether, and to what extent, it does depend on your views of the wrongness of breaking promises. If it's the fact of breaking a promise, as opposed to the consequences of breaking the promise, that moves you, my argument certainly counts against a marriage that involves traditional wedding vows. Of course, you could simply alter the vows to say "or at least make a reasonable effort to do so" or something like it at the end. But, of course, the real reason not to marry without getting as much evidence as you can that you can keep your marriage vows is that the consequences of breaking the vows are bad. Most of these bad consequences apply also to relationships that don't involve explicit promises. A couple who never explicitly promise to remain faithful, and so on, but who nonetheless engage in a stable long-term relationship will suffer most of the same bad effects of a breakdown of that relationship as a formally married couple. The same applies to any children the couple may have. It's possible that some of the bad effects of a breakup are worse with formal marriage, because of extra legal costs and complications (but the status of common-law marriage provides many of the same costs and complications in most states in the United States). Perhaps, then, my argument could be taken to count against engaging in long-term relationships at all. Perhaps. But it's also possible that the benefits of such relationships, even weighed against the risks of the painful breakup, are worth it. What is clear is that it's always better to hold off on making serious commitments, whether implicit or explicit, until you have as much evidence as you can reasonably get that you will fulfill those commitments.

Objections and Replies

(i). *But isn't premarital sex wrong, because god/my pastor/my parents say so?*

Answers: No. If there is a god who is truly benevolent and wise, she doesn't disapprove of premarital sex. How could she? If *I* am able to see what's bad about premarital abstinence, presumably she is too. Her wisdom and benevolence surely exceed mine. But perhaps there's a god who is either not truly benevolent or not particularly wise? Some religious traditions certainly portray god as being very powerful, but not very bright. For example, some traditions that claim that god forbids working on certain days, or forbids referring to her by certain names, also claim that the clever use of technology will somehow fool god or that a slight alteration of the spelling of the word "god" will mollify her. But if god isn't truly benevolent or wise, her wishes concerning premarital sex shouldn't concern us. Perhaps we are afraid of the retributive behavior of a vicious dimwitted deity? Although I can't deductively prove that we shouldn't fear such retribution, the recorded history of humanity provides overwhelming inductive evidence that the morality of an individual's behavior bears pretty much no relation to how well they fare. So what about your pastor? Your pastor isn't a moral authority (just ask the sex worker he paid for sex and drugs). If your pastor can't provide good reasons to refrain from premarital sex (and as I've just demonstrated, "god says so" is not a good reason), there's no reason to follow their guidance. Finally, your parents love you but are incapable of rational objectivity when it comes to your sex life.

(ii). *But what about all those studies coming from all those "think" tanks in Colorado Springs, saying that people who cohabit before marriage have an increased divorce rate over those who don't?*

Answers:

(a) The studies that get all the attention concern cohabitation before marriage, not specifically premarital sex, which is the focus of this chapter. Even if they did show that *cohabitation* increased the risk of divorce, that would tell us nothing about whether *premarital sexual abstinence* reduces the chance of divorce. Lots of premarital sex takes place between couples who are not cohabiting. Studies that ask married (or formerly married) participants about their premarital sexual behavior would be even harder to trust. Everybody lies about sex, even on anonymous surveys. In particular, people lie about doing things they think of as wrong, or even just that they think of as being the kinds of things that others think of as wrong. It's possible that many of the still married participants who claim not to have engaged in premarital sex really did.

(b) Don't confuse correlation with cause. The kinds of factors that make people reluctant to engage in premarital sex (or cohabitation) may well also make them reluctant to divorce. Many people who refrain from premarital sex do so for religious reasons. Many of these religions also frown on, or even outright forbid, divorce.

(c) Relatedly, these studies don't tell us how many premaritally sexually abstaining couples have miserable marriages, blighted by infidelity, and/or ceasing to love, comfort, and honor one another. Bad as divorce is, sometimes it's better than remaining in a bad marriage. Simply not divorcing doesn't mean that you keep your marital vows. In fact, you may break your promises less by divorcing than by staying in a loveless marriage.[3]

(d) Even if these studies focused on premarital sex (as opposed to cohabitation), and obtained similar results, they wouldn't tell us how many bad marriages were prevented by premarital sex. Even if it's true that actual married couples who engage in premarital sex are more likely to get divorced than married couples who don't, and even if we are only concerned with minimizing the risk of divorce (which, obviously, we aren't, given all the other problems with failed, or even suboptimal, marriages), sex before marriage may still reduce the risk of divorce. In order to explain, I'll have to consider a somewhat complicated example, involving numbers and percentages, so bear with me.

For the sake of illustration, suppose that the ratio of married couples who engage in premarital sex to those who don't is 3:1. Suppose further that for every 75 married couples who previously had premarital sex (with each other), 37 of them divorce, a rate of nearly 50 percent. And suppose that for every 25 married couples who abstained, only 10 divorce, a significantly lower rate of 40 percent. It could appear from these figures that the choice to engage in premarital sex increases the risk of divorce by nearly a quarter (from 40 percent to 49.3 percent). But we clearly wouldn't be justified in drawing such a conclusion merely from these figures. Suppose, further, that for every 75 married couples who engage in premarital sex there are another 75 couples who get enough evidence from having sex that they decide not to marry in the first (or rather second) place. And suppose further that had those 75 couples married without first having sex, 65 of them would have divorced. This makes allowance for the fact that it may be harder (both legally and psychologically) to get divorced than to dissolve a less formal partnership. Although all 75 couples would get the same evidence of incompatibility in their marriage as they would have gotten from premarital sex, perhaps 10 of them would have gritted their teeth and stayed together, because of the greater difficulty of dissolving the formal marriage. Suppose further, and overly charitably to the Focus on the Family folks, that had the 75 couples who married after premarital sex married without such evidence gathering, only 30 of them would have divorced (the same 40 percent rate as for the other group). Finally, suppose even further that had the 25 couples who married without premarital sex first engaged in this evidence gathering, 5 of the couples who in fact divorce would have chosen not to marry, but 7 of the remaining 20 would have divorced (the 5 who do in the actual world and 2 others). Now, consider all the 175 couples in this scenario. In the actual world, there are 150 couples who engage in premarital sex, of whom 75 marry, and 37 divorce. There are also 25 premaritally sexual abstinent couples who marry, of whom 10 divorce. So, between these 175 couples, there are 100 marriages and 47 divorces. Had they all chosen to engage in premarital sex, there would have been 5 fewer marriages, and 3 fewer divorces, for totals of 95 and 44. On the other hand, had they all chosen to marry, without first having sex (with each other), there would have been 75 more marriages, and 58 more divorces, for totals of 175 and 105.

So let's look at the difference the choice to engage in premarital sex made to the couples who made the choice and would have made to the couples who didn't. For the 25 couples who chose premarital abstinence, the choice to engage in premarital sex would have prevented 5 of them from entering a failed marriage and would have resulted in 2 marriages failing that didn't actually fail (defining a "failed marriage" narrowly as only being one that ends in divorce). For the 150 couples who chose to have sex with each other before deciding on marriage, that choice prevented 65 of them from entering a failed marriage, prevented a further 10 from entering a successful (as in divorce-free) marriage, and led to a further 7 having a failed marriage that would otherwise have been successful. Unless we value the existence of merely divorce-free marriages as lexically prior to the absence of marriages that end in divorce, or fail in other ways, this imagined scenario is clearly not one in which the weight of reasons opposes premarital sex. Quite the opposite. We don't know the actual numbers, but this illustration shows that the various studies, even if they focused on premarital sex, and even if they could be trusted, don't even come close to showing that premarital sex even increases the overall risk of divorce, let alone leads to worse outcomes. Obviously, similar considerations apply to the question of whether the actual studies, focused on cohabitation, give good reasons for thinking that cohabitation leads to bad outcomes.

(e) Even if you are just interested in the question about cohabitation (as opposed to premarital sex), it's worth noting that there is nothing like a consensus among researchers about the effects of premarital cohabitation on later divorce. For example, two reports, released within a couple of weeks of each other in 2018, came to sharply divergent conclusions from largely the same data. As reported in *The Atlantic*, one study[4] came to the conclusion that "couples who lived together before marriage had a lower divorce rate in their first year of marriage, but had a higher divorce rate after five years."[5] But a report published by the

> Council on Contemporary Families, looking at the same data sets, claimed that the relationship of cohabitation and divorce has reversed over time: "those who were willing to transgress strong social norms to cohabit from the 1950s to 1970 were also more likely to transgress similar social norms about divorce. ... But as cohabitation became more widespread, its association with divorce faded. In fact, since 2000 premarital cohabitation has actually been associated with a lower rate of divorce, once factors such as religiosity, education, and age at co-residence are accounted for."[6]

One more quick point about the studies: Because they are looking specifically at the relationship between premarital cohabitation and later divorce, they don't tell us how many good, loving, stable relationships never involve formal marriage.

(iii). *But doesn't premarital sex increase your risk of disease?*

Answers: Not if done right. Furthermore, many STDs are easily treatable. Finally, it may actually reduce your risk of such diseases contracted from a future cheating spouse. Of course, if we are really that concerned with the prevention of STDs, we could go to the absurd lengths of ruling out all sexual activity that carries even a slight risk of infection (perhaps leaving an exception for specifically procreative sex). Not only would that be an absurd conclusion, it would be irrelevant to my arguments in this paper, because it would apply equally to marital and extramarital sexual activity.

Notes

1. I say "almost certainly," because otherwise, no doubt, some clever dick philosopher will devise an imaginary scenario in which some powerful monster threatens to unleash calamity and suffering on the world unless you promise to love, honor, cherish, and remain faithful to Donald Trump for the rest of your life.
2. Although the focus of this chapter is on premarital sexual abstinence, it should be clear that the arguments extend to close personal contact that goes beyond sexual activity, such as premarital cohabitation. A detailed discussion of the arguments as applied to cohabitation is beyond the scope of the chapter.
3. As I noted above, the studies themselves focus on cohabitation and not premarital sex. But even restricting the focus in this way, in order for what these studies claim to show to provide good reasons to refrain from premarital cohabitation, they would have to be combined with more extensive and intrusive studies that also measured at least (i) the extent to which people break their marriage vows within marriages, (ii) the effect on the well-being of all involved of such breaking of vows, and (iii) the differential effect on well-being of divorcing versus remaining in a failed marriage. It's not clear how such studies could be constructed, but until they are, we learn very little of use from the existing studies.

4. Rosenfeld, Michael and Roesler, Katharina, "Cohabitation Experience and Cohabitation's Association with Marital Dissolution", *Journal of Marriage and Family*, Vol. 81, no. 1, 2019.
5. Ashley Fetters, "So Is Living Together Before Marriage Linked to Divorce or What?", *The Atlantic*, October 24, 2018.
6. Kuperberg, Arielle, "From Countercultural Trend to Strategy for the Financially Insecure: Premarital Cohabitation and Premarital Cohabitors, 1956–2015", *briefing paper prepared for the Council on Contemporary Families*, October 2018.

PART III

Sex and Consent

CHAPTER 15

Sexual Autonomy and Sexual Consent

Shaun Miller

With the cultural reckoning of the #MeToo movement, current conversations are largely revolving around how to give and receive consent properly. Sexual consent is based on the sexual choices and preferences of the people involved. The ability to make choices and exercise one's preferences is intimately connected to autonomy. Therefore, to discuss sexual consent, we must also talk about sexual autonomy. With varying discussions revolving around sexual consent, this chapter distinguishes among three different theories of sexual consent and discusses what these different theories entail. It also shows how these three different ideas of sexual consent correspond to three different types of sexual autonomy. "Procedural Sexual Autonomy and Consensual Minimalism" section discusses the first view: procedural sexual autonomy and consensual minimalism. This view is the starting point of what sexual consent has been, but there have been disadvantages to this view. "Substantive Sexual Autonomy and Consensual Idealism" section discusses substantive sexual autonomy and consensual idealism. This view covers the pitfalls of the previous view of autonomy and consent by making autonomy and consent more robust and having a higher standard. However, there are pitfalls with this view as the standards are too high. Finally, "Weak Substantive Sexual Autonomy and Consensual Realism" section discusses a view of sexual autonomy that is based on competency and consent that is based on the context of the people involved, which I consider superior to the previous two theories.

Before we delve into the topic, there has lately been a rich assortment of articles regarding sexual consent. These articles have different purposes and ranges that involve the nature of consent,[1] case studies,[2] particularities of what

S. Miller (✉)
Salt Lake Community College, Salt Lake City, UT, United States

D. Boonin (ed.), *The Palgrave Handbook of Sexual Ethics*,
https://doi.org/10.1007/978-3-030-87786-6_15

makes sexual consent,[3] sexual negotiation and communication,[4] and presuppositions regarding the language of consent.[5] While these are enriching conversations worthy of discussion, I will not be focusing on those questions here. Rather, I will focus on the conditions that pass for consent, what counts as consent, and what it means to meet those conditions.

Procedural Sexual Autonomy and Consensual Minimalism

Let us take sexual autonomy as an application of personal autonomy. When discussing personal autonomy, there is a distinction between procedural and substantive autonomy. Procedural autonomy is when an agent can freely choose an action. It does not matter *what* the action is or *what* kind of being the agent is. The content of people's desires, values, preferences, and beliefs are irrelevant. Procedural autonomy is based on *how* the decision is made. What matters is that the agent makes choices through, at the very least, some critical reflection and that there are options to choose from.

Procedural sexual autonomy is procedural autonomy applied to the sexual realm. In other words, procedural sexual autonomy is choosing to engage or not engage in some sexual activity, and it does not matter what the action is or what kind of being the agent is. Nor does it matter what the agent desires or values. What matters is *how* the choice is made in that the agent can make sexual choices and that there are options among the sexual choices.

What corresponds to this type of autonomy is what I call consensual minimalism.[6] Under this view, the necessary[7] and sufficient condition for what counts as consent is a voluntary informed agreement. The best representatives of this view are Mappes,[8] Wertheimer,[9] and Steutal and de Ruyter.[10] In short, this position entails that if and only if coercion, deception, or incapacitation occur, then consent is invalid, and the sexual act is unethical. According to Mappes, sex is morally permissible when the participants have made a voluntary informed agreement. Morally impermissible sex is when one person undermines the other's consent or when there is no consent to be undermined, meaning that one coerces, deceives, or takes advantage of the other's desperate situation.

Another way to look at consensual minimalism is to say that it is contractual. Raymond Belliotti argues that

> When two people voluntarily consent to interact sexually they create *obligations* to each other based on their needs and expectations. Every sexual encounter has as its base the needs, desires, and drives of the individuals involved. That we choose to interact sexually is an acknowledgement that none of us is totally self-sufficient. We interact with others in order to fulfill certain desires which we cannot fulfill by ourselves. *This suggests that the basis of the sexual encounter is contractual*; i.e., it is a voluntary agreement on the part of both parties to satisfy the expectation of the other.[11]

The sexual contract entails that we are to help fulfill our sexual needs and desires and, in return, there is an implicit expectation to help fulfill the other person's sexual needs and desires.

Philosophers Robin West and Lois Pineau have argued that consent in this framework is problematic. They both use examples of a woman in a problematic sexual encounter, but under consensual minimalism, she passes the conditions for consent. For simplicity, let us call this woman Monica. West argues that traditional consent has mainly been operating for the benefit of men.[12] She argues that it is possible for a woman to have consensual, non-coercive, non-forceful, non-criminal but harmful sex. These harms, however, may be hard to discover. Many women consent to sex when they do not desire it, and it is usually not pleasurable for them.[13] Through several vignettes, she reveals how engaging in sex multiple times under this context can be harmful: her self-assertion and self-possession are weakened, her integrity is lessened, and, most notably for our discussion, her autonomy is drained. In these experiences, having consensual but unwanted sex over time can take a toll on her. Since it is wrong to act in ways that cause (unjustified) harm to oneself or others, not all sexual activity engaged in under conditions of voluntary informed consent is morally acceptable. Moreover, even if a "yes" was obtained, the "yes" could have been worked out through aggressive tactics.

Pineau discusses a young woman (again, we will call her Monica) who goes on a date with a man.[14] Monica feels attracted to him and believes that he feels the same way. She hopes there will be mutual enjoyment with mutual interest. However, the mutual and reciprocal interests are not realized. She feels immense pressure to have sex with him, though she does not want to have the kind of sex he does. The man uses aggressive, high-pressure tactics to have sex with her, such as various pressures that make it difficult for her to say "no." She has trouble disengaging his body from hers and wishes he would go away, but she feels stuck because she feels afraid to say "no" lest his high-pressure tactics become violent. Instead, she goes along with him just to get it over with. He does not even notice she finds the encounter disagreeable and probably still would not have changed course if he had. He congratulates himself for his aggressive tactics in that they paid off. Monica, however, does not feel quite right with the experience.[15]

The problems that West and Pineau reveal are that consensual minimalism allows men to have the upper hand and that the choices Monica makes are not genuine even if it looks as if there are options. Their target also aims against procedural sexual autonomy, which is similar to the criticisms against procedural autonomy. Procedural autonomy has difficulty explaining away oppressive measures. With internalized oppressive norms, what if society is structured in such a way that we are not really autonomous, but we *think* we are?

According to the criticism, it is difficult to ascertain whether a decision is made through our autonomy or because of our socialization. Thus, the agent's choices could be adaptive preferences. Adaptive preferences are when agents under oppressive social conditions unconsciously adapt to choose and prefer various options within that context. Sometimes they are known as "deformed desires" depending on the context. Performing these adaptive preferences relies on subordination and oppression, which do not count as autonomous. Therefore, procedural sexual autonomy is insufficient to describe what counts as autonomous.

Munro illustrates the problem:

> Imagine, for example, a woman who has sex with her male partner, not so much because she wants to, but because she knows that he wants her to. In the absence of overt coercion or deception, this would be condoned and normalized as an unproblematic instance of consensual sex under a minimalist approach. But under this more ambitious consent-plus model, that conclusion would have to be postponed pending an investigation of the context of, and motivations underpinning, the intercourse. If the woman complied because she loves her partner, values their relationship and knows that responding to his sexual advances is important to its health, this may be a legitimate expression of agency, reflecting her endorsement of the benefits that accrue to her as a result of the exchange. By contrast, if she complied because she fears she cannot survive financially without him or is afraid of his (as yet unthreatened) retribution in the event of rebuttal, her involvement emerges as self-alienating, undertaken in pursuit of an unendorsed benefit, and thus problematic.[16]

Notice that under procedural sexual autonomy, the woman in question is autonomous because she made her decision freely, meaning without external coercion. She also decided with some critical reflection. However, procedural sexual autonomy is insufficient for genuine consent to emerge under oppressive circumstances, even minimally.

Substantive Sexual Autonomy and Consensual Idealism

The substantive form of autonomy is more robust than the procedural view and is packed with a stronger view of what it means to be autonomous. If a choice is made under certain conditions and/or the agent is of a certain type, then the agent is autonomous. The choice is not based on subjective criteria, but on some "external" criteria. Morten Ebbe Juul Nielson explains why:

> formal [procedural] conceptions of autonomy that are meant as action-guiding are said to be so, but it remains unclear *why*. If one launches a formal [procedural] conception of autonomy and adds that "autonomous choice should be respected," we would like to know *why*. If choice is not linked to some sort of value—for instance, to a conception of human flourishing—it is hard to see why we should respect it. Formal conceptions of autonomy, then, stop short of providing us with reasons.[17]

Substantive autonomy enriches the agent because it maximizes autonomy or furthers the conditions of autonomy. The necessary conditions for an action to be consensual are the same as consensual minimalism (i.e., voluntariness, being informed, agreeing to the action). However, for substantive autonomy, the choices must also foster the conditions for flourishing. Moreover, from Nielson, we can say that if people are flourishing, then they are substantively autonomous. And to be substantively autonomous requires that the content of their choice is valuable for which people ought to aim. Therefore, there is something underlying autonomy, something that is objectively and normatively valuable for people to flourish. If Nielson is correct, then the objective and normative values are "providing us with reasons" for autonomy.

Moreover, proponents of a strong substantive autonomy argue that "preferences are autonomous if and only if their contents correspond to morally permissible or correct features of the world."[18] Certain external conditions are necessary for the agents to be autonomous and thereby flourish. One significant external condition is the world being just such that agents can achieve autonomy without their options continuing their oppression. Therefore, to be autonomous is to have the right psychology corresponding to a world without severe constraints. No one would choose to be enslaved or be a deferential wife and still consider themselves autonomous according to substantive autonomy theorists. Here, they solve the problem that challenged procedural autonomists.

By applying substantive autonomy to the sexual realm, there is substantive sexual autonomy. Here the sexual choices agents make *do* matter; their desires and preferences behind their choices must have the right content, which is when the choices further one's sexual autonomy and help the agents sexually flourish which, in turn, would thereby constitute their overall well-being. People cannot make *any* sexual choice they want. After all, certain choices can lower one's well-being. Therefore, some choices are illegitimate because certain choices—certain "deformed desires"—do not lead one to flourish.

The type of consent that corresponds to substantive autonomy is what I call consensual idealism.[19] Under this position, representatives argue that consensual minimalism is necessary but not sufficient for sexual activity to be ethical. Rather, there is a moral requirement that people ought to acknowledge and be responsive to each other's needs, desires, and feelings. A representative of this position is Yolanda Estes. She agrees consent is a necessary condition for moral sexual behavior yet adds two additional criteria: "each sexual partner exhibits concern for the other's interests and needs insofar as their wellbeing includes and extends beyond their sexual wellbeing" and "each sexual partner attend[s] to the other's desires."[20] Starting with the first, without having a concern for the other's interests and needs, the sexual interaction could undermine well-being. She points out that "sex without desire results in sensual or emotional dissatisfaction at best and physical or psychological trauma at worst."[21] Not paying attention and not having any interest in the partner's needs and desires shows a lack of respect. In shorter, non-committed sexual relationships, it becomes more imperative for more transparent, explicit, and specific communication.

For the second criterion, to attend to the other's desire is to know what the other person wants. There must be communication, though not necessarily verbal, which provides the partners enough knowledge to determine whether there is a reasonable, reciprocated consent and an understanding of each other's concern and desire.[22] Communication cannot be rushed or it risks neglecting the other's desires. Estes states:

> We can take time to gain some sexual knowledge of our partner by proceeding cautiously and unhurriedly in the initial stages of a sexual relationship. This increases the chance of correctly interpreting and addressing expressions of consent, expectation, and desire. Before, during, and after sexual interactions, we can solicit more explicit, specific expressions of our partner's thoughts and feelings; observe our partner's reactions carefully; and reflect diligently on what we hear and see. This enhances the possibility of reciprocal consent, concern, and desire while improving our sexual technique and our opportunity for a repeat performance.[23]

"Enthusiastic Consent"

One condition to consider within the framework of substantive autonomy is the popular phrase "enthusiastic consent." What is striking is that there are no investigations of "enthusiastic consent" among philosophers. After a search on philpapers.org and academia.edu, I could not find academics who defend "enthusiastic consent" as a necessary (nor a sufficient) criterion for consent. Those who require enthusiasm as a necessary condition are various sex educators, sex and relationship coaches, and Planned Parenthood.

Project Yes notes that sex must start with enthusiastic consent which means "being as excited and into someone else's enjoyment as we are excited and into our own enjoyment."[24] Planned Parenthood states that "When it comes to sex, you should do stuff you WANT to do, not things people expect you to do. If someone doesn't seem enthusiastic (meaning happy, excited, or energized), stop and check in."[25] In 2014, McGill University held a Forum on Consent and declared that real consent "must be loud and clear. Sex without enthusiastic consent is not sex at all. It's sexual assault or rape."[26] Thus, a weak "yes" or a nonchalant "yes" is not good enough. The "yes" must be "loud and clear."[27] Feminist writer and activist Jacklyn Friedman states that a "yes means yes" philosophy means that only valid consent is enthusiastic consent, which means that "you shouldn't do anything that your partner isn't actively excited about (or at least excited to try)."[28] Even pop icons like actress Jameela Jamil declared that when it comes to consent, "make sure the other person is not just willing, but damn well enthusiastic."[29] And "CONSENT SHOULDN'T BE THE GOLD STANDARD. That should be the basic foundation. Built upon that foundation should be fun, mutual passion, equal arousal, interest and enthusiasm."[30]

Despite the popularity of enthusiastic consent, some thinkers have implicitly questioned it. Quill Kukla notes that we can consent to sex where we feel ambivalent or game to do an activity but not enthusiastic about doing, such as when we know our partners enjoy an activity, but we find the activity neutral.[31] Lily Zheng, a former columnist for *The Stanford Daily*, argues that enthusiastic consent assumes various norms, and these norms play into hegemonic masculinity. Enthusiasm for men, Zheng argues, "manifests itself as confidence, social aptitude and extroversion. So how do shy men express 'enthusiastic consent?'"[32] Introverts may have a harder time expressing their enthusiasm. Moreover, race plays a role. Considering that Asian masculinity is depicted as passive, desexualized, and sexually compliant, is their enthusiasm different from a white man's enthusiasm? Black women are hypersexualized to the point where they are perceived as already enthusiastic, even if they really are not.

Adding to Zheng's analysis, gender also plays a role. Heterosexual, cisgender men are always socialized to desire sex to perform their masculine gender. Thus, enthusiasm to consent—let alone enthusiasm to have sex—comes easy for heterosexual, cisgender men.[33] For heterosexual, cisgender women, they are socialized to not explicitly show their sexuality,[34] let alone show that they desire sex. Enthusiasm may be harder to come by due to their socialization. Moreover, even if they were to desire sex, they may be reluctant to show their enthusiasm, again due to socialization and fear that enthusiasm may give them an unfavorable reputation.

Asexual and queer people do not follow the heteronormative scripts and so their cues on interpreting, giving, and receiving enthusiasm may be different. An asexual person who goes by the name of "Siggy" declares that he cannot imagine shouting "hell yes" to consent.[35] Asexual people sometimes are involved in sexual relationships. They do so because they know their partners enjoy sex, and so some asexual people will have sex with their partners to cement their relationship. Nevertheless, Siggy notes that while they consent to sex in a healthy relationship, asexual people may not be particularly "excited" to have sex, nor do they experience "hotness" when it comes to consent. Nevertheless, they consent to sex because they know their partner enjoys sex.[36]

Finally, while there may be enthusiasm for those at the beginning of their sexual relationship, the enthusiasm lessens over time as the partners get to know each other. "If consent is only granted with a loud, emphatic 'YES,' a coy or shy sexual response becomes a complex question of permission. It's unnecessary. Especially since in many marriages, for example, enthusiasm wanes as the years go on—but that doesn't mean men are abusing their wives."[37] Indeed, no enthusiasm does not mean the sex is unethical; it could mean that the novelty has worn off, but the sex is still pleasurable and enjoyable. We can even imagine a stereotypical long-term happily married couple where one person asks whether to have sex or not, and the other person unenthusiastically responds: "yeah, sure." There is little to no enthusiasm, but this does not mean that there is sexual assault happening. To add to the list, sex workers, those who have sex to maintain the relationships (i.e., maintenance sex), and those who have sex purely for the sake of reproduction, would not meet this criterion.

Beyond enthusiastic consent, there are two problems with consensual idealism. One problem is that some of the features set too high a standard. For example, suppose there is a new couple and they want to have sex. We can even assume that there is strong chemistry—even enthusiasm. However, once they start, there is a bit of awkwardness: they see each other naked for the first time, one of them may be self-conscious, they fumble during sex, which makes the experience sub-par, and perhaps one (or both) are thinking not directly about fulfilling the desires of the other, but more on the mechanics of sex, which detracts from their enjoyment. Both are disappointed when they end. Nevertheless, they want to give it another round at a later time. They did not mutually try to satisfy nor attend to each other's desires, nor did they exhibit concern for the other's sexual well-being. This is not the ideal sex act—which would be where both partners enjoyed the act and exhibited skill—but it does not make the act unethical.

A second problem—which relates to the last—is that people, especially when they are novices or feel uncertain about sex, may not know what they want. They may engage in various activities for curiosity, experimentation, or simply "just to try it out." There is no mutual desire to be had since one is unsure what sort of desires one has. Indeed, young adults and adolescents are still figuring out their own sexual boundaries and values. If consensual idealism is the standard to reach, then very few young people and adolescents actually consent. Going back to enthusiastic consent, they may not enthusiastically say "yes" because they are not sure what they are enthused of, but they still want to have the sexual experience for the reasons mentioned above.[38]

The criticisms of consensual idealism also reveal problems with substantive sexual autonomy. The higher the standard for sexual autonomy, the more stringent the conditions are to be autonomous. And the more stringent the conditions, the more expectations there are to be sexually autonomous. However, if we have too many expectations (such as living without oppressive contexts), then fewer people fit the conditions of being sexually autonomous. Considering that many young people are still figuring out their sexuality, substantive sexual autonomy would have to conclude that young people are rarely sexually autonomous. Indeed, the only people that could be sexually autonomous would be people without any oppressive circumstances. But because we are always in some sort of power structure, no one is purely sexually autonomous. And if no one is sexually autonomous, then no one can truly consent.[39]

Another problem with any ideal theory is that it ignores context and a broad spectrum of cases as falling short in the same way. Elizabeth Sperry uses six cases revealing the problems with substantive autonomy.[40] Influenced by Sperry, I will change the cases a bit. Suppose Ann engages in sexual activities because her husband enjoys them. She wants to please her husband and does what she can to make sure his pleasures are heightened. Her "deformed desires" are so entrenched that she is not even aware of other possibilities. She is unreflective of her desires. Barb also engages in sexual activities with her husband. Some of the activities bother her, but she enjoys watching her husband being pleased and feels a sense of accomplishment that "she did that" such that she feels more attractive and sexy. Barb reflectively decides that the inconvenience is worth it and reflects on her desires, yet bargains with patriarchy. Eve considers the priority of her male partner's sexual pleasures as oppressive and corrupt. In any sexual activity with her male partners, she makes sure that her pleasures will happen and thereby rejects any deformed desires and does not bargain with patriarchy. Proponents of substantive autonomy will endorse Eve as being autonomous and Ann as being heteronomous. But what about Barb? Proponents would argue that since she does not meet the ideal conditions, she is also heteronomous since her desires and behavior align with patriarchal standards. However, she questions them, even if she retains them. Thus, even by retaining these desires, she is more autonomous than Ann. A theory of autonomy should explain *how* Barb is more autonomous than Ann and substantive theory has difficulty explaining that by retaining ideal standards.[41]

Relatedly, another problem is that substantive autonomy is what Serene Khader calls "promiscuous paternalism."[42] People's choices give us clues to what they care about, and not paying attention to what oppressed people care about desensitizes us to their objections to strategies for feminist change.[43] If, however, their choices are replaced by opportunities to increase their well-being, their choices are still diminished. If some choices are illegitimate, then substantive autonomy is too restrictive even if those opportunities are for their well-being. Paternalistic interventions try to improve other's well-being "for their own good" even if it ignores how people *feel* about the intervention. Applied to the sexual realm, some people consent to various sexual practices that are against the norm: BDSM, kink-behavior, fetishes, and role-play just to name a few. Suppose there was a stringent condemnation against these practices because it hinders people's well-being. Therefore, people who choose to engage in these practices are dupes of their own oppression. And so, their choices are invalid. However, restricting people's sexual choices risks ignoring people's reasons for their choices without knowing the context. With the examples of Ann, Barb, and Eve above, let us say that Ann and Barb engage in BDSM where they play the sub, the participant who willingly gives control to the other partner which is usually the dom. This activity has had some contention among feminists because BDSM looks to replicate male domination, especially if the female plays the sub. Ann does so because it is what her husband likes. Barb does so because her husband likes it, and after reflecting on it, she feels neutral about it. There is no harm to her, but she engages in it because she enjoys the pleasure that her husband receives. Eve enjoys playing the sub and receives intense pleasure from the activity. And like above, she demands that her pleasures will happen. As from before, substantive sexual autonomy would have to commit to the idea that all three are oppressed and therefore are not really autonomous in choosing to play the sub. But are they not autonomous in the same way? Even *if* BDSM was oppressive, the context reveals that Eve is *less* heteronomous than Barb and Ann. Ann would be the most heteronomous. And again, a substantive theory that holds to ideal standards would have difficulty explaining why Eve is less heteronomous.

Weak Substantive Sexual Autonomy and Consensual Realism

With the problems of both procedural and substantive sexual autonomy, is there another route that keeps the advantages and discards the disadvantages? Diana T. Meyers offers a route within the procedural account but with an added oomph. Paul Bensen considers this position as weak substantive autonomy.[44] Weak substantive autonomy has normative constraints, not on the contents of people's preferences and values, but more on utilizing certain values that constitute autonomy. For Meyers specifically, her account is a skills-based view of autonomy[45] in that agents must have autonomy competency: a collection of agentic skills and capacities. The collection of agentic skills that Meyers has in mind include introspection, communication, memory, imagination, analytical reasoning, self-nurturing, and resistance to pressures to conform. "Embedded in these skills, I discern a panoply of epistemic and psychological values: perspicacity, resourcefulness, creativity, rationality, self-esteem, stability, resilience, tenacity, and corrigibility, to name a few."[46] These skills build self-definition, self-discovery, and self-direction which Meyers sees as necessary for autonomy. If traditional gender socialization compromises women's capacities to achieve full autonomy and damages their self-respect, this kind of socialization is oppressive.[47]

One of the benefits of Meyers's view of autonomy is that autonomy comes in degrees depending on how the agent gains those capacities and exercises those skills. Moreover, Meyers asks whether all desires deserve the same weight. After all, if desires come about due to their oppressive circumstances, should those desires be given credence? If yes, then we seem to be feeding into the oppression. If not, then we would ignore those who have those desires, which disrespects them. Meyers answers that not all desires have the same weight. If the desires come about autonomously—meaning if the acquisition or endorsement of those desires came about through competent skills—then those desires should be given more weight than desires that have not been critically reflective because they are built into the social norms and expectations.

I endorse Meyers's version of autonomy when applied to sexual autonomy. I will call it weak substantive sexual autonomy. To be sexually autonomous, certain competency skills are needed, such as knowing when one is ready, communication, the courage to say "no," the ability to accept a "no," and emotional intelligence. Sex educatory theorists Michelle Fine and Sarah McClelland point out that sexual agency includes skills such as asking for help, negotiating risk, engaging in critical analysis, and pursuing pleasure.[48] These skills, however, seem to be virtues (either moral or intellectual).[49] However, developing these skills as virtues is beyond the scope of this chapter, but it is worthy of research.

The Complexities of Sexual Consent

What view of consent comes from weak substantive sexual autonomy? The issue becomes complex after considering the latest empirical research discussing sexual responses and how they vary between males and females.[50] Let me briefly bring up three differences and suggest how these differences make consent more complex. The first difference discusses sexual concordance, which is the (mis-)match between one's subjective and one's physiological sexual responses. A plethysmograph machine detects physiological sexual responses (e.g., blood flow, penis erection, vaginal fluid). Meanwhile, people report whether they are subjectively aroused or not. If the subjective and physiological arousal match, then there is a sexual concordance. Empirical research suggests that men typically show a concordance 50 percent of the time, whereas women show a concordance 10 percent of the time. In the case of women, they will say they are not subjectively turned on, but their physiological responses suggest otherwise 90 percent of the time.

The second difference is the dual-control model, the description of modulating sexual arousal. There is the sexual excitation system, which Nagoski calls the sexual accelerator, and the sexual inhibition system, which she calls the sexual brake. Every person has this, and some accelerators and brakes may be more sensitive than others. The sexual excitation system notices relevant information in the environment to motivate arousal (e.g., partner's appearance, ways your partner makes you feel, novelty). The sexual inhibition system notices relevant information in the environment to dissuade arousal (e.g., stress, body image, trauma history, relationship conflict, sleep deprivation, reputation).

On average, men appear to have more sensitive sexual accelerators and women have more sensitive sexual brakes.[51] Nagoski discusses that when we want to turn our partner on, we often think that we just need to press the accelerator more. However, since women are more likely to be attuned to their brakes, they may need to release the brake pedal so that sexual arousal can initiate. Simply turning people on is not merely a matter of touching or caressing, but setting up a context where they are comfortable and already set in a situation where they could be easily aroused. Or, as Nagoski puts it, "arousal is the process of turning on the ons and turning off the offs."[52]

Finally, the third difference involves the genesis of sexual arousal. We often think of sexual arousal happening spontaneously: sexual arousal appears out of nowhere, and we want to have our sexual desires fulfilled, which Nagoski calls *spontaneous sexual arousal*. This narrative is so strong that we assume it is a universal human condition. However, Nagoski points out that spontaneous arousal typically works around 75 percent of men and 15 percent of women.[53] Conversely, other people typically have *response sexual arousal*, which is when arousal arises *after* the accelerator has been pressed and/or the brake pedal has been released. In other words, the person is in a state of arousal *in response* to a context that fosters sexual arousal. This form of arousal occurs in roughly 5 percent of men and 30 percent of women.[54]

With these factors in play, consent becomes complex. For simplicity, let us call this woman who has these typical responses Tonya. She is more likely to engage in sexual relations when the context is set up just right. Moreover, Tonya's sexual desire is responsive rather than spontaneous. Thus, in Tonya's experience, she has difficulty initiating sexual encounters because she may not currently be in the mood. However, she could be, given the right physiology (i.e., possible sexual non-concordance) and the right context (i.e., the release of the brake pedal) from a good external source (i.e., responsive desire from her partner as opposed to someone with aggressive sexual tactics). In this sense, perhaps a playful nudge, a soft persuasion, a positive pressure,[55] or a helpful sway would be ethically permissible. With these complexities and context, there is no problem with consent being *positively* brought out.[56] Indeed, if we could imagine an ideal society without any unethical sexual actions, the helpful sway would be unproblematic to bring forth ethical sexual conduct.

What makes this experience more complex is that of young female adolescents. Most young adolescents are novices when it comes to sex and, for many young women, discussing and thinking about sex, especially sexual pleasure, is still taboo. Thus, many young women may not know their boundaries—what sort of pleasures they have, what they desire, or what they are willing to do. At the beginning of many sexual experiences, Tonya may feel awkward at first. However, she consents to sex because of the context, which enables her arousal mechanisms and thus increases her desire.

With Nagoski's discussion in mind, I claim Tonya's experiences reveal the complexity of sexual encounters and sexual consent, but traditionally most of the literature has focused primarily on women such as Monica, as manipulated actors in the encounter. To show the difference, suppose we had a young woman going on a date with a man. The date seems wonderful, and both parties seem to be enjoying themselves. Later that evening, the man tries to initiate some sexual contact. He is not doing it aggressively, but he is making his intentions known. The young woman is not against having any sexual relations with him, but she does not desire to have sex right then. He displays more arousal techniques. She may experience a mixture of feelings such as awkwardness, stress, vulnerability, self-consciousness, joy, pleasure, detachment, connectedness, and nervousness. Moreover, she may feel unsure because part of her enjoys the experience, and another part of her is hesitant to engage.[57] At some point, his efforts to arouse her succeed in acquiring her permission to initiate the next level of sexual intimacy, even if she remains unsure of herself. During the sexual act, she continues to consent, but she is dissatisfied with the experience: maybe his technique is unimaginative, and to be honest, he half-heartedly focuses on her. It is an awkward experience, but she tells herself that if they continue to see each other, his technique could improve, they will be more familiar with each other's bodies, and overall, things may get better. This is not the worst sexual experience that she has had, but it is undoubtedly lackluster. Eventually, however, this woman starts to feel more at ease and relaxes. She can slowly get into the flow of the sexual experience, even if it was not enjoyable from the beginning. She may not be in the mood or turned on, but she could be if she sees some potential in future encounters, or she may be aroused throughout the sexual encounter. She may consent to the act, even if she is presently not aroused. Or, to make it even more complicated, her lack of experience may mean she does not know what turns her on, so she may be confused as to whether she is aroused or not, but she still consents nevertheless.

Traditionally, the woman I just described is considered Monica's experience. However, the experience easily could have been Tonya's. There may be differences between how men and women desire sex but, as Nagoski's work suggests, these differences seem to have some biological basis.[58] Even if men initiate sex more often than women, there is ethically no *prima facie* wrong happening. Specifically, Nagoski points out that it is normal to feel ambivalence around sex and that a stimulus can hit the accelerator and brake pedal simultaneously, particularly if the person has learned that sexuality is a threat. For example, inexperience could be considered a "threat" of sorts in that people may experience being turned on but not be quite sure if they should continue.[59] We can compare this to inexperienced dancers who may accidentally step on each other's toes, say "sorry," and continue dancing, with everyone still enjoying themselves.

Tonya's and Monica's sexual experiences have been conflated and I suggest three components to show the differences. First, Monica's autonomy and integrity are weakened due to her motivations.[60] A study from Emily Impett and her colleagues shows that the *motivations* for engaging in sex with a partner when one does not specifically desire sex are important. In this study, there are two types of motivations: approach goals and avoidant goals. Approach goals are goals that one pursues to reach a positive outcome, whereas avoidant goals are those one pursues to avoid a negative outcome. In the sexual domain, approach goals could be obtaining pleasure, helping a partner obtain pleasure, and increasing or maintaining relationship satisfaction. Avoidant goals could be avoiding sexual or relationship conflict, a partner's loss of interest, or relationship tension. The study suggests that when someone consistently pursues avoidant goals in their relationship, they are more likely to experience a breakup, find the relationship dissatisfying, or be less satisfied with their sexual experiences over time. In short, consistently pursuing avoidant goals can be detrimental to maintaining relationship satisfaction.[61] Monica's motivation was to avoid a scenario, whereas Tonya wants to approach a scenario.

The second missing component comes from what Ann J. Cahill argues is the nuance between coerced sex and reluctant sex. Cahill considers the sexual experiences motivated by consistent avoidant goals to be "unjust sex." Monica's experience is not explicitly sexual assault because there was consent, but the situation did not present her with options that engage her autonomy skills, let alone promote her well-being. In fact, Monica is presented with a dilemma in which she chooses sex because it is the least bad option. In these scenarios, Monica is a victim of "unjust sex."

So, what is the marker between sexual assault and the "grey area" of unjust sex? Cahill states that "sexual assault entails a sexual interaction where one person (the assailant) either overrides the will of another (the victim) or exploits the situation wherein the victim's will is inoperative."[62] Unjust sex, the sex in the gray area, would be where consent was given reluctantly, where the woman may have had a split will during which she was hesitating and less than willing.[63] Could this description help make a distinction between Monica's and Tonya's experiences? In both scenarios, the women are hesitating, reluctant, and may have a split will. They are both unsure of the experience. However, the difference is that Monica is less than willing, yet feels she has no choice but to consent.[64] Tonya is willing, more than willing, or taking a chance and choosing to engage in the sexual interaction. She may also feel unsure, but she has the option to opt-out if she wishes.[65]

For the third missing component, Monica and Tonya are sexual agents but expressed differently. My influence also comes from Cahill who states:

> A sexual subject does not merely have sexual agency prior to any given sexual interaction, as a kind of freestanding capacity or resource; rather, both the existence and the quality of that sexual agency emanates from sexual (and other) interactions. ... For me, agency in general is both deeply embodied (and thus profoundly affected by bodily interactions with other subjects, specific environments, objects, and discourses of inequality) and fundamentally, not peripherally, intersubjective.[66]

Monica's sexual agency is not intersubjective but is reduced to the man's sexual agency in an unethical way because her contribution to the sexual action is an afterthought, as a way to cross off the checklist of what counts as consent. Again, it is worth quoting from Cahill:

> the woman's agency is deployed *only* to be used against her. Or, to be more precise, the woman's agency can be deployed only to facilitate a specific sexual interaction whose content (that is, the particular acts that will make up the interaction) is predetermined and remains largely unmarked by the specific quality of the woman's sexual subjectivity. Her sexual agency is employed in a weak way, as a mere accreditation of the sexual interaction that is being offered to her. Because her agency is merely providing a kind of ethical cover to the interaction being offered, the interaction itself does not enhance either her sexual agency (that is, it does not empower her to become more knowledgeable or forthright about her sexual needs, desires, and interests in the context of this particular relationship) nor, most likely, does it broaden her sexual subjectivity by creating more possibilities. In this sense, the interaction most likely does not contribute positively to her sexual becoming or flourishing. Thus, her sexual agency is hijacked, used not to forward her interests, but in fact to undermine them, particularly those interests that are related to her always-developing sexual subjectivity.[67]

On the other end of the spectrum, Tonya's interests are not ignored but taken as a factor to consider. Her experiences may be treated with respect, but they may also be treated with reckless indifference or ignorance. There may be confusing moments, but this is primarily due to the ongoing relationship she has with her partner, her familiarity with her body and her partner's body, whether she is comfortable or awkward, whether she is stressed, and so on. Nevertheless, she may continue to act in the hopes that it may get better. Overall, she may be able to flourish sexually. In some cases, she enjoys the act; in others, she may find the experience wanting, but she does not consider the act violent or assaultive, nor does she regret or even consider it "unjust," unlike Monica, who cannot flourish. If, over time, Tonya feels consistently unsure about her sexual actions but still consents, it is possible that she would slowly lose her sexual well-being and that the act would increasingly become "unjust sex," in which case, she may end up like Monica.[68]

To sum up this section, there is a plurality of ways to look at consent, but the addition of looking at how desires and arousals function biologically and looking at the context of the encounter requires us to take this reality into account. Therefore, I call my position "consensual realism."[69] The realism is not looking at the principles of consent first and apply them to people's experiences. Instead, we start by looking at Tonya's experiences, which include her desires, interests, and the context of the sexual encounter. Consensual realism is above the minimum standard and above what Ann Cahill calls "unjust sex" since any sexual activity within that field would be unethical. Moreover, consensual idealism is too rigid in its principles in that it ignores the context and how well the people know each other. With that, my position lies between consensual minimalism and consensual idealism, which is where Tonya's experiences lie.

Adopting from Kukla, autonomy itself is not all or nothing; it comes in degrees. There are paradigmatic examples of being autonomous or not, but there are thresholds when one is *sufficiently* autonomous.[70] Likewise, I suggest that since sexual consent correlates to these different theories of autonomy, there are various conditions when consent takes place. This brings up the question as to whether consent comes in degrees. Since weak substantive sexual autonomy can accommodate degrees, and since consent relates to autonomy, consensual realism also comes in degrees which stipulates that we should have a heavy dialogue to determine the ethics of fringe cases.[71]

Tonya's experience will change depending on the context, but the context will inform whether the interaction is ethical or not. For example, suppose Bob is with Tonya and they are in tune with each other's bodies and can easily read each other's body language. Let us also suppose that they have been in a relationship for a long time; the background of the relationship would give them the experience and context of how to engage in a sexual way that is fun, exciting, and caring. Since Bob and Tonya have known each other for a while and know how to turn each other on, it is almost as if they can do it automatically to gain pleasure for themselves and for each other. Because they know what they are doing, the context suggests that there is a low bar to hurdle. Therefore, they may affirm each other's consent, even to the ideal realm, and they can do so easily.

Now consider Tonya and Jess. They are at the beginning of their sexual relationship, so they have not yet developed the experience of what turns them on, their limits or boundaries on what is appropriate, or what they can do to enhance the experience rather than it leading to awkwardness. Because of this context, there is a higher bar to hurdle. Both Tonya and Jess have to put in extra effort for a mutually enjoyable sexual encounter, which could mean check-in with each other, be more sensitive to the reactions and body language of each other, display a more caring attitude to make everyone more comfortable, and communicate beforehand to ensure each other's boundaries, or at least have a good certainty that pleasure will not be diminished.

Finally, as another example, Tonya may be of two minds about a sexual act, even if she completely trusts a long-term partner. Suppose Tonya has been seeing Kim for quite a while. They trust each other and enjoy being with each other. They have had a sexual relationship that is both satisfying and pleasurable. Kim suggests trying a new activity that is somewhat adventurous and risky (e.g., anal sex, swinging, bondage, or a threesome). Tonya has never thought about the activity except in the abstract. Part of her never thought she would want to participate in it, but she has never categorized the activity as an absolute "no." So when Kim presents an opportunity to try out this new activity, Tonya is not quite sure. The hesitancy, however, is not from her not wanting to do it, but from the idea that she has never thought about wanting to do it. Thus, Tonya is partially intrigued, curious, and game to try it out.[72] At the same time, she is not entirely sure since she is dismayed at the request, worried that the negative consequences could outweigh the benefits, and made to be in a vulnerable position.

Tonya decides to do it even though she is of two minds. Notice that this does not equate to having second thoughts. Having second thoughts entails regret or going along with the activity "just to get it over with." Being in the state of two minds, on the other hand, means she is actively going along with the activity and says "yes" to the activity, but the "yes" is not an excited one. Let me be clear. Tonya is not being coerced and manipulated into the activity. The "no" can easily be used without awkwardness. The "yes" that Tonya states is active and not an acquiescing "yes." The phenomenon of being in two minds comes about by the novelty of the activity, which can have a combination of excitement (incentivizing her to do the activity) and uncertainty (which makes her skeptical about doing the activity). If, on the other hand, this was repeated over time and she said yes just to go along, or if she felt like she had to say "yes" to avoid awkwardness or relieve pressure, then she may become like Monica over time.

Conclusion

Taken altogether, sexual consent is nuanced and more complex than we think. It is not just a "yes" or a "no" for many encounters. Indeed, when most people have sex, they hardly explicitly garner a "yes" or state a "no." Instead, we are searching for clues and giving clues to see if it is ok to proceed, slow down, or stop. Most sexual initiations happen non-verbally. If many sexual interactions happen non-verbally, then we need to teach consent based on giving/receiving clues, and that, of course, depends on the reality of the context.[73] The context is varied and open, but this means that it requires more careful thought and attention to our sexual avenues to determine ethical sexual activities and distinguish between sexual assault, "unjust sex," and permissible sex.

Notes

1. These discussions involve what consent *is*. One view is the attitudinal account, which states consent is given by having a mental attitude. Consent is, therefore, a psychological attitude toward that act. Another view is the performative account, which states consent is given by communicating (either verbally or non-verbally) and intending to do the act.
2. Berit Brogaard (forthcoming). "Sex by Deception." In John M. Doris & Manuel Vargas (eds.), *Oxford Handbook of Moral Psychology*. (Oxford: Oxford University Press).
3. For an account of deception and how that undermines consents, see Tom Dougherty (2013). Sex, Lies, and Consent. *Ethics* 123 (4):717–744.
4. Rebecca Kukla. "That's what she said: The language of sexual negotiation." *Ethics* 129, no. 1 (2018): 70–97.
5. Jonathan Jenkins Ichikawa. "Presupposition and Consent." *Feminist Philosophy Quarterly* 6 no. 4 (2021).
6. There are other names for this type of consent. This would be in the tradition of what Soble has called "thin externalism." See Alan Soble. "Sexual use and what to do about it: Internalist and externalist sexual ethics." *Essays in Philosophy* 2, no. 2 (2001): 37–54. We could also call this Thin Kantian Respect since consensual relations stem from Kant and the concept is "thin" because there is a minimal need as to what counts as consent.
7. I am ignoring various cases where consent is not always necessary for morally permissible sexual contact exemplified by Wertheimer. His cases are interesting, but they are beyond the scope of this essay. Moreover, I am also ignoring whether consensual minimalism or consensual idealism ("Substantive Sexual Autonomy and Consensual Idealism" section) solves the Kantian problem of moral sexual relations.
8. Thomas Mappes. "Sexual Morality and the Concept and the Concept of Using Another Person," in *Today's Moral Issues*, 7th Edition, Ed. by Daniel Bonevac (McGraw-Hill, 2013), 66–72.
9. See Alan Wertheimer. *Consent to sexual relations* (Cambridge University Press, 2003).

10. Jan Steutel and Doret J. De Ruyter. "What should be the moral aims of compulsory sex education?" British journal of educational studies 59, no. 1 (2011): 75–86.
11. Raymond Belliotti. "A Philosophical Analysis of Sexual Ethics." In *Journal of Social Philosophy*, Vol. 10, No. 3 (1979): 8, my emphasis. Belliotti never considers that masturbation is a way to "fulfill certain desires which we cannot fulfill by ourselves," unless he means some sexual interaction with *another* person. Moreover, unless Belliotti is very broad when he says, "satisfy the expectation of the other," he also seems to ignore cases of prostitution, maintenance sex, pity sex, or peace-inducing sex, just to name a few instances of sexual encounters.
12. Robin West. "The Harms of Consensual Sex," in *The Philosophy of Sex*, 6th Edition. Ed. by Nicholas Power, Raja Halwani, Alan Soble. (Lanham, MD: Rowman & Littlefield Publishers, Inc., 2013), 387–394.
13. We can ask why do they engage and consent to sex if they do not consider the sex desirable nor pleasurable? Here are some of the answers I have received: you do not want to offend them, you do not know that you can say no, general insecurity, you have developed the habit of caring more about their pleasure than your own in an abstract sense, it kind of feels good but might be slightly painful because of dryness or minor yeast infection and you do not want to bring it up because it's not cool, you dislike them so much and think the encounter might end faster if they just cum and fall asleep, you have other stuff going on that you are trying to ignore but cannot and you already started having sex but do not want to talk about your deep-seated issues at that time.
14. Lois Pineau. "Date Rape: A Feminist Analysis," in The Philosophy of Sex, 6th Edition. Edited by Nicholas Power, Raja Halwani, Alan Soble, 461–484. (Lanham, MD: Rowman & Littlefield Publishers, Inc., 2013).
15. Cf. Pineau, 465–466.
16. Vanessa E. Munro. "Constructing consent: Legislating freedom and legitimating constraint in the expression of sexual autonomy." *Akron Law Review* 41 (2008): 950–951.
17. Morten Ebbe Juul Nielson. "Safe, Sane, and Consensual—Consent and the Ethics of BDSM," in *International Journal of Applied Philosophy*, 24(2), 2010, 271.
18. Natalie Stoljar. "Feminist Perspectives on Autonomy," The Stanford Encyclopedia of Philosophy (Winter 2018 Edition), Edward N. Zalta (ed.), URL = https://plato.stanford.edu/archives/win2018/entries/feminism-autonomy/.
19. We could perhaps also call this "thick Kantian respect." Munro calls it "consent-plus." This would be in the tradition of what Soble calls "thick externalism."
20. Yolanda Estes. "Mutual Respect and Sexual Morality." *College Sex-Philosophy for Everyone*. Edited by Michael Bruce and Robert M. Stewart. (Wiley-Blackwell, 2010), 212.
21. Estes, 212.
22. Cf. Estes, 213.
23. Estes, 214.
24. Project Respect. "Consent." Accessed October 6, 2020. https://www.yesmeansyes.com/consent/. Does this mean that if we are not excited and into our own enjoyment, then we do not have to be excited and into someone else's enjoyment?

25. Planned Parenthood. "Rape and Sexual Consent | Information for Teens." Accessed October 6, 2020. https://www.plannedparenthood.org/learn/teens/sex/all-about-consent.
26. Robyn Urback, "To McGill Activists, a 'Yes' Doesn't Mean Consent," National Post (National Post, February 27, 2014), https://nationalpost.com/opinion/robyn-urback-to-mcgill-activists-a-yes-doesnt-mean-consent.
27. There are no details on whether consent must be verbal or nonverbal. Presumably, nonverbal consent can be clear, but can it be "loud?" I will leave that to the linguists.
28. Jaclyn Friedman. "'Yes Means Yes' and Enthusiastic Consent." Our Bodies Ourselves, September 12, 2014. https://www.ourbodiesourselves.org/book-excerpts/health-article/yes-means-yes-enthusiastic-consent/.
29. Jameela Jamil, "What We Need to Learn from the Aziz Ansari Clusterfuck." *Jameela Jamil—Diary of a Goon*. January 15, 2018. https://jameelajamil.co.uk/post/169720263620/what-we-need-to-learn-from-the-aziz-ansari.
30. Jamil website.
31. Quill Kukla, "A Nonideal Theory of Consent," A nonideal theory of sexual consent." *Ethics*, 131 no. 2 (2021): 270–292.
32. Lily Zheng. "How to Ace Sex: Why Enthusiastic Consent Doesn't Cut It." *The Stanford Daily*, November 5, 2014. https://www.stanforddaily.com/2014/11/04/how-to-ace-sex-why-enthusiastic-consent-doesnt-cut-it/.
33. It could be the case that it comes easy for all men (as opposed to women), but it comes easier for heterosexual, cisgender men. Gay men and trans men may express enthusiasm easier than women, but we cannot say without quantitative data.
34. However, it is a double bind since they cannot show that they are completely asexual.
35. Siggy. "I'm Not Enthusiastic about Enthusiastic Consent." The Asexual Agenda, March 5, 2019. https://asexualagenda.wordpress.com/2019/03/05/im-not-enthusiastic-about-enthusiastic-consent/.
36. There are many types of asexuals where they feel no sexual attraction, but they still have sexual desires. In which case, they may not mind having sex and still gain pleasure from it. On the other end of the spectrum, some asexuals have total sexual aversion and feel disgusted having sex. Using notes from Siggy, I am mostly discussing the former and not the latter. In the description above, I am also assuming that the sexual relationship is good, all else being equal.
37. Urback.
38. Another model to consider is the affirmative consent model. The default has been "no means no" which states that sex is ethical unless there is a "no" present. Affirmative consent changes the default in that a "yes," rather than a "no" or silence, is seen as necessary for the moral transformation that changes an act of sexual aggression or assault into a permissible engagement with the other sexually. This model has gained traction, and since there is a higher bar to hurdle, it looks as if it is tied up with consensual idealism. However, I do not see this being incompatible with consensual minimalism. One can still obtain a "yes"—even enthusiastically—without explicitly focusing on the other's sexual needs and desires. Consensual minimalism and idealism tell us the content of consent where the focus is on the ethical portion of consent; the affirmation model tells us when it is permissible to proceed by focusing on how a token

consent can be communicated. In short, *consensual minimalism and idealism tell us what consent is; affirmative consent gives us the conditions to make consent valid.* This model does not erase all the problems of whether someone consented or not, but it does get rid of various defenses used by aggressive men to prove there was consent: "She didn't say anything, so it was ok," "She kissed me back, so it was ok to go forward," and so on.

39. Catherine MacKinnon takes this view, but I argue that we should reject the high standard rather than accept MacKinnon's view. This does not mean we should reject her criticisms, which have fair points. However, we can accept her criticisms without choosing these two options.
40. Elizabeth Sperry. (2012). "Dupes of Patriarchy: Feminist Strong Substantive Autonomy's Epistemological Weaknesses." *Hypatia*, 28(4), 887–904. doi:10.1111/hypa.12007. She discusses the problems of feminist strong substantive autonomy specifically in contrast to weak substantive autonomy.
41. Are there reasons substantive autonomists do not say that Barb is more autonomous than Ann? After all, Barb is closer to the ideal than Ann. While true, strong substantive autonomists hold that external circumstances *and* the right psychology are necessary conditions to be autonomous. Since Barb is in a patriarchal circumstance and does not have the right psychology since she "bargains with patriarchy," she is therefore not autonomous. I thank David Boonin for bringing this to my attention.
42. Serene J. Khader. "The Feminist Case Against Relational Autonomy," *Journal of Moral Philosophy* (2020): 2, doi: https://doi.org/10.1163/17455243-20203085.
43. Cf. Khader, 22.
44. To be clear, Meyers is somewhat of an advocate of the "weak substantive" position, but she does not see her position as value-laden that is between the value-neutral position (procedural autonomy) and the value-saturated position (substantive autonomy). Rather, she sees her view as a value-neutral position (hence falling closer to the proceduralist camp), but she invokes an autonomy that utilizes value, namely autonomy competency skills. Indeed, she claims that all theories of autonomy are value-utilizing even if they hold to a value-neutral position. So while her position is nuanced, I will place her in the "weak substantive" camp for brevity. See Diana Tietjens Meyers, "Review of Personal Autonomy in Society by Marina Oshana," *Hypatia* 23:2 (2008): 206 and Diana Tietjens Meyers. "The Feminist Debate over Values in Autonomy Theory." In *Autonomy, Oppression, and Gender*, edited by Andrea Veltman and Mark Piper, 114–140. (Oxford University Press, 2014).
45. Diana Tietjens Meyers, *Self, Society, and Personal Choice* (Columbia University Press, 1989), 56.
46. Meyers, The Feminist Debate over Values in Autonomy Theory," 121.
47. See Catriona Mackenzie and Natalie Stoljar. *Relational Autonomy: Feminist Perspectives on Autonomy, Agency, and the Social Self.* (NY: Oxford University Press, 2000), 18. Indeed, Meyers points out that resistance is necessary for autonomy.
48. Cf. Michelle Fine and Sara McClelland. "Sexuality Education and Desire: Still Missing After All These Years," in *Harvard Educational Review*, 76 (2006): 327.
49. The ability to communicate well could be the virtue of truthfulness, wittiness, and friendliness combined with *phronesis* of what to say at the right time. The

courage to say no is the virtue of courage. The ability to receive a "no" could be the virtue of humility and patience. Emotional intelligence and to know when one is ready is using *phronesis* such that people can display their emotions and sexual interactions at the right place at the right time. Other virtues could be temperance, generosity, and care.

50. I am indebted to Emily Nagoski's *Come as you are: The surprising new science that will transform your sex life.* (Simon and Schuster, 2015) for this discussion from which I am taking most of this information. This is not to say that *all* males will follow a specific sexual response path and that *all* females will follow another specific sexual response path. Nagoski points out that these are averages.

    A disclaimer: the studies that Nagoski provides—and therefore the rest of the essay—will be confined to heterosexual people. There is too little research on whether the sexual desires and functioning are the same for cisgender, heterosexual people, and those in the LGBTQ+ community. They very well could be, but for accuracy, the subjects in the studies are heterosexual. Fortunately, some psychologists taking note of this gap are conducting studies to understand the experiences of the LGBTQ+ community and whether their sexual functioning and well-being are the same as cisgender, heterosexual people.
51. Cf. Nagoski, *Come As You Are*, 60. Hereafter *CAYA*.
52. Nagoski, *CAYA*, 61.
53. Cf. Nagoski, *CAYA*, 225.
54. Cf. Nagoski, *CAYA*, 225. Another note: only about 6 percent of women lack both spontaneous and responsive desire. The remainder of the percentage numbers is context dependent.
55. I am indebted to Patricia Marino's discussion on this in her paper. Patricia Marino. "Affirmative Consent and Female Desire." Presentation, Pacific American Philosophical Association, Seattle, WA, April 10, 2017.
56. Contrast this with how Kate Manne describes men's entitlement to sexual consent where consent is *negatively* brought out. See Kate Manne. *Entitled: How male privilege hurts women* (Crown, 2020), Chapters 3 and 4.
57. It would make sense for someone to feel a wide range, even contrary, emotions. From the dual-control model, sexual interest and desire can be ambiguous.
58. To be clear, even if men typically initiate sex more often than women, this may not be ethically problematic as long as other conditions are in play: following the rules of consent, making sure that both partners are mutually respected, and feeling comfortable to say "no."
59. Cf. Emily Nagoski, *Come As You Are Workbook*, 26.
60. This is not to say that Monica should be *blamed* for her weakened autonomy and integrity. I am offering a descriptive account of what is happening.
61. Cf. Emily A. Impett, Amy Muise, and Natalie O. Rosen. "Is It Good to Be Giving in the Bedroom? A Prosocial Perspective on Sexual Health and Well-Being in Romantic Relationships." *Current Sexual Health Reports* 7, no. 3 (2015): 185. doi:10.1007/s11930-015-0055-9.
62. Cahill, "Unjust Sex vs. Rape," 8.
63. Cf. Cahill, "Unjust Sex. vs. Rape," 8.
64. Cf. Cahill, "Unjust Sex vs. Rape," 9.
65. This is different than experiencing sexual assault where the encounter goes *against* the will of the victim.
66. Cahill, "Unjust Sex vs. Rape, 9.

67. "Cahill, "Unjust Sex vs. Rape," 10.
68. Taking note from Marino's paper where I suggest Tonya may hope that her experiences get better over time, and Monica may hope that things do not get worse.
69. I am indebted to Aubrey Spivey for coming up with this term.
70. Kukla's conditions for a sufficient sexually autonomous person are (1) the choice is not manipulated nor coerced, (2) the action flows from the agent's motivations, and (3) the agent has enough normative responsiveness and reflective capacities to be open to recognize reasons not to do what she is choosing to do. From my account, I take Kukla's conditions as the agent's choices must be voluntary, stemming from the agent's self-definition, and stemming from competency skills.
71. Kukla has a list which they argue is neither necessary nor sufficient for consent. Rather, the list is to scaffold consent and that when some on the list are lacking, we need more of the others. The list includes trust, a concrete ability to exit at will, competent uptake, social contexts that do not undermine the agency of people's sexual choices, avoiding activities that undermine agency, meaningful epistemic agency, redress if their consent was violated, and being socially connected. I consider this a good list as part of the context to see if the agents consented.
72. This fits into another criterion that Meyers considers necessary for autonomy, which is the person's judgment to act on his or her judgments: "Though people may change their minds as a result of making an experiment based on their own deliberations, refusing to try out a creditable possibility compromises autonomy" (Meyers, *Self, Society, and Personal Choice*, 83–84).
73. However, if it happens non-verbally, people also need to be taught how to use their words comfortably when the time comes. See Alice Dreger. *The Talk: Helping Your Kids Navigate Sex in the Real World* (CreateSpace Independent Publishing Platform, 2016), 55.

References

Belliotti, Raymond. 1979. A Philosophical Analysis of Sexual Ethics. *Journal of Social Philosophy* 10 (3): 8–11.

Brogaard, Berit. forthcoming. Sex By Deception. In *Oxford Handbook of Moral Psychology*, ed. John M. Doris and Manuel Vargas. Oxford: Oxford University Press.

Cahill, Ann. 2016. Unjust Sex vs. Rape. *Hypatia* 31 (4): 746–761.

Dougherty, Tom. 2013. Sex, Lies, and Consent. *Ethics* 123 (4): 717–744.

Dreger, Alice. 2016. *The Talk: Helping Your Kids Navigate Sex in the Real World*. CreateSpace Independent Publishing Platform.

Estes, Yolanda. 2010. Mutual Respect and Sexual Morality. In *College Sex-Philosophy for Everyone*, ed. Michael Bruce and Robert M. Stewart, 209–219. Wiley-Blackwell.

Fine, Michelle, and Sara McClelland. 2006. Sexuality Education and Desire: Still Missing After All These Years. *Harvard Educational Review* 76: 297–338.

Friedman, Jaclyn. 2014. "Yes Means Yes" and Enthusiastic Consent. *Our Bodies Ourselves*, September 12. https://www.ourbodiesourselves.org/book-excerpts/health-article/yes-means-yes-enthusiastic-consent/.

Impett, Emily A., Amy Muise, and Natalie O. Rosen. 2015. Is It Good to Be Giving in the Bedroom? A Prosocial Perspective on Sexual Health and Well-Being in Romantic Relationships. *Current Sexual Health Reports* 7 (3): 180–190. https://doi.org/10.1007/s11930-015-0055-9.

Jamil, Jameela. 2018. What We Need to Learn from the Aziz Ansari Clusterfuck. *Jameela Jamil—Diary of a Goon*, January 15. https://jameelajamil.co.uk/post/169720263620/what-we-need-to-learn-from-the-aziz-ansari.

Jenkins Ichikawa, Jonathan. 2020. Presupposition and Consent. *Feminist Philosophy Quarterly* 6 (4): 4.

Kukla, Quill (written as Rebecca). 2018. "That's What She Said": The Language of Sexual Negotiation. *Ethics* 129 (1): 70–97.

———. 2021. A Nonideal Theory of Sexual Consent. *Ethics*, 131(2): 270–292.

Mackenzie, Catriona, and Natalie Stoljar. 2000. *Relational Autonomy: Feminist Perspectives on Autonomy, Agency, and the Social Self.* Oxford University Press.

Manne, Kate. 2020. *Entitled: How Male Privilege Hurts Women.* Crown.

Meyers, Diana Tietjens. 1989. *Self, Society, and Personal Choice.* Columbia University Press.

———. 2008. Review of Personal Autonomy in Society by Marina Oshana. *Hypatia* 23 (2): 202–206.

———. 2014. The Feminist Debate Over Values in Autonomy Theory. In *Autonomy, Oppression, and Gender*, ed. Andrea Veltman and Mark Piper, 114–140. Oxford University Press.

Munro, Vanessa E. 2008. Constructing Consent: Legislating Freedom and Legitimating Constraint in the Expression of Sexual Autonomy. *Akron Law Review* 41: 923–956.

Nagoski, Emily. 2015. *Come as You Are: The Surprising New Science That Will Transform Your Sex Life.* Simon and Schuster.

———. 2019. *Come as You Are Workbook.* Simon and Schuster.

Nielson, Morten Ebbe Juul. 2010. Safe, Sane, and Consensual—Consent and the Ethics of BDSM. *International Journal of Applied Philosophy* 24 (2): 265–288.

Planned Parenthood. 2020. Rape and Sexual Consent | Information for Teens. Accessed October 6, 2020. https://www.plannedparenthood.org/learn/teens/sex/all-about-consent.

Pineau, Lois. 2013. Date Rape: A Feminist Analysis. In *The Philosophy of Sex*, ed. Nicholas Power, Raja Halwani, and Alan Soble, 6th ed., 461–484. Lanham: Rowman & Littlefield Publishers, Inc.

Project Respect. 2020. Consent. Accessed October 6, 2020. https://www.yesmeansyes.com/consent/.

Siggy. 2019. I'm Not Enthusiastic about Enthusiastic Consent. *The Asexual Agenda*, March 5. https://asexualagenda.wordpress.com/2019/03/05/im-not-enthusiastic-about-enthusiastic-consent.

Soble, Alan. 2001. Sexual Use and What to Do About It: Internalist and Externalist Sexual Ethics. *Essays in Philosophy* 2 (2): 37–54.

Sperry, Elizabeth. 2012. Dupes of Patriarchy: Feminist Strong Substantive Autonomy's Epistemological Weaknesses. *Hypatia* 28 (4): 887–904. https://doi.org/10.1111/hypa.12007.

Steutel, Jan, and Doret J. De Ruyter. 2011. What Should Be the Moral Aims of Compulsory Sex Education? *British Journal of Educational Studies* 59 (1): 75–86.

Stoljar, Natalie. 2013. Feminist Perspectives on Autonomy. *Stanford Encyclopedia of Philosophy*, May 02. Accessed October 20, 2020. https://plato.stanford.edu/archives/fall2015/entries/feminism-autonomy/.

Urback, Robyn. 2014. To McGill Activists, a "Yes" Doesn't Mean Consent. *National Post*, February 27. https://nationalpost.com/opinion/robyn-urback-to-mcgill-activists-a-yes-doesnt-mean-consent.

Wertheimer, Alan. 2003. *Consent to Sexual Relations*. Cambridge University Press.

West, Robin. 2013. The Harms of Consensual Sex. In *The Philosophy of Sex*, ed. Nicholas Power, Raja Halwani, and Alan Soble, 6th ed., 387–394. Lanham: Rowman & Littlefield Publishers, Inc.

Zheng, Lily. 2014. How to Ace Sex: Why Enthusiastic Consent Doesn't Cut It. *The Stanford Daily*, November 5. https://www.stanforddaily.com/2014/11/04/how-to-ace-sex-why-enthusiastic-consent-doesnt-cut-it/.

CHAPTER 16

Enthusiastic Consent to Sex

Tom Dougherty

Reluctant sex can strike us as worrisome. Consider how "Rachel" from New York describes some of her reasons for having had sex:

> I hate admitting how much sex I've had because it was "polite" to just let him finish. You read stories of rape and sexual assault but never about your own manners pressuring you into having sex. Sometimes you just don't want to have sex after all the buildup but there is no way to get out of it without coming off as rude or disappointing your partner, who is probably a good person, not some creepy dude in a club.[1]

Similarly, consider the following testimony of "Maria."[2]

> I'm bad at saying no to people […] Like … with men, when they want sex. I don't want to disappoint them, but I'm also not usually into it. […] I mean … the last time I had sex with a guy, he was like, 'I don't have a condom, is that okay?' And I knew it wasn't, but I said yes anyway. […] Because I didn't want him to be mad at me. Or yell at me. And I wasn't sure I didn't want it. I was already there, so I just let it happen.[3]

The sex in these stories doesn't seem great. It's not just that it doesn't sound fun and pleasurable. After all, clumsy sex isn't fun and pleasurable. Instead, there seems something *morally* off about having sex with someone who doesn't

Many thanks to David Boonin for helpful comments on a draft of this chapter.

T. Dougherty (✉)
University of North Carolina at Chapel Hill, Chapel Hill, NC, USA
e-mail: emailtom@unc.edu

D. Boonin (ed.), *The Palgrave Handbook of Sexual Ethics*,
https://doi.org/10.1007/978-3-030-87786-6_16

"want to have sex" but does so to avoid being rude. The same seems true of having unprotected sex with someone who isn't "into it" and thinks it isn't "okay" but who doesn't want to be yelled at.

But if there is something morally defective about these sexual encounters, then how should we characterize the defect? One approach is to focus on how Rachel and Maria are not enthusiastically consenting. Talk of enthusiastic consent is increasingly common nowadays. In popular discourse, we find it advocated in venues from *The Guardian* to *Teen Vogue*:

> don't look for consent in a sexual partner, because that's too low a bar. Look for enthusiasm … the unmistakable sense of not being able to keep your hands off each other.[4]

> But regardless of where you think the *legal* lines should be drawn, we can all agree that we want both ourselves and our partners to be enthusiastic about any sexual encounter.[5]

It has also been espoused by sex educators, such as Gigi Engle:

> Enthusiastic consent is not coerced, pressured, or forced in any way. It's when you're super down to clown, and everyone is extremely excited about it.[6]

We also find the notion in universities' policies and programming:

> Consent is … enthusiastic.[7]

> Only an enthusiastic yes means yes![8]

And the idea has been advocated by politicians around the world. For example, Lily Madigan, the Women's Officer of the United Kingdom's Labour Party, says that enthusiastic consent is

> the idea that consent moves beyond the initial "yes" and is shown by engagement and enthusiasm. It's about checking in with your sexual partner and acknowledging their sexual and emotional signals. It's also about acknowledging that consent can't be given where there is manipulation, pressure, threats, and where both people aren't in the right state of mind.[9]

Meanwhile, Pru Goward, Minister for the Prevention of Domestic Violence and Sexual Assault in New South Wales, Australia, states that:

> You must explicitly ask for permission to have sex. If it's not an enthusiastic yes, then it's a no. … I feel that this is where the law in [New South Wales] needs to go.[10]

While some of these quotations merely advocate that we should aspire to having sex with enthusiastically consenting partners, many imply that it is okay to have sex with someone only if they enthusiastically consent.

To frame our inquiry into enthusiastic consent to sex, let me introduce a threefold taxonomy for the moral effects that consent can have.[11] Consider any two individuals, A and B, who are fully responsible moral agents. On the assumption that A and B are symmetrically related, and both are equally sexual agents, every normative claim that is true of one is true of the other, and so nothing in what follows presupposes that one individual is an active agent and the other a passive recipient of sexual activity. However, for concision I will focus on the possibility that A engages in sexual activity with B. I assume that the moral default is that A gravely wrongs B by engaging in sexual activity with them. This default can be superseded by B giving consent. Under the right conditions, B's consent has the moral effect of making it the case that A does not wrong B with the sexual activity. I call consent that has this effect "fully valid" consent. These conditions include B being suitably free from coercion, being suitably informed, and having suitable decision-making capacities. Suppose that these conditions are not met, B gives consent, and A engages in sexual activity with B. Then, there are at least two moral possibilities. First, A could wrong B exactly as gravely as A would wrong B in the absence of B's consent. In this scenario, B's consent has no moral effect. I call this type of consent "fully invalid" consent. Second, A could wrong B less gravely than A would wrong B in the absence of B's consent. In this scenario, the consent does not prevent a wronging, but it does ameliorate its gravity. I call consent that has this ameliorative effect "partially valid" consent. With these distinctions in mind, I will pose the central question of this chapter as follows:

Enthusiasm Question. Is consent fully valid only if it is enthusiastic?

Given the distinction between partially valid and fully invalid consent, a negative answer to this question does not imply that sex with an unenthusiastically consenting person is as morally grave as sex with someone who does not consent. But a negative answer does imply that an agent wrongs an unenthusiastically consenting person by engaging in sexual activity with them.

To answer this question, we need to do two things. First, we need to consider the different things we might mean by calling consent "enthusiastic." Second, we need to determine whether, on any of these interpretations, enthusiasm is necessary for fully valid sexual consent. Those tasks are the agenda of this chapter. I will ask whether certain conditions must obtain for someone's consent to be fully valid. Must they be eagerly consenting? Must they have made a settled decision to consent? Must they endorse their reasons for consenting? Must they clearly communicate their consent? Must they not have certain reasons for consenting? In addressing these questions, I will defend the following three substantive claims. First, it is false that someone's consent is fully valid only if they are eager, they have a settled motivation for having sex, or they endorse their motivation for having sex. Second, to responsibly have sex with someone, they must clearly communicate their consent. Third, someone's consent is fully valid only if they are not motivated by certain reasons. It is the third claim that explains what is morally amiss in the introductory cases involving Rachel and Maria. In addition to these matters of substance, I will also advocate for a point of terminology: we should not express my second or third substantive claim by saying that consent is fully valid only if the consent is enthusiastic. Combining these points of substance and terminology, my overall conclusion is that we should avoid saying that an agent wrongs an unenthusiastically consenting individual by having sex with them.

This chapter is organized as follows. In the first Section "Enthusiasm as Content-Independent Motivation", I argue that fully valid consent need not be eager consent, that fully valid consent need not be settled consent, and fully valid consent need not be endorsed consent. In the second Section "Enthusiastic Consent as Communication", I argue that to responsibly engage in high-stakes encounters like sex, we must have clear evidence that our partners are willing to do so. This provides a rationale for the requirement that expressions of consent have to be unambiguously positive. But I argue that while this requirement is philosophically defensible, it is not helpfully communicated using the language of enthusiasm. In the third Section "Reasons for Having Sex", I argue that insofar as a consent-giver's motivations undermine their consent, this will concern the *reasons* why they aren't enthusiastic. Consent is not fully valid when it is given for some reasons, for example, to avoid being shouted at. But consent can be fully valid when it is given for other reasons, for example, for procreation. It is the content of these reasons that determines whether the consent is fully valid. That is, some reasons that make people unenthusiastic about consenting to sex can preclude their consent from being fully valid but the lack of enthusiasm itself can't preclude their consent from being fully valid.

Enthusiasm as Content-Independent Motivation

Talk of "enthusiastic consent" needs to be interpreted. The most natural interpretation is psychological: someone's consent is enthusiastic (or not) in virtue of their motivations. I will discuss what I take to be the three most salient ways of fleshing out this idea.

The Eagerness View

We could say that someone gives enthusiastic sexual consent when they eagerly consent. This leads us to the normative view that consent is fully valid only when it is given eagerly. Let us call this the "Eagerness View."

Before we look at the Eagerness View in detail, I want to draw attention to a feature of it in order to introduce some terminology. Consider a non-sexual example in which Taylor eagerly wants to watch the circus. Part of Taylor's motivation concerns the content of their desire—the proposition that they watch the circus. But another part of Taylor's motivation concerns the stance that they take toward this content—they are eager. To mark this difference, we can say that Taylor's eagerness is a "content-independent" feature of their motivation. In due course, we will categorize other views according to whether they construe enthusiasm in terms of content-independent features of motivations.

The Eagerness View has the virtue of explaining why there is something morally problematic about our introductory examples. Assuming Rachel and Maria are consenting reluctantly, the Eagerness View implies that their consent is not fully valid. As well as predicting that result, the Eagerness View seems to provide the right type of explanation of why their consent is not fully valid. What triggers our intuitions that something has gone wrong are the cases' descriptions of Rachel's and Maria's motivations.

However, I think that our judgments about these cases are influenced by assumptions about the type of sexual encounter they involve. Admittedly, the descriptions are brief, but it's natural to read these cases as concerning hook-ups for mutual enjoyment and quite possibly casual encounters between people who do not know each other well. I think that when we limit ourselves to these types of case, a consent-giver's reluctance will often be a warning sign that something else has gone wrong.

But things look different when we pan out to a wider range of cases. People have sex for all sorts of reasons. Sometimes, they have sex without having much intrinsic desire for the sex itself. For example, instead of intrinsically desiring sex, people may have instrumental motivations. They might have sex in order to procreate. They might have sex in order to earn money. (And while sex work under patriarchy is typically marked by oppression, we should not lose sight either of the fact that some actual sex work is unaffected by oppression or of the hypothetical possibility of sex work in a thoroughly egalitarian society.) They may have sex because they are in a relationship characterized by a reciprocal arrangement where partners trade sexual gifts even when they are not particularly inclined to engage in sex for its own sake. Even without such an arrangement, people may have other reasons to offer their partners sexual gifts or to seek sex for other reasons, such as intimacy and bonding. For any of these reasons, they might not have much intrinsic eagerness for the sex, and yet there need not be anything morally defective either about their consent or the ensuing sex. As Quill Kukla notes:

> [W]e can legitimately consent to sex about which we feel ambivalent, or that we are only agreeing to in order to please a partner, for instance. We can consent to sex when we are not really in the mood, just as we can consent to watch a movie we are not especially up for. Asexual people and sex workers have resisted the idea that they can't give legitimate consent to sex despite a lack of felt giddiness about the whole affair. So long as a non-eager individual has non-problematic instrumental motivations for having sex, it is fine for their partner to have sex with them.[12]

An advocate of the Eagerness View might respond by conceding that an intrinsic eagerness for the sex is unnecessary for fully valid consent. Instead, the eagerness might also come from instrumental motivations. According to this response, what matters is whether taking into account all the individual's desires, both intrinsic and instrumental, they are eager to have sex. In other words, in terms of their all things considered motivations, they must eagerly consent in order for their consent to be fully valid.

There are two reasons why this response will not save the Eagerness View. The first reason is that fully valid consent can be given by people who are not all things considered eager to have sex with each other. For example, two people might decide to have sex without having much intrinsic desire for any particular activity—sexual or otherwise. For example, it may be that they live in an isolated location where there is nothing else to do. Perhaps, the idea of having sex doesn't particularly thrill them, but they are thoroughly bored, and all the alternatives seem worse. This is unlikely to be the best sex of their lives, but assuming that they are appropriately responsive to each other's decisions, they would still be treating each other with proper respect and concern. So even though they are not all things considered eager to have sex, their consent could still be fully valid. A similar point could be made about people who are mildly depressed and hence not really enthusiastic about anything. And the same holds true of all the other instrumental reasons we looked at earlier. It may be that someone wants to have procreative sex because they or their partner are ovulating, even though they are tired and not particularly in the mood. As a result, they might marginally all things considered prefer to have sex, even though all things considered they are not enthusiastic about the prospect.

The second reason is that if an advocate of the Eagerness View makes this response, then the view fails to explain what is problematic about the consent given by Rachel and Maria in our introductory examples. Suppose that Rachel is extremely anxious not to come off as rude. This gives her a motivationally powerful reason to engage in sex. Therefore, in terms of her all things considered motivations, she would count as eager to engage in the sex. But this eagerness would not lead us to conclude that Rachel's consent is fully valid. If anything, Rachel's strong desire to avoid being perceived as rude would make us even *more* morally concerned about the encounter. We can generalize this point. Take any case where someone gives consent out of an instrumental motive, such that they do not give fully valid consent. Next imaginatively increase the strength of this motive, so that this person counts as being all things considered eager to have sex. Now we have a case in which someone does not give fully valid consent, even though they are all things considered eager to consent.

The Settled Motivation View

The foregoing shows that we should deny that someone gives fully valid consent only if they consent eagerly—either in terms of their intrinsic desires for the sex or their all things considered preferences. But there remains the possibility that fully valid consent requires a different motivation that is also content-independent in the sense that it does not concern an individual's particular reasons for having sex.

What could this content-independent feature of their motivation be? Let us return to the benign cases, such as the case in which two people decide to have procreative sex without particularly wanting to have sex for its own sake. We might think that their consent is fully valid because it is part of a broader plan and hence based on a settled motivation of each agent. There are different respects in which an agent's motivation could be settled. If they had decided ahead of time to schedule sex, then the motivation would be settled *temporally*. If they would still have decided to have sex in slightly different circumstances, then the motivation would be settled *counterfactually*. If the decision to have sex was a fixed point in their deliberations, such that they are unwilling to change their mind, then it would be settled *resolutely*. We could appeal to any or all of these senses of "settled" and say that someone enthusiastically consents when they have a settled motivation to consent. Let us say that the "Settled Motivation View" is the view that consent is fully valid only if it is enthusiastic in this sense.

We should reject the Settled Motivation View as it has counterexamples. Suppose that a first date has gone well and a couple have had sex and then slept in the same bed. In the morning, they start making out, and this leads them to communicate that they both would enjoy sex. However, sex would make them late for morning brunch dates with friends, and this puts seeds of doubt in their minds. At the spur of the moment, they decide to have sex and clearly communicate their decisions to each other. We can assume that their motivations are not settled in any of the aforementioned senses. Since they made a spur of the moment decision, their motivations are not settled temporally. We can assume that their decision would not be counterfactually settled by supposing that they would have decided differently in slightly different circumstances. For example, if they had been a little less aroused by making out, then they might well have decided instead to get on with their day. We can assume that their motivation would not be resolutely settled by supposing that they were of two minds about whether they were doing the right thing or whether they ought to be setting off to meet their friends. Still, even if their consent is not settled in these ways, their consent could still be fully valid given that they both genuinely want to have sex for its own sake, they have freely decided to have sex, and they have clearly communicated their decisions to each other.

In response to this argument, an advocate of the Settled Motivation View might concede that fully valid consent does not require that a particular motivation to have sex is settled. But they might insist that someone must have a settled policy to be willing to have sex in these types of scenarios. However, this response is implausible. In the brunch case, it may be that both parties are having sex for the first time, and so they have not developed a policy one way or the other as to whether they are willing to forego brunch with friends for sex. But that does not preclude the sex from being morally okay. So I conclude that consent can be fully valid even if the relevant individuals are not consenting from settled motivations.

The Endorsement View

Here is an alternative proposal: someone's sexual consent is enthusiastic when they *endorse* the motivation that leads them to consent. This endorsement would consist in their having a positive higher-order attitude toward this motivation. For example, it might consist in their approving of this motivation as a suitable reason for having sex. Let us say that the "Endorsement View" is the view that someone gives fully valid consent only if they endorse the motivation that leads them to consent.

On the natural interpretations of the cases that we have discussed so far, the Endorsement View yields intuitive verdicts. For example, it is natural to assume that people engage in procreative sex while approving of a desire to have children as a suitable reason to have sex. Similarly, it is natural to assume that the couple missing brunch approve of sexual arousal as a suitable reason for having sex. On the other hand, it seems plausible that Rachel disapproves of wanting to be polite as a suitable reason to have sex. If so, then the Endorsement View implies that her consent is not fully valid.

However, there are two reasons why we should reject the Endorsement View. The first is that there are cases in which people do not approve of their sexual motivations and yet give fully valid consent. Suppose that someone has had an oppressive upbringing that has led them to believe that their own sexual pleasure and desires are shameful. Consequently, they have the higher-order attitude of disapproving of these motivations and so they do not endorse them as suitable reasons for having sex. But suppose they are tempted by pleasure to act against their better judgment, succumbing out of weakness of will to engage in a sexual encounter. (Fill in the rest of the details of the case, so that it is ideal in every other respect.) From this encounter, they come to see the value of hedonistic sex. This leads them to change their mind and see hedonism as a good reason for having sex. All the same, at the point at which they first had sex, they did not endorse their reason for having sex. Yet this is consistent with their consent being fully valid. This case exemplifies a more general point about what to do when our higher-order attitudes do not cohere with our first-order motivations. Sometimes, it is the higher-order attitudes that are the problems. When that is the case, our lack of endorsement of our first-order motivations should not cast shade on acting on these motivations.

The second reason why the Endorsement View fails is that people can form mistaken beliefs about the appropriate way to be treated, and this can lead them to endorse motivations that are morally problematic. Suppose, for example, that Rachel believes that being polite is a good reason or that Maria believes that not wanting to disappoint a partner is a good reason to have sex. If Rachel and Maria have these beliefs, then they would approve of these motivations and hence endorse them. All the same, we would continue to view their consent as not fully valid in light of their motivations. Or to press the point, we could imagine someone brought up to believe that it is their matrimonial duty to have sex whenever their spouse wanted, and it is acceptable for their spouse to enforce this with physical violence. If this individual were fully indoctrinated, then they might endorse their motivation to have sex out of a sense of duty or to avoid abuse. Therefore, insofar as we aim to characterize a feature of Rachel, Maria, and the indoctrinated spouse in virtue of which they do not give fully valid consent, this feature does not concern whether they endorse their motivation for consenting.

Enthusiastic Consent as Communication

In the last section, I considered views that explain the moral force of someone's consent in terms of their motivations. Now, I want to consider a view that explains consent's moral force in terms of features of how someone communicates this consent. Consider the following substantive view:

> Unambiguity View. Someone should have sex with another person only if the latter person has communicated unambiguously that they are willing to have sex.

We might aim to express the Unambiguity View by saying that consent must be enthusiastic. In doing so, we would be understanding enthusiasm in terms of clear communication.

Here are two ways of motivating the Unambiguity View. First, we might argue that consent must be publicly communicated in order to transform people's moral relationships with each other. Elsewhere, I have defended the claim that someone must clearly express their willingness to have sex in order to give sexual consent.[13] A second motivation for the Unambiguity View is that without clear communication, there is a risk that an individual is not consenting. If this risk exists, then it would be negligent for their partner to proceed with sex. This implies that people need each other's clearly communicated consent to responsibly have sex with each other.[14]

Of these two motivations, the first is controversial, and it is denied by people who endorse views according to which consent is a purely mental phenomenon. An example is the view that consent involves intending to forego complaining about another person crossing one's moral boundaries.[15] If that view is true, then consent can be given without any communication, and so a fortiori consent can be given without unambiguous communication. But even if we accept the view that such consent is valid, we can still endorse the second motivation. This is because it is coherent to say that consent is a mental attitude and also to say that responsible sexual behavior involves assuring oneself that one's sexual partner has this attitude. Indeed, it is hard to see any grounds for resisting this thought, and so, I conclude that we have good reason to accept that morally permissible sex involves clearly communicated consent.

Still, let us note four points about the Unambiguity View. First, the absence of clearly communicated consent is not sufficient for the commission of a serious sexual offense. Consider the following case:

> *Ambiguity.* C attempts to communicate that they are willing for D to sexually penetrate C. But because C is inexperienced and nervous, C's attempts to communicate are ambiguous. While D thinks that, most likely, C is willing for D to penetrate C, D is not fully confident about this. However, D is apprehensive about asking, and worried that doing so will indicate inexperience. Moreover, D assumes that C will refuse if C is unwilling. D penetrates C.

As I note elsewhere,[16] there is a respect in which D has wronged C: D has failed to take due care toward them in the respect that D has not done enough to investigate that C is consenting. All the same, this is not one of the most serious sexual wrongs that D can commit against C. C did not have sex against their will. Indeed, we can suppose that C was incredibly eager for the sexual encounter to take place and that they remained glad throughout that it did. In virtue of these features, it is implausible that C has suffered one of the most serious forms of sexual assault.

Second, the Unambiguity View will not help us with our introductory cases of Rachel and Maria. For example, it could be that Rachel is unambiguously clear that she is consenting to sex and yet still be doing so only to avoid seeming rude. Insofar as we are looking for a formulation of an "enthusiastic consent" view that implies that Rachel's consent is not fully valid, it will not come from the Unambiguity View.

Third, we already have a well-known term for expressing the clear communication advocated by the Unambiguity View. This is the term "affirmative consent." This phrase has been popularized in the last decade or so, with the consequence that it has become entrenched in popular discourse and features widely in university campus codes. In light of the popularity of this phrase, there is no need to define a parallel term "enthusiastic consent" with exactly the same meaning.

Fourth, introducing "enthusiastic consent" as another term for affirmative consent problematically creates a risk of misinterpretation. When many people hear the phrase "enthusiastic consent," their minds will not immediately turn to the idea of clear communication. Instead, they are likely to conjure up a psychological interpretation of this phrase. Ultimately, it is an empirical matter how much misinterpretation there would be and I cannot settle this from the armchair. I can only flag the risk and note that this counts against formulating the Unambiguity View in terms of "enthusiastic consent."

We can put these points together as follows. The Unambiguity View will not help us with the cases of Rachel and Maria. Moreover, there is a downside to expressing the Unambiguity View with the slogan that consent must be enthusiastic: it runs the risk of misinterpretation. This downside could be outweighed if there were a significant upside. But there is no upside since we already have the phrase "affirmative consent" with which to formulate the Unambiguity View. I conclude that the Unambiguity View does not give us a good reason to say that only enthusiastic consent is fully valid.

Reasons for Having Sex

I began this essay with the cases of Rachel and Maria. In light of our conclusions in the Section "Enthusiasm as Content-Independent Motivation", we should not explain what is problematic about Rachel's and Maria's consent by appealing to the idea that consent needs to be enthusiastic in the sense that the consent needs to have a certain content-independent feature. Moreover, in the Section "Enthusiastic Consent as Communication", I argued that while it is plausible that someone should have sex only with people who clearly communicate consent, this view cannot explain what is morally problematic about the cases of Rachel and Maria. So the question arises as to what does explain this. My answer is that the problem concerns the *content* of Rachel's and Maria's motivations for having sex. In other words, I think that their consent is not fully valid because of content-dependent features of their motivations.

In the case of Rachel, her motivation is to avoid being thought rude. This fact about the content of her motivation is distinct from facts about whether Rachel is eager, whether her motivations are settled, whether she endorses this motivation, or whether she clearly communicates it. Being thought rude is a type of penalty. It is also an inappropriate penalty: it is not reasonable to see someone as rude simply because they have changed their mind about having sex. As such, we can see Rachel as giving consent because she aims to avoid an inappropriate penalty.[17]

Once we frame Rachel's consent in this light, we can assimilate it to other forms of consent that is not fully valid. Consider someone who gives consent because of coercion. Paradigmatic coercion operates by imposing an illegitimate penalty on someone's decision to refuse to consent. For example, if a perpetrator threatens a victim with violence unless they consent, then the perpetrator is imposing a penalty of violence on refusal, and this is an illegitimate harm. Of course, violence is a worse penalty than being thought rude. In light of this point, we may wish to distinguish penalties of differing levels of unacceptability.[18] Plausibly, these penalties have to reach a certain threshold of severity for it to be the case that a victim's consent is so impaired that it is morally as if they had never consented at all. In other words, the penalties must reach a threshold in order to fully invalidate an individual's consent. But in light of the possibility of partially valid consent, even if a sexual encounter is caused by coercion that is below this threshold, the encounter can still be morally wrongful in a less grave way. One candidate is a sexual encounter that someone participates in to avoid a social sanction like being thought rude. I suggest that it is because we take this candidate seriously that we are inclined to view Rachel's consent as not fully valid.

We can apply a similar analysis to the case of Maria. Maria says that she has had sex to avoid making her partner get mad or yell at her. These are also inappropriate penalties to impose on someone's refusal to have sex. In addition, Maria says that she has had sex to avoid disappointing her partner. This can be taken in two ways. First, it may be that her partner is disappointed *with Maria*. This social reaction is typically a type of social sanction. (Compare the parent who tells their offending child, "I'm not angry, I'm just disappointed with you.") The reaction functions as a social sanction, as it is normal and legitimate for us to seek each other's approval and hence to wish to avoid disappointing others. In this case, it is inappropriate for Maria's partner to impose on her this sanction for refusing sex.

Second, it may be that Maria's partner is disappointed in the sense that they were hoping to have sex, but they did not have sex. So long as this attitude is not targeted at Maria, then this disappointment need not be seen as an inappropriate reaction. It's fine to have got your hopes up and then feel a bit bummed when things didn't go as you wanted. There is then an interesting question about what to say if Maria decided to have unprotected sex to avoid making her partner disappointed in this sense. I think that we may be of two minds about this, but insofar as we are inclined to see Maria's decision as problematic, this will be because we take her to be placing disproportionate weight on her partner's interests compared to her own. She has an interest in avoiding unprotected sex, and it is natural to assume that this derives from a more basic interest in avoiding pregnancy or sexually transmitted infections. These are weighty interests. By contrast, Maria's partner has an interest in the additional pleasure that they would derive from unprotected sex, compared to protected sex. This marginal pleasure is relatively unimportant in the circumstances. If Maria is prioritizing her partner's relatively unimportant interests over her own weighty interests, then she is being excessively selfless. This may be a circumstance in which it is wrong to have sex with someone who is being excessively selfless. If so, then I suspect that this wrong is best characterized in terms of how Maria's sexual interests are set back rather than in terms of her consent not being fully valid, but I will not press this point here.

Note that this is a claim about *excessive* selflessness. As such, it is distinct from the view that appropriate selflessness can be a suitable reason to have sex. Someone is appropriately selfless when they sufficiently respect and value themselves; they are disposed to advocate for their own interests and yet decide to prioritize another person's interests for a good reason, such as the reason that the other person's interests are more weighty. This will often be the case in sexual gifts. Giving a sexual gift may require effort or discomfort on the part of the gift-giver. Yet the gift-giver may judge that their partner would get pleasure that outweighs this effort or discomfort. If so, then they would be acting selflessly with their sexual gift, but not excessively selflessly.

There is room for much further debate about which reasons are such that it is okay to have sex with someone who is motivated by these reasons. I will not attempt to settle this issue here, and I do not assume that my foregoing claims on these topics are beyond reasonable dispute. Instead, the point that I want to insist on is that this is the right way to frame our analysis of the cases of Rachel and Maria. We need to look at the content of their reasons for having sex and ask, "Would one person wrong another person by engaging in sexual activity with them when they are motivated by this reason?" This is a claim about whether, for example, Rachel's partner wrongs Rachel and it does not imply that Rachel is to be criticized for having or acting on these motivations. On the grounds that the content of these reasons determines whether a wronging occurs, we should accept that the relevant features of their consent are content-dependent features.

Since this view is not naturally or helpfully framed as the claim that consent must be enthusiastic, I conclude that this view does not give us reason to say that consent must be enthusiastic. This is significant because the view explains cases like the introductory cases of Rachel and Maria. Since our intuitive verdicts about these cases provide the initial impetus for the hypothesis that fully valid consent must be enthusiastic, the fact that the hypothesis does not explain these verdicts shows that the enthusiastic consent model lacks a good motivation.

Conclusion

My principal concern has been with certain substantive issues. Corresponding to each section of this chapter, I have argued for the following three substantive claims:

(1) We should reject the view that someone's consent is fully valid only if they are eager, the view that someone's is fully valid only if they have a settled motivation for having sex, and the view that someone's consent is fully valid only if they endorse their motivation for having sex.
(2) Someone should have sex with another person only if that person has communicated unambiguously that they want to have sex (The "Unambiguity View").
(3) Whether someone's consent is fully valid does depend (at least in part) on the content of their reasons for giving consent.

While claim (2) is independently plausible on the grounds that responsible sexual agents must be sure of their partner's consent, it is claim (3) that explains why there is something problematic about Rachel's and Maria's consent in our introductory examples.

In addition, I have argued for a position that we should take on the terminology that we use for expressing these substantive claims. I have argued that it is unhelpful and misleading to frame either (2) or (3) as the claim that fully valid consent must be enthusiastic. Regarding claim (2), we already have the language of "affirmative consent" to express this claim, and talk of enthusiasm is likely to bring to mind the consent-giver's motivations. Moreover, since the natural interpretation of "enthusiastic" is content-independent, it is misleading to express (3) by saying that consent must be enthusiastic.

Of course, there is nothing to stop a moral philosopher defining terms of art in stipulative ways that depart from their normal meanings. Similarly, a law or policy can also give specialist definitions of terms. But there has to be a sufficiently good reason for doing this in light of the risk of causing misunderstanding. The risk of misunderstanding arises because people are liable to default to the everyday meanings of these terms. And if we aim for a view to be accepted by a large number of people or for a policy to be applied to a large number of people, then many people will run the risk of misunderstanding. This makes it highly likely that at least some people will actually misunderstand. Given how important it is that people understand sexual ethics and sexual offense policies, we have good reason to avert this misunderstanding. We can do this by avoiding expressing claims (2) and (3) by saying that someone gives fully valid consent only if they enthusiastically consent.

Notes

1. (Bennett and Jones 2019: 82).
2. "Maria" is my pseudonym as no pseudonym is given in the New York Times article containing the testimonies. In this article, Maria is described by "Elise" as a "girl [she] hooked up with." "Elise" is the article's pseudonym.
3. (Bennett and Jones 2019: 68).
4. (Hinsliff 2015).
5. (Vrangalova 2016).
6. (Elizabeth 2019).
7. "Sexual Violence and Title IX: What you need to know at Coastal Carolina University," 2014–2015 accessed November 1 2020 at https://www.coastal.edu/media/studentaffairs/counselingservices/pdfs/SA/20Sexual/20Violence/20booklet/20WEB/20Jul14.pdf
8. This is part of the title of one of the topics of Duke's Women's Center's "interactive, student-facilitated training" accessed November 1 2020 at https://studentaffairs.duke.edu/sexual-misconduct-prevention-and-response/make-a-difference/available-training
9. (Madigan 2018).
10. (Whitbourn 2018).
11. (Dougherty 2021a).
12. (Kukla 2021).
13. (Dougherty 2015, 2021b).
14. Alexander Guerrero (2021) defends affirmative consent policies on the grounds that proceeding without communicated consent would involve negligence. In

(Dougherty 2018), I defend these policies on the grounds that people have duties of due diligence to check that others are consenting. Some people hold that negligence can constitute mens rea for sexual offenses in the sense that they hold it is appropriate to convict people who ought to have been aware that their partners were not consenting. For an analysis of certain cases of incompetent consent in terms of the wrongfulness of taking risks that one's partner does not consent, see (Boonin ms).

15. (Alexander 1996, 2014).
16. (Dougherty 2018: 97, 2021b).
17. I develop this analysis in greater length in (Dougherty ms).
18. (Dougherty 2021a).

References

Alexander, L. 1996. The Moral Magic of Consent II. *Legal Theory 2* (3): 165–174.

———. 2014. The Ontology of Consent. *Analytic Philosophy 55* (1): 102–113.

Australian Broadcasting Corporation. 2018. "'Enthusiastic consent': What is it, how do you prove it, and will it work in court?" https://www.abc.net.au/triplej/programs/hack/how-do-you-prove-enthusiastic-consent-in-court/9740214

Bennett, J., and D. Jones. 2019. 45 Stories of Sex and Consent on Campus. In *Defining Sexual Consent: Where the Law Falls Short*, ed. The New York Times Editorial Staff, 66–82. New York NY: New York Times Educational Publishing.

Boonin, D. Ms. When Yes Means No: Problems of Sexual Consent

Dougherty, T. 2015. Yes Means Yes: Consent as Communication. *Philosophy and Public Affairs* 43: 224–253.

———. 2018. Affirmative Consent and Due Diligence. *Philosophy and Public Affairs* 46: 90–112.

———. 2021a. Sexual Misconduct on a Scale: Gravity, Coercion, and Consent. *Ethics 131* (2): 319–344.

———. 2021b. *The Scope of Consent*, Oxford: Oxford University Press

———. ms. "Social Constraints on Sexual Consent"

Elizabeth, D. 2019. "Enthusiastic Consent is Changing How We Have Sex," MTV News, 3rd April 2019, accessed 6th October 2019 at http://www.mtv.com/news/3119428/how-to-practice-enthusiastic-consent/

Guerrero, A. 2021. "The Epistemology of Consent," in Jennifer Lackey (ed.) *Applied Epistemology*, Oxford University Press

Hinsliff, G. 2015. "Consent is not enough: if you want a sexual partner, look for enthusiasm,". The Guardian, https://www.theguardian.com/commentisfree/2015/jan/29/rape-consent-sexual-partner-enthusiasm,

Kukla, Q. 2021. A Non-Ideal Theory of Consent. *Ethics 131* (2): 270–292.

Madigan, L. 2018. "We need to talk about enthusiastic consent,". *Medium* https://medium.com/@lilymadigan/we-need-to-talk-about-enthusiastic-consent-e23bdc725f55

Vrangalova, Z. 2016. "Everything You Need to Know About Consent That You Never Learned in Sex Ed," Teen Vogue 18th April 2016 accessed 6th October 2019 at https://www.teenvogue.com/story/consent-how-to

Whitbourn, M. 2018. 'Enthusiastic yes': NSW announces review of sexual consent laws," Sydney Morning Herald May 8 2018 accessed 6th October 2019 at https://www.smh.com.au/national/nsw/enthusiastic-yes-nsw-announces-review-of-sexual-consent-laws-20180508-p4zdyn.html

CHAPTER 17

On the Sufficiency of Sexual Consent

Alan Soble

"When it comes to sex, who properly understands their actions?"
Major Crimes, *S5/E2 (2016), "N. S. F. W."*
—[Lawyer Charles Hardin]

"Women need a reason to have sex. Men just need a place."
—Widely attributed to Billy Crystal.

This chapter begins (**I**) with general remarks about the principle of free and informed consent (F&IC). I then examine (**II**) Seiriol Morgan's argument that consent is not sufficient for the morality of sexual activity—which I show to be defective. However (**III**), his argument leads us to recognize a significant feature of F&IC that is slighted even by theorists for whom dialogue or communication before and during sexual activity is taken to be crucial. Knowledge of *why* we desire to have sex with another person should not be ignored during discussion about *what* we want to do. In (**IV**) I express some reservations.

I.

The presence of free (voluntary, uncoerced) and informed (knowledgeable, comprehending) consent (agreement, saying or implying "yes") by a patient is necessary and sufficient for the morality of the performance of a procedure is a central principle in the practice of Western medicine, designed to protect the self-determination and dignity of the patient. It is often argued that *ceteris paribus*

A. Soble (✉)
University of New Orleans (Emeritus), Philadelphia, PA, USA
e-mail: asoble@uno.edu

D. Boonin (ed.), *The Palgrave Handbook of Sexual Ethics*,
https://doi.org/10.1007/978-3-030-87786-6_17

should be appended to "sufficient." Physician-assisted termination of life, intentionally contributing to or causing the death of a suffering, terminally ill patient, may be unjustifiable killing and hence an exception to sufficiency, based on paternalistic considerations, or because euthanasia undermines the reliability of and social trust in medicine (thereby harming nonparticipants), or because interferences with continued life are intrinsically wrong.

Our question is not whether these arguments for curtailing sufficiency are convincing; it is whether analogous maneuvers are convincing for sex. Nonetheless, it is useful to review the territory. The paternalistic objection to euthanasia may be met by emphasizing that the relevant patients, despite suffering, are mentally sound adults. Euthanasia's unwelcome institutional or social effects are not empirically confirmed or invoke slippery slopes. The "intrinsically wrong" objection is contentious; it may even beg the question.

The F&IC principle, as I have believed for decades, governs all interactions among people who are over a specific age and lack mental defects. The principle is not extended from medicine to other areas; historically, it has been extended from the personal, educational, recreational, social, religious, and economic realms to later-developing medical contexts. Sexual acts may be subsumed, depending on the details, within these domains of life. The F&IC principle applies to sexuality, unless it is a unique aspect of life whose axiology is *sui generis.*[1]

The presence of F&IC is nearly always considered to be an unbreakable necessary condition for sexual activity; violations constitute rape or assault. Not surprisingly, many objections have been raised against sufficiency. Paternalistic concerns may be offered to curtail consent to risky sexual practices (BDSM) or other acts contrary to one's self-interest, in particular of women.[2] The occurrence of sadomasochism may have harmful social effects if, say, the legal system faces unmanageable decisions about whether consent was given to inflicted bruises or pain. Adulterous sexual acts may in themselves satisfy the F&IC principle, yet may be rejected (by attaching an appropriate proviso) on account of harmful effects on nonconsenting third parties or because they involve deception or promise breaking. BDSM may be condemned as intrinsically wrong along with homosexuality, as in Natural Law.[3] One attraction of the F&IC principle is that it defuses paternalistic and moralistic restrictions on the variety of human sexual desires. In any event, *ceteris paribus* may without worry be appended to the sufficiency clause, an adjustment made nearly universally and uninterestingly to claims that some *X* is sufficient for some *Y*.

According to Nir Eyal, in an impressively thorough account of the application of the F&IC principle in medicine,[4] F&IC has two distinct components, "free" (voluntary, uncoerced) and "informed" (knowledgeable understanding; the absence of lies, fraud, deception; no important information withheld). Joined together they yield the principle of free and informed consent. Well-known writings about sex by philosophers and legal scholars that claim that rape or other sexual immoralities may be realized by deception accept, innocuously, that in sexual contexts the principle requires consent which is both free and informed.[5] At work is not a principle of "free consent" to which "informed" is welded as an afterthought or which plays no role in creating genuine consent. The power of deception to negate consent is hardly controversial.

II.

Seiriol Morgan has provided an argument that attempts to show that consent, in an interesting, nontrivial way, is not sufficient for the morality of sexual activity.[6] "My aim … is to advance a line of criticism against an influential contemporary view of sexual morality. The view [his main philosophical adversary is Igor Primoratz][7] is that universal participant consent is sufficient for the moral permissibility of any sexual act" (377). "I will be arguing that the complex and frequently dark nature of human sexual desire requires a certain kind of ethical sensitivity from us, and also places upon us obligations not to act on certain sexual impulses we might have, even if there is no-one who fails to consent to their gratification" (378).[8]

Morgan's strategy is to construct a counterexample to "the view" based on Pierre Choderlos de Laclos, *Les Liaisons dangereuses* (1782), the details of which I will not repeat. It is enough to review Morgan's summary of what he believes he has accomplished. "We see in the Vicomte [de Valmont, the book's protagonist] a species of sexual desire which of its nature should not be gratified, whether its gratification is consented to or not" (400). "All in all … the Vicomte presents us with a clear-cut case of sexual behaviour which is consensual and yet contrary to the Kantian conception of duty" (401). Hence consent is not sufficient and "the view" is false. "What is so jaded about the Vicomte is that what he desires is for her [Madame de Tourvel] to want him in spite of herself, for her to be unable to resist the desire he has made her feel, despite her identification with chastity and virtue" (381–82). The Vicomte's "dark desire" is to humiliate Madame by seducing her, who thereby—through her own consent—abandons her sexual values.

In defending the sufficiency of consent, Primoratz criticizes Natural Law, Roger Scruton's "individuating intentionality,"[9] and various "radical" feminists. None, on his view, establishes that factors beyond consent are required for a comprehensive sexual morality. What Primoratz writes about consent makes it imperative that a counterexample to his "consent is sufficient" thesis be a case in which consent, both free and informed, occurs, yet the subsequent sexual activity is morally wrong. The philosophies rebutted by Primoratz "deny legitimacy to sexual interactions based on nothing but mutual consent that is reasonably free and informed and may be motivated by any of a host of reasons."[10] Primoratz's thesis is that free and informed consent is sufficient for the moral permissibility of sexual encounters. The motives sit silently in the philosophical background.

In Morgan's intricate description of the example, the Vicomte engages in extensive deception in his seduction of the Madame. Morgan's counterexample fails because the informed clause of the F&IC principle is violated.[11] The only place he deals with this issue is a footnote (401*n*23):

> [S]omeone holding the infringement of freedom to be the only possible thing wrong with sexually motivated behaviour wouldn't have much trouble producing an argument that what the Vicomte does to Madame de Tourvel in the book is wrong. For one thing, he repeatedly lies to her, breaks promises, intimidates and threatens her. … But to locate the badness of the Vicomte simply in the fact that he does these things is to rather miss the point. The very nature of the desire he feels and the pleasure he anticipates shows the corruption of his character. It would be contrary to his duty of virtue for him to gratify it, even if he cunningly succeeded in doing so without violating her autonomy in this way.

Morgan speaks about "someone holding the infringement of freedom to be the only possible thing wrong with sexually motivated behavior." This "someone" is a straw person and certainly not Primoratz, who does not assert that "free" alone is sufficient. Morgan claims that "someone" could argue that the Vicomte's actions are wrong on the grounds that deception itself is an infringement of "free," but he fails to acknowledge the independent negation of consent by deception. To refute Primoratz's thesis, one must describe a case in which the F&IC principle is satisfied in all ways yet in which the sexual activity is morally wrong. Morgan holds out the possibility that the Vicomte *could* have satisfied his dark desires by seducing the Madame and "cunningly succeeded" in doing so without deception. This suggestion admits that his counterexample has the wrong logical form. Further, it is Morgan's task to construct the proper counterexample by laying out the details of such a case, not merely leave it to agreeable, sympathetic imaginations that the Vicomte could pull it off with thorough honesty. The Vicomte would have to be candid about his motive, which would kill for him the significance of the enterprise and its success.

Raja Halwani has offered another example.[12] "Lisa hates yet desires Nancy; she [Lisa] wants to sexually humiliate her; Nancy knows this but desires Lisa, and she does not care. They have sex during which Lisa heaps verbal abuse on the trembling-with-desire Nancy." His conclusion is similar to Morgan's: "This act has the universal consent of all the relevant parties, yet it is vicious given the motives of both women." Hence "consent is not sufficient for the [vicious] act's goodness, though perhaps it is for its permissibility." (Is this conjunction consistent?)

One virtue of this encounter and the participants is that the women are honest: their dark desires are out in the open. Another virtue is that they proceed voluntarily; they do not coerce each other.[13] So the F&IC principle is satisfied. But does the counterexample have the right form? I'm not convinced that the act is morally wrong or the participants vicious. The women sexually desire each other. They entirely enjoy themselves. They know exactly why each of them is there. Speaking Kantianly (which Morgan and Halwani would bless), they adopt each other's ends and make them their own. Lisa goes that far even though she hates Nancy. We may suppose that after a few shared orgasms, they do not regret the encounter.—Been there, done that. I resist the clever insult that we orchestrated vicious acts. I reject condemning us for not grasping virtue ethics. Maybe *because* the acts manifested F&IC, did not harm anyone, and we were open about our complementary reasons and wants. It helped that we jointly studied Pat Califia.[14]

It may be thought that Lisa's hatred decisively makes her character and Nancy's acquiescence vicious. No. We should not expect or insist that virtuous sexual encounters involve only loving, affectionate, or fond attitudes. That's just a moralizing restriction defused by the F&IC principle. Against a smiley-faced view of humanity is the fact that hatred is a ubiquitous sentiment and exists either consciously or deeply under the surface in everyday interactions, including those that are sexual. That the other person *has what you yearn for* creates a dialectic of antagonism between assertive pursuit and assertive reluctance.[15] It may be that few people execute the elaborate routine of a fictional Vicomte. But sex is darker than the darkness exhibited by the tame, pompous Vicomte, without being irremediably corrupt. And encounters, sexual or not, may be imbued with a complex mixture of many, often conflicting light and dark attitudes (or vanilla and kinky motives). The virtue of the women in Halwani's example is that they are conscious of and (unlike the Vicomte) openly share their darkness. Indeed, exposing and expressing Lisa's hatred in consensual, mutually pleasurable sex may profit them. Reconciliation is likely unreachable. Lisa's hatred, however, may be cathartically tempered and Nancy may come to feel less threatened by and vulnerable to her. True, if Lisa hates Nancy, Lisa might get carried away. But don't we often find ourselves getting carried away in sex? Without hate, sometimes more so with love. Sex is voracious.

III.

"Dark Desires" has a slew of philosophical virtues, but one of them is not that Morgan has delivered a decisive blow to Primoratzian sexual morality. The value of Morgan's essay is that he pushes us to examine motives and how knowledge of them impacts F&IC. Had the Madame known the Vicomte's motives, she would not have acquiesced to his sexual proposal. Here I depart from Primoratz's dismissal of motives or reasons for engaging in sexual activity.

Lois Pineau's famous 1989 essay "Date Rape: A Feminist Analysis"[16] introduced into the philosophical and legal examination of rape, date rape, and sexual assault the idea that parties considering sexual activity must engage in communication before and during the encounter.[17]

> Assuming that each person enters the encounter in order to seek sexual satisfaction, each person engaging in the encounter has an obligation to help the other seek his or her ends. To do otherwise is to risk acting in opposition to what the other desires, and hence to risk acting without the other's consent. … But the obligation to promote the sexual ends of one's partner implies the obligation to know what those ends are, and also the obligation to know how those ends are attained. Thus, the problem comes down to a problem of epistemic responsibility, the responsibility to know. The solution, in my view, lies in the practice of a communicative sexuality, one which combines the appropriate knowledge of the other with respect for the dialectics of desire. [473–74]

My thesis is that the satisfaction of the F&IC principle requires that each person knows their own reasons for the sexual encounter and the reasons of the other person(s). Pineau's communication model of sexual interaction is valuable because it creates space for the parties to investigate their own and the others' motives—the *why*. Pineau, however, seems concerned only with *what* is going to occur between the participants. Notice that the passage quoted above begins, "Assuming that each person enters the encounter in order to seek sexual satisfaction." I do not know why she makes this assumption. Both men and women just need a place?

I hunted for other passages in "Date Rape" which may have raised the *why* question, but found only these vague explanations of the meat of communication: "To be intimate just is to open up in emotional and personal ways, to share personal knowledge, and to be receptive to the openness of the other" (475). And: "the requirement of mutuality means that we must take a communicative approach to discovering the ends of the other" (477). It is not obvious from the context that "ends" here includes "motive." Pineau uses the enticing expression "full possession of knowledge," the lack of which negates consent, but without the details:

> [I]f acquiescence to an encounter counts as consent only if the acquiescence is reasonable, something to which a reasonable person, in full possession of knowledge relevant to the encounter, would agree, then acquiescence to aggressive noncommunicative sex is not reasonable. Hence, acquiescence under such conditions should not count as consent. [477]

If the man does not engage in communication, "he runs the risk of imposing on her what she is not willing to have." *What* is the subject of communication, not *why*. Note that the questions the prosecuting attorney in a trial would ask of the man who is accused of date rape focus on the defendant's knowledge, or lack of it, of *what*.[18] Thus Pineau's "communication model" should be revised and expanded.

Michelle Anderson has more recently than Pineau (who is oddly never mentioned) proposed a "negotiation model" of sexual interactions.[19]

> [T]he law should eliminate the requirement of nonconsent. In its place, the law should recognize the centrality of negotiation, in which individuals would be required to consult with their partners before sexual penetration occurs. Negotiation would not require a verbal contract for penetration. Instead, it would require only what conscientious and humane partners already have: a communicative exchange, before penetration occurs, about whether they want to engage in sexual intercourse. [1407]

The dialogue is thin: "shall we do it?" is its point. There's not a word in Anderson about discussing *why* they want to do whatever they eventually decide to do, whether it be penis-vagina penetration, penis-anus penetration, penis-oral cavity penetration, or a nonpenetrative act. She had the opportunity to do so: "Force, coercion, or misrepresentations by the actor would be evidence of a failure to negotiate"[20] (rather than, on her view, failure of consent) and hence of rape. But emendation is required. Specifically lying about or withholding motives would be evidence that the penetrative act did not fulfill Andersonian negotiation.

Why *why*? Insight into the motives of the other person may well play a crucial role in whether I want any part of this person or sex with them. I will need to extract their reasons accurately and be alert to signs of deception. I must also have insight into my own motives, not only to find out whether I fit well with the other person and the proposed activity, but also because the other person will likewise be circumspect about my reasons and grill me about them. Honesty with ourselves and with the other is mandatory, on pain of violating the F&IC principle.[21] We have now filled in more completely Pineau's "The solution … lies in the practice of a communicative sexuality, one which combines the appropriate knowledge of the other" and ourselves. The parties should know anything about the other and themselves which may influence their decision about the wisdom of the encounter; this includes their reasons for wanting to engage in sex. Of course, both lies about vicious reasons and withholding embarrassing motives violate F&IC.[22]

According to the poet Philip Larkin, "Sexual intercourse began/In nineteen sixty-three" (*Annus Mirabilis*). I had no idea what was going on around me. Indeed, I didn't have a clue until 1972 when, *suddenly*, I was faced with a real, not virtual, cornucopia of sexual partners, largely due (but in the admittedly loose 1970s) to a complete self-designed physical makeover and edging from Vivaldi and Franck to "Layla" and "Do It Again." It all happened surprisingly, this sweetness—*finally*—of touching, smelling, and tasting these bodies and being reciprocally touched, smelled, and tasted. Hearing unanticipated noises and words and seeing what had been modestly or enticingly draped were also sweet. So *this* is what I had been missing, while hanging with Rex Stout and Aldous Huxley, and over which Augustine and Kant got their knickers in a twist. I too briefly asked myself, "*Why*, you neglected, starving fool, did you want them?" I also too briefly asked myself, "*Why* do all these people, each in their own special way a beauty, want me?" The Lord had bailed out a nerd and I was not about to doubt Her decisions, a story which does not, however, enjoy the grandeur of a French novel. I attributed to my partners the motives I thought I had for myself: pleasure, emotional closeness, and escape from *horribilis mundo*. These assumptions were often true but also sometimes naïve. It took me way too long until I learned to be cautious not only about my partners' reasons but also my own. People can be understandably guarded and dreadfully inscrutable. They can also have the benefit of mixed motives or be plagued by them. (Lisa both desires and hates Nancy.) As a result, I didn't always realize what my or their motives were before the encounter. Years later, reflecting on some of these experiences, things began to fall into place. I'm not happy to report that one or both of us may have violated the F&IC principle by partially or wholly withholding reasons.

Among my partners' reasons for having sex with me I detected these: (1) "I was curious about what the person was like in bed." (2) "I wanted to give someone else a sexually transmitted disease." (I was ignorant of this motive for a few days.) (3) "I wanted to hurt/humiliate the person." (I was ignorant of this one for weeks.) (4) "I wanted to end the relationship." (I found out right afterward.) (5) "I wanted the person to love me." (6) "I was tired of being a virgin." (7) "I felt sorry for the person." (8) "I wanted to manipulate him into doing something for me." (9) "I wanted to see if I could get the person into bed." (10) "I wanted to act out a fantasy." (11) "The person was famous [me?!] and I wanted to be able to say I had sex with him." (12) "I wanted to get back at my partner for having cheated on me." (13) "I was on the 'rebound' from another relationship." (14) "I wanted to defy my parents." (15) "I wanted to make money." Not all these reasons are dark. Not all, had I known them, would have blocked my participation. But I should have known them in order to decide, with informed consent, how to proceed.

I did not whimsically make up these reasons; they actually occurred. But I have quoted all 15 from a list assembled by scholars Cindy Meston and David Buss, who "identified 237 distinct reasons why people engage in sexual intercourse."[23] (Heterosexual coitus was not always the event in my 15.) After giving their subjects[24] a questionnaire containing 237 numbered reasons, they asked them to indicate how frequent each reason was for their engaging in sexual intercourse (on a 1–5 scale; 1 = never, 3 = about half the time, 5 = always). Relevant to the Vicomte case is the result that the 9th lowest reason for women was the dark "I wanted to hurt/humiliate the person" (the mean score was **1.04**, the same score women had for "I wanted to feel closer to God"), and the 12th lowest for men was "I wanted to hurt/humiliate the person" (mean score **1.12**, the same score for "I wanted to feel closer to God"). A list of the top seven reasons subjects admitted having for intercourse is evidence that light or vanilla (conventional) motives occupy the minds of Texas undergraduates.[25] Meston and Buss, however, nicely point out a possible limitation of the study. "[T]he low social desirability of some reasons (e.g., to punish a partner; to make money; to give someone a sexually transmitted disease) may have lowered response rates to these reasons, while the high social desirability of other reasons (e.g., to express love; to show affection) may have raised response rates to these reasons."[26] Morgan may still be right about the "frequently dark nature of human sexual desire."

Another interesting feature of this study is that the questions about reasons for engaging in sexual intercourse were altogether aimed at revealing the reasons the subject had: Why did *you*, X, want to engage in sex with Y? The study did not ask how frequently each of the listed reasons led others to have sex with the subject. That is, it did not investigate whether subjects were aware of the reasons their partners wanted to have sexual intercourse with them. Why do you want *me*? Why do you *want* me? (This is why I focused in my 15 cases on my partners' reasons for engaging in sex.) Perhaps the researchers assumed that subjects knew well enough their own reasons for wanting to engage in sex while the reasons of the other person were not reliably reachable. It would have been significant to discover how much agreement there was between the reasons given by the X's for wanting intercourse and the reasons the Y's believed the X's had to copulate with them. Did the Y's take at face value the X's who reported that they "wanted to express my love for the person"?[27] A passage in Meston and Buss's article is relevant here:

> It is important to note that what constitutes a rare reason for the population as a whole might nonetheless constitute a frequent motivation for a subset of individuals within the population. For example, most people are not motivated to have sex in order to "get closer to God." But for some people, this has become their cardinal motivation. Most people are not motivated by the desire to humiliate another or to feel humiliated by another through sex. For some, however, sadism or masochism defines their principal sexual mode of being. [499]

The authors are on the verge of realizing the importance of digging out the reasons that various, uncountably many, X's have for wanting to engage in sex with a *particular* Y.

IV.

Suppose that according to the "informed" portion of the F&IC principle each person must know everyone's motives for the sexual encounter, including their own. I suspect that satisfying the F&IC principle, so understood, is not easy to accomplish. This observation does not entail that we should jettison the principle or cut back on either the necessity or sufficiency of F&IC.

Way back in 1929 Bertrand Russell claimed that sexual activity "should be a mutual delight, entered into solely from the spontaneous impulse of both parties."[28] One of his concerns was the lamentable fact that economic considerations often had a pernicious influence on the course of sexual interactions. The condemnation of social arrangements which incited women to practice prostitution is an obvious application of Russell's "solely from the spontaneous impulse of both parties" principle. But marital sexual activity, neither a "mutual delight" nor free of coercion, motivated by a wife's fear of economic abandonment by her husband was equally his target. Second Wave Feminism elaborated this theme. Robin Morgan, for example, claimed that "sexual intercourse" should occur only when it has been "initiated by the woman, out of her own genuine affection and desire."[29] Intercourse that does not abide by this principle is, for Morgan, rape.

Despite their progressive implications, Russell's "spontaneous impulse" and Morgan's "genuine affection and desire" principles are problematic. How literally we should understand "spontaneous impulse" (spontaneous combustion?), instead of taking it as a poetic description of blissful romance, is unclear. Morgan's principle is ambiguous: Does it require both a woman's "genuine affection" and her "genuine desire," or does Morgan mean "out of her own desire and genuine affection"? In any event, the principles are defied by Pineau's "epistemic responsibility, the responsibility to know"—not about *what*, but *why*.

How does Morgan's woman know whether she is acting "out of her own genuine affection and desire"? And that her partner is *also* acting out of her own genuine affection and desire? On occasions I experienced what seemed to be authentic spontaneous impulses involving apparently genuine affection and desire which motivated me to seek sexual activity. We are often, however, ignorant of many aspects of ourselves, our deeper psychology and unconscious yearnings and resentments, and everyone is susceptible to self-deception and rationalization. (The Vicomte, too.) Knowing that the other is motivated by their own affectionate desire is not always simple. Caresses, grindings, and moans may indicate that she is acting on her own genuine desires, but maybe not; arousal does not necessarily imply either "wanted" or "consensual." If my surface desire for pleasure is masking underlying vicious motives or distress, I cannot share them with my partner. Further, if my deep character has these invisible flaws, could this not also be true of my partner? I was dismayed to find out, eventually, that my partners' motives were not always playfully benign. The "eventually" is the key. Only *ex post facto* could I untangle the threads of "spontaneous impulse" and "genuine desire." Delayed insight, however, does not satisfy the F&IC principle. Communication and negotiation soon before sexual activity are bound to be too skeletal to unearth the buried attitudes and emotions that generate our reasons.[30]

X may say "I love you, Y" sincerely, yet hasn't a clue what is meant or offered as a reason or inducement to engage in sexual activity. X may only be in the grip of sexual desire and arousal, which are genuine enough. Maybe, then, X's safest proclamation is: "Dunno. Just do. Feels good."[31] Or a deadpan "Why [the fuck] not." If Lisa and Nancy were intent on doing a better job of fulfilling their epistemic responsibility, they would explore their psychologies thoroughly and might (like Frasier Crane) refer to Sigmund Freud on "primary masochism." Nancy would have to confront seriously, however, the possibility that her masochism was not primary but secondary and spend months or years on the couch figuring it out.[32] Then what? Psychosis may now be the issue, not morality, in which case she can continue her analytic defense by arguing (or rationalizing?) that her sexual activity is functional by eroticizing secondary masochism, making it palatable rather than distressing, as well as by confining it to a small box inside her life.

Notes

1. Jonathan Webber ("Sex," *Philosophy* 84, No. 328 [2009], 233–50), argues that "sex is not morally special" (236) and falls under the F&IC principle—which, however, does not in general and hence not for sex state a sufficient condition. On *sui generis*, see my "Concealment and Exposure," in Raja Halwani, ed., *Sex and Ethics* (Palgrave, 2007), 229–52.
2. Robin West, "The Harms of Consensual Sex," *American Philosophical Association Newsletters* 94:2 (1995), 52–55, reprinted in *Philosophy of Sex* [*POS*], 6th ed., Nicholas Power, Raja Halwani, and Alan Soble, eds. (Rowman &

Littlefield, 2013), 388–93. I discuss her "paternalism" in *Sexual Investigations* (New York University Press, 1996), 37–40.

3. See my "Kant and Sexual Perversion," *The Monist* 86:1 (2003), 57–92, and "Sexual Use," in *Philosophy of Sex*, 6th ed., 302–33.
4. "Informed Consent" (1/16/2019), in *The Stanford Encyclopedia of Philosophy*, Edward Zalta, ed., https://plato.stanford.edu/entries/informed-consent/.
5. Thomas Mappes, "Sexual Morality and the Concept of Using Another Person" [1985], in *POS*, 6th ed., 272–87 (the criterion is "voluntary informed consent" [273]; and "deception functions to undermine the *informed* character of voluntary consent" [277]); Alan Wertheimer, "Consent and Sexual Relations" [1996], in *POS*, 6th ed., 360–84 ("valid or morally transformative consent must be *informed* consent" [366]); Igor Primoratz, *Ethics and Sex* (Routledge, 1999) and note 10, below; Raja Halwani, *Philosophy of Love, Sex, and Marriage*, 1st ed. (Routledge, 2010): "Sexual acts obtained by deceiving someone or by manipulating her decisions are … obtained without consent" (177). For some feminist jurisprudence, see my review of Joan McGregor, *Is It Rape?* in *Law and Philosophy* 25:6 (2006), 663–72. Interesting is the distinction between "fraud in the factum" and "fraud in the inducement."
6. "Dark Desires," *Ethical Theory and Moral Practice* 6:4 (2003), 377–410.
7. "Sexual Morality: Is Consent Enough?" *Ethical Theory and Moral Practice* 4:3 (2001), 201–18.
8. In his "Sex in the Head" (*Journal of Applied Philosophy* 20:1 [2003], 1–16, in *Philosophy of Sex*, 6th ed., 101–20), Morgan hinted where he was going: "Some people clearly eroticize contempt … and take pleasure in the thought that they are doing something to someone which degrades them. … Such pleasure need not involve violating the autonomy and consent of the other person, since it is quite possible to consent to something that is bad for one, or for which one's partner holds one in contempt" (115).
9. *Sexual Desire* (Weidenfeld and Nicolson 1986).
10. "Sexual Morality," 202; see also 208—hence I do not understand how Morgan overlooked "informed" in Primoratz. In *Sexual Investigations* (33–34) I pointed out, while contrasting Primoratz with Bertrand Russell on prostitution, that Primoratz's principle was "free and informed consent." A year later, the 3rd edition of my *Philosophy of Sex* (1997) included Primoratz's "What's Wrong with Prostitution?" (339–61), in which he writes explicitly about free and informed consent. Igor's essay appeared originally in *Philosophy* 68, No. 264 (1993), 159–82.
11. Laurie Shrage and Robert Stewart (*Philosophizing About Sex* [Broadview, 2015], 36–38) present Morgan's case against sufficiency favorably, concluding that "genuine consent and reasonably good (or the lack of bad) intentions are both necessary for a sexual act to pass the test of minimal decency" (38). They do not mention "informed" here but only later in the book: 66 (BDSM), 156 (pornography), 199 (HIV), and 247 (genital mutilation). Webber, too, is imprecise: "Valmont aims to humiliate Tourvel precisely through winning her consent to a sexual liaison. This liaison seems to be morally reprehensible even though both participants fully consent, as Morgan points out" ("Sex," 249). Fully?
12. "Sex and Sexuality" (10/19/2018), in *The Stanford Encyclopedia of Philosophy*, Edward Zalta, ed., https://plato.stanford.edu/entries/sex-sexuality/, and *Philosophy of Love, Sex, and Marriage*, 2nd ed. (Routledge, 2018), 230.

13. They may be coerced by their desires. Perhaps the Vicomte wasn't able to resist the push of his dark desires, in which case he, not only the Madame, was a victim of desire. See my *Pornography, Sex, and Feminism* (Prometheus, 2002), 182–83.
14. "The things that seem beautiful, inspiring, and life-affirming to me seem ugly, hateful, and ludicrous to most other people" (*Macho Sluts* [Alyson, 1988], 9). Toby equinely concurs: "You may have no way to understand how wonderful a yellow mane is. Well, on Lily it's wonderful. Sometimes at night she will slide along the fence and come close to me, and she will sigh her warm breath on my nose, and I will rub my head against her yellow mane, and the smell will stay with me until morning. She also has a great asshole. Oh, I forgot. You're human and you think that's vulgar. Lily is about the closest thing to God that I've come across" (Steve Martin, "Does God Exist?" *The New Yorker*, 12/7/1998, 100). Available at https://www.pbs.org/wgbh/questionofgod/voices/martin.html.
15. "*Harry*: I love that after I spend the day with you, I can still smell your perfume on my clothes. And I love that you are the last person I want to talk to before I go to sleep at night. … I came here tonight because when you realize you want to spend the rest of your life with somebody, you want the rest of your life to start as soon as possible. *Sally* [furious]: You see? That is just like you, Harry. You say things like that and you make it impossible for me to hate you, and I hate you, Harry. I really hate you. I hate you" (Nora Ephron, *When Harry Met Sally* [Knopf, 1990], 96–97).
16. *Law and Philosophy* 8:2 (1989), 217–43. References are to *POS*, 6th ed., 461–81.
17. Antioch College enacted a Pineauvian hindrance to date rape. See my "Antioch's 'Sexual Offense Policy': A Philosophical Exploration," *Journal of Social Philosophy* 28:1 (1997), 22–36. Reprinted in Ellen Feder, Karmen MacKendrick, and Sybol Cook, eds., *A Passion for Wisdom* (Prentice Hall, 2004), 742–54; David Boonin and Graham Oddie, eds., *What's Wrong? Applied Ethicists and Their Critics*, 2nd ed. (Oxford, 2010), 219–27; and *POS*, 4th ed. (2002), 323–40, 5th ed. (2008), 459–77.
18. "Date Rape," 478.
19. "Negotiating Sex," *Southern California Law Review* 78:6 (2005), 1401–38, available at https://southerncalifornialawreview.com/2005/09/04/negotiating-sex-article-by-michelle-j-anderson/.
20. "Negotiating Sex," 1407.
21. Eyal's "Informed Consent" examines difficult tangles with "informed" yet misses *why*. The motives of a doctor who suggests a procedure and the motives of a patient in agreeing are not always transparent or innocent. I recently worried about a major surgery: Maybe I had a death wish or my doctor primarily wanted to write a case-study paper or sought more experience doing this procedure.
22. Mappes ("Sexual Morality," *POS*, 275–76) gives examples of both kinds of deception.
23. "Why Humans Have Sex," *Archives of Sexual Behavior* 36:4 (2007), 477–507, at 496, 498. Only 142 of these "distinct" reasons are displayed in the article itself (in the appendix); all 237 have been posted at https://www.midss.org/sites/default/files/why_have_sex_reasons.pdf. Webber ("Sex," 243) borrows the authors' language: Meston and Buss "isolated no fewer than two hundred and thirty-seven distinct reasons why people engage in sexual intercourse." The "distinct" is specious. Consider, in the list of 142, these sets of separately num-

bered reasons: (A) I'm addicted to sex; I am a sex addict. (B) The person was too physically attractive to resist; The person was too "hot" (sexy) to resist. (C) I wanted to get back at my partner for having cheated on me; I wanted to even the score with a cheating partner. (D) I wanted to make myself feel better about myself; I wanted to boost my self-esteem. (E) I felt obligated to; I felt like it was my duty. (F) I wanted to keep my partner from straying; I wanted to decrease my partner's desire to have sex with someone else; I was afraid my partner would have an affair if I didn't have sex with him/her. (G) I could brag to other people about my sexual experience; I wanted to impress friends. (H) I wanted to end the relationship; I wanted to break up my relationship. (I) I wanted to have a child; I wanted to reproduce. (Where is "I wanted to get pregnant and then get him to marry me"? That would have been [16] on my list had it been among the questionnaire's 237).

24. The subjects were 1549 undergraduate Introductory Psychology students (UT/Austin, 2000–2003); 96% were aged 18–22. At that age I was still hanging with Rex and Aldous and had a fling with Ayn.
25. "Why Humans Have Sex," 481:

Women	*Men*
1. I was attracted to the person **3.89**	1. I was attracted to the person **4.03**
2. I wanted to experience the physical pleasure **3.75**	2. It feels good **3.96**
3. It feels good **3.59**	3. I wanted to experience the physical pleasure **3.84**
4. I wanted to show my affection to the person **3.58**	4. It's fun **3.57**
5. I wanted to express my love for the person **3.48**	5. I wanted to show my affection to the person **3.46**
6. I was sexually aroused and wanted the release **3.30**	6. I was sexually aroused and wanted the release **3.43**
7. I was "horny" **3.11**	7. I was "horny" **3.38**

Note that #6 suggests that sexual acts of some sort were already underway and the reason given was for proceeding to intercourse. The study was not designed to investigate reasons for having any sexual interaction at all, although subjects may have approached the questionnaire more broadly

26. "Why Humans Have Sex," 501.
27. Falsely proclaiming love is Mappes's first example of deception that makes subsequent sexual activity immoral ("Sexual Morality," 275).
28. *Marriage and Morals* (George Allen & Unwin, 1929), 121.
29. *Going Too Far* (Random House, 1977), 165–66.
30. Meston and Buss conclude, "In summary, human sexuality is motivated by a complex and multifaceted psychology" (502). Yet they had just admitted, "The current study ... dealt with expressed reasons for having sex. As such, it cannot capture motivations that lie outside of awareness" (501).
31. Perhaps X means, "Pleasure is its own bottom" (Richard Mohr, *Gays/Justice* [Columbia, 1988], 113).
32. See https://www.encyclopedia.com/psychology/dictionaries-thesauruses-pictures-and-press-releases/primary-masochism.

CHAPTER 18

Bad Sex and Consent

Elise Woodard

Consent and Its Limits

For the most part, discussions of rape and sexual assault—in legal and academic contexts—focus centrally on consent. In addition, public campaigns and university policies focus on the importance of seeking out and obtaining *affirmative consent*—not mere acquiescence—when proceeding with any sexual activity. Those who fail to do so may be found guilty of sexual assault. This focus on consent is not unjustified. Surely, nonconsensual sex is wrong. For one, by hampering the victim's ability to decide when and with whom to have sex, it violates her autonomy.[1]

At the same time, by focusing almost exclusively on the wrongs involved in nonconsensual sex, we risk overly legitimizing all sex that is consensual. Consent is a minimal standard. It does not necessarily turn "an invasion of privacy into an intimate moment."[2] In some cases, it merely removes one ground for complaint about that invasion. In other words, consent does not always have the robust powers of moral transformation often attributed to it. On this point, I follow Robin West. West insightfully observes that "it is a non sequitur, but a disturbingly common one, to argue from the premise that some act is bad because it is nonconsensual, to the conclusion that the same act, if consensual, is therefore good."[3] In other words, West notes that it is a mistake to think that consensual sex is always sufficient for good—morally or otherwise—sex. The worry here seems to be, in part, that there is a gap between consent and good sex.

E. Woodard (✉)
University of Michigan, Ann Arbor, MI, USA
e-mail: ewoodard@umich.edu

D. Boonin (ed.), *The Palgrave Handbook of Sexual Ethics*,
https://doi.org/10.1007/978-3-030-87786-6_18

Those sympathetic to consent frameworks might attempt to bridge the apparent gap between consensual sex and good sex by amending the standards for consent. More specifically, they might attempt to raise the bar for consent, thereby lowering the bar for nonconsensual sex. For example, they might argue that consent must be either *affirmative* or *enthusiastic*.[4] Unfortunately, this move risks *overcriminalizing* sex that is not clearly criminal. Whatever the merits of a personal policy of only having sex with partners who affirmatively or enthusiastically consent, surely less-than-enthusiastic sex is not necessarily rape. Nor, as we'll see, is it typically experienced as such.

Before raising the bar for consensual sex, or abandoning consent frameworks altogether, we ought to get clearer on what work consent is supposed to do. If consent is a minimal standard—sufficient only for avoiding rape—then it is not necessarily worrisome that consent alone is insufficient for unproblematic sex or good sex. As Quill Kukla, echoing West, notes, "We can consent to all sorts of lousy sex, including demeaning, boring, alienated, unpleasantly painful or otherwise harmful sex."[5] Beyond this, I'll argue that we can also consent to sex even under some pressure to do so. Indeed, some degree of coercion or pressure does not always undermine consent.[6]

These observations motivate theorizing about cases that do not constitute rape and should not be criminalized yet are nonetheless violating to victims.[7] I've motivated this point by looking at the apparent limitations of consent frameworks. However, the point can be framed more generally. Uncoerced or unforced sex alone will never be sufficient for good sex or even unproblematic sex. Our standard for *not-rape* should not also serve as our standard for good sex.

While the point that there are some cases of sex that are not rape yet are nonetheless violating is not necessarily new, this category of violating yet noncriminal sex is radically undertheorized, especially given its prevalence.[8] Even when theorized, it is seldom broken down into sub-categories. As I'll argue below, this is unfortunate, as there are normatively significant differences between and within different sub-categories. These include differences in how victims are harmed or wronged as well as in whether perpetrators are blameworthy. At the same time, the cases are unified in that they all give rise to distinctive agency problems.

The category between rape and just sex goes by many different names. These include 'gray rape' or 'gray sex' (or variations thereof) 'unwanted sexual intercourse,' 'unjust sex,' 'consensual sexual dysphoria,' and 'bad sex' (where 'bad' does not just mean 'disappointing').[9] In this chapter, I'll opt for the term 'Bad Sex.'

One might worry that the term 'Bad Sex' is a poor choice, as it is ambiguous between sex that is merely unpleasurable and sex that is morally problematic or violating.[10] However, there are several strong considerations in favor of this term. First, it is useful to have terms that are not obviously morally loaded—but can become so—available as starting points for classifying our experiences. For example, one might start out thinking that a particular experience was one of merely bad sex and then realize it was more morally loaded. Even extremely coarse-grained and ambiguous labels like Bad Sex can aid in understanding, as they provide a useful starting point for navigating a difficult moral landscape. Indeed, what victims seem to experience is a sense of violation that is difficult to articulate; a vague label like 'bad' is a natural one to reach for. The term Bad Sex, I suggest, facilitates inquiry.

Second, it avoids some problems with more common terms like 'gray rape' or 'gray sex.'[11] As we'll see in §2, the term 'gray rape' in particular has a problematic history. I think this history has made some people unduly hostile to the notion of a sexual gray area. In addition, I think we ought to be open to the possibility that the harms associated with Bad Sex are incommensurable with those of rape. However, the notion of 'gray sex' suggests a simplified model on which Bad Sex is simply a less bad or qualified version of rape. Bad Sex is also more common than the remaining alternatives, which also tend to be more theory-laden.

The aim of this chapter is to get clearer on the category of Bad Sex. I will argue that it is imperative that we develop hermeneutic and linguistic resources for understanding it. The remaining structure of this chapter is as follows. In §2, I offer some background on theorizing about Bad Sex and introduce some examples, starting with the viral story "Cat Person." In §3, I discuss three broad types of Bad Sex, which involve either psychological pressure, social coercion, or epistemic risk. In §4, I consider an objection to my project on which we ought to collapse the categories of Bad Sex and rape. I offer a response in §4.1. §5 concludes.

From 'Gray Rape' to 'Bad Sex'

Many terms have been used to describe sex that is problematic but seems to fall short of the legal definition of rape, with some variation of 'gray rape' or 'gray sex' appearing among the most popular. The term 'gray rape' was popularized in 2007, when an article called "A New Kind of Date Rape" by Laura Sessions Stepp in *Cosmopolitan* went viral.[12] Stepp defined gray rape as "sex that falls somewhere between consent and denial."

The article received intense yet appropriate backlash. The problem was that Stepp had illustrated this concept via examples that seemed to many to constitute rape, not 'gray rape.'[13] For example, one woman had clearly communicated that she did not want to have sex and believed that she was raped; however, she worried that others would not see it that way. Another woman had also communicated lack of consent, though did not identify as raped. Both women worried they were not forceful enough in protesting. In both cases, one could argue that rape myths prevented them from fully recognizing that they had been raped, but that didn't make it such that the case was *gray*. To many, these were clear-cut cases of rape.[14]

The term 'gray rape' has not quite recovered from its complicated history. Those critical of it rightly note that the term has been used to promote the idea that gray rape is merely the result of miscommunication, as when Katie Roiphe says, "there is a gray area in which one person's rape may be another's bad night."[15] Others have criticized the notion of 'gray rape' on the grounds that it seems to imply that "some forms of sexual assault are less serious or trivial."[16] I am sympathetic to this criticism to an extent; we ought to be open to the possibility that the harms and wrongs associated with Bad Sex are incommensurable in some respects with the harms and wrongs associated with rape, especially given that Bad Sex is, as we'll see, both more normal and normalized. Indeed, this is one reason why I prefer the term 'Bad Sex,' as it avoids this implicature.

Over the past few years, there seems to have been renewed interest in understanding and developing a language around "sex that feels violating even when it's not criminal."[17] This interest seems to have been sparked at least in part by the publication of the short story "Cat Person" by Kristen Roupenian in December 2017.[18] The virality of each of these articles is partly explained by the success of the #MeToo hashtag movement in October 2017, which came in the wake of breaking stories about Harvey Weinstein's sordid history of sexually assaulting and harassing dozens of women.[19] What was different about "Cat Person" was that it gave voice to—and illustrations of—sex that was not criminal but was nonetheless 'soul-crushing,'[20] transporting one to 'the place of no return' and offering a 'window into the uncomfortable reality that clouds' consent.[21] For an experience that "we don't really have a language for talking about," these stories gave people something to point to.[22] Indeed, both examples were seen as extremely relatable, with many recounting their own 'cat person stories' online.[23] In response to "Cat Person," "women seem to have been united in a collective cry of recognition and a release of righteous gratitude towards it."[24]

"Cat Person" begins with a 'boy-meet-girl' moment. Margot, a 20-year-old college student, meets 34-year-old Robert at a movie theater where she's working the concession stand. After a series of flirtatious exchanges, Robert asks Margot for her number. A series of witty text exchanges—replete with inside jokes—ensues before Robert asks Margot on a date to the movies. After the movie, he invites her to grab a beer. At the bar, Margot starts imagining what it would be like to have sex with Robert and decides to go to his house.

A disturbing and 'skin-crawling' sexual encounter ensues.[25] Margot is disgusted at Robert's body, recoiling at his "belly thick and soft and covered in hair." But she finds herself unable to summon the tact required to "stop what she had set in motion," worrying that doing so would make her seem "spoiled and capricious." While having sex, the feeling of repulsion overcomes her again but is not enough to disrupt her 'pinned stasis.' Throughout, she seems committed to getting it over with, soothed and delighted only when consumed by thoughts about how excited and mesmerized Robert is by her beauty and youth. Robert moves her through a series of positions brusquely and pornographically until finally collapsing onto her.

Soon after, he drives her home to her dorm. In her dorm room, Margot has complicated feelings about what transpired. On the one hand, she feels "overwhelmed with a skin-crawling loathing." On the other hand, she worries she is being unfair: perhaps Robert "had done nothing wrong except like her, and be bad in bed." In the ensuing days, she tries to determine a tactful way to end things, but her roommate instead sends an abrupt breakup text on her behalf. Robert replies with a series of texts which escalate from apologetic and nostalgic to jealous and hostile. The story ends with a final text from Robert: "whore." Throughout the story, it seems clear that the sex is consensual. Moreover, although Margot found it difficult to communicate that she did not want to have sex, this was not because "she was scared he would try to force her to do something against her will." On the contrary, it was because she thought the norms of politeness dictated that she sees this through "after everything she'd done to push this forward." In this respect, Margot seems compelled in part by social narratives or norms, a point I return to.

The sense that "getting it over with" would be easier than saying 'no' is common among others who describe their own Bad Sex experiences. For example, Jessica Bennett, describing her own "Cat Person" experience, writes that by the time she was sure that she did not want to have sex, she just "let it" happen through "some combination of fear (that I wasn't as mature as he thought), shame (that I had let it get this far), and guilt (would I hurt his feelings?)."[26] Margaret Cho writes, "Often I would initiate the encounter just to get it over with, so it would be behind me, so it would be done."[27] More generally, in her pioneering work on the topic of gray area sex in *Just Sex?: The Cultural Scaffolding of Rape*, Nicola Gavey gives multiple examples of women who felt that having sex was the easiest or quickest means of achieving one of her own needs, such as sleep or tranquility.[28] For example, one woman interviewed named Lee said that on some occasions, she would have sex "in order to both go to sleep and for him to um, to finally relax" after an hour of him coaxing her.[29] Gavey calls these cases *unjust sex*, a double-entendre to indicate that these examples are not '*just* (mere) sex' nor are they '*just* (normatively good) sex.'

In these examples, we've already seen at least two types of Bad Sex, which I'll discuss in more detail in §3. First, we saw cases of what I'll call social coercion in Margot and Bennett's case. Both Margot and Bennett have internalized social norms and narratives, such as norms of politeness and narratives about 'girls who lead people on.' This leads them to have sex or 'get it over with.' In both cases, there is no obvious sign that they were pressured by their sexual partner. Second, and by contrast, we see an example of interpersonal psychological pressure to have sex in Lee's case. Lee's decision to have sex is partly strategic: it is based on a calculation that the costs of proceeding are lower than the costs of refusing or resisting, despite her lack of desire.[30] I discuss psychological and social coercion in §3.1 and 3.2, respectively.

There is, however, a further type of Bad Sex that has yet to be illustrated. These are cases of what I'll call *epistemically unsafe sex*.[31] These are motivated by two observations, the first by psychologist Nicola Gavey and the second by writer Reina Gattuso. In her groundbreaking work on unjust sex, Gavey also describes cases where women consented as a strategic way of avoiding rape: they feared that a 'no' would be ineffective. This fear might take hold even in the absence of psychological or social pressure. Reina Gattuso elaborates on this experience, noting that it can be consistent with consent, even affirmative consent:

> A lot of sex feels like this. Sex where we don't matter. Where we may as well not be there. Sex where we don't say no, because we don't want to say no, sex where we say yes even, where we're even into it, but where we fear—some little voice in us fears—that if we did say no, if we don't like the pressure on our necks or the way they touch us, it wouldn't matter. It wouldn't count, because we don't count. This feeling isn't necessarily assault, but it is certainly on a continuum with it.[32]

In both Gavey's and Gattuso's examples, victims fear finding out what would have happened if they did not consent. More specifically, they worry about finding out that their consent was modally irrelevant: if they had not consented, their partner would have proceeded to have sex with them anyway. This consideration might motivate consent in the first place, as in Gavey's examples, or it might be a haunting afterthought, as in Gattuso's examples. In both cases, one's decision to have or continue with sex is motivated at least in small part by a type of risk aversion: one fears what would happen if one didn't consent.

Of course, the victim's perspective and the perpetrator's perspective can come apart in both directions: the victim might be wrong that her consent is modally irrelevant, or she might not be aware that the perpetrator is in fact indifferent to whether she consented. This observation raises interesting questions about how to delineate the category of Bad Sex. I discuss these cases further in §3.3.

So far, we've seen at least three types of Bad Sex: those involving interpersonal pressure, those involving social pressure (including pressure coming from social narratives and scripts), and those involving epistemic risk. Across these cases, victims may consent for strategic or calculated reasons. In the next section, I'll discuss the dynamics of these cases in more detail.

Varieties of Bad Sex

I'll focus on three broad categories or dimensions of Bad Sex. These distinctions help us classify many of the cases we've already seen and to separate out different threads. I have decided to split these categories up based on victims' motivations for having sex; these categories thereby focus on different narratives of harm to victims.[33] I do not intend these classifications to be exhaustive, nor do I think different types of Bad Sex are mutually exclusive. Moreover, for ease of discussion, I have placed together different types of examples that may deserve further breakdown. I intend this classification to be the first step in getting clearer on different types of Bad Sex, how they harm and potentially wrong victims, and the extent to which perpetrators are blameworthy.[34] My diagnoses are necessarily tentative and schematic, and I emphasize throughout that more work needs to be done to investigate the different dimensions of Bad Sex. Nonetheless, by starting to describe and distinguish between different types of Bad Sex, we can start to make progress on creating hermeneutic and linguistic resources for understanding and articulating the problems with gray area sex.[35]

Psychologically Pressured Sex and Calculated Trade-offs

The first category captures two types of cases that are highly interrelated. The first focuses on cases of psychological pressure. Psychological pressure includes cases involving nonphysical threats or threats of unfavorable emotional responses. Examples include threats of cheating, leaving, or withholding money or other resources or threats of emotional responses like sullenness, sulking, unkindness, or anger.[36] The second type of case captures what I'll call calculated trade-offs. These are cases where an agent decides to have sex because she feels it is the least bad option among the alternatives. This includes many of the remaining cases Gavey describes, such as cases where a woman felt like having sex was the most efficient way of achieving one of her own needs, such as sleep or tranquility (e.g., avoiding tension). It also includes cases that Jessica Bennet describes as "begrudgingly consensual sex, because, you know, you don't really want to do it but it's probably easier to just get it over with."[37]

Why classify these two types of cases together, under one heading? One reason is that calculated trade-offs are often made against a backdrop of psychological pressure. For example, the threat of an unfavorable emotional response may be what renders sex the least bad option. Calculated trade-offs may also be made against a backdrop of *social* pressure, something I'll discuss below. For example, a woman may prefer to have sex rather than feel guilty for 'leading someone on' or being 'prude.' A man might prefer to have sex because he knows that if he refuses, his partner will interpret his refusal as a sign that she is unattractive or undesirable—particularly given the stereotype that men are always interested in having sex. Of course, these situations reflect problematic but operative social scripts and narratives.

That said, calculated trade-offs can be made without any backdrop of pressure. For example, an individual might consent to sex simply to get it over with rather than have to express her preferences, feel the need to explain herself, or say 'no.' On this, Margaret Cho writes, "I said yes because it was too much trouble to say no. I said yes because I didn't want to defend my 'no,' qualify it, justify it—deserve it."[38] Nonetheless, it seems like calculated trade-offs are most obviously problematic when they intersect with some type of pressure—psychological or social. For this reason, I've grouped these cases together.

In cases of psychological pressure, an agent can be made to feel complicit in her own assault (or something akin to it). This is especially true in cases where she consents after consistently having her preferences manipulated and her agency strategically used against her. For example, in *Longing to Tell: Black Women Talk about Sexuality and Intimacy*, Tricia Rose describes the story of 'Sarita,' who was manipulated into having sex by a 50-year-old man whom she trusted when she was in college. Sarita writes: "I wouldn't call it date rape, but I would call it rape by mental force. I was completely manipulated and made to have sex through my own mental shortcomings—by a person I trusted. I still have a lot of shame about it because deep down I feel like I played along with it" (Rose 2004, 32–33). As Linda Alcoff highlights, it is important to discuss cases where agents are made to feel complicit in their own assault. She observes that "part of the manipulation here is to displace causal agency and deflect blame."[39]

Sarita's story and Alcoff's observation help us see one distinction between paradigmatic examples of psychological pressure and of rape. As Ann Cahill argues in her discussion of *unjust sex*, which includes cases of psychological pressure, *unjust sex* and rape differ in the way they involve sexual agency. Whereas rape typically involves overcoming, overwhelming, nullifying, or utterly disregarding the victim's sexual agency,[40] unjust sex implicates a diminished sexual agency. In the latter cases, "women's sexual agency is actively sought, but its efficacy is severely limited in a variety of ways, all of which refer back to the instigator's sexual preferences, interests, and needs, thus resulting in a weak, stunted form of intersubjectivity that does not sufficiently recognize or further women's sexual agency."[41] In other words, unjust sex, but not rape, involves a sort of *split* or ambivalence on the part of the woman. For instance, she may be forced to choose between two selves, a sexual agent or a 'dutiful' wife or partner, or between prioritizing her (lack of) desire or sleep (or simply emotional tranquility).[42]

In short, in paradigmatic cases of psychological pressure, the victim's agency is actively deployed against her in ways discussed above. This is a distinctive type of harm, especially if she is made to feel complicit or like an accomplice in her own assault. Moreover, the perpetrator who pressures the victim is often blameworthy for psychologically pressuring the victim into having sex.[43] However, as noted above, there can be cases where an individual experiences psychological pressure or makes a calculated trade-off, despite no explicit pressure from her partner at the time.

What should we say about cases of calculated trade-offs that occur without any backdrop of pressure? First note that these cases are less likely to involve any blameworthiness. However, they can still be harmful and, indeed, wrongful. Specifically, these cases still reflect or contribute to a diminished or stunted form of sexual agency. In the ideal case, agents base their decision to have sex on their desire to have sex.[44] This desire need not always be physical. Rather, one might desire to have sex with her partner as a form of gift-giving.[45] In the cases at issue, however, one bases her decision to have sex on a desire to avoid something worse, such as exhaustion, emotional distress, or the inconvenience or difficulty of saying 'no.' One does not in fact act on a desire—direct or indirect—to have sex but rather to avoid an alternative.[46]

This discussion is necessarily schematic, and more work would need to be done to flesh out the normative dimensions—including the harms, wrong, and blaming conditions—of these cases. The key point is that the harms or wrongs here plausibly differ from those of nonconsensual sex. For example, it is commonly thought that nonconsensual sex violates the victim's autonomy.[47] In cases of psychological pressure and calculated trade-offs, the victim's autonomy and agency are instead diminished or strategically used against her. It is important to get clearer on this latter, undertheorized type of harm in future work. For now, this suggestion is merely a starting point.[48]

Some might try to argue that these are really just cases of rape rather than Bad Sex. Indeed, consent theorists can claim that psychological coercion can render consent invalid in some cases.[49] It is certainly true that there are ways to fill out the details of some of the cases discussed above to render them rape. However, it seems both implausible and unhelpful to claim that all cases of psychological coercion and trade-offs constitute rape. Moreover, intuitively, there seems to be some morally important difference between the examples described above and rape. The point of this chapter is in large part to argue that we should develop more fine-grained tools for classifying sex that is not morally neutral yet does not constitute rape.[50] That said, the question—both descriptive and normative—of whether we should really classify all or many cases of Bad Sex as rape is an important one, and I return to it in §4.1.

Socially Coerced Sex

We can now turn to social coercion. Social coercion is the pressure induced by cultural expectations and social conventions or scripts, some of which are socialized, internalized, and institutionalized. Certain sexual scripts can also arguably be socially coercive. In the context of marriage, social coercion has historically arisen from social pressure on women to be 'dutiful' wives and thus they sometimes consent to unwanted sex with their husbands.[51] More common today, social coercion can arise from rape myths regarding what women 'owe' to men upon 'leading them on' by, for example, letting them pay for dinner or by wearing provocative clothing.[52]

An example will illustrate. Consider Jessica, who feels like she owes her date, Drew, sex in virtue of going on a date with him and letting him pay. There are operative social scripts, just relatively recently being explicitly called into question, that say she owes him sex. She might think there may be consequences of her not going back to his place after, or going back and not having sex, such as him getting angry with her or spitting vitriol. Even if she thinks he is a 'nice guy,' she might worry about the risks of saying 'no' either based on past experience or how these scripts operate in the public imagination.[53] She might not make these calculations explicitly but simply think 'this is what is supposed to happen next,' as often happens when norms are deeply internalized.[54]

Alternatively, one might worry about what failing to act in accordance with these social norms expresses about *her*. Consider again Margot, from "Cat Person." We are told that Margot allows the encounter to continue not because she "was scared he would try to force her to do something against her will." Rather, she worried that "insisting that they stop now, after everything she'd done to push this forward, would make her seem spoiled and capricious, as if she'd ordered something at a restaurant and then, once the food arrived, had changed her mind and sent it back."[55]

Women are, of course, not the only group to suffer from social norms and scripts. For example, men (both straight and queer-identified) have cited pressure to conform to the stereotype of always being interested in sex.[56] Gender identity, sexual orientation, and race also plausibly intersect with social norms regarding sex. For example, in the context of work on unwanted orgasms, Sara Chadwick et al. have found that queer women who slept with other women felt especially pressured to orgasm due to narratives about women being especially skilled at making other women orgasm, and bisexual men have cited feeling special pressure to "perform" with every partner to prove their sexual identity.[57] It seems plausible that these narratives could intersect not only with pressure to orgasm but also with the question of whether to have sex in the first place, for example, to 'prove' one's sexual orientation or conform to stereotypes about always being interested in sex, lest they offend their partner. Race plausibly exacerbates some of these narratives. For example, black bodies are often hypersexualized, and this plausibly affects social norms or expectations.[58] Social norms can vary in how deeply they are internalized by individuals. More generally, agents can differ in their reasons for conforming to social norms. If a social norm is deeply internalized, it is difficult for the agent to do anything *but* conform to the norm. (Note that internalizing a norm does not entail endorsing it.) In other cases, an agent might conform to a social norm out of fear of the consequences of not doing so or for the expressive value of doing so. Of course, there are positive reasons to conform to social norms as well; for example, they allow us to coordinate. However, as we saw earlier, this coordination often asymmetrically benefits one party (often men) to the detriment of another (often women).

Why might some cases involving social coercion count as Bad Sex? There are at least two reasons. First, if one feels socially coerced into having sex, there is clearly something non-ideal about the situation. More specifically, as in cases of calculated trade-offs discussed in §3.1, it seems that one's sexual agency is stunted or diminished.[59] Second, some paradigmatic examples of Bad Sex involve problematic social norms. We saw this while discussing "Cat Person" and in the case of women consenting to unwanted sex with their husbands. Again, these cases are explicitly described as consensual though they feel violating, and it is implausible that all cases involving social coercion count as rape.

Cases of Bad Sex involving social coercion raise important and underexplored questions about blameworthiness, assuming that socially coerced sex can wrong victims.[60] For illustration, consider again Jessica and Drew. Given social norms and expectations, Jessica felt like she owed Drew sex in virtue of going on a date with him and letting him pay. Is Drew blameworthy for having sex with Jessica, given that she felt socially coerced? That depends on the details. For example, is Drew aware of the operative social norm, and does he actively exploit it? If so, then this seems clearly blameworthy. On the opposite end of the spectrum, if Drew was unaware of this norm through no moral or epistemic fault of his own, then it's difficult to say that he is blameworthy. In between lie cases of motivated ignorance. Another relevant question pertains to the nature of Drew's duties. Is he obligated to attempt to cancel the relevant norm if aware of it? I think the answer is plausibly yes. Yet if Drew tries his best to cancel it, and if Jessica has internalized the norm or doesn't trust his sincerity, then he may still be off the hook morally speaking.

What this discussion highlights is that Bad Sex can occur without any particular person being blameworthy. Indeed, some individuals hesitate to attribute responsibility to their partner despite feeling socially coerced. One woman writes, "I know I was feeling coerced and not doing it willingly most of the time. But I'm not sure it was done by him. It was really my own upbringing and the things that I'd been taught."[61] Indeed, cases of social coercion illuminate the importance of challenging insidious social norms and rape myths on more than just an individual level while educating individuals on them to be more aware. This raises many questions that demand exploration. Again, the discussion here is necessarily schematic. The category of Bad Sex, though underexplored, is ripe for further philosophical and political analysis.

Epistemically Unsafe Sex

The final type of category we'll consider is what I'll call *epistemically unsafe sex*. This includes cases where consent is motivated partially by risk aversion: one does not want to risk finding out that her consent was irrelevant. It also includes cases where one affirmatively consents simply out of a desire for sex yet then fears that her consent was modally irrelevant—as we saw in Gattuso's case. Cases of epistemically unsafe sex are compatible with consent, even affirmative consent, yet they can nonetheless be violating. For example, learning—or suspecting—that one's own consent is modally irrelevant can undermine one's sense of autonomy or sexual agency in deciding whether to have sex. One worries that her "consent wouldn't count, because we don't count."[62]

Suppose the victim is correct that her consent is modally irrelevant. What does this look like from the perpetrator's perspective? On one extreme, a perpetrator ***A*** might be reckless in their decision to have sex with their sexual partner ***B***. For example, they might ignore apparent evidence that ***B*** did not consent (even though ***B*** did). On another extreme, ***A*** might be negligent. For example, ***A*** might fail to seek out any evidence for ***B***'s consent. They may fail to do their due diligence in seeking out ***B***'s consent. (This at least seems problematic in the early stages of a sexual relationship, given the high stakes.) In either of these types of cases, ***A*** might be indifferent to whether ***B*** consented—or, more realistically, ***A*** might fail to care sufficiently about whether ***B*** consented. ***A*** might prefer that ***B*** consent but would still—intentionally or inadvertently—likely have had sex with ***B*** in the absence of their consent.[63] The perpetrator in any of these variations seems blameworthy for failing to care sufficiently, if at all, about whether their partner consents.[64]

Importantly, ***A***'s indifference is compatible with ***B***'s consent. This is most obvious on views where whether ***B*** consented depends entirely on ***B***'s mental state. However, it's also true for views on which consent requires communication and on views where consent must be affirmative. We can imagine that ***B*** affirmatively consents to ***A***, but ***A*** fails to hear it or give it uptake, or ***A*** would've had sex with ***B*** anyway.

Of course, ***B*** might never worry nor realize that her partner was indifferent to whether she consented. Or, she might suspect he was indifferent, but be wrong. Do such cases count as Bad Sex? These are tricky questions, and the answers depend in part on the roles of the term 'Bad Sex.' Insofar as Bad Sex is a large, gerrymandered bucket that captures any sex that is sufficiently similar to but not quite rape, then I think there is some reason to classify both types of cases as Bad Sex. In the first, there is a dangerously close nearby world where ***A*** rapes ***B***; at the very least, ***A*** seems blameworthy. This may seem like a harm, even if ***B*** never realizes this fact. In the second, the victim may still suffer from the (reasonable) suspicion that ***A*** is indifferent and thus be hurt (a subjective notion), even if she is not harmed or wronged. On the other hand, one might prefer Bad Sex to be more restrictive. While I am sympathetic to the former view, I am happy for now to restrict Bad Sex to cases where the victim (at least) truly believes that her consent is modally irrelevant.

Even with this restriction, one might wonder whether this is really a case of Bad Sex. To put the question bluntly, the objector asks: "Is this really that bad?" I think the answer is yes: it can be. Indeed, I think this is a particularly undertheorized kind of Bad Sex. If you justifiably believe or learn that whether or not you consented is irrelevant—or not sufficiently important—to the person having sex with you, then this would likely feel violating and arguably objectifying. In the words of Gattuso, you would feel like you 'don't count,' or that your interest in sex is less important than his pleasure. In more extreme versions of the case, this realization can make one feel frightfully close to being assaulted; had you not wanted to have sex, you would have anyway. Learning that one's consent is modally irrelevant can undermine one's sense of autonomy or sexual agency in deciding whether to have sex. In addition, I think that perpetrators may harm victims by being careless or indifferent toward whether they consent.

* * *

Let us take stock. I briefly described three general types of cases of ***Bad Sex:*** psychological pressure and calculated trade-offs, social coercion, and epistemically unsafe sex. A unifying feature of these cases is that agents have their sexual agency or autonomy diminished or stunted, albeit often in different ways.[65] In cases of psychological pressure, her agency is actively deployed against her. In cases of social coercion and some cases of trade-offs, her agency is stunted in more subtle ways; she is prevented from basing her decision to have sex on her desire to have sex. In cases of epistemically unsafe sex, victims might exercise their agency in deciding to have sex, yet if they find out that their consent is in fact modally irrelevant, this could hinder their sense of agency. I do not claim to have gotten the details or classifications—descriptive or normative—completely right. Nonetheless, I hope to have made some progress toward getting clearer on different dimensions and underexplored choice-points for theorizing about Bad Sex.

Whence Amelioration?

Now that we have gotten clearer on some varieties of Bad Sex, I want to return to an important question alluded to earlier. Why should we theorize this liminal category of Bad Sex? Why not instead try to expand our notion of rape to include such cases?

Indeed, as Catharine MacKinnon notes, there are political reasons for widening the scope of the term 'rape.' She writes:

> Politically, I call it rape whenever a woman has sex and feels violated. You might think that's too broad. I'm not talking about sending all of you men to jail for that. I'm talking about attempting to change the nature of the relations between women and men by having women ask ourselves: 'Did I feel violated?' To me, part of the culture of sexual inequality that makes women not report rape is that the definition of rape is not based on our sense of our violation.[66]

Moreover, what counts as rape has been expanded—often for the better. We have come a long way in identifying acts like marital rape and date rape as *rape*.[67] Why think that the examples of Bad Sex we've seen are any different? In addition, there are narrative and strategic reasons for classifying instances of what I've called Bad Sex as rape. Accusations of rape are taken far more seriously. Of course, this is arguably for good reason. The problem arises when people wrongly infer that just because an act is not rape, it is therefore unproblematic; it is *just sex*, to use Gavey's term. (Indeed, the point of this chapter is in part to push back against this tendency.) Moreover, as Reina Gattuso notes, "The 'gray zone' idea has often functioned as a tactic to minimize or dismiss violence, and therefore evade accountability, by claiming that sex is inherently a murky, illegible realm."[68] Viewing Bad Sex as a gray area, rather than definitively rape, is thus not without its risks.

Another way to press the objection is as follows. The concept *rape* is in desperate need of amelioration. When we ask the question, "What is rape?," we are not merely engaged in a descriptive or conceptual project. That is, we are not merely trying to achieve descriptive adequacy or a satisfactory conceptual analysis. We are also trying to figure out what concept of *rape* we *should* have, given our emancipatory and political ends of combatting sexual inequality.[69] Moreover, rape is an *essentially contested concept*, and thus extending the boundaries of what counts as rape is in fact part of the proper use of the term.[70] Especially given that it's within our rights to expand the notion, perhaps our political ends demand it.

Response: Against Collapse

In response, I'll argue that there are important reasons to make room for victims to be able to make sense of these experiences without facing the binary choice between seeing them as *rape* or *just sex* as usual. First, the experience of 'gray area sex' is extremely common, but currently victims report having few hermeneutic and linguistic resources available for understanding or articulating their experiences. Second, this suggests that ameliorative considerations actually point in favor of theorizing the category of Bad Sex on its own terms. Third, important strategic and narrative considerations point *against* expanding our definition of rape to encompass all instances of Bad Sex. I discuss each of these in turn. By classifying all cases of Bad Sex as rape, we miss out on important moral, legal, and narrative nuance; in other words, we flatten the normative terrain.

Linguistic and Hermeneutic Resources

First, and perhaps most importantly, victims and advocates have expressed a dire need for linguistic and hermeneutical resources for understanding and discussing sex that seems violating but does not rise to the level of rape. For example, Elsie Whittington—a lecturer at the University of Manchester who has a background in sexual health youth education—notes that the gray area is a 'tricky topic' because 'we don't really have a language for talking about it.'[71] Similarly, writer Ashley Ford remarked in a powerful Twitter thread on the need for "more definitive language" about the "spectrum of harm" inflicted on victims as a result of experiences that were consensual but felt violating and harmful. These include years of feeling like one owes their partner sex as a 'compromise' or 'thank you.'[72] Alexandra Brodsky—senior editor of Feministing.com—echoes experiencing a linguistic gap. She writes, "I don't have the language for this substratum of violence we refuse to name as such."[73] She offers examples involving psychological pressure or stealthing (where a man removes a condom during sex). These calls for expanding linguistic and hermeneutical resources are not limited to the United States. For example, in her "Towards a language of sexual gray zones," Lena Karlsson documents a series of similar demands in a Twitter-initiated #talkaboutit campaign in

Sweden, which focused on these 'gray areas.' One author observed that "we need a language, we need to think about our own and others' boundaries, about respect, about desire, lust and fear."[74]

In addition to expressing the absence of language to talk about these gray areas, others note that we also need to challenge the general narrative that just because a particular sexual encounter is consensual, then it is morally and politically unproblematic. Commenting on Aziz Ansari and similar examples, Emma Gray writes:

> If the #MeToo movement is going to amount to sustained culture change—rather than simply a weeding out of the worst actors in a broken system—we need to renegotiate the sexual narratives we've long accepted. And that involves having complicated conversations about sex that is violating but not criminal.[75]

Some advocates reject the notion of gray areas partly because they think that it excuses sexual violence. They worry that it would contribute to "the absurd notion that sexual violence is 'less bad' in some cases" or would risk being victim-blaming as it implicates the victim's agency.[76] The demand for nuance gives way to a putative demand for justice for victims. On the contrary, Linda Alcoff notes that:

> The idea that rape is a simple, straightforward matter actually works to dissuade the many victims from coming forward who feel that their own experience had complexity and ambiguity, and it inhibits the vitally necessary process of being able to discuss one's experience with others. Acknowledging 'gray rape' thus becomes a way to respect the perspective of survivors themselves.[77]

Indeed, terms like 'bad sex' and 'gray sex' often better reflect victims' experiences by their own accounts. For example, the victims surveyed by Gavey repeatedly described their experiences as ambiguous, often expressing ambivalence about how to classify the experience. In Cahill's words, "[t]o attempt to resolve such ambiguity prematurely would risk missing something important about heterosex as currently experienced by heterosexual women."[78] While it's true that we cannot always defer to victims' own descriptions about their experiences—given the phenomenon of unacknowledged rapes—that doesn't mean that we ought to dismiss their experiences as articulated out of hand.

Amelioration

The fact that victims have expressed the need for more fine-grained hermeneutic and linguistic tools offers us a response to the ameliorative considerations for expansion. Even if we accept that the concept of rape is in desperate need of amelioration, I don't think that identifying all instances of Bad Sex as rape actually serves emancipatory ends. Rather, what seems genuinely emancipatory is recognizing Bad Sex as a *problem* to be cared about independently of whether it falls under the extension of rape. As noted, there are good reasons for a more nuanced understanding of the range of sexual harm that can occur to individuals, ranging from rape on the one hand to various forms of sexual harassment. While it's true that appeals to the gray area have often been used to excuse sexual violence, this is not a reason to avoid theorizing this area; instead, it is a reason to advocate for its importance and to correct this misunderstanding.

Moreover, as we've seen, it's not clear that encouraging women to use the term 'rape' to describe cases of Bad Sex adequately fits their experiences. For one, as we've seen, some women consent to sex specifically to *avoid* being raped. Moreover, attempting to expand the term 'rape' in such a way may feel empowering to some but stultifying to others. Indeed, it can be traumatic to identify oneself as a rape victim. Rather, we need a larger vocabulary to describe the types and range of harms we can suffer.

Strategic and Narrative Reasons

Third, there are important strategic and narrative reasons *against* expanding our definition of rape to encompass all instances of Bad Sex. The highly emotionally charged connotations of 'rape' are a double-edged sword. This has consequences for both victims and for broader political struggles. The fact that 'rape' carries these meanings might make some victims of Bad Sex reluctant to describe their experience in that way. In many of the cases Gavey considers, for instance, women struggle to find the vocabulary to convey that they were sexually violated in a way that was not as severe as rape. As we've seen, with few conceptual resources available, they thought that the harm that was done to them was more imaginary than real.

On a political level, expanding the use of the term 'rape' can lead to backlash. While conservative outlets seem more likely to publicly complain that the term 'rape' is being watered down,[79] others privately express their grievances about movements like #MeToo, telling alleged victims to 'grow up.'[80] Of course, some prominent French women have also denounced #MeToo publicly, writing, "Rape is a crime. But insistent or clumsy flirting is not a crime, nor is gallantry a chauvinist aggression."[81] Similarly, Bari Weiss claimed that Aziz Ansari was only guilty of not being a mind reader. She wrote dismissively "I'm apparently the victim of sexual assault. And if you're a sexually active woman in the 21st century, chances are that you are, too."[82] These complaints are not new. For example, Katie Roiphe—a figure we've seen before—remarked in 1993 that the definition of rape has been "stretched" and "the lines between rape and sex begin to blur."[83]

Backlash not only hurts political movements. It also hurts victims of both rape and Bad Sex. It hurts victims of rape who might worry that their claims will not be taken as seriously—that perhaps they mean rape in this 'watered down' sense. It hurts victims of Bad Sex who don't have any other tools at their disposal, as we've seen. It also makes it easier for those accused of rape to claim that they did not actually commit rape by the legal definition, when they did, by saying that the accuser meant the term in a very broad way.[84] The accuser, reluctant to divulge details, may find it harder to 'correct' the record. Of course, we shouldn't merely be motivated by considerations of backlash in determining whether to extend the use of a concept. Indeed, the prospect of backlash is arguably inherent in many attempts for social progress. The point here is simply that in addition to strategic reasons in favor of expanding our use of the term 'rape' to include Bad Sex, there are also costs worth bearing in mind. In addition to these costs, we've also seen a host of positive reasons for classifying many of the cases we've seen as Bad Sex rather than rape. Thus, we ought to think carefully about whether these costs are worth paying when expansion can harm victims and when there is a clear compromise solution, namely, to carve out and articulate this middle ground of conduct that is not criminal—and not even always blameworthy—yet is still harmful to victims.

* * *

Attempts to expand our understanding of what constitutes rape are important. Indeed, much social progress has been made this way. However, attempts to understand cases of Bad Sex on its own terms are also important and, moreover, underexplored. We've now seen three reasons why this project is worth pursuing, appealing to the need for richer linguistic and hermeneutical resources, ameliorative considerations, and strategic reasons.[85] If we were to collapse the distinctions between Bad Sex and rape, we would not only miss out on important moral and legal nuances; we would also, inadvertently, hamper victims' ability to make sense of and communicate their own experiences.

Conclusion

Theorizing about rape has focused centrally on consent. However, by focusing almost exclusively on the wrongs involved in nonconsensual sex, we risk overly legitimizing sex that is consensual. Both proponents and opponents of consent frameworks have failed to fully recognize that consent is, at best, a minimal standard for avoiding rape. On the one hand, this is bad news for consent theorists: it limits consent's powers of moral transformation. On the other hand, it is good news: apparent problems for consent frameworks can be accounted for by acknowledging the existence of Bad Sex. I've argued that Bad Sex includes cases of psychological pressure, social coercion, and epistemic risk, and I have attempted to sketch a descriptive and normative profile of each. Over the past few decades, feminists have helped us make significant progress in understanding the harms of various sorts of encounters, ranging from sexual assault to sexual harassment, and more. By motivating the importance of theorizing about Bad Sex, I hope to have contributed in some small way toward this project.

Acknowledgments For helpful feedback on previous versions of this chapter, I am grateful to David Boonin, Mercedes Corredor, Carolina Flores, Scott Hershovitz, Kathy Hintikka, Josh Hunt, Jonathan Sarnoff, Angela Sun, Jamie Tappenden, and Adam

Waggoner. I am particularly grateful to Scott Hershovitz for encouraging me to pursue this project. I am also thankful to audiences at the Bad Romance Conference at Harvard University and the Michigan Graduate Student Working Group and to participants in Michigan's Candidacy Seminar for their questions and suggestions.

Notes

1. See West (2016) and Archard (2007).
2. Hurd (1996, 123).
3. West (1996, 247).
4. These are popular proposals on college campuses. Philosophically, Dougherty (2018) defends the necessity of affirmative consent.
5. Kukla (2018, 72).
6. See Conly (2004).
7. The term 'victim' has many connotations and may evoke concerns about 'victim culture.' Unfortunately, I do not know of an elegant alternative. Note also that while many of the victims I'll discuss are women, men and gender non-conforming and non-binary individuals are also certainly victims of what I'll call *Bad Sex*. I say more about this point in §3.2. More generally, it's worth noting that Bad Sex is part of a broader picture of sexual violence, which includes things like catcalling, public exposure, 'revenge porn,' and not just rape. (Thanks to Mercy Corredor for this observation.)
8. Prominent exceptions in the academic literature include Gavey (2005), Cahill (2016), and West (2016). As we'll see, much of the literature on Bad Sex is constrained to popular writing.
9. We'll see many uses of the term 'gray rape' or 'gray sex' below. Alcoff (2018) notes the use of the term 'unwanted sexual intercourse' but notes that it is too coarse-grained, as it sounds euphemistic for rape. Gavey (2005) and Cahill (2016), whose work I discuss below, use the term 'unjust sex.' West (2016) uses the term 'consensual sexual dysphoria.' The term 'bad sex' is used (in the relevant way) by Yarmel (2018) and others. Bennett (2017) also uses the term 'bad sex,' noting that 'bad' refers "not to the perceived pleasure of it, but to the way you feel in the aftermath." As we'll see, much of the conversation in terms of 'bad sex' started with the publication of "Cat Person" (see, e.g., Thompson (2018)).
10. Of course, there are lots of ways in which sex between consenting adults might be morally problematic even according to sexual liberals, such as if it involves cheating or deception (assuming these don't undermine consent in the first place). Here I want to set such cases aside; I am instead focused on cases that seem to lie on some continuum with sexual assault.
11. A third, more radical consideration in favor of this label might be that it highlights the continuity between sex that is bad in virtue of being unpleasurable and sex that is bad in virtue of something more serious. If some groups of people (e.g., heterosexual women) disproportionately report finding sex unfulfilling or painful, we ought to wonder if 'merely' bad sex involves failures of considerateness or failures to facilitate the development of sexual agency. These failures might make such cases lie on a continuum with Bad Sex.
12. Stepp (2007).

13. See, for instance, Smith (2019) for discussion.
14. For helpful philosophical discussion of rape myths, see Jenkins (2017).
15. Roiphe (1993a, 54).
16. ConsentEd (2020).
17. Gray (2018).
18. Roupenian (2017). This was followed shortly after with a breaking story about Aziz Ansari in Babe.net in January 2018, which seemed to offer a real-life illustration of "Cat Person's" themes (see Way (2018)). That said, I believe the Ansari case is a bit more problematic, for the victim "Grace" had attempted to communicate that she did not want to have sex, but Ansari failed to pick up on both verbal and non-verbal cues. "Cat Person," by contrast, involved sex that was more clearly consensual and thus is a better illustration of what I'm calling 'Bad Sex.'
19. A more complete history of the #MeToo movement, and events leading up to it, can be found at Nicolaou and Smith (2018).
20. Smith (2017).
21. Bennett (2017).
22. Thompson (2018).
23. Khazan (2017). See also Gray (2018) and Bennett (2017).
24. Parker and Wilkinson (2017).
25. Nicolaou (2019).
26. Bennett (2017).
27. Cho (2019).
28. Gavey (2005).
29. Gavey (2005, 142).
30. Margot and Bennett's cases might also involve some strategic calculation.
31. I have chosen this term for two reasons. First, victims may consent in these cases because it feels like the safest, or least risky, option. In other words, not consenting would seem unsafe. Second, in the cases at issue, the victims fear that their consent may be modally irrelevant: if they hadn't consented, then they would've been forced to have sex anyway. This connects with the notion of *safety* in epistemology. Roughly, one's belief is safe only if one could not easily have been wrong in similar cases (Williamson 2000). Here I'm suggesting that one's consent is safe only if one's consent could not easily have been irrelevant in similar cases.
32. Gattuso (2015).
33. Methodologically, this approach aligns me more with theorists who emphasize the *meaning of rape*—or, in this case, *of Bad Sex*—rather than those who focus on necessary and sufficient conditions for these categories (see Cahill (2016, 751–2) for discussion). That said, other ways of approaching the issue of Bad Sex would also be fruitful. Thanks to Sumeet Patwardhan for raising this methodological question.
34. Joel Feinberg distinguishes harms from wrongs as follows: a harm is a setback to someone's interests; a wrong is an *indefensible* setback to someone's interests (Feinberg 1987). Readers are welcome to translate my claims into wrongs as they see fit. I have chosen to speak primarily in terms of harms, rather than wrongs. This is partly for ease of expression and partly because issues of blameworthiness or whether there is a specific wrongdoer are complicated in the case of Bad Sex, as we'll see. That said, I think many of the examples I'll describe do

involve wrongs to victims, not merely harms. Thanks to Sumeet Patwardhan and Jonathan Sarnoff for prompting me to clarify my focus on harms.

35. Although there are ways of filling in the details of the cases I discuss such that they constitute nonconsensual sex or rape, rather than merely *Bad Sex*, there are surely some versions of the cases that are intuitively consensual and not-rape. My aim here is to understand why these cases, so understood, can still be bad despite not constituting rape.
36. Note that psychological pressure need not always be explicit. As Robin Morgan writes, "The pressure can be there without being 'a knife blade against the throat; it's in his body language, his threat of sulking, his clenched or trembling hands, his self-deprecating humor or angry put-down or silent self-pity at being rejected" (Morgan 1977, 165). Coercive offers and psychological manipulation also arguably fall under or neighbor the category of psychological pressure—at the very least, they seem to involve a similar dynamic. (Thanks to Angela Sun for this point.) Finally, note that although some cases of psychological pressure *will* undermine consent, this is not always the case (see Conly (2004) for discussion).
37. Bennett (2017).
38. Cho (2019, 3). Authors have been rightly critical of the emphasis on refusals and saying 'no' as opposed to non-verbal communication or negotiation. See, for instance, Gavey (2005, 135–136) and Kukla (2018). I should emphasize again that there are certainly cases where one prefers to have sex rather than verbally refuse due to some type of pressure. In addition to cases mentioned above, one might feel uncomfortable refusing due to norms of femininity that prescribe gentleness and nurturance (Gavey 2005, 136).
39. Alcoff (2018).
40. The last possibility comprises cases where someone "exploits a situation wherein the victim's will is inoperative," such as when she is unconscious (Cahill 2016, 753).
41. Cahill (2016, 758).
42. That said, some cases might not involve such ambivalence; rather, the problem is simply that you don't have enough choices for full agency. Thanks very much to Carolina Flores for this observation.
43. There may be extenuating cases where the perpetrator is excused.
44. Cf. MacKinnon (2017, 167): "[W]hat you get out of sex as such is that you are doing it. [...] This is what sex as a human right could look like: the right to have sex that is mutual, so equal that it is its own reward."
45. See Kukla (2018) for helpful discussion of such cases.
46. Even more worrying, two studies have shown that women who consent to unwanted sex are at greater risk of experiencing sexual aggression (see Shotland and Hunter (1995) and Krahé et al. (2000)). However, the causal connection is unclear. One possibility raised by Impett and Peplau (2003) is that "previous experiences with a coercive male partner may lead a woman to develop compliant sexual strategies as a way of avoiding potential conflicts" (p. 96). While making trade-offs in one-off cases may seem relatively unproblematic compared to the harms of rape, doing so can contribute to a problematic pattern where women are more likely to be victimized. Moreover, this fact suggests that the way sexual agency is diminished in cases of Bad Sex is seriously wrongful, not merely non-ideal.

47. See, for example, Archard (2007); West (2016). See also Rubenfeld (2012) who agrees with this claim but uses it to argue against consent frameworks for analyzing rape.
48. See Melo Lopes (forthcoming) for illuminating discussion of how women can be 'half victims, half accomplices.'
49. See, for example, Conly (2004), who offers four conditions for psychological pressure to vitiate consent.
50. This point echoes Conly (2004).
51. In her 1983 book, D. E. H. Russell observed that "many women feel obliged to accommodate their husbands' sexual wishes no matter how repelled they are by them" (Russell 1983, 356). While these norms are less prevalent now in the United States, unfortunately they persist globally. See, for instance, https://www.who.int/violence_injury_prevention/violence/norms.pdf.
52. Gavey (2005) gives several examples of this. See also Jenkins (2017) for a recent philosophical treatment of the effects of rape myths on women's ability to make sense of their experiences.
53. Compare the claim that rape benefits all men (at least indirectly).
54. Thanks to Mercy Corredor for this point.
55. Roupenian (2017).
56. See, for example, Chadwick et al. (2019, 2450).
57. Chadwick et al. (2019).
58. See Rose (2004) for narratives by black women. See Collins (2002)—especially chapter 6—for classic discussion of hypersexualization.
59. Note that, unlike in paradigmatic cases of psychological coercion, Cahill (2016)'s analysis of unjust sex does not seem to apply. In cases of social coercion, there need not be a particular agent seeking out the woman's consent and deploying it against her. A similar point holds for epistemically unsafe sex, which I discuss in §3.3. This illustrates the importance of getting clearer on varieties of Bad Sex before offering general diagnoses. (Of course, Cahill might note that unjust sex is merely a subspecies of Bad Sex, in which case I am sympathetic to her distinction between unjust sex vs. rape.)
60. That said, I think agents can be blameworthy for, for example, failures to care or perform due diligence, even if such failures do not wrong the victims. However, I will not argue this point here. Thanks to Jonathan Sarnoff for raising this issue.
61. Quoted in Finkelhor and Yllö (1985, 86).
62. Gattuso (2015).
63. Indeed, one might think that some cases of sex procured by fraud are similar in that the perpetrators fail to care sufficiently about whether their partner consents.
64. Note that there might be some versions of these cases that count as attempted rape, rather than just Bad Sex. These are cases where the perpetrator intended to rape the victim, and this is the reason the victim's consent is modally irrelevant. Instead, I'm focused here on cases where the perpetrator does not intend to rape the victim, but her consent is nonetheless modally irrelevant in virtue of *A* not being sufficiently sensitive to facts about ***B***'s consent. Thanks to Mercy Corredor and Jonathan Sarnoff for the discussion.
65. The agency problems involved in rape tend to be more severe (Cahill 2016). Moreover, rape is clearly punishable in a way that Bad Sex is not.
66. MacKinnon (1981, 82).
67. For similar examples and discussion, see Reitan (2001).

68. Gattuso (2018). Indeed, we have already seen this when discussing the history of the term 'gray rape.' In addition, so-called libertarian feminists like Katie Roiphe have trivialized the gray area, claiming that "there is a gray area in which one person's rape may be another's bad night" (Roiphe 1993b).
69. See Haslanger (2006) for this three-way distinction between conceptual, descriptive, and ameliorative projects. See Hänel (2018) for an ameliorative proposal for conceptualizing rape. Her view seems to entail an expansion of the concept of *rape*, which she argues is a cluster concept.
70. Reitan (2001).
71. Thompson (2018).
72. Ibid.
73. Brodsky (2014).
74. Karlsson (2019, 211).
75. Gray (2018).
76. ConsentEd (2020).
77. Alcoff (2018).
78. Cahill (2016, 751).
79. See, for instance, https://patriotpost.us/articles/52127-leftists-water-down-the-meaning-of-sexual-assault-2017-10-30
80. Merkin (2018).
81. Safronova (2018).
82. Weiss (2018).
83. Roiphe (1993b).
84. If this sounds far-fetched, I can at least report first-hand experience of this tactic.
85. See Cahill (2016, 751–752) for further considerations against collapse.

Works Cited

Alcoff, Linda Martín. 2018. *Rape and Resistance*. London: John Wiley & Sons.

Archard, David. 2007. The Wrong of Rape. *The Philosophical Quarterly* 57 (228): 374–393.

Bennett, Jessica. When Saying 'Yes' Is Easier Than Saying 'No', 2017. https://www.nytimes.com/2017/12/16/sunday-review/when-saying-yes-is-easier-than-saying-no.html.

Brodsky, Alexandra. That Bad, 2014. http://feministing.com/2014/06/02/that-bad/.

Cahill, Ann J. 2016. Unjust Sex vs. Rape. *Hypatia* 31 (4): 746–761.

Chadwick, Sara B., Miriam Francisco, and Sari M. van Anders. 2019. When Orgasms Do Not Equal Pleasure: Accounts of "Bad" Orgasm Experiences During Consensual Sexual Encounters. *Archives of Sexual Behavior* 48 (8): 2435–2459.

Cho, Margaret. 2019. Foreword. In *Yes Means Yes!: Visions of Female Sexual Power and a World Without Rape*, ed. Jaclyn Friedman and Jessica Valenti, 1–4. New York: Seal Press.

Collins, Patricia Hill. 2002. *Black Feminist Thought: Knowledge, Consciousness, and the Politics of Empowerment*. London: Routledge.

Conly, Sarah. 2004. Seduction, Rape, and Coercion. *Ethics* 115 (1): 96–121.

ConsentEd. Myth: Grey rape, 2020. http://www.consented.ca/myths/grey-rape/.

Dougherty, Tom. 2018. Affirmative Consent and Due Diligence. *Philosophy and Public Affairs* 46 (1): 90–112. https://doi.org/10.1111/papa.12114.

Feinberg, Joel. 1987. *The Moral Limits of the Criminal Law: Volume 1: Harm to Others*. Oxford: Oxford University Press.

Finkelhor, David, and Kersti Yllö. 1985. *License to Rape: Sexual Abuse of Wives*. New York: Simon and Schuster.

Gattuso, Reina. Rape Culture Is a Contract We Never Actually Signed, 2015. http://feministing.com/2015/05/26/rape-culture-is-a-contract-we-never-actually-signed/.

———. Is There a "gray zone" in Consent?, 2018. http://feministing.com/2018/05/11/is-there-a-gray-zone-in-consent/.

Gavey, Nicola. 2005. *Just Sex?: The Cultural Scaffolding of Rape*. London: Routledge.

Gray, Emma. On Aziz Ansari and Sex That Feels Violating Even When It's Not Criminal, 2018. https://www.huffpost.com/entry/aziz-ansari-sex-violating-but-not-criminal_n_5a5e445de4b0106b7f65b346.

Hänel, Hilkje Charlotte. *What Is Rape?: Social Theory and Conceptual Analysis*. transcript Verlag, 2018.

Haslanger, Sally. What Good Are Our Intuitions? *Proceedings of the Aristotelian Society, Supplementary Volumes*, pages 89–143, 2006.

Hurd, Heidi M. 1996. The Moral Magic of Consent. *Legal Theory* 2 (2): 121–146.

Impett, Emily A., and Letitia A. Peplau. 2003. Sexual Compliance: Gender, Motivational, and Relationship Perspectives. *Journal of Sex Research* 40 (1): 87–100.

Jenkins, Katharine. 2017. Rape Myths and Domestic Abuse Myths as Hermeneutical Injustices. *Journal of Applied Philosophy* 34 (2): 191–205.

Karlsson, Lena. 2019. Towards a Language of Sexual Gray Zones: Feminist Collective Knowledge Building Through Autobiographical Multimedia Storytelling. *Feminist Media Studies* 19 (2): 210–224.

Khazan, Olga. A Viral Short Story for the MeToo Moment, 2017. https://www.theatlantic.com/technology/archive/2017/12/a-viral-short-story-for-the-metoo-moment/548009/.

Krahé, Barbara, Renate Scheinberger-Olwig, and Susanne Kolpin. 2000. Ambiguous Communication of Sexual Intentions as a Risk Marker of Sexual Aggression. *Sex Roles* 42 (5): 313–337. https://doi.org/10.1023/A:1007080303569.

Kukla, R. 2018. That's What She Said: The Language of Sexual Negotiation. *Ethics* 129 (1): 70–97. https://doi.org/10.1086/698733.

MacKinnon, Catharine A. 1981. A Rally Against Rape. *Feminism Unmodified: Discourses on Life and Law* 1: 81–88.

———. 2017. Trafficking, Prostitution, and Equality. In *Butterfly Politics*. Harvard: Harvard University Press.

Melo Lopes, Filipa. forthcoming. 'Half Victim, Half Accomplice': Cat Person and Narcissism. *Ergo: An Open Access Journal of Philosophy*.

Merkin, Daphne. Publicly, we say #MeToo. Privately, We Have Misgivings, 2018. https://www.nytimes.com/1993/06/13/magazine/date-rape-s-other-victim.html.

Morgan, Robin. Going Too Far: The Personal Chronicle of a Feminist. 1977.

Nicolaou, Elena. Kristen Roupenian on 'Cat Person', #MeToo and Going Viral, 2019. https://www.ft.com/content/95d4456e-36b6-11e9-bd3a-8b2a211d90d5.

Nicolaou, Elena, and Courtney E. Smith & how far we need to go, 2018. https://www.refinery29.com/en-us/2018/10/212801/me-too-movement-history-timeline-year-weinstein.

Parker, Sam, and Abi Wilkinson. 'Cat Person': A Male and Female Take on the Most Talked- About Short Story of 2017, 2017. https://www.esquire.com/uk/culture/

a14414351/cat-person-a-male-and-female-take-on-the-most-talked-about-short-story-of-2017/.
Reitan, Eric. 2001. Rape as An Essentially Contested Concept. *Hypatia* 16 (2): 43–66.
Roiphe, Katie. 1993a. *The Morning After: Sex, Fear, and Feminism on Campus*. Boston: New.
———. Date Rape's Other Victim, 1993b. https://www.nytimes.com/1993/06/13/magazine/date-rape-s-other-victim.html.
Rose, Tricia. 2004. *Longing to Tell: Black Women Talk About Sexuality and Intimacy*. New York: Farrar, Straus and Giroux.
Roupenian, Kristen. Cat Person, 2017. https://www.newyorker.com/magazine/2017/12/11/cat-person.
Rubenfeld, Jed. 2012. The Riddle of Rape-by-deception and the Myth of Sexual Autonomy. *Yale LJ* 122: 1372.
Russell, Diana E.H. 1983. *Rape in Marriage*. London: Macmillan Publishing Company.
Safronova, Valeriya. Catherine Deneuve and Others Denounce the Me Too Movement, 2018. https://www.nytimes.com/2018/01/09/movies/catherine-deneuve-and-others-denounce-the-metoo-movement.html.
Shotland, R. Lance, and Barbara A. Hunter. 1995. Women's "token resistant" and Compliant Sexual Behaviors Are Related to Uncertain Sexual Intentions and Rape. *Personality and Social Psychology Bulletin* 21 (3): 226–236. https://doi.org/10.1177/0146167295213004.
Smith, Kyle. Dear Cat-person Girl, 2017. https://www.nationalreview.com/2017/12/pervasive-culture-sex-drunkeness-regret/.
Smith, Erika. What Do We Mean by "grey rape"?, 2019. https://www.refinery29.com/en-us/2019/10/8515346/what-is-grey-rape-consent.
Stepp, Laura Sessions. A new kind of date rape, 2007. https://www.cosmopolitan.com/sex-love/advice/a1912/new-kind-of-date-rape/.
Thompson, Rachel. We Urgently Need to Talk About the Grey Areas of Bad Sexual Encounters, 2018. https://mashable.com/2018/01/24/grey-areas-conversation/#wu.iqNcxcmqA.
Way, Katie. I Went on a Date with Aziz Ansari. It Turned into the Worst Night of My Life, 2018. https://babe.net/2018/01/13/aziz-ansari-28355.
Weiss, Bari. Aziz Ansari Is Guilty. Of Not Being a Mind Reader, 2018. https://www.nytimes.com/2018/01/15/opinion/aziz-ansari-babe-sexual-harassment.html.
West, Robin. 1996. A Comment on Consent, Sex, and Rape. *Legal Theory* 2 (3): 233–251.
———. 2016. Consensual Sexual Dysphoria: A Challenge for Campus Life. *J. Legal Educ.* 66: 804.
Williamson, Timothy. 2000. *Knowledge and Its Limits*. Oxford: Oxford University Press.
Yarmel, Aaron J.. Bad Sex (and what to do instead), 2018. https://medium.com/@aaronyarmel/bad-sex-and-what-to-do-instead-9cc59ec8eebe.

CHAPTER 19

"Respect Women": Thinking Beyond Consent After #MeToo

Jordan Pascoe

"For the past two years I have been interviewing high school and college-age men for a book on their experience of physical and emotional intimacy," wrote Peggy Orenstein in the *New York Times* the weekend after Christine Blasey Ford testified before the Senate Judiciary Committee.[1] "At some point in our conversation, these young men usually referred to themselves as 'good guys,' and mostly, I would say, they were. They had also all been duly admonished by some adult in their lives—a parent, a coach—to 'respect women.' But that, along with 'don't get anyone pregnant,' was pretty much the totality of their sex education. As one college sophomore said to me, 'That's kind of like telling someone who's learning to drive not to run over any little old ladies and then handing him the car keys. Well, of course, you think you're not going to run over an old lady. But you still don't know how to drive.'"

This chapter asks: what does it mean to "respect women" in a sexual context? I ask this question in an era in which claims to "respect women" have often been deployed by men accused of sexual misconduct in their own defense: "I have great respect for women," Donald Trump told Anderson Cooper after the video leaked in which he bragged about grabbing women "by the pussy."[2] "I have always treated women with dignity and respect," Brett Kavanaugh said on Fox News the week before Christine Blasey Ford testified about the night he allegedly sexually assaulted her in high school.[3] In these locutions, "I respect women" is offered by men as a reason *not* to "believe women"—it is offered, in fact, as an assertion of the man's entitlement to our respect and our belief.[4]

J. Pascoe (✉)
Manhattan College, Bronx, NY, USA
e-mail: jordan.pascoe@manhattan.edu

D. Boonin (ed.), *The Palgrave Handbook of Sexual Ethics*,
https://doi.org/10.1007/978-3-030-87786-6_19

"Respect women" is a curious formulation. If we assume that we already have a generalized duty to respect persons—as rational beings, or as individuals with their own preferences, or as human beings—then why do we need to *specify* this respect, in the case of women? What does it mean to respect women *qua* women? Is it to signal—as a popular meme suggests[5]—that women require some special, heightened version of respect? Is respecting women *different* from respecting men, requiring some special mode of respect? Or is it a reminder, that since one would not *normally* respect women, one should *try* to—and be patted on the back for doing so (much as the imperative to "believe women" specifically names our pattern of *not* believing women)?

In this sense, a formal requirement to "respect women" is likely to trigger misogynist assumptions in men about which women are deserving of respect, and why. The command to "respect women," uttered so that the very otherness of women is encoded into the command, positions this duty in this larger pattern of what Kate Manne has identified as the misogynist system of reward and punishment.[6] Respect is one of the rewards women can expect when they act in accordance with male expectations and entitlements; the corollary, of course, is that women who resist or contravene these expectations are not deserving of respect *qua* women. "Respect women" designates women not as rational beings, but as a particular subset whose purpose is to serve and support men; "respect women" is then a command to uphold this patriarchal structure, a reminder to young men that sex itself is one of the means of asserting and maintaining male dominance. In this formulation, the assumption is that sex is not an activity to be suffused with respect, but one of degradation; though sex is generally a tool for asserting male dominance over women, the command asks men to resist this "natural" order of things and respect women anyway. The fact that one needs a reminder to "respect women," that doing so requires a special command, rather than being part of one's general duties as a decent person, reveals the misogynist dynamics at work. "Respect women" as a vague, formal, but specific command is a form of misogynist discipline that reminds men of their duty to uphold and participate in the logics of misogyny.

In this chapter, I explore what "respect women" might mean in sexual situations under conditions of patriarchy. I will argue that when we understand "respect women" not as a gendered subset of a formal duty to respect rational beings that signals the discipline of misogyny, but as a material duty to know and respect the ends chosen by others, the duty carries with it a range of useful epistemic and communicative requirements.

In making this argument, I explore "respect women" explicitly within the logics of patriarchal heterosex, and consider the operative definition of sexual consent under conditions of misogyny.[7] My goal is not to reinforce the gendered discipline of heterosex, nor to define sexual consent in exclusionary terms, but first, to examine how our conceptions of sexual consent and respect both assume and reproduce the assumptions of patriarchal heterosex, and second, to develop an account of sexual respect that is capable of resisting, instead of reinforcing, the logic of misogyny.

Respect in a Sexual Context

What does it mean to "respect women" in a sexual context? In contemporary contexts, this will often simply be taken to mean respect for sexual consent. To respect women, then, is to respect their "yes"es and their "no"s, to understand that "yes means yes" and "no means no."

But as feminists since the 1970s have insisted, this is a rather impoverished conception of both respect and agency.[8] By insisting on the importance of women's right to say "no" or "yes" (depending on your framework), consent can function to silence other kinds of sexual speech.[9] It reduces our sexual scripts to a programmed pattern of asking permission, a dynamic in which one party proposes and the other assents, in which the proposer is attuned to their duty to respect the consent of their partner, often at the expense of attentiveness to that partner herself. And in doing so, it operates to place the definition of sex itself in the hands of the proposer. In the normative gendered dynamic of heterosex, this means that men define and propose sex, and women consent.[10]

In her analysis of date rape, Lois Pineau proposes an alternative to the consent framework, suggesting that we think of sex as more akin to a conversation than a contract.[11] Sex, she argues, is a dynamic more like a friendship than a business relationship; instead of thinking of sex as something fixed by the terms of a predetermined contract, we should understand it as a cooperative act, cocreated by the people engaging in it. Sex educator Al Vernacchio articulates this idea by describing negotiations about sexual encounters as more like ordering a pizza than playing a game of baseball.[12] Both parties agree they want pizza, but they still have to decide what kind of pizza, and they have to navigate each person's preferences; "pizza" could be broadly or narrowly interpreted. We don't simply consent to "pizza"; beyond an agreement to have pizza, we need to negotiate one another's limits and preferences, respecting one another's vision for what having "pizza" should look like.

The communicative negotiation that Pineau maps and Vernacchio describes pushes us beyond consent to consider what sexual respect involves. For Pineau, this model is explicitly Kantian.[13] On a Kantian account, rational beings are deserving of respect because they are end-setting beings; end-setting is how we create concrete pathways of respect. To respect me is not to treat me merely with a mild, general form of respect, but to respect my right to set my own ends.

In many contexts, respecting me means simply respecting my right to set and pursue ends, without treating me merely as a means; to treat me merely as a means is not simply to treat me with disrespect, but to block my ability to pursue concrete ends of my own, such as by deceit, manipulation, or coercion. In most contexts, I am not required to know and further the ends of others, but merely to act in ways that treat them as persons with ends of their own.

But when we find ourselves in a relationship with someone, organized by contract or affection, we may have a duty to help others to pursue their ends. Part of what involvement in someone else's life means is an awareness of and attentiveness to their ends—or at least the subset of those ends relevant to your relationship or contract with them. So, my employer need not know the full range of my private ends, but our employment contract may require them to know a particular set of my professional ends, and our contract may require them to help me attain these ends, even as I help them attain those ends that pertain to me. My employer will also have general duties not to thwart me as an end-setting being—such as through coercion or deceit—but can be held accountable only for *helping* me pursue those ends relevant to our contract. In a friendship, the ends I must know and respect will be less formally designated, but friendship carries with it clear duties to pay attention to what one's friends want; to know them is often to know their ends. To support a friend is often not merely to not thwart their ends, but to actively help them toward them. This entails an epistemic duty: respect, in relationships, requires us to learn and understand the ends of our employees or our friends in order to help them fulfill those ends.

Sexual encounters, because of their embodied nature, require a higher burden of respect. Since sex involves the use of one's body, and since bodies are the mechanism through which we can pursue ends in the world, to share one's body must be to share one's ends, since failure to do so—to grant someone access to your body *against* your ends—does grave violence to you as an end-setting, embodied being. For this reason, Kant argues, we should be wary of consent, particularly if we think that consent can be *conclusive*, even if we subsequently change our minds. In lieu of consent, Kant proposes that we think of permissible sexual relations as a shared activity involving a "shared community of ends."[14] To respect someone in a sexual encounter means to seek out and know their specific sexual ends, and to allow their ends to transform your own. As Vernacchio's pizza model illustrates, this account of sexual respect takes sex to be a *shared activity* that requires parties to *share* ends.

But if sexual respect is concrete respect for one's partner's particular sexual ends, this requires both that one *knows* one's partner's sexual ends—an epistemic duty—and that one's partner can set sexual ends of her own. If one's partner does not have—or does not know that they can have—sexual ends, then sex cannot be a shared community of ends. If there is only one set of ends, then what we have is merely consent.

The Epistemology of Consent

Sexual consent, as it emerges within heterosex, is a sexual script that insists upon women's sexual speech: that her "yes" or "no" determines the permissibility of sex. This means that women's sexual speech is scripted: women's speech registers only when they say "yes" or "no."

This communicative dynamic shapes what Kristie Dotson has called "testimonial quieting," which occurs when a person's speech cannot be given uptake by an interlocutor, often because of "controlling images" that prevent the listener from recognizing the speaker as a knower.[15] We could, of course, say much about the kinds of controlling images—for example, pornographic images—that shape sexual communicative dynamics. But there is also a straightforward sense in which consent's (hetero)sexual script itself operates as a controlling image, training men to listen only for women's "yes" or "no"—and training them, therefore, to ignore other speech acts, such as expressions of sexual ends, preferences, or limits. Consent's normative discipline of testimonial quieting produces significant epistemic challenges for men's capacity to respect women in sexual contexts.

In doing so, consent's communicative dynamics produce what Dotson calls "testimonial smothering," which occurs when a speaker, believing that their words cannot be heard and understood by their interlocutor, silences themselves.[16] Consent creates a pattern in which women cannot reasonably expect their ends to gain uptake in a communicative exchange. Dotson argues that testimonial smothering may occur when a speaker recognizes that their testimony is risky or unsafe, or when pernicious ignorance has rendered their interlocutor incompetent to recognize their speech as authoritative.[17] Testimonial quieting and testimonial smothering are forms of epistemic violence: they undermine both participation in, and the development of, our shared epistemic resources. Because of consent's testimonial quieting, men may not know—and hence cannot respect—women's sexual ends; because of consent's testimonial smothering, men—and women—may not consider women as end-setting beings in sexual contexts.

This dynamic excuses men's ignorance about what their sexual partners want in bed: in following the script of consent, how could they be expected to know what has not been scripted? Given that this is the sexual dynamic that is affirmed and valorized on high school and college campuses, as well as in our culture at large, it should be unsurprising that young women often do not think of themselves as interlocutors in determining a sexual encounter, and that men often do not notice when their sexual partners express no desires or preferences of their own. Consent's script enforces the epistemic violence of misogyny, mapping men as sexual "knowers" and women as sexual "givers."[18]

Thus, the shift required in order to make "respect women" a functional, material command is not merely communicative, but epistemic. Feminist theorists and sex educators have emphasized the shifts in communicative practices that must occur for sex to be truly reciprocal.[19] Within the context of consent culture, this requires an epistemic shift: men must come to see women as sexual knowers, agents with ends of their own, whose articulations of those ends must be seen as authoritative. The challenge is not merely to create communicative dynamics in which these ends can be expressed, but to create conditions in which partners could reasonably be said to know one another's sexual ends.

Within the framework of patriarchal heterosex, then, cultivating mutual respect in sexual contexts will require men and women to grapple with distinct epistemic challenges. I turn first to the epistemic transformations required of men to counter the persistent testimonial quieting that consent cultivates, before exploring the epistemic challenges consent produces for women as sexual end-setters.

Respect and Male Sexual Entitlement

I have argued that respect in sexual contexts means respect for your partner's specific, concrete sexual ends. This is, as Lois Pineau argues, both a moral duty and an epistemic one.[20] The epistemic dimension of this duty is apparent in cases when your partner's specific sexual ends may come as a surprise to you; they may seem inconsistent with other things you know about your partner as a person; they may feel inconsistent with what you generally think "respect" entails. Respecting your partner as an end-setting being means respecting their freedom to set ends *of their own*, not the ends you might expect or prefer them to set. This is particularly true in the sexual domain, precisely because our sexual ends are often resistant to rational deliberation, grounded in animality and desire.[21] When we set ends, we choose the desires we will act on, but we cannot fully determine our sexual desires. To respect someone as a sexual end-setting being means respecting the sexual ends they *choose* for themselves; it means prioritizing their sexual subjectivity as they share it with you. Recognizing a person as a sexual *subject*, as one capable of choosing and determining their sexual ends, means paying attention to what they tell you they want, not what, for example, you take their body to be saying.

But most of what men will have been taught about sex—from the silencing of women built into the consent framework, to the objectification of women baked into porn culture, to the general ignorance of women's sexual physiology, to the pervasive entitlement that organizes male understandings of sex—will undermine the epistemic practices necessary for respecting women in sexual contexts. Taken together, these will tend to produce what Jose Medina has described as the "epistemic vices" of privilege: men will fail to participate productively in communicative exchanges with women about sex because of epistemic arrogance, epistemic laziness, and closed-mindedness.[22] Even in moments where men are desperate to seduce or please a sexual partner, or are aware of their own limitations in understanding what their partner wants, their practiced closed-mindedness about what sex is, epistemic laziness in seeking out and paying attention to women's specific desires and ends, and epistemic arrogance in assuming that they are entitled to a sexual encounter that fits their own ends will undermine their ability to successfully participate in a communicative exchange about sex. These practices mean that even in moments where a woman does communicate her sexual ends, men may fail to perceive this as knowledge, may fail to perceive it as testimony that should outweigh their own faulty perceptions of body language or arousal, or may fail to perceive it as pertaining to them.

But sexual epistemic duties do not stop there; even in a case where a woman successfully communicates her sexual ends, men may fail to respect them. We fail to respect sexual ends when we fail to understand these ends as acting on their own ends in a concrete and transformative way. Respecting sexual ends thus means having what Miranda Fricker has called *agential epistemic follow-through*.[23] Fricker identifies three distinct epistemic features of this kind of transformative knowledge. First, respecting someone's ends involves *inferential follow-through*: how do your ends change the way I think about my ends, and about what this sexual encounter is? If your partner's sexual ends do not include intercourse, but do include orgasms of her own, then how do those ends change the way in which you think about what this encounter is? How does knowing these ends transform the ends, expectations, and entitlements you bring into the encounter?

Second, respecting someone's ends means having *motivational follow-through*: it means recognizing these ends as *a reason to act accordingly*. It is not a mere passive, or interesting, fact that your sexual partner sets certain limits, or would prefer a particular position or kind of touch. Respecting those sexual ends means taking them seriously, as limits on, or reorientation of, your own ends. It means rethinking, and transforming, your own ends for the sake of this sexual encounter; it means acting *as if* your partner's ends were your own. And finally, this means having *practical follow-through*—it means in fact acting accordingly: the sexual ends you act on in this encounter will be those that have been subject to this process of reconsideration in light of the ends of your partner. The ends you act on will be those ends consistent with those of your partner, in a shared community of ends.

This epistemic analysis makes clear that the duty to "respect women" is not merely a feature of men's sexual duties but is connected to male epistemic practices in all domains of life. The closed-mindedness, arrogance, and laziness that characterizes men's perception of women's sexual testimony are the same epistemic patterns that uphold gender hierarchies across social spaces, and that orient men to assume that it is other men, not women, who must be prioritized as knowers and interlocutors. Medina argues that the only way to undermine these poor epistemic habits is through epistemic friction, or interactions between epistemic worlds that produce discomfort and tension, and force those with poor epistemic practices to work harder as knowers.[24] Holding men accountable for poor epistemic practices means creating conditions for male epistemic discomfort that undermine the practices of entitlement that organize male sexuality.[25]

The #metoo era offers a cultural masterclass in ways in which the consent framework fails to disrupt men's sense of entitlement to sex, undermining the epistemic practices that might make "respect women" possible. Take, for example, Les Moonves, the former head of CBS, who defended himself against charges of sexual harassment, saying "I always understood and respected—and abided by the principle—that 'no' means 'no,'"[26]; Louis C.K. opened his response to allegations of sexual assault by saying, "At the time, I said to myself that what I did was O.K. because I never showed a woman my dick without asking first, which is also true."[27] Both Moonves and C.K. defend their behavior as organized by the duty of asking for consent; neither seems able to understand how they can have violated someone whose consent they sought.

These stories illustrate how empty and formal the mere practice of asking for consent can be. Moonves' defends the aggressive pursuit of women *prior* to asking their consent; that behavior, he suggests, cannot have been wrong, since the women in question had not yet had the opportunity to say no. C.K. defends masturbating in front of women after he asked for permission—even when that permission was not granted. For C.K., *asking* for consent was the full extent of the requirement; for Moonves, all behavior was permissible until "no" was explicitly said. Consent here is merely a game, a requirement to be simultaneously fulfilled and circumvented; attention to the game of consent becomes a mechanism for *ignoring* the experience and ends of the woman in question.

Thus, understanding "respect women" as involving concrete epistemic, communicative, and moral practices means identifying the multiple ways in which one can fail to fulfill this duty. Under consent's sexual scripts, men may persistently fail to hear women's articulations of their own sexual ends; when they *do* hear them, they may fail to understand it as a *reason to act accordingly*, may fail to interrogate and transform their own ends accordingly, and may thus fail to act accordingly.[28]

Consent itself undermines the epistemic transformations required to make respecting women possible. And so we are badly miseducating men when we tell them that all that is required of them is obtaining consent: we are setting them up for a lifetime of poor epistemic and communicative practices around sexuality, and we are doing nothing to undermine their sense of sexual and gendered entitlement. As our shared epistemic resources for identifying sexual violations reshape our sexual credibility economy, men trained to think only in terms of consent will find themselves guilty of and complicit in assault and harassment, precisely because they have not developed the epistemic attunements necessary to recognize women as sexual end-setters, and continue, therefore, to treat them as sexual givers. Men who are attuned only to questions of consent will inevitably fail in the material requirement to "respect women," and in doing so, they will likely find themselves confused, like Moonves and C.K., about how their practices of seeking consent could land them in a sexual assault scandal.

The Epistemology of Sexual Self-Respect

Thus far, I have argued that respect in a sexual context requires partners to reciprocally respect one another's concrete, specific sexual ends; this, in turn, requires partners to develop the epistemic practices that make them competent to *know* their partner's concrete and specific sexual ends. But this model necessarily depends upon a prior duty: the duty to set sexual ends in the first place.

If we respect persons as end-setting beings, then setting ends makes respect possible. Respect for persons is cashed out through respect for the ends they have set for themselves: by setting ends, I determine what it will mean for someone to respect me. And so, we might best understand sexual end-setting as a project of self-respect, a way of first respecting myself as an end-setting being, and second, making it possible for others to respect me. Thus, the disciplinary epistemic silencing built into consent poses significant and gendered problems for sexual respect, by creating epistemic conditions in which women learn *not* to share sexual ends, and thus, not to set them in the first place.

Consent is structured to undermine not just men's ability to "respect women" but women's ability to respect themselves. Kant, too, understood all sex as a violation of our duties of self-respect. Through sex, I treat myself as a mere means to my own pleasure, or my own desirability.[29] Through consent, I agree to someone else treating me as a mere means to their ends—rather than as a rational being with ends of my own. In this way, he thought, sex inevitably undermines our rational capacities, and our ability to treat one another with respect. He thought the only solution was marriage—a lifelong, all-encompassing community of shared ends—but I, like recent feminist Kantians, think that we can envision a limited community of shared ends: a sharing of *sexual* ends.[30]

End-setting is part of the project of being a creature with wants and desires; it is the intentional process of critically filtering our desires and inclinations, and deciding which of these we will make a part of our life projects. In the case of sexual ends, we filter our desires and inclinations—which is not to say that we are in control of these desires and inclinations—and we decide which will be part of our sexual projects. Thus, end-setting is the process by which we choose our lives and, by extension, constitute our identity. Having particular ends, connected to particular life projects, is part of what makes us distinctive persons, and so respecting me as a particular, embodied person means not just respecting my abstract humanity, but respecting the particular, concrete ends I have chosen for myself.[31]

This process of choosing and setting ends has an epistemic dimension, so that patterns of oppression or epistemic violence that undermine our rational capacities may damage our capacity to set sexual ends. We set ends in socially constructed contexts that may unknowingly inform both our desires and preferences, and our end-setting projects. As Carol Hay has pointed out, there are a range of ways that oppression can harm a person's ability to choose valuable ends for themselves, either because they do not think they deserve valuable ends, or have been conditioned to prefer less valuable ends, or because they cannot imagine that such valuable ends are real choices for them.[32]

In sexual contexts, my sense of what ends are possible for me may be double burdened, since sex is a shared activity. The sexual ends that are possible for me are those I can imagine successfully sharing: my choice of ends may be further circumscribed by the kinds of ends I believe I can successfully communicate in the contexts in which I find myself. If I have no reasonable belief that a given end could gain "uptake" in a communicative exchange, then I have no reason to believe that that end can be shared, and this may further dissuade me from setting such an end. As we have seen, consent can be such a context for women; setting sexual ends within a culture organized through sexual consent will therefore pose pernicious challenges not only for women, but for all those who experience persistent oppression that undermines the capacity to choose valuable ends for oneself.

Thus, the project of setting sexual ends for oneself must be recognized as an example of what Hay has identified as the duty to resist one's own oppression.[33] Setting sexual ends under conditions of heterosexual patriarchy is, for women and LGBTQ2 folk, a practice of self-respect, and of resisting one's oppression by making it possible for others to respect us in a sexual context.

But in a patriarchal world shaped by male sexual entitlement, the duty to resist one's own oppression will often cash out through a defensive awareness of and resistance to the sexual expectations of others, which may further undermine one's own capacity to set sexual ends for oneself. We train women to avoid sexual assault and harassment through attentiveness to men's sexual ends; by defining women's sexual agency as their capacity to consent to these ends, consent disciplines women's epistemic attunement to men's sexual ends, rather than to their own. Thus, the duty to resist our own oppression may in practice often conflict with the duty to set our own sexual ends. Or, the duty to resist our own oppression may take over our sexual end-setting practices, so that the sexual ends we set may simply be the end of not being raped, or not being degraded.

Sexual ends are meant to be the end result of a process in which I know what I want and am able to imaginatively engage with what sort of desires drive me, what sorts of actions might fulfill those desires, and in which I feel entitled to the sorts of ends and actions suggested by those desires. Imagination, especially in the form of sexual desire and fantasy, is essential to the capacity to set sexual ends. In conditions of persistent gendered oppression, the duty to set sexual ends will require women to develop what Jose Medina has called a "resistant imagination" grounded in sexual self-knowledge and resistant epistemic practices.[34]

It is worth noting that it does not follow from this that we are entitled to set and pursue *any* sexual ends. Our sexual ends are limited by the requirement that, to act on them, we must be able to share them. They are thus limited by their intersubjectivity: they must be *materially* but not necessarily *formally* permissible. Setting my own sexual ends obligates me to reciprocally respect the right of others to set ends of their own. In other words, I can't set sexual ends that preclude me from recognizing others as sexual end-setters themselves. I can't set a sexual end of rape, for example, because that would, by definition, mean that my sexual end precludes my partner from having sexual ends of their own.

Beyond the material requirement that I respect the rights of others to set sexual ends, the Kantian shared sexual ends model is agnostic about what kinds of ends are permissible, or what sexual activities are consistent with sexual respect. This is, in part, because Kant thought all sex was morally dangerous and frankly, scarily kinky, and so his framework is designed to be amenable to kinky sex.[35] Sexual ends thus encompass a fun, kinky contradiction: I must set ends that respect my partner as an equal sexual end-setter, but my ends may also objectify my partner, or envision their submission, or my own. Indeed, the BDSM community's practices of sharing one's respective kink in order to create a shared vision for a scenario or encounter hew more closely to Kant's model than to the normative sexual consent framework. A resistant practice of sexual end-setting, then, might orient us toward the tools developed by kink communities, rather than our normative, patriarchal conceptions of what "sexual respect" might involve.

Conclusion

This chapter has conceived of "respect women" through a set of resistant epistemic practices, cashed out through the concrete duty to treat women as sexual end-setters. I have argued that the epistemic structure of consent within heterosex, which disciplines women's silence and enforces men's ignorance, cannot produce a communicative context in which sexual respect is possible. Instead, drawing on feminist elaborations of Immanuel Kant, I propose that we rethink sexual respect as the work of knowing and sharing one's partners' concrete, communicated sexual ends.

In this framework, adopting new ends, in the face of new desires, or new end-sharing projects, may require both men and women to gradually shift or transform our existing projects.[36] To do this, we will need to map the epistemic transformations required to rethink permissible sex in resistance to the gendered logics of consent culture, where women will often have experienced persistent sexual epistemic silencing that has hampered their capacity to set and express sexual ends of their own, and men will have internalized persistent practices of pernicious sexual ignorance. If permissible sexual encounters require us to do more that obtain another's consent to our proposal for sex, then they require all participants to enter the encounter with ends of their own; if one partner enters a sexual negotiation without sexual ends of her own, the communicative process reverts to a one-sided consent model. Thus, alternatives to sexual consent depend on both a duty to *set* sexual ends, and on a duty to treat one's partner's epistemic ends as *a reason to act accordingly*.

Notes

1. Orenstein, Peggy. 2018. "Why We Can't Just Let Boys Be Boys" *The New York Times* September 29, 2018. Accessed October 9, 2020: https://www.nytimes.com/2018/09/29/opinion/sunday/sex-education-ethics-assault-boys.html.
2. Paquette, Danielle. 2016. "Anderson Cooper told Trump 'That was sexual assault.' The Justice Department agrees." *The Washington Post.* October 9, 2016. Accessed October 2, 2020: https://www.washingtonpost.com/news/wonk/wp/2016/10/09/anderson-cooper-told-trump-that-is-sexual-assault-the-justice-department-agrees/.
3. Blake, Aaron. 2018. "Brett Kavanaugh's Fox News interview transcript, annotated." *The Washington Post.* September 24, 2018. Accessed October 9, 2020: https://www.washingtonpost.com/politics/2018/09/24/brett-kavanaughs-fox-news-interview-transcript-annotated/.
4. Manne, Kate. 2020. *Entitled: How male privilege hurts women.* Crown.

5. Feldman, Brian. 2017. "On the Web, Respecting Women Is a Full-Time Job." *New York Magazine* August 23, 2017. Accessed October 9, 2020: https://nymag.com/intelligencer/2017/08/respecting-women-memes-explained.html.
6. Manne, Kate. 2017. *Down girl: The logic of misogyny*. Oxford University Press.
7. Haslanger, Sally. 2012. *Resisting reality: Social construction and social critique*. Oxford University Press., pp. 92–94; Gavey, Nicola. 2005. *Just Sex? The Cultural Scaffolding of Rape*. New York: Routledge, pp. 11–12.
8. MacKinnon, Catherine. 1989. *Toward a Feminist Theory of the State*. Harvard University Press. pp. 129–134.
9. Kukla, Rebecca. 2018. That's what she said: The language of sexual negotiation. *Ethics*, *129*(1), pp. 70–97.
10. MacKinnon (1989, p. 137).
11. Pineau, Lois. 1989. Date rape: A feminist analysis. *Law and Philosophy*, *8*(2), pp. 217–243.
12. Vernacchio, Al. 2013. "You're Out, Baseball!": A Healthier, More Equitable, Satisfying and Safer Model for Sexual Activity. *American Journal of Sexuality Education*, *8*(1–2), pp. 80–96.
13. Pineau develops her Kantian analysis through Onora O'Neill's analysis of Kantian sexual consent. O'Neill, Onora. 1985. Between consenting adults. *Philosophy & Public Affairs*, pp. 252–277.
14. Kant, Immanuel. 1996. *The Metaphysics of Morals*. Translated by Mary Gregor. Cambridge: Cambridge University Press. 6: 278–279.
15. Dotson, Kristie. 2011. "Tracking epistemic violence, tracking practices of silencing." *Hypatia* 26.2: 236–257, p. 242.
16. Dotson (2011, p. 244).
17. Dotson (2011, p. 244).
18. Manne (2017, p. 110).
19. *Cf.* Pineau (1989); Vernacchio (2013).
20. Pineau 1989, p. 234.
21. Varden, Helga. 2020. *Sex, Love, and Gender: A Kantian Theory*. Oxford University Press, USA, p. 125.
22. Medina, José. 2013. *The Epistemology of Resistance: Gender and racial oppression, epistemic injustice, and the social imagination*. Oxford University Press, pp. 30–40.
23. Fricker, Miranda. 2018. "Knowing, Not-Knowing, and Half Knowing." Public Talk at "Me Too and Epistemic Oppression" Conference, City University of New York, October 6, 2018.
24. Medina (2013, p. 29).
25. Manne (2020, pp. 30–74).
26. Cauterucci, Christina. 2018. "Les Moonves and the Limitations of No Means No." *Slate* Magazine, July 30, 2018. Accessed October 9, 2020: https://slate.com/human-interest/2018/07/les-moonves-and-the-limitations-of-no-means-no.html.
27. C.K., Louis. 2017. "Louis C.K. Responds to Accusations: 'These Stories Are True.'" *The New York* Times, November 10, 2017. Accessed October 9, 2020: https://www.nytimes.com/2017/11/10/arts/television/louis-ck-statement.html.

28. These failures are unlikely to be mitigated by reforms that insist upon "enthusiastic consent," which places emphasis on the *kind* of consent that counts, rather than on the epistemic practices embedded in the structure of consent.
29. Kant (1996). 6:278–279 and 6:359. While Kant emphasizes pleasure, Lisa Wade argues that often, the sexual end that motivates young women is not sexual pleasure, but the experience of sexual desirability, of being wanted by men. Wade, Lisa. 2017. *American Hookup: The New Culture of Sex on Campus.* W.W. Norton., pp. 186–188.
30. O'Neill (1985); Denis, Lara. 2001. "From friendship to marriage: Revising Kant." *Philosophy and phenomenological research* 63.1: 1–28; Papadaki, Lina. 2010. "Kantian marriage and beyond: Why it is worth thinking about Kant on marriage." *Hypatia* 25.2: 276–294.
31. Herman, Barbara. 1993. *The Practice of Moral Judgment.* Cambridge: Harvard University Press, pp. 228–9; O'Neill (1985, p. 120).
32. Hay, Carol. 2011. "The Obligation to Resist Oppression." *Journal of Social Philosophy* 24:1, p. 26.
33. Hay (2011).
34. Medina (2013, pp. 256–257).
35. Pascoe, Jordan. 2012. "Hunger for you: Kant and Kinky Sex." *What Philosophy Can Tell You About Your Lover*, pp. 25–36.
36. Herman (1993, p. 196); Korsgaard, C. M. (1996). *Creating the kingdom of ends.* Cambridge: Cambridge University Press, 180.

Works Cited

Blake, Aaron. 2018. Brett Kavanaugh's Fox News Interview Transcript, Annotated. *The Washington Post*, September 24, 2018. Accessed October 9, 2020. https://www.washingtonpost.com/politics/2018/09/24/brett-kavanaughs-fox-news-interview-transcript-annotated/.

Cauterucci, Christina. 2018. Les Moonves and the Limitations of No Means No. *Slate Magazine*, July 30, 2018. Accessed October 9, 2020. https://slate.com/human-interest/2018/07/les-moonves-and-the-limitations-of-no-means-no.html.

Denis, Lara. 2001. From friendship to marriage: Revising Kant. *Philosophy and Phenomenological Research* 63 (1): 1–28.

Dotson, Kristie. 2011. Tracking epistemic violence, tracking practices of silencing. *Hypatia* 26 (2): 236–257.

Feldman, Brian. 2017. On the Web, Respecting Women Is a Full-Time Job. *New York Magazine*, August 23, 2017. Accessed October 9, 2020. https://nymag.com/intelligencer/2017/08/respecting-women-memes-explained.html

Fricker, Miranda. 2018. "Knowing, Not-Knowing, and Half Knowing." Public Talk at "Me Too and Epistemic Oppression" Conference, City University of New York, October 6, 2018.

Gavey, Nicola. 2005. *Just Sex? The Cultural Scaffolding of Rape.* New York: Routledge.

Haslanger, Sally. 2012. *Resisting reality: Social construction and social critique.* Oxford University Press.

Hay, Carol. 2011. The Obligation to Resist Oppression. *Journal of Social Philosophy* 42 (1): 21–45.

Herman, Barbara. 1993. *The Practice of Moral Judgment*. Cambridge: Harvard University Press.
Kant, Immanuel. 1996. *The Metaphysics of Morals*. Translated by Mary Gregor. Cambridge: Cambridge University Press.
Korsgaard, C.M. 1996. *Creating the Kingdom of Ends*. Cambridge: Cambridge University Press.
Kukla, Rebecca. 2018. That's What She Said: The Language of Sexual Negotiation. *Ethics 129* (1): 70–97.
Louis C.K. 2017. Louis C.K. Responds to Accusations: 'These Stories Are True'. *The New York* Times, November 10, 2017. Accessed October 9, 2020. https://www.nytimes.com/2017/11/10/arts/television/louis-ck-statement.html.
MacKinnon, Catherine. 1989. Pornography: On Morality and Politics. In *Toward a Feminist Theory of the State*. Harvard University Press.
Manne, Kate. 2017. *Down girl: The Logic of Misogyny*. Oxford University Press.
———. 2020. *Entitled: How Male Privilege Hurts Women*. Crown.
Medina, José. 2013. *The Epistemology of Resistance: Gender and Racial Oppression, Epistemic Injustice, and the Social Imagination*. Oxford University Press.
O'Neill, Onora. 1985. Between Consenting Adults. *Philosophy & Public Affairs* 14 (3): 252–277.
Orenstein, Peggy. 2018. Why We Can't Just Let Boys Be Boys. *The New York Times*, September 29, 2018. Accessed October 9, 2020: https://www.nytimes.com/2018/09/29/opinion/sunday/sex-education-ethics-assault-boys.html
Papadaki, Lina. 2010. Kantian Marriage and Beyond: Why It Is Worth Thinking about Kant on Marriage. *Hypatia* 25 (2): 276–294.
Paquette, Danielle. 2016. Anderson Cooper Told Trump 'That Was Sexual Assault.' The Justice Department Agrees. *The Washington Post*, October 9, 2016. Accessed October 2, 2020. https://www.washingtonpost.com/news/wonk/wp/2016/10/09/anderson-cooper-told-trump-that-is-sexual-assault-the-justice-department-agrees/.
Pascoe, Jordan. 2012. Hunger for You: Kant and Kinky Sex. In *What Philosophy Can Tell You About Your Lover*. Open Court Press.
Pineau, Lois. 1989. Date Rape: A Feminist Analysis. *Law and Philosophy 8* (2): 217–243.
Varden, Helga. 2020. *Sex, Love, and Gender: A Kantian Theory*. Oxford University Press.
Vernacchio, Al. 2013. 'You're Out, Baseball!': A Healthier, More Equitable, Satisfying and Safer Model for Sexual Activity. *American Journal of Sexuality Education 8* (1–2): 80–96.
Wade, Lisa. 2017. *American Hookup: The New Culture of Sex on Campus*. W.W. Norton.

CHAPTER 20

Should Statutory Rape be a Crime?

Stephen Kershnar

Part One: Background

Statutory rape is a crime that occurs when people—specifically postpubescent individuals—have sexual relations that would be legal were it not for one or more of the participants' ages. That is, it is non-forcible sexual intercourse in which at least one person is younger than the statutory age of consent.[1] Most jurisdictions do not label statutory rape as "statutory rape." Among the names they use are carnal knowledge of a minor, child molestation, corruption of a minor, sexual misconduct, statutory sexual assault, statutory sexual seduction, and unlawful carnal knowledge.

In the United States and elsewhere, statutory rape is often a felony. Sentencing can, and often does, include a lengthy prison term and the offender's being classified as a sex offender. Sex-offender status can have severe implications. It can, and often does, affect where an offender may live, what jobs he may hold, and the community activities in which he may participate. It often results in an offender being ostracized and sometimes in his being attacked.

In the United States, thirteen states make the age of sexual consent 18. The thirteen include the three largest states (California, Texas, and Florida). These three states contain 28% of Americans and probably a larger percentage of American adolescents. Six other states make the age of sexual consent 17, including the fourth largest state (New York). Thirty-one states make it 16.

Various laws modify the age-of-consent threshold. Romeo and Juliet Laws reduce or eliminate the penalty for statutory rape when the participants are

S. Kershnar (✉)
Department of Philosophy, State University of New York at Fredonia, Fredonia, NY, USA
e-mail: stephen.kershnar@fredonia.edu

D. Boonin (ed.), *The Palgrave Handbook of Sexual Ethics*,
https://doi.org/10.1007/978-3-030-87786-6_20

above a minimum and close enough in age. Eighteen states have such laws. The close-enough period varies from two to four years. The exception does not apply in some states to adulterous or incestuous sex.[2] Some states do not recognize the age of consent if it is below 18 for cases in which one of the participants is related by blood, a guardian, or occupies a role of authority with regard to the minor. A majority of states and the federal government exempt married couples from statutory rape laws.[3]

Because statutory rape is the name of a crime and pejorative, I will use the term "age-different sex" to refer to the sex with which statutory rape laws are concerned. Age-different sex is sex in which at least one individual is younger than the statutory age of consent. For the purposes of this chapter, I will assume the youngest individual is 15 years old or older and consider whether the state should criminalize such sex.

There are a number of purported justifications for criminalizing age-different sex. Different purported justifications include that criminalizing age-different sex prevents sex to which valid consent has not been given, prevents adults from manipulating or pressuring teens into having sex, reduces the harm and negative externalities that result from such sex (e.g., out-of-wedlock births and single parenting), reduces forcible rape of minors by making such cases easier to prosecute, and prevents young people—especially women—from losing their social eligibility for marriage.[4]

Part Two: Theses

In this chapter, I defend the following theses.

Thesis #1: Not a Crime. Age-different sex should not be a crime.

The argument for this position rests in part on a general rule of thumb that asserts that the side that wants to criminalize something should bear the burden of proving its case.

Thesis #2: Practical Recommendation. As a practical matter, age-different sex should be the bare minimum crime found across states today.

The bare minimum addresses the age of sexual consent, Romeo and Juliet Law, and penalty. It should be 16 years old, 3-year age gap, and misdemeanor/no sex-offender status, respectively.

The motivation for eliminating punishment for age-different sex can be seen in some trilemmas that seem to characterize statutory rape law in the US. Consider the trilemma concerning the seemingly inconsistent manner in which the states punish age-different sex. First, when prosecuted, the punishment for age-different sex is often severe.[5] Second, such sex is infrequently punished and, when it is punished, courts sentence offenders in a haphazard and inconsistent manner. Third, if an activity is often severely punished, then it should be systematically and consistently punished.

To see how infrequently it is punished, note that prior to the 1990s, statutory rape was largely ignored. It is estimated that in the US, there are 7.5 million instances of statutory rape each year and yet only 15,700 statutory rapes in the US were reported to law enforcement agencies.[6] This is 0.2% of statutory rapes and this figure likely overestimates the percentage.

To see the haphazard and inconsistent way in which it is punished, note that in New Jersey 66% of female teachers convicted for having sex with underage students did not even go to prison.[7] Female teachers who had sex with underage students in Kansas, Kentucky, and New York were not imprisoned or got minimal sentences (e.g., 90 days).[8] Other statutory rape convictions result in significant sentences. In New Jersey, the average male teacher convicted of statutory rape of students spent 2.4 years in prison.[9] In contrast, in the US, the average time (median) served for car theft, drug trafficking, and rape are 1.0 years, 1.5 years, and 4.2 years respectively.[10] Such teachers were often labeled as sex offenders, a label that often has enormous personal and professional cost.

Consider, next, the trilemma concerning our intuitions regarding criminalizing age-different sex. First, a state should criminalize age-different sex. Second, a state should criminalize something only if we know that it involves a right-infringement or significant harm. Third, we do not know whether criminalizing age-different sex involves a right-infringement or significant harm.

Here are some oddities about statutory rape and other laws that suggest we do not know whether the sex is right-infringing or significantly harmful.

- Because of differences in statutory age of consent, Romeo and Juliet laws, and marriage exceptions, sex between the same two people can be a felony in one state, a misdemeanor in a second, and not a crime in a third.[11]
- Federal law criminalizes videotaping of otherwise legal sex when a participant is younger than 18 even in states in which the sex is legal because the younger participant is old enough to consent. That is, he or she is 16 or 17, depending on the state. The same thing is true if the age-different sex is legal because the couple is married.[12]
- In some states, two underage individuals can be convicted for having sex with one another.[13]
- A male victim of statutory rape can be liable for child support that comes about as a result of pregnancy that comes about from the rape.[14]

This suggests that states differ noticeably in how they treat age-different sex.

Part Three: Argument

This chapter argues that states should not punish age-different sex. The argument revolves around a three-part test for when a state should criminalize an activity followed by an argument that age-different sex does not satisfy the test.

Here is the argument in general terms. If an activity fails one or more parts of the three-part test, then a state should not criminalize it. By "state," I mean "a government." Age-different sex fails one or more parts of the three-part test. Hence, age-different sex should not be a crime.[15] In what follows, I defend the three-part test and then apply it.

Again, for simplicity, I will assume that the people involved in age-different sex are 15 or older. I will also sidestep the concern that a very minor crime that is more akin to regulation or taxation than punishment need not satisfy the three-part test.

Here is the three-part test. The state should punish an activity only if the activity infringes a moral right and the punishment satisfies a cost-benefit analysis and satisfies intermediate scrutiny. Let us consider these in turn.

(1) **Right-Infringement.** The activity infringes a moral right.

This condition requires that there be a victim of injustice. One underlying idea is that justice depends on, and only on, individuals' moral rights. Goals that are distinct from justice (e.g., efficiency, expressing moral ideas, and paternalism) do not by themselves warrant the coercion and suffering that characterizes the criminal justice system.

(2) **Cost-Benefit Analysis.** State punishment of the activity satisfies a cost-benefit analysis.

A punishment-practice should make the world a better place. More specifically, it should make individuals' lives go better. A cost-benefit analysis is the most reliable measure of whether a practice makes individuals' lives go better. It is the best practical way we have to handle tradeoffs. A cost-benefit analysis considers a practice's opportunity cost. This is done by comparing it to refraining from affecting people's behavior and affecting it through other means. Other means might include government incentives (e.g., fines, subsidies, and taxes) and private civil suits.

(3) **Intermediate Scrutiny.** State punishment of that activity satisfies intermediate scrutiny.[16]

Intermediate scrutiny allows the state to punish a type of action only if the state has a substantive (legitimate-and-important) state interest, punishment directly advances that interest, and the combination of what is punished and the severity of punishment are the least restrictive alternative. This condition requires that the punishment be consistent with what makes the government legitimate.

The argument for the first premise (*If an activity fails one or more parts of the three-part test, then a state should not criminalize it*) rests on the notion that there are three plausible candidate-justifications of the state. If successful, a candidate-justification makes the state legitimate and, if this is different, justifies a duty to obey the law. The three candidate-justifications are: consent, consequentialism, and the duty of fair play.[17] Each of these candidate-justifications supports the three-part test.

If consent justifies the state, people likely consent to the state's use of coercion and force only when it meets these conditions. If the consent is actual, then this is an empirical claim. If the consent is hypothetical, then this is a claim about what people in a hypothetical choosing condition would choose. Consider, for example, what rational people would choose in a free-and-fair choosing scenario. In such a scenario, they would likely require that people's body and property rights be respected and that, as a general matter, the state punish only when such force makes people's lives go better and the punishment is within the government's purview.

If consequentialism justifies the state, the best consequences likely come about only if these three requirements apply. States that punish activities without direct victims tend to make its people worse off. Consider, for example, the way in which communist and fascist countries' restrictions on freedom of association, contract, religion, and speech made their people worse off. The best way to discover whether punishing an activity makes people's lives go better is through a cost-benefit analysis. Intermediate scrutiny relates in part to the notion that a government should act within the scope of its authorized role. A government that stays within its authorized role will likely be more just and democratically accountable than when it goes outside it. Here, justice and democratic accountability are good consequences that count in justifying a government and its actions on consequentialist grounds.

Intermediate scrutiny also relates to the notion that other things being equal, a less restrictive prohibition of something will have better results than a more restrictive prohibition of it. Consider, for example, prohibition and the drug war. If the goal of these policies were to prevent violence and prevent children from using them, it would have been better for the state to have targeted these things rather than all recreational enjoyment of alcohol and drugs. This is in part because of the cost of the wider targeting. The cost includes more organized crime, pricey government enforcement apparatus, significant loss of liberty, and suffering from mass incarceration.

If the duty of fair play justifies the state, then an individual has a duty to support a just cooperative project in which he voluntarily accepts the project's benefits. The just-condition is filled out by the right- and intermediate-scrutiny-conditions. They ensure that the punishment is not unjust and that the state stays within its authorized boundaries. It is less clear whether this basis justifies the cost-benefit-analysis condition. If it does not, then the fairness justification does not support this condition.

Here, then, is a summary of the argument for the first premise.

(1)	If the state is justified, then an activity should be a crime only if it satisfies the three-part test.	**Assumption #1a: State-Justification.** There are three plausible justifications of state legitimacy and the duty to obey the law: consent, consequentialism, and the duty of fair play. **Assumption #1b: Three-Part Test.** Each of the plausible justifications supports the three-part test.
(2)	If (1), then if an activity fails one or more parts of the three-part test, it should not be a crime.	Trivial
(3)	Hence, if an activity fails one or more parts of the three-part test, then it should not be a crime.	[(1), (2)]

On some theories that are not so closely tied to the state, punishment involves coercion or violence, and these require a strong justification. If consequentialism justifies punishment, then the practice of punishing an activity must make the world better than it would be if that practice were not in place. That is, the second condition must be satisfied. If non-consequentialism justifies punishment, then such punishment must respect people's rights. That is, the first condition must be satisfied. If role-based considerations justify punishment, then punishment must be consistent with the punisher's legitimate role. That is, the third condition must be satisfied. Hence, on accounts not focused on the state, punishment must satisfy at least one part of the three-part test. Hence, if punishing an activity were to satisfy none of the parts of the test, it would be wrong.

On some accounts, but not mine, it matters for rightness whether we know that a punishment-practice passes the three-part test. I think this is mistaken because belief or knowledge about an act makes an agent praiseworthy or blameworthy for doing it but does not make his act right or wrong. This is in part the result of the intuition that a person can do the right thing for the wrong (incorrect) reason and the wrong thing for the right (correct) reason. Consider, for example, a case in which a husband shoots a man chasing his wife around the room with a machete only to discover later that the man wielding the machete and his wife were practicing a scene for a play in which they were planning to appear. Intuitively, the husband did the wrong thing, but for the right reason. This is also in part because rights and consequences focus on an act's effects on individuals' autonomy or interests and these are largely, if not entirely, independent of an agent's belief or knowledge about what he is doing.

Still, a blame-based argument applies to the state's (or state-relevant actors') knowledge regarding age-different sex. If the relevant state actors do not know whether punishing an activity fails one or more parts of the three-part test, then they are blameworthy for making it a crime. Below I argue that the relevant state actors do not know whether age-different sex fails one or more parts of the three-part test. Hence, the actors are blameworthy for making age-different sex a crime. If belief or knowledge goes to an act's wrongness, then a parallel argument would show that criminalizing age-different sex is wrong.

The argument for the second premise thus goes as follows.

(1)	Criminalizing age-different sex fails all three parts of the three-part test.	See below.
(2)	If (1), then age-different sex fails one or more parts of the three-part test.	Trivial
(3)	Hence, age-different sex fails one or more parts of the three-part test.	[(1), (2)]

Consider the argument for premise (1). First, consider rights. A person 15 years old or older can validly consent to a matter equal to or more important than age-different sex and, if she can do so, then she can validly consent to age-different sex. If she can give valid consent to age-different sex, then it does not infringe her right. This is because valid consent waives a right.

Intuitively, we think that a person 15 years old or older can give valid consent to same-age sex. Sex between minors is equal in importance to different-age sex. After all, it is the same type of act and involves the same risks. We think there is a big difference between a couple of 16-year-olds having voluntary sex and one of them forcing himself on the other. The difference is valid consent. Intuitively, we also think that a person 15 years old or older can give valid consent to surgery. Surgery is usually more important than age-different sex. The law aligns with these intuitions as can be seen in Romeo and Juliet Laws and laws requiring that emancipated teens give informed consent to surgery.[18] In addition, there is some empirical evidence that mid-teens are competent decision-makers when making decisions about sexual and medical matters.[19]

Another way to see that a person 15 years old or older can validly consent is to imagine a world in which all people have the mental level that 15-year-olds have today. We would think they could consent to commercial transactions, employment, marriage, medical treatment, and sex. Two people's rights regarding one another do not come into or go out of existence depending on the people who surround them. If individuals in this imagined world can validly consent, and intuitively they can, then, in the actual world, people 15 years old or older can give valid consent.

On some theories, a child has a right to an open future.[20] This strikes me as implausible as it does not intuitively seem that others have a correlative positive duty to promote the child's open future. If the correlative duty is negative, merely not to interfere with the child's body or property, then the right is better seen as a body- or property-right, rather than a right to an open future. Even if there were such a right, we would still need to discover whether age-different sex lessens or eliminates a teenager's open future.

Second, consider people's interest in age-different sex. We do not know whether such sex is good, bad, or neither good nor bad for individuals and, if it is bad, whether this justifies criminalizing it.

Let us begin by noting that large numbers of young people have sex in general and age-different sex in particular. Roughly, 50% of US teenagers 16 years old or younger have had sex.[21] In addition, there is some reason to believe that adolescent sexuality is beneficial. In particular, it is thought to promote independence and prepare teens for adult roles.[22] Of this sex, roughly 7.5 million were incidents of statutory rape per year.[23] Much of the statutory rape sex (roughly 30%) involved people in an ongoing relationship.[24]

Most of the reported incidents of statutory rape are heterosexual. 99% of female victims and 94% of male victims had relations with a member of the opposite sex. As a result, I shall focus on heterosexual cases.

Consider first cases in which adolescent males have sex with adult females.

The evidence on adolescent males is poor.[25] A noticeable number of men report that as adolescents they had a sexual relationship with an adult woman (5%). Male adolescents appear to have positive attitudes to such sex after the relationship ended. The majority of men who as adolescents had sexual relations with an adult female had a positive attitude. Less than 5% had a negative view toward it. The studies were done after the age-different sexual relationships were over.[26]

Some studies show that women who as adolescents had sexual relations with an adult male had a negative attitude toward it; others show a positive reaction. Again, the studies were done after the age-different sexual relationships were over. Even if the studies were not mixed, and they are, in the absence of a baseline on what females in general think about their previous romantic or sexual relationships, it is hard to interpret the dissatisfaction. I cannot locate such a baseline.

We also do not know how much of the dissatisfaction results from society disapproving of the sex or the lower quality of male participants that results from social disapproval. That is, we do not know whether adolescent females' negative feelings toward the sex, if they have negative feelings, results from a bad-making or wrong-making feature of it. By analogy, it is unclear if a white mother's interracial relationship should affect her custody claim to a child when she cohabitates with a black man and there is social stigma regarding her relationship that might affect her child's well-being.[27] The concern in this case and in the studies is that negative feelings and risk to the child result from stigma rather than from the relationship being objectionable.

The studies differ on whether women who had age-different sex are more likely to have psychological or behavioral problems. Different studies show negative effects, positive effects, and neutral effects.[28] Other studies show that adolescents who had age-different sex have substantial problems that precede the sexual relationship, including antisocial behaviors, family dysfunction, and poverty. Other studies show that female adolescents who had age-different sex have a higher risk of pregnancy, single parenthood, and psychosocial problems.[29] Still, it is not known whether the sex caused the problems or the problems caused the minors to have sex.

Even if age-different sex were to cause psychological or behavioral problems in some minor participants, this cost has to be weighed against the benefit that comes about because minors enjoy age-different sex and, perhaps also, the dating and socialization tied to it. These factors are likely significant. 95% of Americans have premarital sex and this is likely due in part to the fact that they enjoy it.[30] The same is likely true for many people having age-different sex.

Even if the studies showed that age-different sex was harmful, it still needs to be shown that this harm justifies criminalizing it. This is because criminalization is costly in terms of money and suffering. There is also a cost in liberty. Consider, for example, overincarceration in America and the increasing aspects of American life that has been criminalized.[31] A concern here is that when the state uses the criminal justice system to deter or prevent people from freely transacting with one another, it is very costly in terms of liberty, money, and suffering. Consider, for example, the drug war and prohibition. It would likely be very costly were the state to try to strictly enforce prohibitions on adultery, internet gambling, and prostitution rather than looking the other way.

In short, the evidence here is poor and does not allow us to draw a conclusion as to whether such sex is harmful. Some correlations involving adolescent females and adult men are troubling, but it is unclear whether the sex is causing the females to have problems or whether their problems are causing them to have sex. There is also a concern about whether stigma causes the purported harm. In terms of retrospective attitude toward the sex and relationship, there appears to be a gender difference. Even if such sex were harmful to many minors, and this is unclear, we still do not know whether the sex is beneficial overall once the pleasure it generates is factored in.[32] In addition, the purported harm, accompanying pleasure, or cost of criminalization cannot be accurately assessed, let alone compared to one another. We simply do not know enough to do a cost-benefit analysis of criminalizing such sex. In any case, such an analysis has not been done.

Third, consider intermediate scrutiny. Intermediate scrutiny permits a state to punish a type of action only if the state has a substantive (legitimate-and-important) state interest, punishment directly advances that interest, and the combination of what is punished and the severity of punishment is the least restrictive alternative.

On some theories of the limits of government coercion, the state does not have a substantive state interest in criminalizing something unless it involves a right-infringement. On these theories, criminalizing some economically inefficient activities is wrong because the prohibited or discouraged activity does not involve a right-infringement. Consider, for example, criminalization of out-of-wedlock or dysgenic procreation, or obesity.

Still, the justification of such limits on government coercion is unclear. The harm principle in morality or government is hard to formulate.[33] Preventing other-things-being-equal harm justifies too much coercion because nearly every activity causes some harm. Consider, for example, that free speech on controversial topics frequently upsets people. Preventing all-things-considered harm has a different problem in that if harm is understood in a counterfactual comparative sense, then the harm principle is implausible.[34] This is because every diminishment of others' well-being becomes a harm. Consider that, on the counterfactual comparative account of harm, a good teacher harms a student when the teacher retires because his teaching would have benefitted the student. Similarly, on this account, an adult woman harms a 16-year-old male when she refuses to have sex with him if the sex would have been good for him. Even if there were no such problems in formulating and understanding the harm principle, there seems to be little argument outside of rule-consequentialism for it.

Whether punishment directly advances the state's legitimate-and-important interest in preventing harm to minors and does so via the least restrictive alternative is also unknown. Criminalizing statutory rape must be compared to allowing it to occur unhindered or reining it in via civil law. This comparison depends on the specifics of criminalization. The problem here is that there are many different approaches to criminalization. States differ in their rules (e.g., age of consent, Romeo and Juliet Law, severity of punishment, and sex-offender registries) and the way they go about enforcing them (e.g., actively pursuing cases or mostly looking the other way). Without specifying a specific set of rules and way the rules are enforced, it is hard to compare criminalization to allowing such sex to occur or providing civil, but not criminal, liability for it.[35] Other ways of discouraging such sex, at least as related to preventing to out-of-wedlock births, are available. These other ways include subsidizing abortion and birth control and reducing the amount or types of welfare single mothers receive.[36]

If age-different sex with adults were to harm minor females but not minor males, the least restrictive alternative might suggest that the state criminalize age-different sex involving the former but not the latter. Other constraints on laws might be relevant here, such as the requirement that the sexes be treated equally. Still, the different constraints might conflict and, thus, have to be weighed against one another.[37]

In short, it is unclear whether age-different sex passes intermediate scrutiny. It is unclear whether the state has a legitimate-and-important interest in preventing such sex given that it does not involve a right-infringement. Even if there were such an interest, it is hard to assess whether criminalization promotes this interest and whether it does so in the least restrictive way. This is in part because the assessment depends on a particular set of penalties and way of enforcing them.

Here, then, is a summary of the argument for decriminalizing age-different sex.

(P1)	If an activity fails one or more parts of the three-part test, then it should not be a crime.	**Assumption #1a: State-Justification.** There are three plausible justifications of state legitimacy and the duty to obey the law (consent, consequentialism, and the duty of fair play). **Assumption #1b: Three-Part Test.** All three justifications support the three-part test.
(P2)	Age-different sex fails one or more parts of the three-part test.	**Assumption #2a: Rights.** Age-different sex is not a right-infringement because minors can validly consent to an activity that is as important and an activity that is more important than age-different sex. **Assumption #2b: Cost-Benefit Analysis.** We do not know whether age-different sex passes a cost-benefit analysis. This is in part because we do not know whether, in the aggregate, the sex makes people's lives go better. **Assumption #2c: Intermediate Scrutiny.** Age-different sex probably fails intermediate scrutiny because the state does not have a legitimate-and-important interest in protecting individuals against harm in the absence of a right-infringement. Without a specific criminal justice model, we do not know whether criminalizing age-different sex directly advances the state's interest in preventing harm to minors and whether criminalization is the least restrictive way to prevent it.
(C1)	Hence, age-different sex should not be a crime.	[(P1), (P2)]

Part Four: Practical Recommendation

As a general rule of thumb, the side that wants to criminalize something should bear the burden of proving its case. This is because it is better to err on the side of liberty. It is better to err on the side of liberty in part because liberty tends to bring about the best results and in part because it is morally worse to bring about harm or injustice than to fail to prevent it. If this is correct, then the proponents of criminalizing age-different sex have not proven their case. Specifically, they have not shown that such sex infringes on minors' rights or that criminalization makes people's lives go worse and passes intermediate scrutiny. Hence, given the burden of proof, it should be decriminalized. Given that this is not likely to happen, a practical recommendation is that states adopt the bare minimum requirements for criminalizing it. The bare minimum addresses the age of sexual consent, Romeo and Juliet Law, and penalty. It should be 16 years old, 3-year age gap, and a misdemeanor that does not result in sex-offender status, respectively.

Here is a summary of this argument.

(C1)	Age-different sex should not be a crime.	[(P1), (P2)]
(P3)	If (C1), then, as a practical matter, age-different sex should be the bare minimum crime found today across states.	**Assumption #3a: Not Decriminalize.** In the near future, states will not decriminalize age-different sex. **Assumption #3b: Closer.** The bare minimum across states is closer to decriminalization. **Assumption #3c: Duty to Get Closer.** If (C1), #3a, and #3b, then, as a practical matter, age-different sex should be the bare minimum crime found across states today.
(C2)	Hence, as a practical matter, age-different sex should be the bare minimum crime found across states today.	[(C1), (P3)]

Part Five: Objections

Let us now consider some objections. An objector might argue that the three-part test is the wrong test. She might claim that the state need only meet one or two parts of the three-part test. Alternatively, she might argue that the state should criminalize something because it results from a fair procedure (see, e.g., democracy), maximizes aggregate weighted preferences (see, e.g., an auction), or satisfies a non-rights-based principle of justice (see, e.g., Rawls' Difference Principle). The problem with this objection is that in the absence of an argument for these bases for criminalization, there is no reason to accept this position. Here I merely assert without argument that these alternative positions are implausible.

If a fair procedure or maximizing aggregate preference were a sufficient moral reason to criminalize something, then a vote or auction would justify criminalizing gay sex, interracial sex, or sex between adult Jews if enough people voted for or preferred it. If there is something that prevents a mere procedure- or preference-based reason from providing a sufficient moral reason to criminalize a type of sex (e.g., a moral right), then this also applies to age-different sex.

A second objector might argue that even if the three-part test were the correct test, if the state does not know whether criminalizing an activity passes it, the state has moral discretion as to whether to criminalize it. The underlying principle might be the following: *If the government does not know whether a practice is right or good, and it is not at fault for not knowing it, then it has discretion whether to institute it.* The problem with this objection is that there appears to be a good reason to favor liberty by putting the burden of proof on the side advocating criminalizing an activity. We should favor liberty because, in general, it tends to make people's lives go better and to make them wealthier.[38] Alternatively, we should favor liberty because it is harder to justify causing harm (via criminalization and punishment) than allowing it to happen (via decriminalization).

A third objector might argue that age-different sex is wrong because it violates parental rights. The problem with this is that if parents have a right over their child's sexuality, whether as owner or guardian, then they could sell it or waive it. This would not produce across-the-board criminalization of age-different sex. In addition, if a parent were to have such a right directly, rather than via guardianship, then she would seem to own the teen. She would then be within her rights to treat her child as a slave or have sex with him. If the parent had such a right, and the right cannot be waived, then the case for a Romeo and Juliet Law would weaken considerably.

A fourth objector might argue that we know that premarital age-different sex is wrong or bad for minors, especially minor females. The objector argues that such sex might be intrinsically wrong or bad. Alternatively, she might argue, we know certain things without needing an empirical study. For example, we know that being a prostitute or performing in pornography is bad for 17-year-old women. We do not need a study to tell us these things. The same is true for premarital age-different sex. The objector notes that when making policy decisions, we need not leave morality or common sense at the door.

Contra the objector, premarital age-different sex is neither intrinsically wrong nor intrinsically bad. It is implausible that marriage itself makes an otherwise intrinsically wrong sex act permissible unless we make strong religious assumptions. In any case, premarital sex is not limited to age-different sex. Similarly, having premarital sex does not intuitively seem to be an objective-list bad, that is, something that makes a person's life go worse independent of its effects on pleasure and desire fulfillment. In addition, the sex does not necessarily involve degradation, disrespect, exploitation, right-infringement, and so on. There are at least possible worlds in which age-different sex does not have these features. This is likely true for some of the sex that characterized our ancestors over the tens of thousands of years of human history. Consider, for example, hunter-gatherer tribes. In discovering whether age-different sex is degrading, disrespectful, exploitative, etc., especially when the minor female wants and initiates the sex, we have to look at the world. We cannot know the moral and psychological features of such sex across large numbers of different people in very different situations without looking at the world in a systematic manner.

Even if premarital age-different sex were wrong, bad, or harmful, it does not automatically follow that the state should criminalize it. There might be good reasons not to do so. For example, it might be wrong, but not in a right-infringing way. Alternatively, criminalizing such sex might produce more harm than not criminalizing it. Thus, even if the objector succeeds in showing that such sex is wrong, bad, or harmful, she still needs a further argument to show that it should be criminalized. By analogy, even if cheating on one's boyfriend or girlfriend is bad or wrong, a further argument is needed to show that the state should criminalize such cheating.

Here, then, is a summary of the objections and responses.

Objection #1: Wrong Test. The three-part test is not the correct test for criminalization. The correct test supports criminalization.	**Response #1a: Correct Test.** The objector needs an argument to show that this is not the correct test. **Response #1b: Fails New Test.** If only one or two conditions are required for criminalization (e.g., right-infringement or passing a cost-benefit analysis), criminalizing such sex likely fails this more limited test.
Objection #2: Knowledge. Even if the three-part test is the correct test for criminalization, if the government does not know whether an activity passes it, and is not at fault for not knowing this, then it has discretion over whether the activity should be criminalized.	**Response #2: Liberty.** The side that wants to criminalize an activity should bear the evidentiary burden. The government cannot currently meet it.
Objection #3: Parental Rights. The state should criminalize such sex in order to protect parental rights.	**Response #3a: Criminalization.** If a parent were to have such a right, then she could waive it. Hence, this position would not support criminalization. **Response #3b: Implausible.** A parent does not have the right to control an adolescent's sexuality because she does not own the adolescent.
Objection #4: Knowledge Again. We know premarital age-different sex is wrong, bad, or harmful because of its intrinsic features or through common sense. Hence, the state should criminalize it.	**Response #4a: No Knowledge.** We do not know that such sex is wrong, bad, or harmful. **Response #4b: Criminalization.** Even if we were to know that such sex is wrong, bad, or harmful, this feature alone would not justify criminalizing it.

Part Six: Conclusion

In this chapter, I have argued that, in principle, age-different sex should not be a crime. The argument is that if an activity fails one or more parts of the three-part test then it should not be a crime and that age-different sex fails one or more parts. Specifically, I argued that such sex does not infringe on the minor's right, we do not know whether the sex passes a cost-benefit analysis, and criminalization fails intermediate scrutiny. Perhaps this is unsurprising. States are increasingly recognizing the right of teenagers 16 and older to have sex with one another without criminal penalty and there does not seem to be a significant moral difference between a 16-year-old having sex with a peer and with an adult. I also argued that because age-different sex will not be completely decriminalized, as a practical matter, age-different sex should be the bare minimum crime found across states today.[39]

Notes

1. For a similar analysis, see, also, U.S. Department of Justice, "Statutory Rape Known to Law Enforcement," August 2005, https://www.ncjrs.gov/pdffiles1/ojjdp/208803.pdf, p. 2.
2. For a discussion of the different issues in statutory rape law, see Noy Davis and Jennifer Twombly, *State Legislators' Handbook for Statutory Rape Issues* (Washington, DC: American Bar Association Center on Children and the Law, 2000).
3. See Equality Now, "5 Things You Should Know about Child Marriage and The Law In the United States," May 24, 2019, https://www.equalitynow.org/5_things_you_should_know_about_child_marriage_the_us.
4. See Daryl Olszewski, "Statutory Rape in Wisconsin: History, Rationale, and the Need for Reform," *Marquette Law Review* 89 (2006): 693–719, esp. 698–700 and Meredith Cohen, "No Child Left Behind Bars: The Need to Combat Cruel and Unusual Punishment of State Statutory Rape Laws," *Journal of Law and Policy* 16:2 (2008): 717–756, esp. 727–728.
5. Such crimes are often felonies. Twenty-nine states require some statutory rape offenders to register as sex offenders for age-different sex. In eleven states, there are no Romeo and Juliet Laws. See Cohen, "No Child Left Behind Bars: The Need to Combat Cruel and Unusual Punishment of State Statutory Rape Laws," 738–739.
6. See Michelle Oberman, "Regulating Consensual Sex with Minors: Defining a Role for Statutory Rape," *Buffalo Law Review* 48:3 (2000): 703–784, esp. 703–704 citing Charles Warren et al., "Sexual Behavior Among U.S. High School Students, 1990–1995," *Family Planning Perspectives* 30:4 (1998): 170–172. See, also, U.S. Department of Justice, "Statutory Rape Known to Law Enforcement."
7. See Julia Terruso, "In teacher-student sex cases, men average longer jail terms, newspaper analysis reveals," *NJ.com*, March 30, 2019, https://www.nj.com/news/2013/04/a_look_at_teacher-student_sex.html#:~:text=News-,In%20teacher%2Dstudent%20sex%20cases%2C%20men%20average%20longer,jail%20terms%2C%20newspaper%20analysis%20reveals&text=A%20Star%2DLedger%20analysis%20of,receive%20longer%20sentences%20than%20women.
8. See Amanda Cairns, "Hot for Teacher: Gender Bias in Sentencing of Teachers that have Sex with Their Students," Campbell Law Observer, October 2, 2012, http://campbelllawobserver.com/hot-for-teacher-gender-bias-in-sentencing-of-teachers-that-have-sex-with-their-students/#:~:text=Campbell%20Law%20Observer-,Hot%20For%20Teacher%3A%20Gender%20Bias%20in%20Sentencing%20of%20Teachers,have%20Sex%20with%20Their%20Students&text=A%20female%20teacher%20from%20Texas,male%20students%20in%20her%20home.&text=She%20could%20have%20faced%20up%20to%20five%20years%20in%20prison.
9. See Terruso, "In teacher-student sex cases, men average longer jail terms, newspaper analysis reveals."
10. See Danielle Kaeble, "Time Served in State Prison, 2016," U.S. Department of Justice, November 2018, https://www.bjs.gov/content/pub/pdf/tssp16.pdf.

11. Consider a case in which a 22-year-old has sex with a 17-year-old. This is not a crime in Georgia and a felony or misdemeanor in California. See O.C.G.A. 16-6-3 (2010) 16-6-3. Statutory rape and CA Penal Code 261.5 and
12. See the Protect Act section 503 of 1992 (18 U.S.C. 2251–2260.
13. In eleven states, there are no Romeo and Juliet Laws. See Cohen, "No Child Left Behind Bars: The Need to Combat Cruel and Unusual Punishment of State Statutory Rape Laws," 738–739
14. See *Hermesmann v. Seyer*, 847 P.2d 1273 (1993).
15. There are other arguments for this conclusion. If retributivism is true and adults do not deserve punishment for having sex with minors, then the state should not punish age-different sex. See Michael Moore, *Placing Blame: A General Theory of the Criminal Law* (Oxford: Oxford University Press, 1998). If, instead, the state should not punish at all, then it should not punish age-different sex. For a general argument against punishment, see David Boonin, *The Problem of Punishment* (New York: Cambridge, 2008) and Randy Barnett, "Restitution: A New Paradigm for Criminal Justice," *Ethics* 87 (1977): 279–301.
16. In law, intermediate scrutiny comes from *Craig v. Boren*, 429 U.S. 190 (1976). The idea of its being a necessary condition for criminalization comes from Douglas Husak, *Overcriminalization: The Limits of the Criminal Law* (New York: Oxford University Press, 2008), 55–177.
17. For the consent justification, see John Locke, *Two Treatises of Government*, Peter Laslett, ed.

    (New York: Cambridge University Press, 1988). For discussion of the consequentialist

    justification, see M. B. E. Smith, "Is There a Prima Facie Obligation to Obey the Law," *Yale Law Journal* 82 (1973): 950–76. For the duty of fair play, see John Rawls, "Legal Obligation and the Duty of Fair Play," in Sidney Hook, ed., *Law and Philosophy* (New York: New York University Press, 1964), 3–18.
18. For the idea that emancipated minors must provide informed consent to surgery, see Jessica Feierman et al., *Teenagers, Health Care the Law: A Guide to the Law on Minors' Rights in New York State* 2nd ed. (New York: New York Civil Liberties Union, 2018), accessed August 31, 2020, https://www.nyclu.org/sites/default/files/thl.pdf. In general, though, the law is murky with regard to need for minors to consent to medical treatment. See Nina Shapiro, "What Rights Do Teens Have When It Comes To Their Health?" *Forbes*, March 28, 2019, https://www.forbes.com/sites/ninashapiro/2019/03/28/what-rights-do-teens-have-when-it-comes-to-their-health/#39af39186065.
19. See Cohen, 731, Lewis Bossing, "Now Sixteen Could Get You Life: Statutory Rape, Meaningful Consent, and the Implications for Federal Sentence Enhancement," *New York University Law Review* 73 (1998): 1205–1250, esp. 1227–1228, and Gail Wyatt et al., *Sexual Abuse and Consensual Sex: Women's Developmental Patterns and Outcomes* (Washington, DC: Sage Publication, 1993). As a matter of Constitutional law, the Supreme Court has held that minors are less blameworthy than adults for crimes. See *Roper v. Simmons*, 543 U.S. 551 (2005) and *Thompson v Oklahoma*, 487 U.S. 815 (1988).
20. See Joel Feinberg, "The Child's Right to an Open Future," in W. Aiken and H. LaFollette, eds., *Whose Child?* (Totowa, NJ: Rowman & Littlefield, 1980), 124–53.
21. See Oberman, 703.

22. See Susan Moore and Doreen Rosenthal, *Sexuality in Adolescence: Current trends* 2nd ed. (East Sussex, UK: Psychology Press, 1993).
23. See Oberman, 703–704.
24. See U.S. Department of Justice, "Statutory Rape Known to Law Enforcement," 1.
25. See Denise Hines and David Finkelhor, "Statutory sex crime relationship between juveniles and adults: A review of the social scientific research," *Aggression and Violent Behavior* 12 (2007): 300–314, esp. 302.
26. See Paul Okami, "Self-reports of "positive" childhood and adolescent sexual contacts with older persons: An exploratory study," *Archives of Sexual Behavior* 20 (1991): 437–457, Sylvia Condy et al., "Parameters of sexual contact of boys with women," *Archives of Sexual Behavior* 16:5 (1987): 379–394, and M. E. Fromuth and B. R. Burkhart, "Childhood sexual victimization among college men: Definition and methodological issues," *Violence and Victims* 2:4 (1987): 241–253.
27. The Supreme Court held that the Constitution did not allow race-based social stigmatization to given weight in a custody battle. See *Palmore v. Sidoti*, 466 US 429 (1984).
28. See Harold Leitenberg and Heidi Saltzman, "College women who had sexual intercourse when they were underage minors (13–15): Age of their male partners, relation to current adjustment, and statutory rape implications," *Sexual Abuse: A Journal of Research and Treatment* 15:2 (2003): 135–147 and Harold Leitenberg and Heidi Saltzman, "A statewide survey of age at first intercourse for adolescent females and age of their male partner: Relation to other risk behaviors and statutory rape implications," *Archives of Sexual Behavior* 29 (2000): 203–215, Joanna Higgenson, "Defining, excusing, and justifying deviance: Teen mothers' accounts for statutory rape," *Symbolic Interaction* 22 (1999): 25–46, Okami, "Self-reports of "positive" childhood and adolescent sexual contacts with older persons: An exploratory study," 437–457, and Allie Kilpatrick, "Some correlates of women's childhood sexual experiences: A retrospective study," *Journal of Sex Research* 22 (1986): 221–242. For criticisms of the studies that found positive or neutral effects on women, see Hines and Finkelhor, "Statutory sex crime relationship between juveniles and adults: A review of the social scientific research,"307.
29. For support for one or both of the following empirical claims, see Jennifer Manlove et al., "Sex between young teens and older individuals: A demographic portrait, Child Trends," September 2005, https://www.childtrends.org/wp-content/uploads/2005/09/Sex-Between-Young-and-Old.pdf, Barbara Marin et al., "Youth with older boyfriend and girlfriends: Associations with sexual risk," in B. Albert et al., eds., *14 and younger: The sexual behavior of young adolescents* (Washington, DC: National Campaign to Prevent Teen Pregnancy, 2003), 83–90, Jacquelin Darroch et al., "Age differences between sexual partners in the United States," *Family Planning Perspectives* 31 (1999): 16–167, S. G. Elstein and N. Davis, *Sexual relationships between adult males and young teenage girls (Final Report)* (Chicago: American Bar Association Center on Children and the Law, 1997), K. S. Miller et al., "Sexual initiation with older male partners and subsequent HIV risk behavior among female adolescents," *Family Planning Perspective* 29 (1997): 212–214, and Michael Lamb, "Behavioral profiles of adolescent mothers and partners with varying intracouple age differences," *Journal of Adolescent Research* 1 (1986): 399–408.

30. See Lawrence Finer, "Trends in Premarital Sex in the United States, 1954–2003," *Public Health Reports* 122:1 (2007): 73–78.
31. Roughly, one-third of African-American men have a felony conviction. See Sarah Shannon et al., "The Growth, Scope, and Spatial Distribution of People With Felony Records in the United States, 1948–2010," *Demography* 54 (2017): 1795–1818. The US has 4.4 % of the world's population and 22 % of its prisoners. See Sean Gorman, "Webb says U.S. has 5 percent of world's population, 25 percent of its "known" prisoners," *Politifact*, December 15, 2014, https://www.politifact.com/factchecks/2014/dec/15/jim-webb/webb-says-us-has-5-percent-worlds-population-25-pe/. For the increasing criminalization of American life, see Harvey Silverglate, *Three Felonies A Day: How the Feds Target the Innocent* (New York: Encounter Books, 2011).
32. For a similar move with regard to laws prohibiting recreational drug use, see Douglas Husak, "Recreational Drugs and Paternalism," *Law and Philosophy* 8:3 (1989): 353–381.
33. For the classic statement of the harm principle, see John Stuart Mill, *On Liberty* (New York: Dover Publications, 2002).
34. The counterfactual comparative account of harm says that an event harms a person if and only if he would have done better were the event not to have occurred. This theory clashes with intuitions with regard to overdetermination, preemption, and refraining from benefitting.
35. It is unclear whether some criminal-law-related programs—for example, sex offender registries—make the public safer. One study found that "[P]roponents of these laws are not able to point to convincing evidence of public safety gains from them." See Human Rights Watch, "No Easy Answers: Sex Offender Laws in the US," September 11, 2007, https://www.hrw.org/report/2007/09/11/no-easy-answers/sex-offender-laws-us. Let us set aside this type of consideration because it is not tied to age-different sex in particular. Such a consideration, though, is relevant.
36. The relation of welfare benefits to marriage and out-of-wedlock births is controversial with some studies finding a weak if any relation. One study found that at the time women are receiving welfare benefits, they are less likely to get married. See Julien Teitler et al., "Effects of Welfare Participation on Marriage," *Journal of Marriage and Family* 71:4 (2009: 878–891.
37. In *Michael M. v. Superior Court of Sonoma County*, 450 U.S. 464 (1981), the Supreme Court found that a statutory rape law that discriminated on the basis of gender was constitutional.
38. Citizens in countries that have more economic freedom (i.e., more capitalist) are wealthier, healthier, more educated, and politically freer. See Peter Leeson, "Two Cheers for Capitalism?" *Sociology* 47 (2010): 227–233. For the correlation between economic freedom and per capita income, see Terry Miller et al., *2020 Index of Economic Freedom* (Washington, DC: Heritage Foundation, 2020), https://www.heritage.org/index/book/chapter-4. These patterns suggest that, other things being equal, greater freedom—whether economic or social—correlates with happiness and wealth.
39. I am grateful to David Boonin and Bob Kelly for their extremely helpful comments and criticisms of this paper.

CHAPTER 21

Sexual Consent, Dementia, and Well-Being

Andria Bianchi

Introduction

In April 2015, Henry Rayhons, a former Iowa State Representative, faced a felony charge of sexual abuse for having sex with his wife, Donna, in her nursing home.[1] The initial accusation occurred in 2014 and was motivated by the claim that Donna was unable to consent to sex and that Henry had thereby violated the law. Although consensual sexual activities are allowed to occur at the nursing home in which she resided, Donna had severe Alzheimer's at the time of the accusation, scoring zero on a cognitive test designed to assess memory and orientation. Consequently, staff at the nursing home, including her primary care physician, determined that Donna was incapable of consenting to sex. Although Donna was deemed incapable of consenting, she and Henry were known to have a loving relationship, she was always pleased to see her husband, and no allegation was made to suggest that she resisted sexual contact with him. In fact, Henry testified that Donna frequently initiated sex while living with dementia.[2]

The case of Henry and Donna Rayhons is just one of several ethically complex and controversial cases involving sex and people with dementia. At least part of the reason that these cases are complex is because consent is a significant part of what it typically means to engage in lawful and ethical sex, and some people with dementia may want to have sex but be unable to consent in traditionally accepted ways.

The purpose of this chapter is to consider the question of whether, and if so, under what circumstances, it may be ethically permissible for people with

A. Bianchi (✉)
University of Toronto, Toronto, ON, Canada
e-mail: andria.bianchi@utoronto.ca

D. Boonin (ed.), *The Palgrave Handbook of Sexual Ethics*,
https://doi.org/10.1007/978-3-030-87786-6_21

dementia to have sex. In the first section, I explain what it means to have dementia. I subsequently describe some cases that demonstrate the types and prevalence of circumstances related to the topic. I also explain that people with dementia may be considered unable to consent to sex in accordance with conventional consent models. Consequently, one may argue that people with dementia cannot pursue ethically permissible sex and, as such, should not be permitted to engage in sexual acts; this is not the argument that I endorse. Instead, I suggest that we ought to consider enabling people with dementia to pursue activities that promote their well-being, including sexual acts, even in cases where they cannot give clear consent to them. I ultimately propose a novel framework to respond to cases of sex and dementia called "a framework of prioritizing well-being." The framework is based on the idea that sexual acts for people with dementia may be ethical insofar as they contribute to their well-being; a person with dementia's well-being ought to be prioritized above other considerations, such as autonomy and consent.

Dementia and Sex

Dementia is a term that is used to classify several brain disorders that cause a person's memory and cognitive functioning to decline. There are many conditions that cause dementia; two of the most common are Alzheimer's disease and vascular dementia.[3] Vascular dementia occurs when one's brain fails to receive sufficient oxygen and nutrients because blood vessels are blocked[4] and Alzheimer's is when brain cells are destroyed, causing other cells to "lose their ability to do their jobs and, eventually die, causing irreversible changes in the brain."[5] There is currently no cure. Although different diseases can cause dementia, many of the symptoms are similar. Some of these symptoms are: language and communication challenges, an inability to concentrate, problem-solving difficulties, and memory loss (Alzheimer Society of Canada 2020, What is dementia?). As a result of these symptoms, people with dementia often struggle to—and eventually may be completely unable to—perform regular activities, such as dressing oneself, shopping, toileting, driving, etc. Dementia is typically described as a progressive disease where moments of lucidity will decrease over time.[6] Consequently, those affected by the disease become increasingly dependent on others. While most people diagnosed with some type of dementia are older than sixty-five, 2-8% of diagnoses occur amongst people who are younger, with some being in their 30s, 40s, or 50s.[7]

At the time of writing this chapter, there are an estimated fifty million people living with dementia across the globe.[8] That number is expected to increase to eighty-two million within the next ten years (i.e., by 2030) and to 152 million by the year 2050.[9] The increasing pervasiveness of dementia led the World Health Organization (WHO) to adopt a global action plan in 2017, where 146 countries (i.e., 75% of WHO Member States) were tasked with developing national plans to reduce the stigmatization of, and to help better understand, prevent, and treat dementia by the year 2025.[10]

Although one of the action items in the WHO global plan says that dementia stigmatization ought to be addressed, neither the global plan nor any of the national plans mention the topic of sex; silence surrounding the stigmatization of and response to the topic of sex and dementia, however, is not uncommon. In a 2015 interview about sex and nursing homes, Dr. Cheryl Phillips (an advocate for LeadingAge[11]) said: '"We'll ask [people with dementia] about their religion, the music they like, what kind of food they want to eat. We don't dream of asking them about their preferences around sexuality and intimacy."'[12] Reporter Bryan Gruley has also said that nursing homes and family members are not ready to manage cases of sex and dementia.[13] Relatedly, in her discussion of sexual consent and elderly people with neurocognitive disorders, Stephanie Tang said that "governmental and societal attitudes reflecting general discomfort with the idea of cognitively impaired elders engaging in sexual activity have resulted in an overall lack of laws, regulations, and general guidelines on the subject."[14]

One of the primary reasons that silence surrounding sex and dementia is problematic is that, as demonstrated below, people with dementia may want to engage in sexual activity and it may contribute to their happiness and well-being. The well-being of persons with dementia is important to consider since, ultimately, dementia is an incurable disease that will eventually lead to death. Additionally, the disease typically prevents people from being able to participate in activities that they may have previously enjoyed (e.g. cooking, doing crosswords, dancing, driving, etc.). Prioritizing quality of life makes sense because of the many losses associated with dementia. If sexual acts may add to a person's quality of life, then enabling them to occur in unharmful ways ought to be considered.

Cases of Sex and Dementia

Cases involving sex and people with dementia have made news headlines, which seems to suggest that the topic warrants increased attention. I will highlight a few of these cases below.

In addition to the case of Henry and Donna Rayhons, another case took place at Windmill Manor nursing home in Coralville, Iowa in 2009. In this case, two people with dementia, a 78-year-old man who was divorced and an 87-year-old married woman, were found having sex. Upon learning of the sexual interaction, the nursing staff stopped them, and the woman responded by "kicking, screaming, and biting the nurses."[15] The woman was more content and calmer when the male resident was with her; forcing them to separate caused distress.

In 2010, the case of R v. Adcock concerned an incident in a care home for senior citizens (Bartlett, p. 140). This case involved the husband of a resident, Peter Adcock, who was found engaging in non-penetrative sexual acts with a different woman (i.e., not his wife). Adcock's sexual partner had dementia and a history of sexual disinhibitedness. The staff who discovered the sexual interaction noticed that both Adcock and the resident were smiling and seemed to be enjoying themselves. There was no suggestion of physical or emotional harm. Adcock was charged, however, and pleaded guilty to a sexual offense for impeding the choice of a person with a mental disability. He was sentenced to a maximum of four years in prison, which was reduced to three years on appeal. The relatively light sentence was influenced by the fact that Adcock had a previously innocent character, did not penetrate the resident, and "that his crime was one of acquiescence to the [uninformed] sexual request of the victim."[16]

A case involving former Supreme Court Justice Sandra Day O'Connor and her husband John O'Connor also made headlines in 2008. When John was diagnosed with Alzheimer's, his cognition rapidly deteriorated. He could hardly remember his family and became depressed and introverted.[17] When he went to live in a nursing home, however, John fell in love with another resident, identified as Kay. Because this new relationship positively influenced his happiness and well-being, Justice O'Connor allowed her husband to pursue it.[18] According to the nursing home manager, this is not an isolated case; people with dementia sometimes want to have new intimate relationships and families are responsible for deciding what to do in such cases.

A similar case to John's was outlined in a 2017 interview. In the interview, the husband of a person with dementia, Bob Best, and his daughter were asked about the topic of sex and dementia in relation to his wife. Bob's wife, Karen, was living in a nursing home with early onset Alzheimer's and frontal lobe dementia. Bob said that "he received a call from Karen's [nursing] home notifying him that they found her in bed with another man—who also has dementia—with no pants on."[19] Moreover, "[i]t was one of these things that the home felt that every time they caught or saw or found my wife in a 'compromising' position, they would have to call myself or my daughter. And, it got to be a fair number of times. I remember one weekend there were four phone calls."[20] Although one cannot consent to sex on behalf of another person, Bob and his daughter supported Karen's intimate activities since they contributed to her happiness. In his interview, Bob posed the following: "the bottom line is, is she happy? Is she having a meaningful life right now? Because [i]f we took this [intimate relationship] away from her, what would she have?"[21]

Sex and Dementia: Why Not?

The above cases show that some people with dementia may want to participate in sexual acts, but they are seldom permitted to do so and/or to do so without substantial interference. And while there are important differences in each of the above cases, the range of responses demonstrates that there is no consistent and agreed-upon way to manage them.

At least part of the reason that sexual acts involving people with dementia may be seen as unacceptable is because of widespread ageist stereotypes that suggest that older people are not interested in sex. And given that most people with dementia are above the age of sixty-five, this stereotype likely applies to persons with dementia as well. Many studies exist, however, that explicitly dispel this claim. For instance, a 2003 study completed quality of life assessments and in-depth interviews with forty-four participants between the ages of fifty and ninety-two.[22] The study concluded that sex is a substantial part of having a strong emotional relationship. One seventy-six-year-old participant stated that "[i]t's physical problems that make your sex life less really, not the actual needs and wanting."[23] Also, a study from 2007 included 1,384 participants and learned that more than one quarter of the couples aged seventy-five and older had sex at least once a month.[24] Another study explored 319 elderly men's interests in maintaining their sexual capacity, of which 83% of men said that it was very important.[25] While all of these studies support the conclusion that older individuals may be interested in sex, ageist stereotype are still prevalent.

In addition to the inaccurate descriptive stereotype that older persons are non-sexual, a normative claim also exists, where it is thought that (older) people with dementia *should not* have sex because they may be unable to consent in accordance with conventional consent frameworks.[26] The types of consent frameworks that I have in mind are, for instance, "no means no" and affirmative consent, both of which require sexual participants to have certain capabilities. For example, according to at least one model of affirmative consent, an ongoing positive response (either verbal or non-verbal) from all people participating in sex is required in order for the act to be consensual; verbal and non-verbal cues must be agreed upon before having sex.[27] It seems plausible, however, that many people with dementia may be unable to agree on, remember, and/or effectively communicate an agreed upon positive response to a sexual act. Beyond that of communication, an additional concern is whether people with dementia would be able to sufficiently understand what they are consenting to in an informed and meaningful way. If a person with dementia cannot understand the sexual act to which they are consenting, then any communication of consent would be uninformed and invalid. As such, persons with dementia may be considered unable to consent. It may be the case, however, that a person with dementia can *assent* to sex even if they cannot *consent*; this important distinction will be described in the next section.

Insofar as at least some people with dementia may be interested in having sex, yet unable to consent in accordance with consent frameworks, it seems that we can respond in a few different ways. One way to respond would be to consider it legally and ethically illicit for people with dementia to engage in sexual acts; this seems to be the way that most people and/or organizations respond. Another response may be to develop alternative consent frameworks for the dementia population that take their symptoms into account; this is a task that I previously pursued.[28] Finally, one may respond by considering whether something other than consent per se may deserve to be prioritized in specific circumstances and for the dementia population in particular. Ultimately, I argue that since sex may play an important and desired role in individuals' lives, and since some older people with dementia may want to have sex, an alternative to consent is worth considering. In the section below, I explore what an alternative framework to consent may look like.

Prioritizing Well-Being

As mentioned, people with dementia may be unable to consent to sex in accordance with conventional consent frameworks; this may lead some to the conclusion that persons with dementia cannot engage in ethical sex. While this conclusion may be motivated by good intentions (e.g. the desire to protect vulnerable populations from the harms that may result from unwanted sex[29]), it fails to acknowledge that sex may benefit individuals in certain circumstances. As such, I suggest that we may want to respond to cases of sex and dementia by considering another framework, one that does not always consider consent to be the primary determining factor of whether a sexual act is ethical. I refer to this framework as "a framework of prioritizing well-being."[30]

A framework of prioritizing well-being responds to the question "*if well-being is important for persons with dementia then how do we enable it?*". I suggest that there is a prima facie reason to believe that satisfying desires, experiencing pleasure, and being happy can contribute to a person's well-being,[31] where each element can be independently fulfilled in the sexual domain. Through a framework of prioritizing well-being, I ultimately propose that it may be ethically permissible for people with dementia to have sex if their well-being is enhanced, even in cases where they cannot give clear consent. I am using the term "desire" to define a state in which a person has a disposition to act.[32] Some theories say that desire is when one has a disposition to act to achieve some specific end (e.g. pleasure, goodness, etc.), but I am using a more open-ended definition. The term "pleasure" refers to a satisfying or enjoyable experience; it is defined as "the affective positivity of all joy, gladness, liking, and enjoyment—all our feeling good or happy."[33] Finally, I am using the term "happy" to describe a mental state that often includes "life satisfaction, pleasure, or a positive emotional condition."[34] A person's desire, pleasure, and happiness are determined by the person experiencing a particular sexual act and I am using the term "well-being" to refer to a state in which at least one of these elements is fulfilled.

A framework that prioritizes well-being over something like autonomy or consent may be considered a hedonistic approach to sex and dementia. Hedonism typically considers pleasure to be the only thing that is "intrinsically prudentially good"[35], though some "contemporary proposals [of hedonism] have replaced [pleasure] with notions of enjoyment, happiness or satisfaction, and the absence of suffering."[36] From a hedonistic perspective, it is important to increase happiness and pleasure and minimize unpleasant experiences. In his book on the ethics of aging, Frits de Lange suggests that a focus on increasing pleasure and decreasing suffering "seems to be a trustworthy criterion for judging the quality of life, especially that of elderly patients with dementia."[37] This is the primary reason that this framework is being proposed, i.e., to improve people with dementia's well-being by allowing them to pursue relevant sexual acts even if they cannot clearly consent to them. Although I am not a hedonist, there may be some reason to focus on pleasure and happiness when it comes to people with dementia specifically, as discussed above.

In the subsequent sections I consider what it may be like for someone with dementia to experience desire, pleasure, and happiness in accordance with a framework of prioritizing well-being. If none of the elements are satisfied then the likelihood that someone with dementia may experience undue harm (from engaging in undesired sex, for instance) may increase.

Desire

As mentioned, one of the elements that may contribute to a person's well-being is the fulfilment of desires. Satisfying one's desires is, prima facie, a good thing. There exist some exceptions to satisfying particular desires (e.g. a recovering alcoholic satisfying their desire for alcohol), but overall the satisfaction of desires is typically positive.[38] Some desires are developed based on autonomous life choices, whereas others arise because of innate human needs. For instance, if a person wants to present their work at a particular conference and is successful, then the satisfaction of this desire would be positive, and it may contribute to their well-being at least temporarily. If a person is thirsty and desires a beverage, then giving them water may, similarly, contribute to their well-being in some way. Turning to the sexual domain, then, if satisfying sexual desires for people with dementia can lead to some positive consequence(s) that may contribute to their well-being, then the satisfaction of the desire would be seen as ethically licit in accordance with this framework; in the absence of any complicating factors (e.g. a recovering alcoholic), satisfying desires is good. It seems that satisfying a person's desire to engage in sex may contribute to their well-being in two possible ways—either *just because* the desire was satisfied in and of itself, or because it creates a pleasurable experience and/or happiness, which also leads to well-being.

In order to satisfy a person's sexual desires, evidence of their desires must be determined.[39] This may seem problematic when thinking about people with dementia since some people may be unable to explicitly communicate a desire to have sex because of certain dementia symptoms (e.g. language and communication challenges). If a person with dementia cannot explicitly communicate their desires, then the potential for them to experience undue harm (and/or a lack of benefit and well-being) may increase.

I suggest that one way to respond to this concern is by implementing a practice that is built on responding to assent. In discussing sexual consent, David Archard says that while consent and assent are different concepts, they are closely related. He says that "consent is essential agreement *to* something, [whereas] assent is essential agreement *with* something."[40] The concept of assent is normally used to describe a person's active willingness to engage in an act, specifically when they may lack the ability to fully comprehend it (e.g. the benefits and risks).

Recognizing and responding to assent is a practice that is often used in the medical domain when it comes to people who are incapable of providing informed consent (e.g. some people with dementia, children, etc.). For instance, if a person with dementia expresses a willingness to participate in an activity or to pursue a particular treatment then this would be evidence of assent. Assent is important in medical contexts because it provides a way for healthcare providers and caregivers to involve an (incapable) person in decision-making about their care, even if the person cannot fully comprehend relevant decision-making factors.

Since assent is already used in the medical domain as a way for people with dementia to communicate their desires and have their desires (at times) respected, I argue that it could and, perhaps should, also be applied to the sexual domain. Sexual assent could be determined through a person's initiation of a sexual activity and/or through their active participation in response to a sexual request. Based on this practice of sexual assent, some of the individuals in the above-mentioned cases would probably be seen as desiring to have sex since they expressed a willingness to participate. Although sexual assent may not necessarily mean that one is going to experience pleasure or happiness upon engaging in a sexual interaction, it could be indicative of one's sexual desire.[41]

One of the primary differences between how assent is used in a framework for well-being versus the medical domain is that in medical contexts, a person with dementia who assents to a treatment is not the final decision-maker (at least in most jurisdictions); (an)other capable individual(s) has to consent on the person's behalf. For instance, suppose a person with dementia requires some type of surgical intervention. In this case, assent may be a way for the person to contribute to the decision (e.g. they may agree to proceed even if they do not fully understand potential benefits/risks), however, a substitute decision-maker would still need to provide informed consent.[42]

Under a framework of well-being, a practice of assent for sexual acts would *not* require others to consent on a person's behalf. The purpose of sexual assent would be to indicate a person's desire to have sex so that their desire can be satisfied; formal consent would not be necessary. The reason that I diverge from the medical domain is twofold. First, others may be unable to fully understand and/or appreciate the reason that a person wants to have sex—a person's sexual desire is deeply personal. Second, there may be barriers to involving other individuals since sex is a value-laden and moralized topic. If a person with dementia assents to have sex and if another party needs to consent in addition to the person's assent, then it seems plausible that the other party may impose their own values and possibly halt or alter the desired activity. People have individual ideas about when and what kinds of sexual desires ought to be pursued, which may be detrimental when it comes to satisfying the desires of people with dementia.[43]

One problem that ought to be considered when it comes to assent and satisfying sexual desires is that people are sometimes unaware of what they want. As mentioned by de Lange, "[d]esires can be disordered or wrongly expressed: I think I desire one thing, but I am actually longing for something else."[44] This problem may reflect a lack of self-understanding and/or a difference between first and second-order desires, where a person with dementia may not have access to their second-order preferences and therefore assent to sex based on first-order desires.[45] Ultimately, I suggest that enabling the act to which a person with dementia assents may be ethically defensible from the perspective of prioritizing well-being even if it is only a first-order preference. The justification for prioritizing well-being in this manner is based on the assumption that a person with dementia's quality of life is of the upmost importance, where second-order desires may be less relevant to consider.

The possibility that a person's expressed desire may be inaccurately interpreted is the primary problem with this part of the overall framework. One way to resolve this may be to determine if a person communicates their desire(s) in a consistent manner in the present moment. If a consistent desire is expressed, then we may be able to more confidently assert that their desire is authentic and decrease the possibility of misinterpretation.

Pleasure

Another way to contribute to a person with dementia's well-being may be for them to have pleasurable sexual experiences. Pleasure is different from desire satisfaction since a person may experience pleasure from activities that were not initially desired. Conversely, satisfying desires will not necessarily result in pleasure. A pleasurable experience may contribute to a person's well-being by itself and/or it may cause one to be happy and consequently lead to well-being—this latter possibility will be described below. Under a framework of well-being, if there is good reason to believe that a sexual act may result in pleasure, then that would be a reason in favor of allowing a person with dementia to pursue it.

In *Sexual Consent*, Archard contemplates why consent matters. He suggests that consent is important for a few reasons, one of which is its relation to pleasure. He says:

> In the case of sexual activity… it is plausible to think that individuals will normally only consent to what they believe they will find pleasurable and only find pleasurable what they consent to. In the first place, sex is mainly engaged in for its pleasure. Sex may be freely practised out of duty, habit, boredom, curiosity, defiance, revenge, the impulse to procreate, and, almost certainly, many more motives. But it is chiefly done for and in the expectation of pleasure… [t]he point is simply that it is unlikely that individuals should always voluntarily engage in sexual activity with no aim or hope of deriving pleasure from doing so."[46]

This idea may be oversimplified, but if it is at least somewhat true—if pleasure is at least one of the main reasons that people engage in/want to have sex—then a person with dementia may want to and/or continue to participate for the same reason.

From a historical perspective, pleasure has often been depicted as "a lesser good, a sin, a sickness, and a perversion"[47] rather than something that is valuable or relevant to well-being. The idea of sexual pleasure as a sin was influenced by Greek philosophical ideas that were upheld by the Roman Catholic Church, where "the assumption that the genitalia are an intrinsically inferior part of the body, much lower and less holy than the mind, the "soul," the "heart," or even the upper part of the digestive system (the status of the excretory organs is close to that of the genitalia)"[48] was prevalent. Additionally, the medical model of sexuality (which developed in the late 18th and early 19th centuries) recognized non-procreative sex as a sign of sickness; the potential for experiencing and/or benefiting from sexual pleasure was ignored.[49] In his discussion of sex and pleasure, Mitchell Tepper says that "[t]he pleasurable aspect of sex in our culture has been largely ignored, vilified, or exploited."[50] Furthermore, and relevant to this chapter, Tepper argues that "[s]exuality as a source of pleasure… is not readily recognized for populations that have been traditionally marginalized in our society… [such as] people who are older, who are larger, who are darker, who are gayer, [and/or] who are mentally or physically disabled."[51] Sexual pleasure is rarely depicted as something that can contribute to well-being in general, never mind for (older) persons with dementia.

The topic of sex and pleasure also has a gendered component. Legal scholar Katherine Franke says that sex and sexuality are considered in relation to two principal concerns: dependency and danger.[52] She says that a primary focus in discussions about sex is to prevent women from engaging in risky sexual acts. And although this is important, Franke argues that it risks making "women's actual [sexual] experiences with pleasure invisible."[53] Although Franke does not discuss intersectional challenges for people with dementia specifically, perhaps a similar claim could apply to all persons with dementia since they are a vulnerable population who are more likely to experience harm. While it is important to protect people with dementia from harm, a complete focus on harm prevention could risk making pleasurable experiences invisible.

Ultimately, if sex can be a pleasurable activity, and if pleasure is relevant to well-being, then sex would be deemed ethical under this framework when pleasure is achieved. In order for pleasure to be a part of a framework of prioritizing well-being, we need to be able to determine whether someone with dementia is experiencing it. Some people with dementia may be able to effectively articulate whether they are experiencing pleasure, but for people who are unable to do so because of dementia symptoms, alternative methods of assessing pleasure need to be considered.

One suggestion for determining if a person is having a pleasurable experience may be to consider their physiological response(s) to sex. If a person's body is responding to a sexual interaction in a certain way (e.g. through an erection and/or vaginal lubrication) then this may be one way to determine whether they are experiencing pleasure. The challenge with this recommendation, however, is that physiological responses are not consistently reliable. As stated by sex educator Emily Nagoski, a person's body may naturally respond to a sexual interaction in a certain way, but this may not mean that a person is experiencing pleasure. For instance, a woman's vagina may naturally lubricate during sex, and it is typically thought that this means that a woman is experiencing pleasure. However, Nagoski says that this is just her body's natural response to a sexual activity; she may or may not be experiencing pleasure at all.[54]

Upon considering this option, it seems that, again, the best way of determining whether a person is experiencing pleasure may be to consider whether assent occurs during a sexual interaction; it is plausible that sexual assent may be indicative of both desire *and* pleasure. If a person expresses their desire to have sex via assent and then actively participates in a sexual act (as opposed to passively receiving sexual favors) then this may reflect their desire to continue participating *because* it is a pleasurable experience; a desire to continue to participate in a sex may be based on pleasure and expressed via assent.

As mentioned above, however, a person may experience pleasure from participating in acts that they do not (at least initially) desire. In response to this possibility, I suggest that if there is good reason to believe that a person with dementia will experience sexual pleasure, then it may be ethically permissible under limited and careful circumstances to commence a sexual interaction in order to see whether pleasure is experienced/whether sex is subsequently desired. In her work on scientific information related to sex and sexuality, Nagoski says that many women have a responsive sexual desire rather than a spontaneous sexual desire. A responsive desire is when a person's desire to have sex is awakened *during* a sexual act.[55] If a person develops a desire to have sex after their partner pursues them and they commence a sexual activity, then they may have a responsive sexuality.

The circumstances under which it may be permissible to commence a sexual act to see if pleasure is experienced and sex is desired are those under which a person's responsive sexuality is known *and* only insofar as a person does not dissent. For instance, if Henry Rayhons knew that Donna previously: (1) seldom initiated sex because she has a responsive sexual desire, and (2) always enjoyed sex and wanted to participate once a sexual act started, then it may be permissible for Henry to commence sex to learn if Donna experiences pleasure and wants to continue; Donna's pleasure and desire to continue to have sex would be communicated through assent. It would be impermissible for Henry to initiate a sexual act if Donna clearly dissented, such as saying "no" or pushing Henry away. Ultimately, sexual assent may alleviate some concerns regarding an individual with dementia's inability to explicitly convey whether they are having a pleasurable time. Assent not only considers a person's physiological response to sex as a determination of pleasure, but also whether a person is actively expressing a willingness to participate. This may be a more reliable way to determine whether a person with dementia is experiencing pleasure, and consequently, enhanced well-being.[56]

Happiness

Happiness is the third and final factor that is prima facie relevant to well-being.[57] It seems possible that sex may cause happiness in at least one of three ways. First, if a person satisfies a desire to have sex and/or experiences sexual pleasure, then it may create happiness. Second, a person who is in a long-term relationship may agree to have sex *not* because they desire it or experience pleasure, but solely because their partner wants to have sex with them, and satisfying their partner's desire may make the person happy. Sex may also contribute to a person's happiness just because of the satisfaction gained from being close to another person—not necessarily because of any satisfied desire or pleasure.[58]

There has been minimal research on the idea that sex may cause happiness, so it is difficult to know if and/or which of the above suggestions may most accurately reflect a link between sex and happiness in all circumstances. Three studies, however, do suggest that there is a link. One study from 2004 included 909 employed women who retrospectively rated sex as the activity that contributed to the most enjoyment and happiness (in a list of nineteen activities).[59] A similar study explored the link between sex and happiness by surveying approximately 16,000 randomly selected Americans. The main finding was that sex is, indeed, linked with happiness as the dependent variable: "[t]he more sex, the happier the person."[60] Moreover, they say that "[t]he effect of sex on happiness is statistically well-determined, monotonic and large"[61] for both men and women, and for people both under and over forty years of age. In their analysis, the researchers also note that "individuals who have no sexual activity are less happy than average."[62] Another study was done on 3800 adults in China. This study also concluded that people who frequently engage in better quality sex (i.e., pleasurable sex) and with one primary partner are more likely to be happy. Furthermore, they found that "having unwanted sex or having sex simply to satisfy one's partner is associated with lower happiness."[63]

Many people with dementia may be unable to effectively describe whether they are happy after sex. There is no reason to believe, however, that the link between sex and happiness would differ when it comes to people with dementia or other cognitive impairments. Based on the available studies, it seems plausible that if a person with dementia satisfies a sexual desire and/or experiences sexual pleasure (as determined by a practice of assent), then we may be able to infer that they will experience happiness as well.

Conclusion

This chapter introduced a framework of prioritizing well-being as a new way to respond to cases involving sex and people with dementia. Since many people with dementia may be unable to consent to sex in accordance with conventional consent models (because of symptoms associated with dementia), I aimed to prioritize a person with dementia's well-being by satisfying desires, sexual pleasure, and happiness even when a person cannot clearly consent.

In order to determine whether a person with dementia desires to have sex, a practice of responding to assent was proposed. Assent is often used in medical contexts so that people who are deemed incapable of consenting to their own treatment decisions can still be included in the decision-making process to a certain extent. Assent is when a person expresses a willingness to participate in an activity even if they cannot critically reflect upon potential risks and benefits. Assent may be helpful in the sexual domain so that a person's sexual desire to sex can be determined. Assent may also be helpful when it comes to the second element discussed in this chapter, i.e., for evidence of pleasure. If a person assents to continue to participate in a sexual act, then there is some reason to believe that this may be because they are experiencing pleasure. Finally, empirical evidence suggests that sex often leads to happiness for people who engage in sex that is desired and pleasurable. Insofar as at least one of these three elements (i.e., satisfied desire, pleasure, and/or happiness) is attained when it comes to sex and persons with dementia, then the sexual act may contribute to increased well-being. And, under this framework, the sexual act would therefore be deemed ethical.

Overall, this chapter was based on the idea that honoring a person with dementia's well-being is important when it comes to approaching sexual acts. I recognize that determining if a sexual act is ethically licit by prioritizing well-being may be seen as counterintuitive and/or entirely inappropriate given the significance of consent in Western societies. I argue that it ought to be considered for people with dementia, however, if it is important to promote well-being.

Notes

1. Pam Belluck, "Sex, Dementia and a Husband on Trial at Age 78," *The New York Times*, April 13, 2015, https://www.nytimes.com/2015/04/14/health/sex-dementia-and-a-husband-henry-rayhons-on-trial-at-age-78.html
2. Belluck, "Sex, Dementia and a Husband on Trial at Age 78."
3. Alzheimer's Association, "What Is Dementia?," 2018, accessed October 15, 2020, https://www.alz.org/what-is-dementia.asp
4. Mayo Clinic. "Vascular Dementia," 2020, accessed October 15, 2020, https://www.mayoclinic.org/diseases-conditions/vascular-dementia/symptoms-causes/syc-20378793
5. Alzheimer's Assosciation, "What is Dementia?"
6. The reason that I say that dementia is typically described as progressive is because other views exist which consider dementia to be a different kind of experience.

For instance, in their discussion of the troubling discourse that is used to describe dementia, Gail J. Mitchell, Sherry L. Dupuis, and Pia C. Kontos say that dementia is a "dynamic and fluid phenomenon that is not, as portrayed in the medical and societal discourses, a progressive, irreversible decline of one's humanity." The authors argue that the experiences of people with dementia regularly fluctuate, and so defining dementia as a progressive disease is misleading.

7. Alzheimer Society of Canada, "Young onset dementia," 2020, accessed October 20, 2020, https://alzheimer.ca/en/about-dementia/other-types-dementia/young-onset-dementia
8. World Health Organization, "Dementia," September 21, 2020, accessed October 13, 2020, https://www.who.int/news-room/fact-sheets/detail/dementia
9. World Health Organization, "Dementia"
10. World Health Organization, "Global action plan on the public health response to dementia 2017—2025," 2017, accessed November 5, 2020, https://www.who.int/mental_health/neurology/dementia/action_plan_2017_2025/en/
11. LeadingAge is a non-profit organization whose mission is to help combat ageism. One of their strategic goals is to improve care for people with dementia.
12. Eliza Gray, "Why Nursing Homes Need to Have Sex Policies," *Time Health,* April 23, 2015, http://time.com/3833358/nursing-home-sex/
13. Melissa Block, "Can Elderly Patients With Dementia Consent To Sex?," *National Public Radio, Inc.*, July 22, 2013, https://www.npr.org/templates/story/story.php?storyId=204580868
14. Stephanie L. Tang, "When Yes might Mean No: Standardizing State Criteria to Evaluate the Capacity to Consent to Sexual Activity for Elderly with Neurocognitive Disorders," *Elder Law Journal* 22 (2015): 451.
15. Block, "Can Elderly Patients With Dementia Consent To Sex?"
16. Peter Bartlett, "Sex, Dementia, Capacity, and Care Homes," *Liverpool Law Review* 31, (2010): 137–154; Bartlett refers to this population as people with learning disabilities, but the term intellectual disability may better capture the broader community that he is referring to.
17. Philip Sherwell, "Judge lost husband to Alzheimer's—and love," *The Telegraph,* January 27, 2008, https://www.telegraph.co.uk/news/uknews/1576716/Judge-lost-husband-to-Alzheimers-and-love.html.
18. The reason that Justice O'Connor had to *approve* of John's relationship was because adultery was still a crime in twenty-three states at the time of John's case. Consequently, any outside-of-marriage relations had to be approved by one's married spouse so that one would not be convicted of adultery. Now, adultery is a crime in 21 states; this is discussed by: Deborah L. Rhode, "Why is adultery still a crime?," *LA Times*, May 2, 2016, http://www.latimes.com/opinion/op-ed/la-oe-rhode-decriminalize-adultery-20160429-story.html; Available information about this case discusses John and his new partner kissing and hand-holding, though if other intimate acts occurred that also contributed to her husband's happiness then perhaps Justice O'Connor may have approved.
19. Bob Best and Cassandra Trach, "What does consent look like when you have dementia?" interview by *Out in the Open,* Canadian Broadcasting Corporation, May 21, 2017, http://www.cbc.ca/radio/outintheopen/sexual-consent-1.4123222/what-does-consent-look-like-when-you-have-dementia-1.4123272.

20. Best and Trach, "What does consent look like when you have dementia?"
21. Best and Trach, "What does consent look like when you have dementia?"
22. Merryn Gott and Sharron Hinchliff, "How important is sex in later life? The views of older people," *Social Science & Medicine* 56, (2003): 1617.
23. Gott and Hinchcliff, "How important is sex in later life? The views of older people," 1623.
24. John DeLamater and Sara M. Moorman, "Sexual Behavior in Later Life," *Journal of Aging and Health* 19, no. 6 (2007): 922.
25. Asgeir R. Helgason et al., "Sexual Desire, Erection, Orgasm and Ejaculatory Functions and Their Importance to Elderly Swedish Men: A Population-based Study," *Age and Ageing* 25, (1996): 285.
26. Alexander A. Boni-Saenz, "Sexuality and Incapacity," *Ohio State Law Journal* 76, no. 6 (2015); Andria Bianchi, *Sex, Dementia, and Consent.* (doctoral thesis). 2018; Andria Bianchi, "Sex, Aging, and Dementia," In *Routledge Companion to the Philosophy of Sex*, edited by Brian D. Earp, Clare Chambers, Lori Watson, Routledge, forthcoming.
27. Lois Pineau, "Date rape: A feminist analysis," *Law and Philosophy* 8, no. 2 (1989): 217-243.
28. Boni-Saenz, "Sexuality and Incapacity"; Bianchi, "Sex, Aging, and Dementia"; Bianchi, *Sex, Dementia, and Consent.*
29. Stephen J. Schulhofer, *Unwanted Sex: The Culture of Intimidation and the Failure of Law* (Massachusetts: Harvard University Press, 1998).
30. It is important to highlight that I am not suggesting that conceptions of sexual consent for people *without* dementia need to be reconsidered and I recognize that there are some potentially serious consequences of this framework that will need to be taken into account.
31. Economists David G. Blanchflower and Andrew J. Oswald differentiate between two types of well-being. They say that well-being can be associated with a single area of a person's life and one's entire life. I am specifically interested in focusing on a single area/context of a person's life, namely, that which focuses on sexual acts David G. Blanchflower and Andrew J. Oswald, "Money, sex and happiness: an empirical study," *The Scandinavian Journal of Economics* 106, no. 3 (2004): 395.
32. Tim Schroeder, "Desire," *The Stanford Encyclopedia of Philosophy* (Summer 2017 Edition), ed. Edward N. Zalta, https://plato.stanford.edu/archives/sum2017/entries/desire/.
33. Leonard D. Katz, "Pleasure," *The Stanford Encyclopedia of Philosophy* (Winter 2016 Edition), ed. Edward N. Zalta, https://plato.stanford.edu/archives/win2016/entries/pleasure/.
34. Dan Haybron, "Happiness," *The Stanford Encyclopedia of Philosophy* (Fall 2011 Edition), ed. Edward N. Zalta, https://plato.stanford.edu/archives/fall2011/entries/happiness/; Different theorists have different conceptions of desire, pleasure, and happiness. These differences should not matter when it comes to considering a framework of well-being, however, it may be an area that could be researched in the future.
35. Maartje Schermer, "In search of 'the good life' for demented elderly," *Medicine, Health Care and Philosophy* 6 (2003): 36; Matthew Silverstein, "In Defense of Happiness: A Response to the Experience Machine," *Social Theory & Practice* 26, no. 2 (2000).

36. Schermer, "In search of 'the good life' for demented elderly," 36.
37. Frits de Lange, *Loving Later Life:an ethics of aging* (Michigan: William B. Eerdmans Publishing Company, 2015, 15.
38. We typically explain why satisfying certain desires is bad by referring to the effect that they have on the satisfaction of other desires. For instance, we might say that satisfying the alcoholic's desire for a drink is bad because it makes it more challenging for them to satisfy other central desires (e.g., autonomy, personal relationships, health, financial stability, etc.).
39. Desires that require other people's involvement will need the desire to be understood in some way.
40. Archard, *Sexual Consent,* 5
41. In general, it seems that one of the primary reasons that a person with dementia may assent to an activity is in order to experience something that they want. For instance, a person with dementia may assent to eating a type of food by opening their mouth if they desire to eat/desire that type of food, whereas they may not assent if they lack the desire.
42. *Health Care Consent Act,* 1996, c. 2, Sched. A, s. 20 (1), https://www.canlii.org/en/on/laws/stat/so-1996-c-2-sch-a/latest/so-1996-c-2-sch-a.html.
43. For instance, Perlin describes the case of an owner of a care facility who only allows their residents to engage in sexual acts on Saturdays from 6:00-8:00pm because this is when he and his wife engage in sex (*Sexuality, Disability, and the Law: Beyond the Last Frontier?,* 2).
44. de Lange, *Loving Later Life,* 15.
45. In his discussion of free will and personhood, Harry Frankfurt discusses first-order and second-order desires. First-order desires are those that influence a person to act in a certain way—these actions are guided by a person's first-order *volition* or *will.* For instance, one's desire to yell at an intoxicated person who is disturbing one's sleep would be a first-order desire. Second-order desires are the desires that one has about one's first order desires. So, one's desire to *not* act on one's desire to yell at the intoxicated individual would be a second-order desire. Second-order desires are sometimes considered to be intrinsic (e.g. the desire to not yell is desirable *for its own sake*) and sometimes considered to be instrumental (e.g. one's second order desire to not yell at the person is because of the instrumental desire to not have rocks thrown at one's window after yelling).
46. Archard, *Sexual Consent,* 20.
47. Mitchell S. Tepper, "Sexuality and Disability: The Missing Discourse of Pleasure," *Sexuality and Disability* 18, no. 4 (2000): 286.
48. Rubin, *Deviations,* 148; Shrage and Stewart, *Philosophizing about Sex,* 74-75.
49. Tepper, "Sexuality and Disability: The Missing Discourse of Pleasure," 286.
50. Tepper, "Sexuality and Disability: The Missing Discourse of Pleasure," 284.
51. Tepper, "Sexuality and Disability: The Missing Discourse of Pleasure," 285.
52. Katherine M. Franke. "Theorizing Yes: An Essay on Feminism, Law, and Desire," *Columbia Law Review* 101, no. 1 (January 2001): 182.
53. Franke, "Theorizing Yes: An Essay on Feminism, Law, and Desire," 200-201.
54. Nagoski, *Come As You Are,* 192.
55. Nagoski, *Come as You Are,* 227 and 307.
56. While sexual assent may be a more reliable way to determine if a person is experiencing sexual pleasure, I recognize that the potential to misinterpret a person's willingness to participate in sex is a risk.

57. Daniel Kahneman et al., "Toward National Well-Being Accounts," *American Economic Review* 94, no. 4 (2004): 429-434.
58. Doll, *Sexuality and Long-Term Care*, 4 and 104.
59. Kahneman, "Toward National Well-Being Accounts," 432.
60. Blanchflower and Oswald, "Money, sex and happiness: an empirical study," 400.
61. Blanchflower and Oswald, "Money, sex and happiness: an empirical study," 411; the authors do note, however, that the results may be influenced by individuals' undetectable biases, the desire to boast about their sex lives, discomfort speaking about sex, etc.
62. Blanchflower and Oswald, "Money, sex and happiness: an empirical study," 400.
63. Zhiming Cheng and Russell Smyth, "Sex and Happiness," *Journal of Economic Behavior and Organization* 112, (2015): 26; the authors' latter point which says that having sex to please one's partner lowers happiness shows that my second proposed reason that sexual acts may result in happiness (namely, for the purpose of pleasing one's partner) may be wrong. Also, similar to Blanchflower and Oswald, the authors note that the results may be influenced by biases.

References

Alzheimer Society of Canada. "Young Onset Dementia." 2020. Accessed November 8, 2020. http://www.alzheimer.ca/en/About-dementia/Dementias/young-onset-dementia.

Alzheimer's Association. "What Is Dementia?" 2018. Accessed October 15, 2020. https://www.alz.org/what-is-dementia.asp.

Archard, David. 1998. *Sexual Consent*. Colorado: Westview Press.

Bartlett, Peter. 2010. Sex, Dementia, Capacity, and Care Homes. *Liverpool Law Review* 31: 137–154.

Belluck, Pam. "Sex, Dementia and a Husband on Trial at Age 78." *The New York Times*. April 13, 2015. https://www.nytimes.com/2015/04/14/health/sex-dementia-and-a-husband-henry-rayhons-on-trial-at-age-78.html.

Best, Bob, and Cassandra Trach, "What does consent look like when you have dementia?" interview by *Out in the Open, Canadian Broadcasting Corporation*. May 21, 2017. http://www.cbc.ca/radio/outintheopen/sexual-consent-1.4123222/what-does-consent-look-like-when-you-have-dementia-1.4123272.

Bianchi, Andria. Sex, Dementia, and Consent. (doctoral thesis). 2018. University of Waterloo, Waterloo, Ontario, Canada.

———. "Sex, Aging, and Dementia." In *Routledge Companion to the Philosophy of Sex*, edited by Brian D. Earp, Clare Chambers, Lori Watson, Routledge, forthcoming.

Blanchflower, David G., and Andrew J. Oswald. 2004. Money, Sex and Happiness: An Empirical Study. *The Scandinavian Journal of Economics* 106 (3): 393–415.

Block, Melissa. "Can Elderly Patients With Dementia Consent To Sex?" *National Public Radio, Inc.* July 22, 2013. https://www.npr.org/templates/story/story.php?storyId=204580868.

Boni-Saenz, Alexander A. 2015. Sexuality and Incapacity. *Ohio State Law Journal* 76 (6): 1201–1253.

Cheng, Zhiming, and Russell Smyth. 2015. Sex and Happiness. *Journal of Economic Behavior and Organization* 112: 26–32.

DeLamater, John, and Sara Moorman. 2007. Sexual Behavior in Later Life. *Journal of Aging and Health* 19 (6): 921–945.

Doll, Gayle Appel. 2012. *Sexuality and Long-Term Care*. Baltimore: Health Professions Press.

Franke, Katherine M. 2001. Theorizing Yes: An Essay on Feminism, Law, and Desire. *Columbia Law Review* 101 (1): 181–208.

Frankfurt, Harry G. 1971. Freedom of the Will and the Concept of a Person. *The Journal of Philosophy* 68 (1): 5–20.

Gott, Merryn, and Sharron Hinchliff. 2003. How Important is Sex in Later Life? The Views of Older People. *Social Science and Medicine* 56 (8): 1617–1628.

Gray, Eliza. "Why Nursing Homes Need to Have Sex Policies." *Time Health*. April 23, 2015. http://time.com/3833358/nursing-home-sex/.

Haybron, Dan. "Happiness," *The Stanford Encyclopedia of Philosophy*, ed. Edward N. Zalta (Fall 2011 Edition). Accessed April 6, 2018. https://plato.stanford.edu/archives/fall2011/entries/happiness/.

Helgason, Asgeir R., Jan Adolfsson, Paul Dickman, Stefan Arver, Mats Fredrikson, Marianne Gothberg, and Gunnar Steineck. 1996. Sexual Desire, Erection, Orgasm and Ejaculatory Functions and their Importance to Elderly Swedish Men: A Population-Based Study. *Age and Ageing* 25 (4): 285–291.

Kahneman, Daniel, Alan B. Krueger, David Schkade, Norbert Schwarz, and Arthur Stone. 2004. Toward National Well-being Accounts. *The American Economic Review* 94 (2): 429–434.

Katz, Leonard D. "Pleasure." *The Stanford Encyclopedia of Philosophy*, ed. Edward N. Zalta (Winter 2016 Edition). Accessed April 6, 2018. https://plato.stanford.edu/archives/win2016/entries/pleasure/.

de Lange, Frits. 2015. *Loving Later Life: An Ethics of Aging*. Grand Rapids, Michigan: William B. Eerdmans Publishing Company.

Mayo Clinic. "Vascular Dementia," 2020, accessed October 15, 2020, https://www.mayoclinic.org/diseases-conditions/vascular-dementia/symptoms-causes/syc-20378793

Mitchell, Gail Joyce, Sherry L. Dupuis, and Pia Kontos. "Dementia Discourse: From Imposed Suffering to Knowing Other-Wise." (2013).

Nagoski, Emily. 2015. *Come As You Are*. New York: Simon & Schuster Paperbacks.

Perlin, Michael L., and Alison J. Lynch. 2016. *Sexuality, Disability, and the Law: Beyond the Last Frontier?* New York, NY: Palgrave Macmillan.

Rhode, Deborah L. "Why is adultery still a crime?" *LA Times*. May 2, 2016. http://www.latimes.com/opinion/op-ed/la-oe-rhode-decriminalize-adultery-20160429-story.html.

Rubin, Gayle. 2012. *Deviations a Gayle Rubin Reader*. Durham, NC: Duke University Press.

Schermer, Maartje. 2003. In Search of`the Good Life' for Demented Elderly. *Medicine, Health Care and Philosophy* 6 (1): 35–44.

Schroeder, Tim. "Desire." *The Stanford Encyclopedia of Philosophy, ed. Edward N. Zalta* (Summer 2017 Edition). https://plato.stanford.edu/archives/sum2017/entries/desire/.

Schulhofer, Stephen J. 1998. *Unwanted Sex: The Culture of Intimidation and the Failure of Law*. Cambridge: Harvard University Press.

Sherwell, Philip. "Judge lost husband to Alzheimer's—and love." *The Telegraph*. January 27, 2008. https://www.telegraph.co.uk/news/uknews/1576716/Judge-lost-husband-to-Alzheimers-and-love.html.

Shrage, Laurie, and Robert Scott Stewart. 2015. *Philosophizing about Sex*. Peterborough, ON: Broadview Press.

Silverstein, Matthew. 2000. In Defense of Happiness: A Response to the Experience Machine. *Social Theory and Practice* 26 (2): 279–300.

Substitute Decisions Act, 1992, *S.O. 1992, c. 30,* https://www.ontario.ca/laws/statute/92s30.

Tang, Stephanie L. 2015. When Yes might Mean no: Standardizing State Criteria to Evaluate the Capacity to Consent to Sexual Activity for Elderly with Neurocognitive Disorders Notes. *The Elder Law Journal* 22 (2): 450–490.

Tepper, Mitchell S. 2000. Sexuality and Disability: The Missing Discourse of Pleasure. *Sexuality and Disability* 18 (4): 283–290.

World Health Organization. "Global action plan on the public health response to dementia 2017—2025." 2017. Accessed November 5, 2020, https://www.who.int/mental_health/neurology/dementia/action_plan_2017_2025/en/

———. Dementia. 2020. Accessed October 13, 2020. http://www.who.int/mediacentre/factsheets/fs362/en/.

CHAPTER 22

Exploitation and Sexual Consent

David Boonin

Alice is an anxious person. Barbara is a psychologist whose specialties include the treatment of anxiety, and Alice has been going to sessions with Barbara every other week for several months now. Alice has clearly made progress, and during this time she has developed feelings of deference, trust, and affection for Barbara as a result of their professional interactions. One afternoon, after their session has concluded, Barbara tells Alice that she finds her attractive and asks if she would like to have sex with her. Alice says yes, and she has sex with Barbara in her office.

The case of Barbara and Alice involves no apparent coercion. Barbara doesn't threaten to retaliate in any way if Alice says no. The case also involves no deception. Barbara doesn't try to trick Alice into having sex with her, for example, by convincing her that sex with her would somehow help to cure her anxiety. And we can assume that while Alice is an anxious person, she is clearly a competent adult. If Alice was considering having sex with anyone other than Barbara, there would be no question that she was capable of giving valid consent to it, the kind of consent that can make an otherwise impermissible act permissible. The question I want to address in this chapter is whether there's something about Alice's relationship to Barbara in particular that could render Alice's consent to have sex with Barbara invalid in this case despite the fact that Alice is a competent adult and that Barbara's proposal involves no apparent coercion or deception. If there is something about their relationship that renders Alice's consent invalid in this case, it presumably has something to do with the unequal positions they occupy within it. The relationship between a therapist and their patient is a paradigm case of positional inequality. It's also the example focused

D. Boonin (✉)
Department of Philosophy, University of Colorado Boulder, Boulder, CO, USA
e-mail: david.boonin@colorado.edu

D. Boonin (ed.), *The Palgrave Handbook of Sexual Ethics*,
https://doi.org/10.1007/978-3-030-87786-6_22

on by David Archard, the philosopher who has made the strongest case for the claim that this kind of inequality can invalidate consent even in the absence of coercion, deception, and incompetence.[1] I will therefore follow Archard in focusing on this example in most of this chapter, though what I have to say here should apply to other cases involving positional inequality, and I return to such cases toward the very end.

Archard's Argument

On Archard's account, Barbara's proposal involves a certain kind of exploitation and the exploitive nature of Barbara's request doesn't just make her behavior sleazy or unprofessional. It makes her guilty of having sex with someone without their valid consent. The basic idea behind Archard's argument is that Barbara is taking unfair advantage of Alice's feelings of deference, trust, and affection toward her and that the unfairness of her doing so renders Alice's consent to have sex with her invalid. Before we can subject this argument to scrutiny, it's important to say a bit more about why Archard thinks this.

Archard identifies five features that combine to generate cases of what he refers to as "exploited consent." Put in terms of the case of Alice and Barbara, the five features are as follows.[2] First, the relationship between a therapist and their patient involves fundamental structural inequality. Alice is vulnerable to Barbara in a way that Barbara isn't vulnerable to Alice, Alice is dependent on Barbara's expertise while Barbara isn't dependent on Alice's, Alice needs Barbara's help, but Barbara doesn't need Alice's, it's Alice who looks up to Barbara, not the other way around, and so on. Second, this fundamental inequality in the dynamics of their relationship is necessary for the relationship to function in the way that it's meant to function. Barbara is the expert and Alice is the one who needs her help, so Alice must defer to Barbara. If Alice simply chatted with Barbara on equal terms, as she might with a friend when they were exchanging advice with each other, Barbara would not be able to successfully treat her condition. Third, the scope of the activity that's appropriate within this professional relationship is determined by the goals of the relationship. If a psychologist were to give their patient financial advice, for example, they would be acting outside the boundaries established by their professional relationship, just as a financial advisor would be if they gave their client dating advice.

Fourth, in the kinds of cases Archard is concerned with, there's good reason to believe that Alice would not consent to having sex with Barbara were it not for the positional inequality in their relationship. Alice has developed feelings of deference, trust, and affection toward Barbara, that is, precisely because certain features that typically characterize the patient-therapist relationship dispose patients to develop such feelings. This might be because Barbara is available to listen to Alice talk about herself and her problems in a way that no one else is. It might be that Barbara genuinely seems to care about Alice and to really "get" her in a way that Alice's friends and family don't seem able to do.

It might be because certain parallels between the patient-therapist relationship and the child-parent relationship lead Alice to transfer onto Barbara some of the feelings of deference, trust, and affection she has toward her parents. It might simply be that Barbara is a smarter and more impressive person than those Alice typically gets to spend time with. It might be some combination of these factors or some harder to characterize feature of the relationship that renders Alice particularly vulnerable to developing these kinds of feelings. The particular reasons don't matter for the purposes of the argument. What matters is simply that it seems to be a sufficiently common, well-established, and widespread phenomenon that patients develop such feelings in the therapeutic context.

Finally, Barbara appreciates, or can reasonably be expected to appreciate, the four features of the case already identified. As a trained and experienced therapist, Barbara understands the positional inequality in her relationship with her patients, understands that it plays an essential role in the success of that relationship, and that the scope of the professional relationship is limited to treating the issues the patient is there to be treated for, and she knows from her training, and perhaps from prior experience, that patients are often led to develop feelings of deference, trust, and affection toward their therapist as a result of these distinctive dynamics.

With all these features of the case in mind, Archard's idea seems to be as follows. It's okay for the unequal power dynamics to influence the interactions between Barbara and Alice within the scope of their professional relationship because those inequalities are essential to the success of that relationship and the success of that relationship is beneficial to them both. But when Barbara takes advantage of the feelings of deference, trust, and affection that their professional relationship generates in Alice and uses them as leverage to get Alice to consent to entering a sexual relationship with her, Barbara exploits a vulnerability that's justified within the professional context for her own selfish pleasure outside the bounds of that relationship. Alice, too, can certainly take pleasure from their sexual relationship, of course, but because of the unequal dynamics that have already been established between them, Barbara is going to be the one who calls the shots, controlling when and where they meet and how long the sexual relationship lasts, and Archard cites evidence to the effect that people in positions relevantly similar to Alice's are likely to regret their choice once the sexual relationship has ended.[3]

That last claim is an empirical one, of course, but let's suppose for the sake of the argument that it's true. And even though we could never be certain that this is so in any particular case, let's go ahead and simply stipulate that Alice only has the feelings that lead her to say yes to Barbara's proposal because they have been generated by the structural inequality in their professional relationship. If Barbara and Alice had met on equal terms in any other context—at a bar, for example, or at a party—let's simply assume that Alice would not have developed a willingness to have sex with Barbara. If these assumptions are accepted, then when Barbara gets Alice to consent to having sex with her,

Barbara takes advantage of the vulnerability generated by her non-exploitive professional relationship with Alice to get Alice to agree to take part in a different kind of relationship that is both sexual and exploitive. It's exploitive because Barbara is getting the real benefits of that relationship at Alice's expense. If we assume all this is true, that might be enough to make Barbara's behavior objectionable, but would it be enough to render Alice's consent to have sex with Barbara invalid?

Objecting to Archard's Argument

I don't think so. There are two parts to Archard's case against the validity of exploited consent: Alice only says yes to sex with Barbara because their professional relationship has made her feel deference, trust, and affection toward Barbara, and the agreement she makes with Barbara is ultimately good for Barbara but bad for Alice. But the claim that these considerations are enough to render Alice's consent invalid has unacceptable implications. This can be seen most clearly by starting with a case of consent that has just one of these features and then working our way up to the case where Barbara asks Alice to have sex with her. Doing so will also help us see why we might originally be led to think that the positional inequality in their relationship makes Alice's consent to have sex with Barbara invalid.

So let's start with the fact that Alice only says yes to sex with Barbara because their professional relationship has made her feel deference, trust, and affection toward Barbara, and consider a case where Barbara instead proposes that they increase the number of sessions she holds with Alice from every two weeks to every week. Suppose that Alice's insurance fully covers all her sessions, that Alice would in fact benefit from meeting with Barbara once a week rather than every other week, and that while Barbara would prefer to fill the empty slots on her schedule with a new patient because she would find it more rewarding to increase the diversity of cases she handles, she offers to fill the slots with additional sessions for Alice instead out of a sense of obligation to Alice. Alice, though, is skeptical. If anyone else told her she should go to therapy once a week instead of every other week, she'd say no. It's only because her professional relationship with Barbara has made her feel deference, trust, and affection toward Barbara that she says yes to Barbara's proposal.

If it's reasonable to think that patients often become disposed to say yes to sex with their therapists only because their professional relationship inclines them to feel deference, trust, and affection toward them, then it's surely reasonable to think that patients often become disposed to say yes to changes to their therapy schedule only because their professional relationship inclines them to feel deference, trust, and affection toward them. This means that if the mere fact that Alice says yes to sex with Barbara only because their professional relationship has made her feel deference, trust, and affection toward Barbara is enough to invalidate her consent to have sex with Barbara, then the mere fact that she says yes to Barbara's proposal to change her therapy schedule only because their professional relationship has made her feel deference, trust, and affection toward Barbara is enough to invalidate her consent to change her therapy schedule. Because Archard's argument depends on the claim that patients develop such feelings frequently enough to make it unreasonable for a therapist to ever think the consent to sex they receive from a patient is untainted by it, this would mean that therapists would have to say the same about cases where their patients consented to a change in their therapy schedules. This would include cases where a patient consents to continue going to therapy only because they have developed such feelings for their therapist. This, in turn, would mean that patients can never give valid consent to continue to be treated by their therapists. And that is plainly an unacceptable result.

There are, of course, some notable differences between the case where Barbara proposes to change Alice's therapy schedule and the case where Barbara proposes to have sex with Alice. But I don't think any of them can overcome this problem with Archard's argument. First, in the schedule changing case, Barbara is generously proposing something that's good for Alice and bad for Barbara. But, if we accept Archard's empirical claims, when Barbara asks Alice to have sex with her, she's instead proposing something that's good for Barbara and bad for Alice. We can test the claim that this difference makes a difference to the validity of Alice's consent by changing the scheduling case to one where the change in scheduling would be good for Barbara and bad for Alice. So suppose Alice's insurance doesn't cover her treatment, that she'll have to pay out of pocket for the additional sessions, and that Barbara has no other potential patients to add to the currently empty slots in her schedule and will lose out on some valuable extra income if Alice doesn't take those empty slots. We can assume that Alice will get a little bit of extra benefit from meeting weekly rather than every other week, but that the modest amount of extra benefit won't be worth the considerable additional expense she'll have to incur. This version of the case is like the first version in that Barbara proposes to change their sessions from every other week to every week, but in this version this is a good deal for Barbara and a bad deal for Alice while in the first version it was a good deal for Alice and a bad deal for Barbara. This change to the story might affect our evaluation of Barbara's character, but it's implausible to think it could affect the validity of Alice's consent to the schedule change. To say that Alice's consent to the schedule change is valid when the proposal is a good deal for her but invalid when the very same proposal is a bad deal for her would mean that people can't validly consent to bad deals. Anytime someone paid more for a product than the product was really worth, for example, their consent to buy the product would be invalid for that very reason. This result strikes me as clearly unacceptable. So if we agree that Alice's consent to the schedule change is valid when it's a good deal for her, we should agree that her consent to the schedule change is also valid even when it's a bad deal for her.

A second difference between the schedule changing case and the sexual consent case is that the schedule changing case takes place within their professional relationship while the sexual consent case does not. Part of Archard's argument depends on the claim that this distinction, too, is relevant. But even when this part of the argument is accounted for, Archard's position still yields implausible results. Suppose, for example, that Alice needs to buy a car. She can't afford a new one, but she doesn't trust used car dealers and she's wary of buying a car from a stranger trying to sell one online. Then one day she comes across a car ad on a local website and discovers that the seller of the car is Barbara. It turns out that Barbara has been trying to sell her car for several months now, and while she had originally been asking for a thousand dollars more than the car's market value, she's now asking for a few thousand dollars less than market value. It's a very good price and a very good deal. Alice normally wouldn't buy a used car from someone selling it through a website like this, but because she trusts Barbara, she sends her a message and agrees to pay her asking price. It seems clear that Alice's consent to pay Barbara for the car is valid in this case even though this exchange takes place outside their professional relationship and even though Alice would not have agreed to the transaction were it not for the trust she developed in Barbara as a result of their professional relationship.

We've now considered a case where Alice agrees to a proposal of Barbara's that's a bad deal for Alice and a case where Alice agrees to a proposal of Barbara's that lies outside the scope of their professional relationship. In both cases, it turned out to be implausible to treat the feature in question as one that could render Alice's consent to Barbara's proposal invalid. But what about a case that has both features? This would make the case just like the case of sexual consent in terms of the two features that Archard takes to be relevant. We can construct such a case by simply supposing that Alice had come across Barbara's used car ad a few months earlier, when Barbara was still asking for considerably more than the car's market value rather than a few months later, when she was asking for considerably less than its market value. This change to the story makes Alice's discovery of the ad less fortunate for her, but it's implausible to think it could affect the validity of her consent to pay the asking price. To say that her consent is valid when agreeing to Barbara's asking price is a good deal for her but invalid when agreeing to Barbara's asking price is a bad deal for her would again mean that people can't validly consent to bad deals. Anytime someone paid more for a product than the product was worth, their consent to buy the product would be invalid for that very reason. So if we agree that Alice's consent to buy Barbara's car is valid when Barbara is asking for less than market value, we should agree that Alice's consent to buy Barbara's car is also valid when Barbara is asking for more than market value. This seems to be the correct result in any event since it seems clear, at least to me, that Alice's consent to buy the car is indeed still valid even if Barbara is asking for more than the car is really worth.

If we agree that Alice's consent to buy Barbara's car is valid when Barbara is asking for more than the car's market value, though, then we should agree that if Alice consents to a proposal of Barbara's, her consent is not rendered invalid by the combination of the facts that (1) the proposal lies outside the scope of their professional relationship, (2) the proposal is a bad deal for Alice and a good deal for Barbara, and (3) the proposal is one that Alice would not have said yes to were it not for the positive attitude she developed toward Barbara as a result of their professional interactions. And if we agree that the combination of these facts is not enough to invalidate Alice's consent to buy Barbara's car, then we should agree that it is not enough to invalidate Alice's consent to have sex with Barbara even if all three facts hold in that case as well. Archard's argument should therefore be rejected.

Assessing the Objection

We might have the nagging suspicion, though, that even if Alice's consent to buy the car from Barbara is valid, her consent to have sex with Barbara isn't. If this is so, it can't be because having sex with Barbara is outside the scope of their professional relationship, is a bad deal for Alice, and is something she wouldn't have agreed to were it not for the trust she's come to have for Barbara as a result of their professional relationship, because those things are also true of her buying the car from Barbara at above market value. This might lead us to suspect that there's something distinctive about the sexual domain that renders Alice's consent invalid in that context while leaving it valid in other contexts that lie outside her professional relationship with Barbara, and this suspicion might seem to help vindicate Archard's position.

If you find it sufficiently clear that Alice's consent is valid in the car case but plausible that her consent is invalid when she agrees to have sex with Barbara, though, I think a better explanation of your different reactions is available, one that provides no support for Archard's position. This is because when I described the case where Alice consents to have sex with Barbara, Barbara tells Alice to her face that she wants to have sex with Alice, and Alice says yes in response to Barbara's personal request. But in the case where Alice consents to buy Barbara's car, Barbara simply posts an impersonal notice on a website inviting offers from anyone who's interested in buying her car, and Alice simply happens to come across the ad and decides to respond to her invitation. This difference between the two cases means that if Alice decides not to buy Barbara's car, Barbara will never know that Alice decided not to buy it. Barbara will have no way of knowing that Alice noticed her offer online in the first place and so Alice can decline to buy Barbara's car without feeling pressured into buying it or running the risk of disappointing Barbara if she doesn't buy it. But if Alice decides not to have sex with Barbara in the story we began with, she'll have to tell Barbara that to her face, knowing that Barbara will be disappointed. This may very well lead Alice to feel pressured into saying yes to Barbara and concerned that Barbara might retaliate against her in some way if she doesn't. This, I suggest, is why we might still feel uncomfortable treating Alice's consent to having sex with Barbara as valid even if we find it clear that Alice's consent to buying the car from Barbara is valid. If that's right, then the fact that one case involves sex and one doesn't has nothing to do with our having different responses to the two cases and the positional inequality between Barbara and Alice does not by itself render Alice's consent to sex invalid.

We can test this hypothesis in one of two ways, changing the car case to make it more like the sex case or changing the sex case to make it more like the car case. Both tests, I believe, will serve to confirm my suggestion that it's the pressure and plausible fear of retaliation that generates whatever remaining doubts we might still have about the validity of Alice's consent to sex with Barbara in the version of the case I began with and not the mere fact of the positional inequality between Alice and Barbara within their professional relationship that Archard's argument depends on. This result is important both as a way of vindicating my objection to Archard's argument in the therapeutic context in particular and as a way of accounting for the similar sorts of concerns that might seem to arise in cases involving other kinds of positional inequalities.

So suppose first that instead of taking out an ad in which Barbara tries to sell her car to anyone who's interested, Barbara tells Alice right after one of their sessions that she really wants Alice, in particular, to buy her car because she really likes the idea of Alice, in particular, being her car's next owner. In this version of the story, Alice can't decline to buy Barbara's car without Barbara becoming aware of this and being disappointed by it, and I suspect that you'll find it considerably less clear that Alice's consent to buy the car would be valid in this case. Changing the sex case to make it more like the car case might require a bit more creativity, but it too can be done, and I think the results will point in the same direction. So suppose that Alice has decided that she'd like to have a one-night stand with someone and she's visiting a website that features photos of women who are looking to have a one-night stand with someone like her. She's interested in many of the women she sees pictures of, but she doesn't feel sufficiently confident that she can trust any of them. Then, unexpectedly, Barbara's picture turns up. Because she trusts Barbara, she clicks on her profile and sends her a message agreeing to have sex with her. She would not have done so had their professional interactions not lead her to trust Barbara. In this version of the story, Barbara would have no way of knowing that Alice went on the website if she decided not to have sex with her, and Barbara wasn't looking to have sex with Alice in particular in the first place. As a result, I suspect that you'll find it much easier to accept that Alice's consent to have sex with Barbara is valid in this version of the case than you may have found it to be in the original version.

If you respond to these variations on the two cases in the way I think you will, finding it harder to accept Alice's consent as valid in the revised car case than in the original and easier to accept Alice's consent as valid in the revised sex case than in the original, this provides strong evidence that whatever doubts you might have had about the validity of Alice's consent to have sex with Barbara in the original version of the case were due to contingent facts about that case that go beyond the features that Archard's argument appeals to. As a practical matter, these contingent facts are surely important. They suggest that in many, perhaps most, perhaps even virtually all real-life cases, a patient can't give valid consent if their therapist asks to have sex with them because in many, perhaps most, perhaps even virtually all real-life cases, they would reasonably be concerned about potential retaliation if they said no. But as a theoretical matter, they nonetheless help to show that Archard's argument should be rejected. If a reasonable fear of retaliation is what invalidates Alice's consent to have sex with Barbara, after all, then it's coercion that invalidates her consent. And the question this chapter sought to consider was whether there's something about relationships like those between a psychologist and their patient that could render Alice's consent to sex with Barbara invalid even if Alice is a competent adult and Barbara's proposal involves no coercion or deception. Archard's argument aimed to show that the answer to this question is yes. But the conclusion we have reached here is no.

Implications

This theoretical conclusion, in turn, has two important practical implications. First, if Archard's argument were successful, then Alice's consent to have sex with Barbara would be invalid regardless of whether Barbara asked Alice to have sex with her or Alice asked Barbara to have sex with her. In either version of the case, after all, the conditions Archard thinks are sufficient to render sexual consent invalid would be present. But if my objection to Archard's argument has been successful, then even if in most real-life cases it's still true for other reasons that a patient can't give valid consent to sex when their therapist asks to have sex with them, there's no reason to think a patient can't give valid consent to sex if they ask their therapist to have sex with them. If Barbara asks Alice to have sex with her, Alice might reasonably fear retaliation from Barbara if she says no, but if Alice is the one who asks Barbara to have sex with her, there's no reason to think Barbara has illicitly pressured Alice into having sex with her if Barbara says yes to Alice's proposal.

The second practical implication has to do with applying Archard's analysis to other cases of positional inequalities. Assuming that both parties involved are competent adults, these could include cases like a teacher and their student, a coach and their player, a clergy person and their congregant, or an agent and an actor they represent. If Archard's analysis can be successfully extended beyond cases of professional-client relationships, it might also be applied to other forms of positional inequality, like a senator and a constituent, a cop and a citizen in their precinct, maybe even a rock star and a long-time fan. Perhaps it could also be extended to cover cases involving social hierarchies like those involving race or gender, and other kinds of inequality, too.

If my critique of Archard's argument here has been successful, I think we should say the same thing about these other cases, too. In many, or most, or perhaps even virtually all real-life cases, a sexual proposition from a teacher to a student, or an agent to a client, or a cop to a citizen on their beat, might prove unable to elicit valid consent because the recipient might reasonably fear some form of unjust retaliation if they were to say no. But there's no reason to think their consent to sex would be invalid if the proposition ran in the other direction, from student to teacher, client to agent, or citizen to cop. This is not to say that there aren't other good reasons to prohibit sexual relationships in some of these cases. Surely there are. But even if it's wrong for a therapist to have sex with a patient who asks to have sex with them, and wrong for a teacher to have sex with a student who asks to have sex with them, the wrongs they commit do not include the wrong of having sex with someone without their valid consent.

Notes

1. Archard, David. 1994. "Exploited Consent." *Journal of Social Philosophy*, vol. 25, no. 3, pp. 92–101.
2. Archard, p. 95. I have changed the order of presentation of the five points and in elaborating on them draw in part on clarifications that Archard makes later in the article.
3. Archard, p. 97.

CHAPTER 23

A Solution to the Problem of Rape by Fraud

Laurie Shrage

Introduction

The concept of "rape by deception" or "rape by fraud"[1] exposes the fault lines in our contemporary understandings of rape. In many jurisdictions in the U.S. and elsewhere, rape, a sub-category of sexual assault, is defined as sexual intercourse or penetration without consent.[2] There are three main conditions for consent to be genuine: it must be given without coercion,[3] by someone who is competent or not incapacitated, and it must not be induced by misrepresentation. If one or more of these conditions is not met, a person's consent is vitiated.

While the first two conditions for consensual sex are reasonably well understood and acknowledged, the third condition seems to open a can of worms. For example, what information is material to sexual consent, and can it vary from person to person? Should there be mandatory disclosure requirements for some kinds of information, such as one's sexually transmitted illness (STI) status? Is there a morally significant difference between deceiving someone about oneself or intentions in order to coax consent to sex and deceiving someone about the sexual nature of an act in order to obtain consent? And, when consent to sex is invalidated because of deception (rather than coercion or incapacitation), is criminal prosecution and punishment appropriate? In this chapter, I address this last question.

Many kinds of cases can be classified as sex-by-deception, both real and hypothetical, which illustrates the expansiveness of this concept. There is the case of a man who misrepresented his religion and ethnicity to obtain consent

L. Shrage (✉)
Florida International University, Miami, FL, USA
e-mail: lshrage@fiu.edu

D. Boonin (ed.), *The Palgrave Handbook of Sexual Ethics*,
https://doi.org/10.1007/978-3-030-87786-6_23

to sex which otherwise would have been withheld; the doctor who misrepresented a sexual act as a medical procedure; the man who entered a woman's bedroom and impersonated her husband in order to gain her consent; the person who promises money or expensive gifts in order to obtain consent with no intention of providing these; the HIV+ person whose medication makes it virtually impossible to transmit the virus and who does not disclose their HIV status; the man who impersonates a celebrity in order to induce consent to sex; the 17-year-old who lies about her age in order to induce an otherwise cautious 25-year-old to consent to sex (making the latter both a victim of sex-by-deception and a statutory rapist); the man who is deceived by a third party into forcing sex on a unwilling woman as part of an allegedly voluntary BDSM rape scene (making this man both a victim of sex-by-deception and a rapist);[4] the woman who does not inform her partner that she has XY chromosomes;[5] and so on. Courts in different countries have ruled that some of these cases constitute rape.[6]

An Israeli court convicted a man of rape by deception for lying about his religion and ethnicity, as these facts were material to his partner's willingness to consent to sex.[7] Some courts treat only fraud "in the factum" (misrepresenting the nature of an action or activity in order to obtain someone's consent) and not "in the inducement" (making an insincere promise or offer in order to obtain someone's consent) as sufficient to invalidate consent to sex.[8] For example, fraud in the factum occurred in a case in which a doctor sexually penetrated a patient after representing the act as a nonsexual medical procedure.[9] Fraud in the inducement occurred in a case in which a doctor falsely told a patient that she needed to be sexually penetrated by him to cure her ailment.[10] A person can be criminally prosecuted for sex-by-deception in the U.S. if they fail to disclose that they are HIV+ to a partner, even if their treatment reduces their viral load to negligible amounts and when no transmission has occurred. However, someone who may be infectious, but doesn't know it because they have not sought testing or treatment, is not likely to be criminally charged.

U.S. courts are more likely to prosecute a man who impersonates a woman's husband or fiancé in order to procure sex than one who impersonates a celebrity.[11] Is the woman who thinks she is having sex with her husband when she is actually having sex with someone else more victimized than the woman who thinks she is having sex with an admired celebrity when she is only having sex with an imposter? Moreover, the courts generally have not prosecuted those who make deceitful promises, such as offers of financial support or marriage, in return for sex. Yet, if someone takes possession of another's property or accepts a service without following through on a promise to compensate the provider, they are typically guilty of theft. These rulings suggest that judgments about a woman's virtuousness still carry more weight than facts about fraudulent behaviors in rape cases.[12]

If consent obtained without deceit is required to legitimize a sexual encounter, then more kinds of cases will count as criminal rape. While courts have prosecuted some cases of rape by fraud, in general, adults who procure sex by fraud with other adults are rarely prosecuted. Some legal scholars find the reluctance of courts to prosecute rape by deception cases to be problematic, while others attempt to explain how sex procured through deception differs from sex procured through coercion or incapacitation.[13] Philosophers who have written about this topic generally conclude that many acts described as "rape by deception" involve serious moral wrongs for which perpetrators should be held more accountable.[14]

The failure of courts to prosecute most rape by fraud cases calls into question the contemporary consensus that the purpose of antirape law is to protect a person's sexual freedom or autonomy. While this conception of the basis of antirape law represents a significant advance over legal regimes that treat rape as a form of damage to a man's property, some legal scholars are developing new ways to conceive of the aims of antirape law. In this chapter, I investigate two proposals for reconceiving the moral and civil purpose of antirape statutes: that rape violates a person's fundamental right to bodily possession or control, and that rape violates a person's fundamental right to be treated with dignity and respect. I also consider an argument that defends the current consensus that protecting sexual autonomy should be the primary aim of antirape laws.

I ultimately argue for a pluralist approach to antirape law and endorse a proposal to treat rape as a cluster or family resemblance concept. There is no need to assume that there is only one human right or interest that all rapes violate and that antirape laws protect, such as sexual autonomy or bodily control. Sexual transactions can involve different kinds of violations: denying a victim the freedom to choose when and with whom to have sex, depriving a person of the right to maintain control over her bodily activities and capacities, and ignoring another's agency and subjectivity. Moreover, sexual assaults cover a variety of behaviors or methods: from coercion, incapacitation, deception, physical force, and egregious forms of sexual objectification. In some cases, the right to choose when and with whom to have sex may be salient, and in others, the right to bodily control or to be treated with dignity may be more salient. Antirape statutes can be written to recognize different kinds of sexual assault, and to mandate different kinds of punishment. My proposal is compatible with efforts to use both civil and criminal law to address sexual assaults. And finally, to address the complexities of sexual assault law, I endorse a recent innovation involving "sexual offense courts" which, like drug courts, would be able to consider a range of sanctions, including mandatory counseling or medical treatment.

Rape as a Violation of the Right to Self-possession

Jed Rubenfeld claims that

> rape law has a serious problem…If rape law were really to eliminate the force requirement—as so many argue it should, as many statutes have already seemingly done, and as courts have begun to do—then sex-by-deception would and

> should be rape, because the legal definition of rape would then be sex without consent, and a defrauded "consent," like a coerced one, is no consent at all.[15]

Nevertheless, as Rubenfeld points out, American courts, legislators, and legal scholars are generally reluctant to extend existing rape law to most rape by fraud cases. Rubenfeld identifies three possible ways to solve what he calls "the riddle of rape by deception." The first involves keeping the force requirement; the second involves accepting the logical consequences of defining rape as nonconsensual sex and treating sex by fraud as rape; and the third involves defining coercive sex as rape, but not deceptive sex. Rubenfeld ultimately argues for a modified version of the first option. He defends the intuition that most cases of sex by fraud are not rape, and then rejects the idea that rape is simply nonconsensual sex, as well as the idea that the purpose of proscribing rape is to protect a fundamental right to sexual autonomy. Instead, he argues that rape violates a fundamental right to self-possession, much like slavery and torture.[16]

Rubenfeld explains the right of self-possession as

> the possession of one's own body…self-possession is not a property right. You don't own your body the same way you own a car. Rather, bodily possession is a matter (like most forms of possession) of physical control…almost all of us enjoy a basic integration of mind and body that gives us an irreducible measure of physical governance over our bodies and makes our bodies our own.[17]

Both slavery and torture violate a person's right to self-possession because, in each practice, one person takes physical control over another's body. Both the slave master and the torturer use another person's body to serve their own ends or to inflict pain and terror, so that "the victim's own bodily self-governance is nullified…rape shares with slavery and torture the same fundamental violation. The victim's body is utterly wrested from her control, mastered, possessed by another."[18]

Defining rape as a violation of one's right to bodily control recognizes the traumatizing, terrorizing, and dehumanizing aspects of such physical assaults. Even if the victim is not fully conscious, once she learns what has happened to her, she is likely to be traumatized when she tries to envision it. For example, Chanel Miller, the former Stanford student who was sexually assaulted after passing out from drinking at a party, wrote: "I don't want my body anymore. I was terrified of it, I didn't know what had been in it, if it had been contaminated, who had touched it. I wanted to take off my body like a jacket and leave it at the hospital with everything else."[19] Nonsexual physical assaults typically do not cause the disintegration of mind and body, which is one way that sexual assaults differ from ordinary physical assaults and battery.[20] According to Rubenfeld, rape violates a person's basic right to bodily self-governance, and in this way, it can seriously injure our sense of selfhood or personhood. It does not, he emphasizes, damage a person's virtue or character, and it need not even involve serious physical injuries. It is a crime against something integral to selfhood—bodily self-governance—which is suspended when a person's body is used by another.

Slavery, torture, and rape are forms of involuntary servitude, commanded by the use or threat of force, or through legal coercion.[21] Rubenfeld reincorporates the force requirement back into definitions of rape, because it is through force or the threat of force that someone takes control of another's body. While acts of deception may be psychologically manipulative, they do not deprive another of bodily self-governance. Rubenfeld writes "A fraud victim retains his basic bodily self-governance"[22] unlike the victim of rape, slavery, or torture. We don't hear people talk about "slavery by deception" because this is an incoherent phrase, as the phenomenon it purports to describe is simply labor procured by fraud (e.g., false promises of compensation, working conditions, and so on). Importantly, the fraud victim can quit or leave, once she recognizes she has been deceived, unlike a person subject to slavery or physical torture. For Rubenfeld, the concept of "rape by deception" makes no more sense than the concept of "slavery by deception," because the victim of sex by deception retains control over her body and can retreat. In this way, Rubenfeld solves "the riddle of rape by deception" (the reluctance of courts to prosecute such acts as rape) by arguing that sex by deception is not rape, because rape is a crime against bodily self-governance and not autonomy.

Rubenfeld's analysis does not preclude prosecuting some cases of rape by deception because, as he states, it offers a floor and not a ceiling: "States are perfectly free to criminalize forms of sexual imposition other than forcible rape."[23] He concludes that while his account explains (and justifies) the infrequent prosecution of sex by deception, he suggests that such acts could be legally proscribed and sanctioned in ways appropriate to acts that undermine our autonomy or agency. It may be a virtue of his account that it doesn't collapse different kinds of "sexual imposition" into the single crime of "rape." Legal rape includes violent sexual assaults (ones that involve the use or threat of physical force and bodily harm), while sex obtained by deception or coercion (but are not physically violent) can be recognized as a kind of wrongdoing that does not amount to rape.[24] This allows room for legislators and courts to devise different rules and sanctions for different kinds of sexual offenses. Wrongdoing that violates a person's right to self-possession (rape, enslavement, torture) could be criminalized, making the wrongdoer subject to imprisonment, while wrongdoing that undermines autonomy in different realms (sexuality, reproduction, work, marriage) could be handled by either the criminal or civil courts, depending upon the nature of the harm done or the sanctions deemed necessary.

As we increasingly recognize that criminalization and incarceration are not the answer to every social problem, Rubenfeld's analysis offers a way to distinguish sexual assaults that should be criminalized and punished with incarceration from those that could be addressed in other ways, such as civil lawsuits for damages, or mandatory counseling, treatment, and public service. Violent rapes and the sexual molestation of children fall into the first category, while other types of sexual manipulation involving adults may require a different societal response. In other words, by expanding the category of criminal sexual assaults to include rape by deception, we may do more harm than good. By criminalizing an act, we authorize law enforcement personnel to arrest and prosecute anyone suspected of committing that act, and such powers can be misused to harass people who are poor or lack political influence, or who are the victims of prejudice. Criminalization does not necessarily serve as a deterrent to morally abhorrent behaviors, and it rarely solves the underlying social and psychological problems that lead to such behaviors. In our quest to protect the rights of those who have been sexually manipulated, we risk harming the dignity, privacy, and freedom of anyone who might be accused of sexual manipulation. The case involving the prosecution of an Israeli Arab man for "rape by deception," and those involving the prosecution of individuals who are transgender, should set off alarm bells regarding the potential for criminal prosecution to devolve into persecution.[25]

One worry some may have about Rubenfeld's analysis is that he reserves the term "rape" only for forcible rape, which may lead to the inference that sexual manipulation accomplished without physical force is not "real rape" or a serious violation. Another alternative would be to distinguish different kinds of rape: for example, rape by the use or threat of physical violence, rape by coercion, rape by deception, and so on, and design appropriate laws and sanctions for each. Or we could drop the culturally and emotionally charged term "rape" from our laws, and instead use terms, such as "sex imposed by physical force," "sex imposed by coercion," and so on. Before deciding which way forward is best, I consider a response to Rubenfeld's proposal that dissolves the riddle of rape by deception while holding onto the idea that protecting sexual autonomy is the basic principle behind rape law.

Rape as Coerced Consent

Luis Chiesa writes

> the riddle of rape-by-deception is based on a misunderstanding of the kind of autonomy that lies at the heart of modern rape reform statues. Properly understood, the chief goal of contemporary rape laws is to neutralize the coercion inherent in sexual relationships that take place in a male-dominated society. Since minimizing deception is only tangentially related to this goal, respect for the kind of sexual autonomy that rape law is primarily designed to protect is compatible with selective criminalization of rape-by-deception.[26]

Chiesa opts for the third path Rubenfeld identified for solving the tension between the infrequent prosecution of sex-by-deception cases and the notion that protecting sexual autonomy is the primary aim of antirape laws. Rubenfeld finds this solution logically inconsistent because both coercion and deception undermine consent and autonomy, and consistency thus requires that we either punish both sex-by-coercion and sex-by-deception, or neither.[27] However, Chiesa finds a way to get around this internal contradiction.

Chiesa argues that consent is both multidimensional and scalar. It can be weak along one dimension (e.g., the noncoercive dimension), while strong along another dimension (e.g., the competence or informational dimension). Our laws and the courts give different weightages to these different dimensions in different contexts, so that consent in theft law is generally more robust along the informational dimension than consent in rape law, but less robust along the noncoercive dimension. Chiesa notes, for instance, that we permit medical providers to sell goods to their patients, and we allow minors to lawfully transfer property to adults, even those with some authority over the minor, such as a teacher. Such transactions are not considered "theft" even though they may be somewhat coercive. By contrast, failures of disclosure of material information in property transactions typically invalidate consent.[28] The form of autonomy or dimension of consent that is salient is context-sensitive, and our laws do take into account many background historical, social, and practical features of transactions, according to Chiesa. In short, property exchanges may be valid in situations that modern rape law would find coercive, while sexual transactions may be permissible in situations that modern theft law would find fraudulent.

In considering the historical and social context of rape law reform, Chiesa writes "the feminist rape reform movement was primarily about securing additional freedom from coercion in the context of sexual relations….while leaving the competency and information dimensions largely untouched."[29] Feminists sought to draw attention to how men, in a male-dominated society, often impose sex on women against their will without the use of physical force or violence—typically by exploiting women's weaker social position and economic dependency on men. They also focused on the distortion of women's sexuality to serve male interests, so that a woman's refusal to consent to sex is interpreted by misogynists as coy behavior, or so that "no" means "yes." Also, feminists objected to how the courts acquitted predatory and even violent sexual behaviors of men if a woman was thought "promiscuous," or to dress in sexually attractive ways, or entered a man's apartment, and so on. The focus of rape reform was to take away men's license to impose sex on women because, due to various rape myths and the social stigmatization of women, men were only prosecuted for rape in a very narrow range of cases (violent stranger rape of a woman thought to be virtuous) and, even then, most prosecutions failed.[30]

Chiesa concludes that "Rubenfeld's failure to account for the scalar and multidimensional nature of consent and autonomy prevents him from appreciating that we can selectively criminalize rape-by-deception without calling into question the kind of sexual autonomy (and consent) that lies at the core of rape offenses. More specifically, he fails to appreciate that rape offenses principally seek to safeguard the non-coercive dimension of autonomy. Since modern rape laws are primarily about preventing coercion in sexual encounters, rather than deception, there is nothing perplexing about rape laws that only selectively criminalize deception."[31] The prioritization of coercion over deception in modern rape law is a societal response to a historical context in which there are vast imbalances of social power between men and women, but less commonly imbalances of knowledge or information.

Chiesa considers whether there are independent reasons for modern rape law's focus on coercion, other than that this happened to be what feminist rape law reformers sought, and he concludes that there are. First, he argues that coercion is more wrongful than deception, and that this is implicit in our criminal justice system, which punishes the latter less frequently. Second, he argues that it is appropriate for our legal system to address entrenched social disparities, such as differences of power or information. For example, the asymmetrical access to information or knowledge shapes the focus of medical malpractice laws and real estate law, just as gender disparities in social and economic power shape the focus of rape law.[32]

Although Chiesa dissolves the "riddle of rape-by-deception," his account leaves open a number of questions. First, for Chiesa, non-coerciveness is one dimension of consent and coercion is also a matter of degree. But Chiesa does not give any guidance, for example, regarding when the degree of coercion vitiates consent so that an imposition of sex counts as rape. For example, compare the case of a woman who is coerced because her victimizer has the power to take away her means of support or her children, or to deport her, with the case of a woman who agrees to sex in order to gain an undeserved advantage, such as a promotion or raise, or a good grade. Both situations involve an imbalance of power between the parties which our criminal justice system could be fine-tuned to address, or it may be that the latter kind of case is best handled through sexual harassment policies and the civil law. Chiesa does not offer a way to measure the degree of coercion involved, or how to handle cases where there is significant coercion, but the consequences of refusing consent are not severe.

Moreover, Chiesa's coerced-consent approach retains consent and autonomy as the guiding principles of modern rape law, unlike Rubenfeld's, and he does allow the law to treat deception as invalidating consent in some instances. For example, he offers the example of mandatory HIV disclosure laws as allowable under his account, but he does not explain why the informational aspect of consent should be given weight in such cases. Today, people who have access to treatment for HIV are rarely infectious, and it is unclear why they should be criminally prosecuted as rapists. In short, Chiesa fails to explain what circumstances justify giving weight to the informational dimension of consent as well as the coercive dimension.

In sexual transactions between adults, one could argue that there is no entrenched imbalance of knowledge or information: that is, person A generally has as much access to information about person B as B has to information about A. But in another sense, there is a built-in asymmetry: person A has access to much more information about A than B, and person B has access to much more information about B than A. How should our laws address such systemic imbalances, especially when setting standards of materiality? For example, what may be material to A's consent is likely to be different from what is material to B's consent. If it is reasonable for our laws to give priority to the informational dimension of consent when there are characteristic asymmetries in access to information (as Chiesa contends), how should such imbalances be addressed in cases of sexual consent? Chiesa's account fails to offer guidance on these questions.

Rape as the Failure to Respect Human Dignity

Ben McJunkin argues that the narrow prosecution of rape by fraud cases reflects a culture that values and protects a practice of manipulative seduction, whereby men increase their social standing through the sexual conquest of women. Our contemporary norms of masculinity valorize men who treat sex as a game or sport, and who use both coercion and deception to overcome women's resistance to having sex with them. Feminist law reformers should be, and are, just as concerned to change dominant social norms that treat women as sexual prey to be caught, or sexual trophies to be acquired, by men.[33] Changing these norms requires that men be held more accountable for sex obtained by fraud.

McJunkin, like Rubenfeld, questions whether protecting sexual autonomy, or the right to choose, should be the guiding principle behind rape law. Paradoxically, the more the law protects a woman's right to choose, the more valuable it is for men, socially, to seduce women into choosing them as sexual partners. McJunkin writes "As rape law reforms to valorize individual choice, being chosen becomes the designation of masculine status. Meanwhile, rape by fraud laws remain entrenched against reform by a discourse of 'seduction,' employed by normative masculinity to eroticize, as sexual conquest, the permissible transgression of women's autonomous resistance."[34] For McJunkin, rape law reform is shaped by, and also shapes, the ways that men use women to raise their masculine status. He continues: "Contemporary rape law represents a continuation of the structures of male dominance that were once explicitly embraced by the property regime. Rape law still exists to protect the claims of men against other men, and it still contributes to the objectification and abuse of women."[35] In short, the transition from the property rationale to the autonomy principle has not proved adequate to hold men accountable for imposing sex on women.

McJunkin holds that there is ample justification for criminally prosecuting sex obtained by fraud, not because such acts violate a person's sexual autonomy, but because they fail to respect a person's right to be treated with dignity. Treating women like sexual playthings to be manipulated in order to enhance one's masculinity ranking fails to treat women with dignity. McJunkin's account, in this way, solves "the riddle of rape-by-deception" by taking Rubenfeld's second path: treating sex by deception as *rape*. Unlike Chiesa, he does not attempt to defend sexual autonomy as the guiding principle of rape law, but instead to build a new foundation for rape law—the protection of human dignity.

McJunkin contends that a focus on dignity compels us to respect persons *qua* persons, and would allow us to focus on the social meaning of sex acts, as well as their meaning to the participants involved. Moreover, treating others with dignity requires respect for them as subjects and, in particular, acknowledging women's subjectivity. The principle of dignity, McJunkin notes, encompasses the Kantian idea that we are morally required to treat others as ends in themselves and not as mere instruments for one's own ends.[36] Furthermore, he points out that one important practical benefit of introducing dignity as the moral and legal standard behind rape law is that it shifts the focus of legal inquiry from the mental state and conduct of the victim (whether she consented) to the behavior of the defendant.[37]

Although McJunkin's account attempts to justify the broader criminalization of rape by fraud, he does not explain just how broadly the law should reach. Some "fraud in the factum" cases seem appropriate for criminal prosecution, such as the 1962 case involving a doctor who misrepresented his sexual penetration of his patients as a medical procedure,[38] and of the course the now-notorious case of the USA Gymnastics team doctor Larry Nassar, who misrepresented his sexual molestation of numerous child and adolescent female athletes as necessary medical examinations or therapies. Some cases of impersonation may also be appropriate for criminal prosecution. Yet broadly prosecuting "fraud in the inducement" cases could require that courts treat as material any information valued by the complainant, even when their desire for such information is based on ignorance or bigotry, such as information about a sexual partner's racial or ethnic identity, birth sex, HIV status,[39] and so on. Given the potential harm done by prosecuting such cases (encouraging racial bigotry, homophobia, transphobia, and violations of medical privacy, sexual privacy, more incarceration for nonviolent offenses, and so on), the remedy of criminalization would likely do more harm than good. In many cases involving sex obtained by some kind of morally troubling deception, pursuing restitution through the civil courts would be more appropriate.

Sex by fraud can cause many different kinds of harm, including the loss of trust, humiliation, contracting an STI, lasting trauma, infliction of emotional distress, and reputational damage. For example, women who are used by men for sexual conquest are often subject to public ridicule, which compounds the harm done. They may be regarded as "easy" targets or damaged goods. In the logic of male conquest that McJunkin describes, a rape by a fraud victim may come to be viewed as less sexually desirable because overcoming her will is seen as less challenging, and therefore less enhancing of a man's masculinity. In short, sex by fraud enhances a man's social standing while lowering that of his female victims. McJunkin claims a dignity approach can take account of the social meaning of an act in order to illuminate the kind of harm done in sex by fraud cases. However, if what is desired is changing the background conditions that tolerate and reward men who treat women as targets for sexual conquest, then what is needed is not incarceration or other harsh penalties, which would result from broader criminalization, but instead compensation for damages and perhaps mandatory public service and counseling.

Rape as a Cluster Concept

Hilkje Hänel proposes treating rape as a cluster concept or family resemblance term (à la Wittgenstein). On this model, there are no necessary and sufficient criteria that an act must meet in order to count as rape, but rather there is a set of features, and any act that fits some of the features will fall under the concept. There are many features that different cases of rape share, but few cases have every feature. The more features in the cluster a particular act has, the more securely the act fits within the category than at its borders. The cluster concept model recognizes that there are different kinds of contexts in which rapes take place, different methods used, and different outcomes. Hänel argues that the criteria in the cluster should track how rape is experienced by its victims and how it is embedded in heterosexist social practices.

Hänel includes the following ten dimensions in her definition of "rape": sexual activity, violence, means of physical coercion, means of psychological coercion, ability of resistance, lack of consent, capacity to consent, interpersonal relationality, context, as well as gender inequality and heteronormativity.[40] Each dimension in the cluster represents a spectrum, so that an act may share this feature to a higher or lower degree. She explains how this set of features fits different kinds of rape, including stranger and acquaintance rape, and genocidal rape (mass rape used as a biological weapon in a militarized conflict). Hänel aims to make her definition of "rape" ameliorative, in the sense that it can contribute to resisting and changing problematic social practices.[41] For example, new understandings and definitions of the category "woman" can contribute to ending the oppression of people who are trans or nonbinary, and thereby advance the feminist goal of ending the discrimination against those who fail to conform to restrictive gender norms.

The cluster concept model can also be deployed for developing an approach to the selective criminalization of rape by deception, and other kinds of rape. Not every act that falls under the category "rape" should be treated as a criminal offense punishable by incarceration. We need to identify other ways to hold men accountable for acts of rape that have fewer features in the category, or possess these features to a lower degree, such as using the civil court system to sue for damages, or to mandate public service or counseling. A rape by deception, for example, in which the perpetrator exploits a high degree of asymmetry in information and knowledge between him and his victim, such as the case of a doctor who misrepresents a sexual act as a medical procedure, and who abuses the authority and trust placed in him, deserves criminal punishment. Such predators pose a substantial threat to society, and their acts violate a victim's right to bodily self-sovereignty and to be treated with dignity. Their acts have the potential to do a great deal of harm, such as causing life-long trauma and depression, an inability to establish relationships based on trust, low self-esteem, and so on. Such rapes generally cause substantial harm to a person's sense of self, as others have pointed out, even though they do not involve the use or threat of physical violence. If the victim is a minor, or some form of coercion is used, such as a threat to deprive someone of medical services or insurance, then the act shares more features of the category "rape" and deserves more severe punishment.

A rape by deception in which one adult misrepresents their religion to another adult, fearing that such information could be a deal-breaker, involves sexual wrongdoing, but not of the same degree as the doctor's case. If the case does not also involve exploiting an asymmetry in access to information or knowledge about the act or persons involved, the abuse of authority and trust, coercion, or physical force or incapacitation, then it falls more along the border or edges of the concept, and should be addressed accordingly. If the victim can show that this act caused a significant amount of harm (perhaps her religious community now shuns her), then the perpetrator could be required to compensate her through a transfer of money or property to her. He might also be required to do some kind of public service or receive counseling that could lead him to understand better how deceiving someone to obtain sex can harm them. But the victim should not be able to use the law to persecute someone who is a member of a minority or stigmatized religious group.

These examples serve to illustrate the cluster of features that should be included in a definition of "rape" and that acts falling into the category should share, more or less. Many of these will overlap with those Hänel includes in her definition: sex obtained by coercion, violence, or deception; sex that exploits asymmetries of power and/or information, or economic and other forms of dependency; sex that involves a victim who is incompetent due to age or mental condition; sex that carries the risk of a substantial harm to one party who is not warned of the risk; sex in which one party covertly intends to degrade another person; sex that aims to impose an unwanted pregnancy in order to genetically or socially "contaminate" a group of people (genocidal rape); and so on. This is not an exhaustive list, but it captures many of the core features of acts that should be included in an "ameliorative definition" of "rape." By ameliorative here, I mean a definition that could undergird legal rules for prosecuting rape and meting out appropriate sanctions, that would hold men (the vast majority of perpetrators) more accountable for imposing sex on others, and that would alter the background conditions that reward men for aggressive sexual behaviors.

Lastly, there need not be a single moral standard for grounding rape law; instead, the acts we call "rape" can violate a number of basic moral rights, such as a right to bodily self-governance, sexual self-determination, and to be treated with dignity. Some of these rights may be more salient in some cases than in others, but all of them should guide societal responses to acts of rape. Instead of grounding rape on one of these moral concepts in order to draw some lines around the category, a better approach is to treat rape as cluster concept with somewhat fuzzy boundaries. Those cases that fall more along the edges of the concept need not be handled with the full force of our criminal law system, but the perpetrators still deserve to be held accountable and to provide restitution for injuries. The cluster concept model provides more flexibility for prosecuting rape charges, and devising appropriate punishments, including those that address the causes and not only the effects of this sexual practice.

Recently the state of New York created "Sex Offense Courts" to address many types of sexual offenses, similar to the drug court model.[42] Sex Offense courts are part of a larger system of "Problem-solving courts" that "look to the underlying issues that bring people into the court system, and employ innovative approaches to address those issues."[43] These approaches include having judges and other court personnel who have expertise and experience dealing with a particular category of offenses, as well as more tools at their disposal for imposing punishments on defendants and providing services to victims. This model holds much promise for addressing rape, given the variety of circumstances, methods, and consequences involved in such acts. Moreover, rape law reformers aim to change the underlying social conditions that condone male sexual aggression, as well as hold individual men more accountable for such behavior. A court system designed to address the underlying social problems "with innovative approaches" is likely to be more helpful with this project than one designed only to react punitively to the effects of various social problems. South Africa is similarly experimenting with sexual offenses courts in order to achieve greater justice for victims as well as address high levels of sexual assault in various communities.[44]

Conclusion

The debate over the narrow criminalization of rape by fraud or deception cases opens questions about the principles that underlie rape law, and how they are, or should be, put in practice. I have argued that we can address many apparent inconsistencies by treating rape as a cluster concept, with a variety of features that acts in the category may possess more or less. My account does not force a choice between competing moral standards and their operationalization in order to define "rape." It also recognizes that cases falling along the edges of the category (i.e., that possess fewer features and to a lesser degree) can and should be dealt with by means other than criminalization and incarceration, such as with civil lawsuits, so that our laws don't do more harm than good. In the U.S., rape laws have been used to persecute Black men, and some of the first rape by fraud cases show the potential of expanded definitions of "rape" to be used to persecute socially stigmatized ethnic and gender minorities. Cases of sex by deception can involve the infliction of extensive and lasting injuries, substantial asymmetries in information and knowledge, and abuses of authority and trust, and those that have these features to a strong degree should be treated differently than those that lack them. In this way, sex by fraud is similar to sex by coercion or sex by the use or threat of violence, and therefore such acts should not be ignored by our criminal justice system, but they should be selectively criminalized.[45]

Notes

1. Ben McJunkin defines 'rape by fraud' as "those instances of deception that are legally sufficient to invalidate a token of consent" (p. 2, note #3). Under U.S. criminal law, the conditions or circumstances that are legally sufficient to vitiate consent vary from state to state. Ben A. McJunkin, "Deconstructing Rape By Fraud," *Columbia Journal of Gender and Law* 28, no. 1 (2014): 1–47.
2. In the not too distant past, American courts understood rape as sex by physical force. Feminist scholars and activists, in the 1970s and later, led efforts to reform this understanding of rape, as it did not cover cases of rape involving coercion (such as a threat of exposure, public humiliation, loss of income), incapacitation, manipulation, the abuse of power, date rape, and so on. What is common to these kinds of situations, as well as rape by physical force, is the lack of genuine consent.
3. "Coerced sex" generally refers to acts that do not involve physical force or violence, but typically involve people in unequal social positions, such as sexual predation by someone in a position of authority or power, or quid pro quo demands for sex, such as making sex a condition for a job benefit, financial or emotional support, or some desperately needed help such as immigration services.
4. Jed Rubenfeld imagines such double-rape cases in order to test our intuitions regarding sex by deception (1414–15), in "The Riddle of Rape-by-Deception, and the Myth of Sexual Autonomy," *The Yale Law Journal* 122 (2013): 1372–1443.
5. Rape by deception laws have been used to prosecute people who are transgender, and could possibly be used to prosecute people with various intersex conditions. See Sophie Wilkinson, "The Troubling Case of a UK woman Convicted of 'Gender Fraud'," *Vice*, December 22, 2017: https://www.vice.com/en_ca/article/43qvz9/the-troubling-case-of-a-uk-woman-convicted-of-gender-fraud.
6. For more imagined and actual cases, see Caroline Lowbridge, "Sex, lies and legal consent: can deceit turn sex into rape?" *BBC News*, September 26, 2019: https://www.bbc.com/news/uk-england-49127545.
7. Paraszczuk, Joanna, "Court cuts Arab-Israeli rape-by-deception sentence," *The Jerusalem Post*, January 27, 2012: https://www.jpost.com/National-News/Court-cuts-Arab-Israeli-rape-by-deception-sentence.
8. McJunkin, p. 9.
9. *People v. Minkowski* (1962): https://law.justia.com/cases/california/court-of-appeal/2d/204/832.html.
10. *Boro v. Superior Court* (1985): https://law.justia.com/cases/california/court-of-appeal/3d/163/1224.html.
11. McJunkin, p. 13. According to McJunkin, one rationale that some legal commentators offer for treating such cases differently is that, in the husband impostor scenario, the woman believes she is having non-adulterous sex when in fact she is allegedly committing adultery. That is, this case involves fraud "in the factum." However, given that adultery has been decriminalized in most states, why should deception regarding the adulterous nature of the act be legally sufficient to invalidate consent, but not deception regarding other aspects of the act? For example, "sex with Brad Pitt" is different than "sex with a Brad Pitt

imposter." Abby Ellin reports that some states have prosecuted men for impersonating a woman's boyfriend, and an old high school friend. "Is Sex by Deception a Form of Rape?" *The New York Times*, April 23, 2019: https://www.nytimes.com/2019/04/23/well/mind/is-sex-by-deception-a-form-of-rape.html.

12. Jeffrie Murphie argues that if the inducement provides evidence that the victim "does not value sexuality in the way characteristic of the norms we seek to protect," such as "a woman trading sex for the promise of a mink coat," then we need not treat such cases as rape when the inducement is fraudulent. Quoted in McJunkin, p. 19.
13. Some recent psychological studies show that many people believe an individual's consent is valid even when they have been deceived by the person asking for consent. See: Sommers, Roseanna, "You Were Duped Into Saying Yes. Is That Still Consent?" *The New York Times*, March 5, 2021: https://www.nytimes.com/2021/03/05/opinion/sexual-consent.html.
14. See, for example, Tom Dougherty, "Sex, Lies, and Consent," *Ethics* 123 (July 2013): 717–744; Danielle Bromwich and Joseph Millum, "Lies, Control, and Consent: A Response to Dougherty and Manson," *Ethics* 128 (January 2018): 446–461; Jennifer Matey, "Sexual Consent and Lying About One's Self," *Philosophy and Phenomenological Research* (2019, early view online); Robert Sparrow and Lauren Karas, "Teledildonics and rape by deception," *Law, Innovation and Technology* 12, no. 1 (2020): 175–204.
15. Jed Rubenfeld, p. 1378.
16. Rubenfeld, p. 1380.
17. Rubenfeld, pp. 1425–26.
18. Rubenfeld, p. 1427.
19. "Stanford sexual assault case: victim impact statement in full," *The Guardian*, June 6, 2016: https://www.theguardian.com/us-news/2016/jun/06/stanford-sexual-assault-case-victim-impact-statement-in-full.
20. Slavery and torture also involve abduction or capture, and imprisonment, which even without slavery and torture violate a person's right to self-possession.
21. Rubenfeld, pp. 1430–33. One can add to this list prison labor and human trafficking. Rubenfeld includes compulsory pregnancy as a form of forced servitude, and argues that access to abortion protects the basic right to bodily self-governance, not merely reproductive autonomy or the right to choose when and with whom to have children.
22. Rubenfeld, p. 1432.
23. Rubenfeld, p. 1435.
24. Rubenfeld recognizes that it can be difficult to define "force," and stipulates that "sex is rape whenever exacted through the kind of force that turns labor into slavery: roughly speaking, physical incapacitation, whether through restraint or imprisonment, or serious physical assault (or the threat of either). Absent such force, sex under conditions of power imbalance, material want, or psychological pressure isn't rape." p. 1436.
25. For an interesting discussion of how rape by deception laws can undermine the right to privacy, see Florence Ashley, "Genderfucking Non-Disclosure: Sexual Fraud, Transgender Bodies, and Messy Identities," *Dalhousie Law Journal* 41, no.2 (2018): 339–77.

26. Luis E. Chiesa, "Solving the Riddle of Rape-by-Deception," *Yale Law and Policy Review* 35 (2017): 410.
27. Rubenfeld writes "Coercion is objectionable because a coerced 'yes' does not reflect a valid or genuine consent. But the same is true of a deceived 'yes.' An anti-coercion principle is attractive because coerced sex is unconsented-to sex. But if unconsented-to sex is rape law's target, then *deceptive sex ought to be punished as well*" (italics in original). Rubenfeld, p. 1412.
28. Chiesa, pp. 428–9.
29. Chiesa, pp. 429–30.
30. Chiesa writes: "There is something more pernicious and more fundamental about sex in a patriarchal society than its simply being misinformed. A male-dominated social order does not need misinformation in order to make women submit to sex. Given the coercion inherent in this kind of social arrangement, women regularly give in to men's aggressive sexual proposals even when they are informed of all the relevant facts." Chiesa, p. 438.
31. Chiesa, pp. 442–43.
32. Chiesa, pp. 450–51.
33. McJunkin, pp. 21–33.
34. McJunkin, pp. 39–40.
35. McJunkin, p. 40.
36. McJunkin, pp. 42–3.
37. McJunkin, pp. 43–4.
38. People v. Minkowski, https://law.justia.com/cases/california/court-of-appeal/2d/204/832.html.
39. Regarding HIV or any incurable STI, what should matter is whether someone knows they are infectious (i.e., positive and not under treatment), or hides from a sexual partner that they engage in risky behaviors that put them at high risk for being infectious (such as needle sharing, being sexually active with multiple partners and not using condoms, not taking their medications, etc.). People who have tested positive but are under treatment do not put their sexual partners at risk for contracting HIV, and should not be treated as criminals. Treating them as criminals amplifies the stigma that surrounds this disease. One could argue that criminal prosecution is justified for nondisclosure of information important for assessing the risk of transmission of an illness only in cases involving disregard for the risk of harm, or the intent to harm on the part of the defendant. Undergoing testing and treatment for HIV shows that a person is not negligent in this way, given that current methods of treatment make the risk of transmission highly improbable.
40. Hilkje Charlotte Hänel, *What is Rape?: Social Theory and Conceptual Analysis* (Bielefeld: Verlag, 2018): p. 188.
41. On the idea of ameliorative conceptual analysis and definition, see Sally Haslanger, "What Are We Talking About? The Semantics and Politics of Social Kinds," *Hypatia* 20, no. 4 (2005): 10–26.
42. http://ww2.nycourts.gov/courts/problem_solving/so/mission_goals.shtml.
43. http://ww2.nycourts.ov/COURTS/problem_solving/index.shtml.
44. Jen Thorpe, "Part 2: The Roll Out of the First Sexual Offences Courts," *Rape Crisis*: https://rapecrisis.org.za/part-2-the-roll-out-of-the-first-sexual-offences-courts/; "Part 4: The Future of Sexual Offences Courts in

South Africa," *Rape Crisis*: https://rapecrisis.org.za/part-4-the-future-of-sexual-offences-courts-in-south-africa/; Luvuyo Mdeni, "South Africa's sexual offenses courts," *SABC News*, April 21, 2020: https://www.sabcnews.com/sabcnews/map-south-africas-sexual-offences-courts/.

45. *Alternative accounts of 'rape'*:

Right that laws protect	*Self-possession*	*Autonomy*	*Dignity*	*Cluster of rights*
Criterion for assessing rights violation	Use or threat of force	Nonconsensual	Objectification	Multiple criteria
Sanctions	Selective criminalization	Selective criminalization	Broad criminalization	Selective criminalization, along with increased use of civil court system

PART IV

Sex, Discrimination, and Exclusion

CHAPTER 24

Sexual Racism

Sonu Bedi

We often discuss racism that takes place in the political or public sphere. This chapter draws on the contribution of feminist political theory that the personal is the political to enlarge the boundary of racial justice and discuss sexual racism, racism that occurs in our personal, intimate lives.

This chapter is divided into three parts. First, it shows how feminist political theory enlarges the boundary of justice by recognizing that justice is not simply about what happens in our public or professional lives but also what happens in our personal, intimate ones. Second, it enlarges the boundary of racial justice by discussing two kinds of sexual racism: racial discrimination and racial stereotyping. Third, it discusses the way in which sexism and racism in private inform an intersectional analysis. This chapter draws from and builds on my previous work on sexual racism.[1]

The Personal Is the Political

A central contribution of feminist political theory is the idea that the personal is the political. This idea sees the family, the intimate sphere, as within the boundary of justice. Feminist political theory shows that justice is not simply about what happens in the public or political sphere but also what happens in the private one. As a matter of justice, we ought to care what happens in our personal, intimate lives not just in our political or professional ones. Susan Okin's *Justice, Gender and the Family* (1989) is one of the primary texts that makes this contribution; so this essay will draw from it.

S. Bedi (✉)
Dartmouth College, Hanover, NH, USA
e-mail: Sonu.S.Bedi@dartmouth.edu

D. Boonin (ed.), *The Palgrave Handbook of Sexual Ethics*,
https://doi.org/10.1007/978-3-030-87786-6_24

At the outset, Okin makes clear that even if there was:

> a high and uniform standard of public education and the provision of equal social services—including health care, employment training, job opportunities, drug rehabilitation, and decent housing—for all who need them...we would still not attain equal opportunity for all. This is because what has not been recognized as an equal opportunity problem, except in feminist literature and circles, is the disparity within the family, the fact that its gender structure is itself a major obstacle to equality of opportunity. This is very important in itself, since one of the factors with most influence on our opportunities in life is the social significance attributed to our sex.[2]

The boundary of justice does not stop simply at the public or professional sphere. It's important, as Okin writes, to enlarge this boundary and include the family and the intimate sphere within its ambit. Doing so recognizes the way in which gender often structures our personal, intimate lives, including "the personal sphere of sexuality, of housework, of child care and family life."[3] This sphere "*is* [also] political."[4]

As Okin goes on to say, feminists:

> of different political leanings and in a variety of academic disciplines have revealed and analyzed the multiple interconnections between women's domestic roles and their inequality and segregation in the workplace, and between their socialization in gendered families and the psychological aspects of their oppression.[5]

By focusing in particular on inequality and stereotyping that occur in the intimate sphere, feminist theory enlarges the boundary of justice. Just as women can experience sexism in their public or professional lives, so too can they experience it in their personal, intimate lives.

The personal is the political argument seeks to:

> expose and elucidate the problems of a dichotomy that has been accepted as fundamental to liberal thought so far: that between the "public" world of political life and the marketplace and the "private" domestic world of family life and personal relations.[6]

Central to this insight is the idea that "what happens in domestic and personal life is not immune from the dynamic of *power*, which has typically been seen as the distinguishing feature of the political."[7] This kind of power differential between men and women can lead to inequality, where, for instance, women may find it more difficult to take on the role of breadwinner and easier to take on the role of caregiver. The fact that this takes place, as Okin says, "within the family"[8] does not make a difference. This is why "Carole Pateman has said, 'the dichotomy between the public and the private... is, ultimately, what the feminist movement is about.'"[9]

This is a core contribution of feminist political theory. It rejects the idea that "circumstances of family life are such that justice is not an appropriate standard to apply to them."[10] Okin realizes that:

> The morality that often prevails in communities or associations that are governed in large part by affection, generosity, or other virtues morally superior to justice is a form of supererogation; individuals' narrowly construed interests give way to their concern for common ends or the ends of others they care about a great deal.[11]

She goes on to say that:

> Nevertheless, it is essential that such higher moral sentiments and actions, within the family as well as in society at large, be underwritten by a foundation of justice. Justice is needed as the primary, meaning most fundamental, moral virtue even in social groupings in which aims are largely common and affection frequently prevails.[12]

Feminist political theory says that our personal and intimate relations are within, not outside, the boundary of justice.

> Thus feminists have turned their attention to the politics of what had previously been regarded—and, as I have shown, still is seen by most political theorists—as paradigmatically nonpolitical. That the personal sphere of sexuality, of housework, of child care and family life is political became the under-pinning of most feminist thought.... We have strongly and persistently challenged the long-standing underlying assumption of almost all political theories: that the sphere of family and personal life is so separate and distinct from the rest of social life that such theories can justifiably assume but ignore it.[13]

By discussing sexism that happens in our private lives, not just in our political and professional ones, feminist political theory enlarges the boundary of justice. It reveals that gender inequality and stereotype can characterize the intimate, familial sphere. Gendered norms operate in our private, intimate lives as well.

Enlarging the Boundary of Racial Justice

Sexual racism is racism that happens in our personal, intimate lives. This also shows how the personal is the political. This enlarges the boundary of racial justice to include the intimate or personal sphere. In her book *Inequalities of Love*, sociologist Averil Y. Clarke uses the personal narratives of college-educated black women to show that in the realm of love, romance and family they are "disadvantage[d] relative to other women at [their] age and station in life."[14] In fact, she concludes that "degreed black women's experience of deprivation in romantic partnership and marriage develops and intensifies over time."[15] Christian Rudder, the creator of OkCupid, a popular free online dating site founded in 2004, published statistical results that reveal this disparity. Rudder stated that "[W]hen you're looking at how two American strangers behave in a romantic context, race is the ultimate confounding factor."[16] His results show that white users are more likely to be messaged or responded to than their nonwhite counterparts. They show, in particular, that black women and Asian men are the least likely to be messaged or responded to.[17] For racial minorities, the personal, intimate sphere can also be a political one. Recognizing this enlarges the boundary of racial justice. This essay discusses two kinds of sexual racism: racial discrimination and racial stereotyping.

Racial Discrimination

Racial minorities are often discriminated against simply for not being white. Consider that OkCupid, one of the most popular dating platforms, has over 50 million registered users since its launch.[18] In 2009, Christian Rudder published results titled "How Race Affects The Messages You Get."[19] OkCupid has a distinctive percentage matching system. The user answers a list of questions (the more questions answered the more "accurate" the match percentage) and then can search for compatible matches. The OkCupid match/compatibility algorithm does not consider appearance including the user's racial identification. It is self-described as considering the users' "inner selves."[20]Consequently, the match score of individuals across different racial groups were roughly even, in the same way that the match scores of individuals across different zodiac signs were roughly even.[21]

But although the data found that reply rates did not vary among those with different zodiac signs, such rates vary significantly based on whether the user is white.[22] For instance, OkCupid concluded that among those who identify as straight, white men get a greater percentage of responses than any other racial group. "White women prefer white men to the exclusion of everyone else—and Asian and Hispanic women prefer them even more exclusively." [23] So women of all racial groups prefer men who are white, findings that persisted in a more recent analysis of the data from OkCupid.[24] In fact, this disparity in racial matching was similar for gay users as well.[25] This discrimination makes it harder for many who are not white to find intimacy. Racial minorities have fewer opportunities to find intimacy than similarly situated white people.

We need not look just to OkCupid. Recent empirical work also reveals that individuals discriminate against racial minorities, often preferring intimacy with white people. This work primarily uses online dating profiles and response rates as well as more conventional survey data to reveal the relationship between race and dating preferences. With regards to heterosexuals, these studies conclude that: females prefer whites over nonwhites;[26] white men and women are more likely to seek out potential dates with whites rather than with blacks;[27] white women are less willing to date nonwhites than white men;[28] college students are more likely to exclude blacks as potential dates;[29] whites are least likely to date outside of their race and Asians and Latinos are least likely to date blacks;[30] blacks were 10 times more likely to contact whites than whites were to contact blacks;[31] and controlling for appearance, Arabs suffered an ethnic penalty in one of Sweden's largest online dating sites.[32]

These studies also reveal that gay men discriminate against racial minorities:[33] Asian men are the least desired in a sample of online profiles of urban males seeking sex with men;[34] interviews reveal sexual marginalization of Asian men in the gay community;[35] an ethnic hierarchy exists with whites and Latinos as the most desirable racial groups in the online gay male community;[36] and gay men of all races prefer to date whites over nonwhites.[37]

The data show that just as racial minorities have difficulty in finding employment,[38] housing,[39] and credit,[40] they also have difficulty in finding intimacy, because they are not white. Racial discrimination occurs not just in our political and professional lives but also in our personal ones. This discrimination is not simply a proxy for other factors such as socioeconomic status or ideology. If that were the case, individuals would be discriminating not on the basis of race but on the basis of these other characteristics. The data and studies reject that conclusion, showing that racial minorities are often discriminated against simply and only on the basis of their race. Again, OkCupid's data on race preferences controlled for match percentage. That is, it considered the response rate for pairs of individuals that had a similar match score—the only difference being their race. The match score considers a wide variety of compatibility questions relating to education, income, expectations, and likes/dislikes. Even when compatibility is the same across racial groups, individuals still discriminate against those who are not white.

Or consider that the studies above conclude, in part, that: white women prefer to date Latinos and blacks over Asians and East Indians, even though the Latinos and blacks have lower than average economic statuses;[41] college students exclude blacks as potential dates based just on physical attractiveness;[42] randomizing occupation on one of the largest Swedish online dating sites still generates an ethnic penalty for Arabs;[43] and education does not mediate the observed racial preference among white Internet daters.[44] This suggests that race is not simply a proxy for compatibility in terms of socioeconomic status or education. The data suggest that sexual racism is exactly what is going on.

We can see this on gay dating sites as well.[45] Consider racism on Grinder, perhaps the most popular platform for gay men to find intimacy. It is not surprising to see profiles on Grinder that read: "No Asians" or "Whites only."[46] These monikers express unambiguous instances of intentional racial discrimination in our private, intimate lives. These are not isolated instances of discrimination. Studies show that among gay men, racial minorities have a more difficult time finding intimacy as compared to those who are white.[47] In fact, in some of these studies researchers discovered that racial discrimination persists even using the same photo, where the study altered only the name to suggest a different race or explicitly stated that the individual was of another race.[48]

This suggests that individuals discriminate against a potential romantic or intimate partner because of not in spite of their race. Just as individuals discriminate in this way, the state or government has also discriminated because of not simply in spite of someone's race.

Racial Stereotyping

Racial stereotyping also characterizes our personal lives. Here, individuals may prefer those who are not white but only on the condition that the person fulfill a certain sexual or romantic stereotype. Robin Zheng describes this as a racial fetish. She defines a fetish as a "person's exclusive or near-exclusive preference for sexual intimacy with others belonging to a specific racial *out-group*."[49] Zheng's article titled "Why Yellow Fever Isn't Flattering" focuses on the stereotype that Asian and Asian-American women are often seen as the "docile, domestic 'Lotus Blossom' or the seductive, treacherous 'Dragon Lady.'"[50] Or what Sheridan Prasso calls the "Asian Mystique."[51] These racial stereotypes of Asian women often view them as submissive and sexual. According to Zheng, individuals may prefer Asian women over women of other races as long as they fulfill this kind of stereotype.

Zheng argues that this kind of fetish limits the opportunity of individuals to enjoy intimacy. She draws on testimony from Asian and Asian-American women, whose voices are often at the "margins of contemporary academic philosophy but whose unique racial vantage point complicates a racial discourse overwhelmingly centered on the Black/White binary."[52] Zheng argues that this kind of stereotype poses a "psychological burden" on these women, constituting a "form of racial disadvantage."[53] Zheng argues that even

> those Asian/American women who ultimately resolve doubts about yellow fever by being unbothered by it or by embracing it or by undergoing the relational work required to establish definitively that their partners are not homogenizing and otherizing them were still saddled, in the first place, with a problem that others are not forced to deal with; very few Asian/American women, if any, can be wholly oblivious to or untouched by it. And it is worth noting that women reporting more positive experiences of yellow fever are often quickly disillusioned, conflicted, or accept the stereotypes themselves.[54]

This imposes a disadvantage on Asian women, a difficulty that does not exist for white women. This can make it difficult for these women to find intimacy that is not subject to this kind of caricature or stereotype. This is why according to Zheng some Asian American women will simply "embrace" the stereotype. Whatever they decide to do, this burden arises only because of their race. Zheng argues that the "racial depersonalization inherent in yellow fever threatens Asian/American women with doubts as to whether they are or can be loved as individuals rather than as objects in a category."[55] This makes it more difficult for Asian women to access intimacy fully, because they must navigate this racial stereotype in their intimate lives.

This kind of racial stereotype also affects Asian males.[56] It deems Asian men effeminate and asexual (what David Eng calls a kind of "racial castration"[57]), furthering discrimination against them simply on the basis of their race. According to sociologist Chong-suk Han, this is why gay Asian men must sometimes also assume this "submissive" intimate role.[58] Doing so may be the only way to find intimacy given the powerful force of this kind of racial stereotype against Asians.

Asians are not the only racial group that faces racial stereotypes. Black men also face difficulties in finding romantic or sexual relationships. Here too racial stereotypes can impose obstacles to intimacy. Black men and, in particular, black gay men must contend with the stereotype that they are aggressive, sexually dominant, and physically well endowed.[59] Drawing our attention to this kind of stereotyping, Nathaniel Coleman argues that:

> "The African American male who isn't sexually well endowed or isn't as sexually aggressive as both homosexual and heterosexual society say he must be, may feel inferior."[60] Such a black MSM [men who have sex with men] is likely to consider himself to have lessened bartering power in the market-place for sex with MSM. He is likely to have lessened confidence in his ability to attract sexual approbation from, and to achieve sexual activity with other MSM. He is likely to have low sexual self-confidence. Young black MSM, fraught with the insecurities of adolescence and of "coming out of the closet," are especially likely to find themselves with low self-confidence.[61]

Gay black men, especially those who are just "coming out," will face this difficulty in finding intimacy, difficulties that white gay men will not face. In fact, Coleman suggests that because of sexual racism, black men are often forced to take on the sexually aggressive role. So, like some Asian men and women, they too may have to "embrace" the racial stereotype in order to find intimacy.

Intersectionality

The fact that racism occurs in our sexual lives is another way that the personal is the political. It shows that both sexism and racism can occur in our private, intimate lives. The core contribution of intersectionality is that these often occur simultaneously and alongside other kinds of social disadvantage. Kimberlé Crenshaw, whose work is seminal in the field of critical race theory, defines intersectionality to include the experience of double discrimination, where racial minorities experience racial discrimination in the context of other kinds of social disadvantage. She draws on the way in which black women can experience this kind of double discrimination in finding employment. As Crenshaw points out, they often experience racism and sexism.

> Black women can experience discrimination in ways that are both similar to and different from those experienced by white women and Black men. Black women sometimes experience discrimination in ways similar to white women's experi-

> ences; sometimes they share very similar experiences with Black men. Yet often they experience double-discrimination-the combined effects of practices which discriminate on the basis of race, and on the basis of sex. And sometimes, they experience discrimination as Black women-not the sum of race and sex discrimination, but as Black women.[62]

This happens in our private lives as well. As the examples of sexual racism make clear, racial minorities can experience this "double discrimination" based on their gender as well as their sexual orientation. Asian women experience discrimination on the basis of their race and their gender. Gay black men do so on the basis of their race, gender, and sexual orientation. These are instances of intersectionality.

Racism in our intimate lives also means that we often experience racial bias along with a bias against how tall we are or our body shape, our socioeconomic status, our professional status, educational pedigree, or family history. These too are instances of intersectionality, showing the way in which race can intersect with any number of social categories or hierarchies. This reveals how racism intersects with class-ism and look-ism where racial minorities face "double-discrimination" in our intimate lives based on our race and, at the same time, how we look or what we do for work.

Women can also experience "double discrimination" based on their gender *and* their socioeconomic status. As Okin writes:

> Those who discuss the family without paying attention to the inequalities between the sexes are blind to the fact that the gendered family radically limits the equality of opportunity of women and girls of all classes—as well as that of poor and working-class children of both sexes. Nor do they see that the vulnerability of women that results from the patriarchal structure and practices of the family *exacerbates* the problem that the inequality of families poses for children's equality of opportunity.[63]

According to Okin, by failing to acknowledge sexism in our private, intimate lives, we fail to see how this works as a "double discrimination" on those women who are poor or socially disadvantaged. They will be even more vulnerable to sexism in our private, intimate lives.

White males may face discrimination based on their socioeconomic status or their looks but generally do not face discrimination on the basis of their race or gender in their private lives. This is why Iris Marion Young notes, referencing intersectionality: "White males… in so far as they escape group marking, can be individuals."[64] White males are often seen as individuals in our private lives, escaping the racial discrimination and stereotype that face racial minorities or the gendered norms that face women.

In this way, racial minorities and women often experience their intimate and family lives as political ones. This informs Catharine MacKinnon's famous claim that:

> The private is the public for those for whom the personal is the political. In this sense, there is no private, either normatively or empirically.[65]

MacKinnon may only have had sexism in mind but this insight applies to racism as well. Racial minorities and women face racism and sexism in their personal lives. We are hard pressed to find a "private" space where racism or sexism does not occur. If racism and sexism happen even in our personal, intimate lives, the private sphere is also a political one. This challenges the idea that we can normatively or even empirically distinguish the public from the private in some categorical or definitive way. Sexual racism enlarges the boundary of racial justice by recognizing that the personal is the political.

Once we enlarge this boundary, we face the question of what society can do about it. Russell Robinson suggests "[l]awmakers might consider regulating web site design decisions that produce, exacerbate, or facilitate racial preferences."[66] So, one option is to consider banning dating websites from permitting users to screen and filter on the basis of race. After all, most major dating websites and apps, including OkCupid and Match.com, permit users both to identify their race and screen potential dates on the basis of race. These sites do not just ask users for their name or to upload a photograph, they ask users to disclose their race.[67] In fact, in 2020 Grindr removed the filter that allowed its users to filter on the basis of race.[68] Such measures enlarge the boundary of racial justice to include our sexual lives by recognizing that personal is indeed the political.

Notes

1. Bedi 2015, 2019a, 2019b
2. Okin 1989: 16
3. Okin 1989: 125
4. Okin 1989: 125
5. Okin 1989: 125
6. Okin 1989: 111
7. Okin 1989: 128
8. Okin 1989: 128
9. Okin 1989: 111, quoting Pateman 1983
10. Okin 1989:
11. Okin 1989: 29
12. Okin 1989: 29
13. Okin 1989: 125-126
14. Clarke 2011: 15
15. Clarke 2011: 117
16. Rudder 2014: 99
17. Rudder 2014: 102-113
18. https://mashable.com/review/okcupid-dating-review/
19. http://blog.okcupid.com/index.php/your-race-affects-whether-people-write-you-back/ (October 5, 2009).

20. Rudder 2014: 101
21. Rudder 2014: 101
22. http://blog.okcupid.com/index.php/your-race-affects-whether-people-write-you-back/ (October 5, 2009).
23. http://blog.okcupid.com/index.php/your-race-affects-whether-people-write-you-back/ (October 5, 2009).
24. Rudder 2014: 102-109
25. Rudder 2014: 243; *see also* Robinson and Frost 2018
26. Tsunokai et al. 2013
27. Herman and Campbell 2012; Lin and Lundquist 2013
28. Hwang 2013
29. Bany et al. 2014; McClintock 2010
30. Robnett and Feliciano 2011
31. Mendelsohn et al. 2014
32. Jakobsson and Lindholm 2014
33. *See generally* Han and Smith 2019
34. White et al. 2014
35. Han 2008b
36. Brown 2003; Robinson 2007
37. Tsunokai et al. 2013. There is research that shows that white lesbians seeking partners are more likely to prefer women without children, which disparately impacts women of color. (Rafalow and Kizer 2018).
38. *See, e.g.* Bertrand and Mullainathan 2004
39. *See, e.g.* Hanson and Hawley 2011; Ewens et al. 2014
40. *See, e.g.* Kau et al. 2012; Ladd 1998
41. Feliciano et al. 2009
42. Bany et al. 2014
43. Jakobsson and Lindholm 2014
44. Lin and Lundquist 2013
45. The analysis focuses on gay men rather than lesbians, because the studies usually focus on gay men. That being said, there is research that shows that white lesbians seeking partners are more likely to prefer women without children, which disparately impacts women of color. (Rafalow and Kizer 2018).
46. https://www.advocate.com/commentary/2018/7/19/no-asians-not-preference-its-racism
47. *See* Brown 2003; Robinson 2007; Robinson and Frost 2018; White et al. 2014.
48. *See, e.g.,* Jakobsson and Lindholm 2014; Robinson 2007
49. Zheng 2016: 401
50. Zheng 2016: 405
51. Prasso 2005
52. Zheng 2016: 401
53. Zheng 2016: 407
54. Zheng 2016: 409
55. Zheng 2016: 408
56. Robnett and Feliciano 2011
57. Eng 2001
58. Han 2008a
59. *See, e.g.,* Coleman 2011; Smith 2014
60. Coleman 2011: 13, quoting Brown 2003: 24

61. Coleman 2011: 13
62. Crenshaw 1989: 149
63. Okin 1989: 134-135
64. Young 1990: 59
65. MacKinnon 1987: 100
66. Robinson 2007: 2794
67. *See* Emens 2009: 1322
68. https://www.inputmag.com/culture/grindr-said-it-wouldnt-allow-filtering-by-ethnicity-but-still-does

References

Bany, J.A., B. Robnett, and C. Feliciano. 2014. Gendered Black Exclusion: The Persistence of Racial Stereotypes Among Daters. *Race and Social Problems* 6 (3): 201–213.

Bedi, Sonu. 2015. Sexual Racism: Intimacy as a Matter of Justice. *Journal of Politics* 77 (4): 1.

———. 2019a. *Private Racism*. New York: Cambridge University Press.

———. 2019b. "Online Dating Sites as Public Accommodations: Facilitating Racial Discrimination," in *Free Speech in the Digital Age* (Oxford University Press) (eds., Susan Brison and Katharine Gelber)

Bertrand, Marianne, and Sendhil Mullainathan. 2004. Are Emily and Greg More Employable than Lakisha and Jamal? A Field Experiment on Labor Market Discrimination. *The American Economic Review* 94 (4): 991–1013.

Brown, William I.I.I. 2003. Discrimination dot com: Racially Biased Interaction in the Online Gay Male Community. *The McNair Scholars Journal of the University of California, Davis* 6: 22–29.

Clarke, Averil Y. 2011. *Inequalities of Love: College-educated Black Women and the Barriers to Romance and Family*. New York: Duke University Press.

Coleman, Nathaniel. 2011. What? What? In the (black) butt. *Newsletter on philosophy and lesbian, gay, bisexual, and transgender issues* 11 (1): 1.

Crenshaw, Kimberlé. 1989. Demarginalizing the Intersection of Race and Sex: A Black Feminist Critique of Antidiscrimination Doctrine, Feminist Theory and Antiracist Politics. *University of Chicago Legal Forum* 1 (8): 1.

Emens, Elizabeth F. 2009. Intimate Discrimination: The State's Role in the Accidents of Sex and Love. *Harvard Law Review* 122: 1307.

Eng, David L. 2001. *Racial Castration: Managing Masculinity in Asian America*. Durham, NC: Duke University Press.

Ewens, Michael, Bryan Tomlin, and Liang Choon Wang. 2014. Statistical Discrimination or Prejudice? A Large Sample Field Experiment. *Review of Economics and Statistics* 96 (1): 119–134.

Feliciano, Cynthia, Belinda Robnett, and Golnaz Komaie. 2009. Gendered Racial Exclusion among White Internet Daters. *Social Science Research* 38 (1): 39–54.

Han, Chong-suk. 2008a. A qualitative exploration of the relationship between racism and unsafe sex among Asian Pacific Islander gay men. *Archives of Sexual Behavior* 37 (5): 827–837.

———. 2008b. No fats, femmes, or Asians: The utility of critical race theory in examining the role of gay stock stories in the marginalization of gay Asian men. *Contemporary Justice Review* 11 (1): 11–22.

Han, C. Winter, and Jesús Gregorio Smith, eds. 2019. *Home and Community for Queer Men of Color: The Intersection of Race and Sexuality*. Lanham: Lexington Books.

Hanson, Andrew, and Zackary Hawley. 2011. Do landlords discriminate in the rental housing market? Evidence from an internet field experiment in US cities. *Journal of Urban Economics* 70 (2): 99–114.

Herman, Melissa R., and Mary E. Campbell. 2012. I wouldn't, but you can: Attitudes toward interracial relationships. *Social Science Research* 41: 343–358.

Hwang, Wei-Chin. 2013. Who are People Willing to Date? Ethnic and Gender Patterns in Online Dating. *Race and Social Problems* 5 (1): 28–40.

Jakobsson, Niklas, and Henrik Lindholm. 2014. Ethnic Preferences in Internet Dating: A Field Experiment. *Marriage & Family Review* 50 (4): 307–317.

Kau, James B., Donald C. Keenan, and Henry J. Munneke. 2012. Racial discrimination and mortgage lending. *The Journal of Real Estate Finance and Economics* 45 (2): 289–304.

Ladd, Helen F. 1998. Evidence on discrimination in mortgage lending. *The Journal of Economic Perspectives* 12 (2): 41–62.

Lin, Ken-Hou, and Jennifer Lundquist. 2013. Mate Selection in Cyberspace: The Intersection of Race, Gender, and Education. *American Journal of Sociology* 119 (1): 183–215.

MacKinnon, Catharine A. 1987. *Feminism Unmodified: Discourses on Life and Law*. Cambridge, MA: Harvard University Press.

McClintock, Elizabeth Aura. 2010. When does race matter? Race, sex, and dating at an elite university. *Journal of Marriage and Family* 72 (1): 45–72.

Mendelsohn, Gerald A., Lindsay Shaw Taylor, Andrew T. Fiore, and Coye Cheshire. 2014. Black/White dating online: Interracial courtship in the 21st century. *Psychology of Popular Media Culture* 3 (1).

Okin, Susan Moller. 1989. *Justice, Gender and the Family*. New York: Basic Books.

Pateman, Carole. 1983. *"Feminist Critiques of the Public/Private Dichotomy" in Private and Public in Social Life, ed. Stanley Benn and Gerald Gaus*. London: Croom Helm.

Prasso, Sheridan. 2005. *The Asian Mystique: Dragon ladies, Geisha girls, & our fantasies of the exotic Orient*. New York: Public Affairs.

Rafalow, Matthew H., and Jessica M. Kizer. 2018. Mommy markets: Racial differences in lesbians' dating preferences for women with children. *Journal of Lesbian Studies* 22 (3): 297–312.

Robinson, Russell K. 2007. Structural Dimensions of Romantic Preferences. *Fordham Law Review* 76: 2787.

Robinson, Russell K., and David M. Frost. 2018. LGBT Equality and Sexual Racism. *Fordham Law Review* 86: 2739.

Robnett, Belinda, and Cynthia Feliciano. 2011. Patterns of racial-ethnic exclusion by internet daters. *Social Forces* 89 (3): 807–828.

Rudder, Christian. 2014. *Dataclysm*. New York: Crown Publishers.

Smith, Jesus. 2014. Getting Off Online: Race, Gender, and Sexuality in Cyberspace. In *Illuminating How Identities, Stereotypes and Inequalities Matter through Gender Studies*, ed. Demetrea Nicole Farris, Mary Ann Davis, and R. D'Lane. Netherlands: Springer.

Tsunokai, Glenn T., Allison R. McGrath, and Jillian K. Kavanagh. 2013. Online dating preferences of Asian Americans. *Journal of Social and Personal Relationships* 31 (6): 796–814.

White, Jaclyn M., Sari L. Reisner, Emilia Dunham, and Matthew J. Mimiaga. 2014. Race-Based Sexual Preferences in a Sample of Online Profiles of Urban Men Seeking Sex with Men. *Journal of Urban Health* 91 (4): 1–8.

Young, Iris Marion. 1990. *Justice and the Politics of Difference*. Princeton: Princeton University Press.

Zheng, Robin. 2016. Why Yellow Fever Isn't Flattering: A Case Against Racial Fetishes. *Journal of the American Philosophical Association* 2016: 400–419.

CHAPTER 25

Racialized Sexual Discrimination: A Moral Right or Morally Wrong?

Cheryl Abbate

Introduction

It's generally assumed that we have a "right" to date and have sex with whomever we want (assuming the "whomever" are willing parties) and thus that racialized sexual discrimination isn't morally wrong. Call this the *Personal Preference Thesis* (PPT).[1] The basic idea behind this thesis is that sexual preferences, including race-based sexual preferences, are merely personal preferences, which are thought to be beyond the scope of moral evaluation. According to the PPT, people aren't immoral for having or acting on racialized sexual preferences.

Against the PPT, some claim that acting on racialized sexual preferences is *sexual racism*, a term coined by Charles Herbert Stember who defines sexual racism as "the sexual rejection of the racial minority, the conscious attempt on the part of the majority to prevent interracial cohabitation."[2] While I agree that acting on certain racialized sexual preferences is morally problematic, I suspect that it rarely amounts to racism. Arguably, few people who have racialized sexual preferences make a *conscious* attempt to prevent interracial cohabitation. For instance, many white people with racialized sexual preferences against Black people *subconsciously* avoid sexual encounters with Black people and never stop to examine why they never consider them as potential sexual partners. Indeed, many white people who claim not to have a same-race sexual preference act as if they do, which indicates an unconscious racial bias.[3]

C. Abbate (✉)
University of Nevada, Las Vegas, NV, USA
e-mail: cheryl.abbate@unlv.edu

D. Boonin (ed.), *The Palgrave Handbook of Sexual Ethics*,
https://doi.org/10.1007/978-3-030-87786-6_25

Consequently, if Stember is right that sexual racism involves the conscious attempt to prevent interracial cohabitation, then few people who act on racialized sexual preferences are sexual racists. But, still, they often act wrongly.

The view I will defend in this chapter is that, special considerations aside, acting on certain (anti-Black) racialized sexual preferences is neither morally permissible nor sexual racism. Rather, as I argue, acting on certain (anti-Black) racialized sexual preferences is a form of *wrongful discrimination*. In particular, when white people act on racialized sexual preferences against Black people, they violate moral norms of respect and equality. Consequently, white people do not have a "right" to engage in racialized sexual discrimination against Black people.

Preliminary Remarks

The focus of this chapter is the morality of a certain kind of racialized sexual discrimination. In particular, I am concerned with racialized sexual preferences that are expressed and acted upon both explicitly and implicitly during the use of what is commonly referred to as "intimate platforms," that is, online dating and hookup platforms.[4]

Sometimes, what appears to be racialized sexual discrimination is not morally problematic. For example, if a Black woman acts on her desire for only Black people, this is not problematic. For one, the woman's reason for desiring only Black people may actually be a form of interest-based discrimination, which is morally unproblematic and distinct from racialized sexual discrimination. For instance, some Black people may desire to be in relationships with only Black people *because there is a shared cultural experience and background and/or shared experience of oppression and discrimination*. In this case, white people are not excluded on the basis of their race; rather, they are excluded because they lack certain experiences. Moreover, the meaning behind Black discrimination against white people in the dating context does not have the same problematic meaning (which I discuss in Section 8) that white discrimination against Black people in the dating context does. As Charles Mills suggests, the case for Black endogamy is more respectable than the case for white endogamy.[5]

In this chapter, I focus exclusively on white people who exclude Black people from their sexual lives. I suspect that other kinds of racialized sexual preferences are problematic, such as the strong preference some white people have for members of certain ethnic groups, due to some kind of racial fetish. However, I do not have space to sufficiently address every kind of immoral racialized sexual preference. I will note, though, that racial fetishization is morally problematic, as it is often derived from and reinforces racial stereotypes and reduces ethnic minorities to racialized aspects of their appearance.[6] As one Latina woman puts it, when you are racially fetishized, "you immediately feel like you are no longer a personality—you are just a thing."[7]

The Ethics of Intimate Platform Use

There are various ways in which people can engage in racialized sexual discrimination while using intimate platforms. First, they might display their exclusionary racialized preferences on their profiles. They might, for instance, explicitly state on their profiles that they aren't interested in members of particular races, such as when users specify that they are "only interested in white people; no Blacks." They might also directly communicate this to other users through private messaging. Call this *visible* racialized sexual discrimination.

Note that *visible* racialized sexual discrimination often causes direct offense to those who encounter such profiles. For instance, an Asian man who encountered profiles that say something along the lines of "I don't date Asians—sorry, not sorry" reported that such encounters were disheartening and hurt his self-esteem.[8] *Continual* and explicit rejection based on one's race is surely hurtful and likely makes one feel undesirable. Indeed, users report that *repeatedly* being devalued and rejected on the basis of race in the dating context had significant adverse effects, such as reduced self-esteem, a reduced sense of self-worth, and increased feelings of shame, humiliation, and inferiority.[9] Moreover, it's been noted that experience of racialized sexual discrimination is associated with higher chances of alcohol use.[10] Consequently, persistent visible racialized discrimination in the sexual context adversely affects the psychosocial health of members of racial and ethnic minority groups.[11]

There is another kind of racialized sexual discrimination that is neither verbalized nor visible, which is the focus of this chapter. I will refer to this as *invisible racialized sexual discrimination*. This includes opting to use race-filters that aren't visible on one's profile and simply ruling out profiles on the basis of the users' race, such as when white users, at the early stage of profile selection, skip over the profiles of Black people after seeing their photos. With invisible racialized sexual discrimination, the discriminator engages in race-based rejection in a way that does not *directly* cause emotional harm. So, if any kind of racialized sexual discrimination is justified, it is the invisible kind. Nevertheless, as I will argue, white invisible racialized sexual discrimination against Black people is wrong.

Deconstructing the Personal Preference Thesis

Those who defend some version of the PPT often move from the claim that "sexual preferences for people of certain races are personal preferences," to the claim that "we have the right to have racialized sexual preferences." Note, though, that I am not concerned merely with the possession of racialized sexual preferences. Rather, I am primarily concerned with the morality of *acting* on these preferences. When it comes to either visible or invisible racialized discrimination on intimate platforms, users *act* on their racialized preferences. And to defend this behavior, defenders of the PPT appeal to some version of the following argument:

P1) Racialized sexual preferences are personal preferences.
P2) We have the right to *act* on personal preferences.
C1) Therefore, we have the right to *act* on racialized sexual preferences.
P3) Those who engage in racialized sexual discrimination act on their racialized sexual preferences.
C2) Therefore, we have a right to engage in racialized sexual discrimination.

The remainder of this chapter is a response to P2, which assumes that we have a right to act on our personal preferences.[12] There are three distinct reasons someone might give in defense of this premise. First, they might say that acting on personal preferences never causes harm. According to this view, acting on personal preferences is always *benign*. Second, one might say that even if acting on personal preferences causes harm, people who act on personal preferences don't act wrongly because they cannot control them. Third, one might say that even if acting on personal preferences causes harm, people who act on these preferences don't act wrongly because their "right" to exercise their personal preferences outweighs any harm that is consequently produced.

Against these, claims I will argue that (1) it is wrong to act on personal preferences when doing so causes harm, (2) when white people act on anti-Black sexual preferences, they cause harm, (3) we can, to some degree, control our racialized sexual preferences, and (4) white people don't have even a prima facie right to act on anti-Black sexual preferences, thus there is no individual "right" that outweighs the harm that is produced when white people act on anti-Black sexual preferences.

Some Harms of Invisible Racialized Sexual Discrimination

The usual defense of racialized sexual discrimination is centered upon the notion of "personal preferences," which are thought to be beyond our control. Stephen Kershnar, for instance, claims that "it is not clear that an individual or even a population can control their sexual preferences and it is not wrong to think a certain way if you can't avoid it."[13] But even if Kershnar is right that we can't control our sexual preferences, we must remember that we can control whether we *act* on those preferences. Consider, for instance, that while an adult might, beyond his control, have a sexual preference for children, he can control whether he acts on this preference. And he certainly acts wrongly if he acts on this sexual preference. Clearly, sexual assaulting children is not morally defensible just because the assaulter was acting on a personal preference he couldn't control *having*.

In response, defenders of the PPT could revise P2 of the argument to say something like this: *we have a prima facie right to act on personal preferences.* This claim, which assumes that the right to act on personal preferences can, under certain conditions, be overridden, is compatible with the view that it's immoral to act on personal preferences when doing so would seriously harm others. And defenders of the PPT might claim that either acting on *racialized sexual preferences* usually doesn't harm others, or that even if acting on racialized sexual preferences does cause harm, the harm is usually relatively minor and is thus usually outweighed by the "rights" of individuals to act on their sexual preferences.

It's peculiar that many people think that moral norms don't apply when it comes to our dating preferences, given that dating almost always involves harms and benefits. As Nathaniel Adam Tobias Coleman notes, in the dating context, there are critical goods at stake, including sexual approbation and sexual activity. There is, for instance, "the good of being considered sexually attractive and the good of participating in sex with another human being."[14] So, if some people are continually passed over on dating sites because of their race, they are *unequally* deprived of the opportunity to access these important goods. Arguably, they are harmed in the deprivational sense.

Coleman points out that many people suffer felt harms from the deprivation of sexual goods. For instance, a limited "sexual network" leads to frustration and low sexual self-confidence, and having a limited sexual network may put one at higher risk for contracting HIV & STIs. Moreover, Black men in the men who have sex with men (MSM) community have a limited sexual network due to "sexual racism," which means that "the sexual networks of Black MSM are constrained to smaller numbers and are therefore potentially more highly interconnected than other groups. Once HIV enters one part of such [a] tightly connected network, it is likely to spread rapidly throughout." Indeed, a similar phenomenon is sometimes observed in general Black populations, often referred to as the "intraracial network effect."[15] There is thus a connection between even invisible racialized sexual discrimination and negative public health outcomes.[16]

It seems clear that even invisible racialized sexual discrimination causes, or at least contributes to, either felt harms or harms of deprivation, and this is of moral concern. Perhaps, though, defenders of the PPT might grant that even if acting on anti-Black sexual preferences causes harm, the harm is outweighed by the "right" to act on personal sexual preferences, including anti-Black preferences. In what follows, I argue that white people don't possess even a prima facie "right" to act on anti-Black sexual preferences.

On Moral Rights

Although we have a general right to liberty, this doesn't mean that we have the right to do whatever we want to whomever we want. We don't, for instance, have a "right" to murder others. And it's not just that our prima facie "right" to murder is overridden by the right to life of others. Rather, there is no prima facie "right" to murder to override in the first place. As Michael Huemer suggests, we lack even a prima facie right to perform acts that cause unnecessary harm to others, treat others as mere means, or use others without their valid consent.[17]

So, what prima facie rights are derivable from the right to liberty? To answer this question, I'll first say something about the function of moral rights, which is to protect the (objective) *welfare interests* of individuals. (Objective) welfare interests are distinct from *preference interests*, although many of our preference interests overlap with our (objective) welfare interests. So, while these "overlap" preference interests are protected by prima facie moral rights, not all preference interests enjoy such protection.

To better understand the distinction between (objective) welfare interests and preference interests, first consider that individuals have two kinds of welfare: *subjective* welfare and *objective* welfare. Our subjective welfare level is determined by the degree to which our subjective experiences are positive, and positive subjective experiences include positive physical experiences (pleasure) and positive mental experiences (happiness or satisfaction), including the positive mental experience we often enjoy when our desires are satisfied. Our objective welfare level corresponds to the degree to which we fulfill our potential to be the kind of beings we are. It is thus promoted when we act and are viewed and treated in ways that reflect the kind of beings we are, and it is reduced when we act and are treated and viewed in ways that don't reflect the kind of beings we are. For instance, my subjective welfare level is increased when I am in a romantic relationship that brings me more positive subjective experiences (e.g., pleasure and positive emotions) than negative subjective experiences (e.g., pain and frustration). But my objective welfare level is lessened if my romantic partner cheats on me, even if I never find out about his infidelity. After all, if my partner cheats on me, he fails to treat me as the kind of creature I am: an autonomous creature who is worthy of respectful treatment.[18]

All sentient beings have the same kind of subjective welfare, insofar as, for *all* sentient beings, it is promoted in the same way: by increasing positive subjective experiences. But different species have different forms of objective welfare. There is, for instance, "human objective welfare" and "feline objective welfare." As humans, people like you and I not only have the capacity to suffer and be happy, but we also are the kind of creatures with the capacity to love, think, make free and informed choices, socialize, and act morally, and it is our species norm to engage in loving, social, intelligent, and moral behavior. In other words, it is the human-species norm to use certain capabilities—human-normal capabilities— to engage in human-normal behavior in the pursuit of human-normal goods, such as love, friendship, wisdom, and moral goodness. Humans thus have a(n) (objective) welfare interest in exercising human-normal capabilities in the pursuit of human-normal goods. This is to say that it is in the (objective) welfare interest of humans to love, think, socialize, act morally, and so forth, and to be treated and viewed in ways that reflect that we are the kind of beings who possess these human-normal capabilities.

We are harmed when our objective welfare level is reduced and are benefited when our objective welfare level is increased, but we aren't always harmed when our subjective welfare level is reduced and we aren't always benefited when our subjective welfare level is increased. To see why this is so, consider Tom Regan's example of a contented drug addict.[19] Say there is a person who has been, since she was a child, forcibly addicted to drugs. Say this person spends her entire life in one room, continually consuming drugs that give her a lot of subjective welfare. And say that the total positive subjective experiences of this person is much greater than it would be if she were to live a normal human life. After all, if she were to live a normal life, she would face constant struggles and painful disappointments, such as heartache, failure, grief from the loss of loved ones, and so forth.

Even if the drug addict's subjective welfare level is as high as it could possibly be, she is harmed insofar as her objective welfare level is much lower than what it could be. In other words, the drug addict's life is not going as well as it could be, insofar as her overall well-being would be increased if she were to live a normal life, with all the struggles and disappointments that come with it. This is because human objective welfare should always be prioritized over human subjective welfare. After all, a subjective experience benefits us *only* if it doesn't reduce our objective welfare level. Indeed, pleasure is good for us primarily because it motivates us to achieve intrinsically valuable human goods. Consider, for instance, that when we first form friendships and sexual relationships, we feel deep kinds of pleasures. But, over time, friendships and sexual relationships become less pleasurable, and even difficult. But we continue to maintain them because once we are invested in these kinds of relationships, we come to understand that they are deeply valuable, even if they don't enhance our subjective welfare. And, arguably, if we didn't initially enjoy some kind of positive subjective experience regarding these kinds of human goods, we wouldn't seek them out. Positive subjective experiences, then, are instrumentally good insofar as they motivate us to pursue intrinsic human goods, but positive subjective experiences are not the ultimate end of human life, and they don't always benefit us.

Essentially, there are "preference-(objective)welfare interests," "non-preference (objective) welfare interests," and "non-(objective)welfare preference interests." There are things that we prefer that increase our objective welfare level, there are things we prefer that reduce our objective welfare level, and there are things we don't prefer that increase our objective welfare level. Non-(objective)welfare preference interests are desires that, if fulfilled, would reduce our objective welfare level by impairing our ability to achieve intrinsically valuable human goods. Because morality is concerned with promoting a person's overall wellbeing and because objective welfare is more important than subjective welfare in overall wellbeing, morality is primarily concerned with promoting objective wellbeing. This means that the only preference interests that are protected by moral rights are those that do not reduce our objective welfare level.

To make this clear, consider again the desire to murder. Even if murdering another would make us happy, this doesn't matter from the moral point of view, as murdering others always reduces our objective welfare level, insofar as murderers fail to exercise their human-normal capacity for morally good conduct and thus fail to achieve the human-normal good of moral virtue. Consequently, even if we have a preference interest in murdering another, we don't have even a prima facie right to murder others because we don't have a(n) (objective) welfare interest in doing so. Relatedly, if spending all day on drugs would make a drug addict have more positive subjective experiences but less objective welfare, the drug addict doesn't have a right to use the drugs.[20]

Likewise, although some white people might prefer to date only white people, acting on this preference reduces their objective welfare level, and thus harms them, because this preference conflicts with an important human (objective) welfare interest: *the interest in being a morally good actor*. As I will argue in the next section, white people who engage in racialized sexual discrimination against Black people act in morally corrupt ways, given that their behavior disrespects the personhood of and demeans Black people, which is always morally wrong. This in turn reduces the objective welfare level of Black people, and thereby harms them. And because acting on anti-Black sexual preferences is morally corrupt and thus reduces the objective welfare level of white people who act on these preferences, the anti-Black sexual preferences of white people are non-(objective) welfare preference interests, which don't enjoy the protection of even prima facie moral rights.

On Respect for Persons

When people join intimate platforms, they have the potential to browse through thousands of profiles of potential "matches." Surely, then, users should have some method for quickly deciding which profiles to "rule out" and which to seriously consider. That is, users should, for prudential reasons, engage in some form of *permissible* sexual discrimination.

Elsewhere, I draw a distinction between *appearance-based discrimination* (often called "lookism") and *interest-based discrimination* in dating.[21] One engages in interest-based discrimination when one rules out potential matches based on their having interests or values that deeply conflict with one's own. For instance, I am a vegan, an avid hiker, a skier, and I despise football. If I come across a dating profile of a man who claims to be a meat-eating, couch-potato, football fanatic, it is permissible for me to rule out this man from further consideration. After all, by pausing to read his biographical information in his profile and considering his interests and values, I view and treat him as a person. I reject his profile not on the basis of his appearance, but because I realize that his preferences and values are in stark contrast to mine.

Ruling out potential dating partners from further consideration because of their looks is another moral story. Appearance-based discrimination is a form of *wrongful* discrimination insofar as it fails to view people *as people*, and thus it fails to view and treat them with the respect they are due *as persons*. Andrew Davis, for instance, compellingly argues that lookism is a form of stereotyping, which fails to view people as sources of actions with moral significance and as potential possessors of moral virtues.[22] To fail to view and treat people as individuals who have values, beliefs, and preferences is to fail to view and treat them as the kind of creatures they are: creatures with unique identities, beliefs, preferences, and value systems.

The basic idea is that you are morally required to treat the people you encounter, including those you encounter on intimate platforms, as the kind of creatures they are—that is, *as persons*, who have unique interests, identities, and values. "Ruling out" a person from social interaction, including an intimate platform interaction, because of her personal appearance, fails to view and treat her as the kind of creature she is. Essentially, it is tantamount to viewing and treating her like a mere vessel of physical characteristics. And this reduces her objective welfare level, and thus harms her. And because it is always wrong to disrespect the personhood of others, this moreover wrongs her.[23]

On Demeaning Persons

All forms of appearance-based discrimination are wrong because they violate the moral norm of respect, thereby reducing the objective welfare level of, and thus harming, those who are discriminated against. Yet, as Xiaofei Liu argues, there is a moral distinction between "simple lookism" and "racial lookism."[24] "Racial lookism" is, from the moral point of view, more problematic than "simple lookism." This is because, in addition to violating the moral norm of *respect for persons*, racial lookism violates the moral norm of *equality*.

As Deborah Hellman compellingly argues, to determine how wrong an instance of wrongful discrimination is, we must look at the meaning of the discriminatory act, which often stems from the context in which it is performed.[25] Consider, for instance, that we all act within a racially charged context. Black people have historically been subject to atrocious treatment, including slavery, lynching, and redlining. To this day, harmful racial discrimination, including race-based violence, is prevalent throughout our society. And because of our racially unjust history and current social situation, drawing distinctions on the basis of race often has a *demeaning* meaning.

Consider, for instance, the moral and legal difference between an employer who refuses to hire people with blonde hair and an employer who refuses to hire people who are Black. While both employers act wrongly, the second acts in an especially wrongful manner, insofar as the distinction he draws between Black people and white people has a troubling meaning that the distinction-drawing between blondes and non-blondes lacks—and this is true *even if the employer discriminates in private*. After all, racial distinctions have been, and are still currently, *systematically* used to disadvantage Black people in the workplace and job market. The same cannot be said for blondes. Indeed, employers who discriminate against Black people *demean*—denigrate and put down— Black people because the meaning of this employer's race-based discrimination is that Black people are inferior to white people, aren't entitled to employment, and so forth.[26]

Like employers, people who use intimate platforms act within a uniquely racially charged context, given the deep historical connection between racism and sex. Consequently, drawing race-based distinctions in the dating context is socially significant and often has a *demeaning* meaning. Consider, for instance, that in the post-Civil War decades, Black men were often castrated and lynched when they were accused of raping white women.[27] Moreover, consider that state bans on interracial marriage, referred to as anti-miscegenation laws, were widespread and not deemed unconstitutional until 1967. Consider, too, that we continue to see the effects of anti-miscegenation laws, given that Black-white couples represent only 1 percent of marriages.[28] This is, in part, because 17 percent of people still (consciously) believe that interracial marriage is morally wrong, and even those who claim to accept racial marriage often elicit disgust when shown photos of interracial couples.[29] Finally, consider that Black-white relationships have been categorized as the "last (interracial) taboo" and a sign of "final (interracial) intimacy."[30] Indeed, acceptance of Black-white relationships is said to be a sign of whether white people have genuinely overcome their racist upbringings.[31]

Because of this painful history involving Black-white couples that has a continued effect today, when white people reject Black people because of their race, white people demean Black people insofar as they not only express a devaluation of Black people because of their race, but they moreover express the view that Black people are not the kind of people white people should date, thereby denying the fact that Black people are, like white people, the kind of creatures who are capable of loving and meaningful relationships. These white people, in a sense, symbolically stamp all Black people with, as Hellman might put it, a "badge of inferiority," thereby violating moral norms of respect *and* equality, *even if the discrimination is private or invisible*. And in doing so, white people not only reduce the objective welfare level of, and thereby harm, the Black people whose profiles they disregard, but they also commit a dignitary harm against all Black people. Indeed, white racialized sexual discrimination against Black people is an affront to the dignity of Black people everywhere. And insofar as white people who act on anti-Black sexual preferences act wrongly by, *without good reason*, reducing the objective welfare level of (i.e., harming) the Black people they discriminate against, they essentially act on a preference that reduces *their own* objective welfare level. Essentially, they act wrongly, in part, because *they wrong themselves* by demeaning Black people and viewing and treating them disrespectfully.

The basic idea is this: if white people act on sexual preferences against Black people, they reduce their own objective welfare level. After all, humans have a (objective) welfare interest in living a morally good life, and it's always morally wrong to demean and disrespect the personhood of others. When white people act on sexual preferences against Black people, they demean and disrespect the personhood of Black people, and thus they harm *themselves*. And because it's morally wrong to harm oneself (without morally good reason), white people who act on sexual preferences against Black people don't just harm themselves; they moreover *wrong* themselves.

The Sources of Racialized Attraction

It's often assumed that we don't have control over our sexual preferences, but research indicates that sexual preferences are somewhat fluid and are, in part, shaped by unexpected encounters.[32] And if we explore the root of attraction, we can determine why we are attracted to certain people (and not others) and how we might change our attraction patterns.

Because Eurocentric beauty ideals, which equate attractiveness with light skin, are so widespread, it's assumed that this "white racist stereotype of beauty" primarily influences our sexual preferences.[33] Yet, research shows that our sexual preferences are influenced by more than just the physical characteristics that we find desirable (such as light skin). Consider, for instance, that we are attracted to people who are *familiar* to us; the more exposure we have to a person (or thing) the more attractive it becomes to us.[34] This is known as the mere-exposure effect. And consider, for instance, that we are attracted to people who are *similar* to us. This is especially true when it comes to reproductive partners (as opposed to mere "hook up" partners), who tend to exhibit similarities regarding background, political and religious beliefs, social and economic status, intelligence, attractiveness, and personality characteristics, a phenomenon often referred to as assortative mating.[35] These considerations suggest that a white person's anti-Black sexual preference is likely attributed to feelings of *unfamiliarity* and *dissimilarity*, and not just a preference for light-skinned people.

Arguably, one reason why many white people skip over the dating profiles of Black people is that white people haven't had many, if any, quality interactions with Black people, and thus white people perceive Black people to be, in some sense, unfamiliar. This sense of unfamiliarity is, in large part, a product of racial segregation (e.g., residential and occupational) that still persists throughout our society today.[36] And if white people see Black people as unfamiliar due to a lack of personal interactions, they likely will not perceive Black people to be "similar" to them.

Moreover, given the powerful influence of media, it's also likely that our media exposure, in part, determines whom we find familiar and similar to us (and thus whom we find attractive). Consider, for instance, that since there is a lack of racial diversity in popular films and shows and a lack of interracial relationships depicted in movies and television, the media likely perpetuates in white people a sense of unfamiliarity, including a sexual unfamiliarity, with Black people. And because Black actors are often cast into roles that represent uniquely minority experiences (such as in the television series *Black-ish*), their "otherness" is emphasized, which reinforces the belief and feeling that Black people are *dissimilar* to white people. Relatedly, popular culture frequently depicts Black men as "thugs," "criminals," "persons of interests," or "criminal suspects," while the news media overrepresents Black people in its coverage of crime, which further perpetuates the belief that Black people are more threatening, and thus dissimilar to, white people.[37]

The Ethics of Intimate Platform Design

Arguably, dating platforms are in an ideal position to challenge systemic racism, as evident from the fact that online dating is the "main driver" behind the recent rise of interracial marriages in the United States.[38] Yet, some intimate platforms seem to promote wrongful discrimination, such as those that provide users with the option of filtering searches not only by location, age, and sexual orientation, but also by *ethnicity*. While not all intimate platforms offer ethnicity filter options, many use problematic race-based algorithms. For instance, many intimate platforms use machine-based learning, which results in users being shown profiles of certain types of people, based on their user history. So, users who continually demonstrate an interest in only white users eventually will be shown only or mostly white users.

Intimate platforms, as businesses that provide dating services to the general public, have, at the very least, a moral duty not to perpetuate or facilitate the exclusion of certain ethnic groups from their services.[39] So, at the very least, designers ought not to include ethnicity filters in their intimate platform-designs, as this clearly promotes racial exclusion. Relatedly, because there is a high risk that intimate platforms will be misused in a way that promotes wrongful discrimination, designers of intimate platforms ought to implement measures to guard against this. For instance, they could include a non-discrimination pledge that potential users are required to sign before registering with their service. At the very least, designers should include rules against problematic forms of racial discrimination in their community policies and ban users who violate these policies, such as people who use prejudicial language on their profiles. And perhaps they ought to counter wrongful racial discrimination by signaling to white users that they are compatible with Black people, such as by regularly suggesting the profiles of Black users as promising "matches" for white users.

While I don't have sufficient space in this chapter to thoroughly explore the design-based moral obligations of intimate platforms, I highlight these design-based concerns, as we should be cognizant of the unique power intimate platforms have to both perpetuate and combat wrongful racial discrimination.[40] So, I encourage readers to keep in mind that individual users aren't the only ones at fault here; the design of some intimate platforms arguably contribute to misconduct in the dating context.

Concluding Remarks

Surely we should, at the very least, strive to cultivate sexual preferences that are in line with moral norms of respect and equality. But how can white people effectively challenge the feelings of unfamiliarity and dissimilarity they likely experience when they encounter Black people on dating websites?

To begin with, white people should attempt to build meaningful interracial personal and social relationships of all kinds by increasing not just the quantity but also the *quality* of interracial contact. Research shows that the majority of interracial interactions are brief, superficial encounters, such as when Black people and white people cross paths at shopping malls and sporting events.[41] Because adults tend to develop friendships with their neighbors, coworkers, and fellow churchgoers, white people can develop meaningful interracial friendships by seeking out residences in racially mixed neighborhoods, seeking out employment at racially diverse businesses, seeking out racially diverse colleges and post-graduate schools, and seeking out houses of worship that have multiracial congregations. Building quality interracial relationships, including platonic friendships, will increase feelings of familiarity and similarity between Black people and white people. And if white intimate platform users make the easy effort to view and treat the Black users they encounter as people by seeking to learn more about their interests and values instead of automatically ruling them out because of their race, white people will find that many potential Black dating partners have backgrounds, experiences, preferences, values, and behavioral and interpersonal characteristics that are similar to their own.

Finally, white people should, at the very least, educate themselves about harmful racial stereotypes and the misrepresentation of Black people in the media. White people, for instance, can read articles and books about the misrepresentation of Black people in the media and misguided racial stereotypes that wrongfully "other" Black people, watch documentaries that debunk the "Black criminal myth" (such as the documentary *13th*), join anti-racist educational coalitions and groups, attend educational webinars, and so forth. Essentially, white people can cultivate a realistic understanding of the lived experiences of Black people, which will ultimately challenge the "othering" stereotypes they've been exposed to by the media. For instance, by making an effort to understand the lives and realities of Black men, white women might, instead of thinking of Black men as "de-victimized" criminals who are so dissimilar to them, realize that they have much in common with Black men, insofar as they are both victims of white-male superiority and oppression in an incredibly unjust world. Indeed, seeking a more accurate understanding of the lives and realities of Black people is something *all* white people, at the very least, can and should do.

Notes

1. For a defense of this position, see Kershnar (2018). Halwani (2017), too, suggests that people with racialized sexual desires are like people with other sexual preferences.
2. The term sexual racism appears in Callander et al. 2015 and Coleman 2011. See Stember 1978, pg. (xi) for his definition of sexual racism.
3. See Anderson et al. 2014.
4. For the use of this terminology, see Hutson et al. 2018.
5. See Mills 1994 for his discussion of white endogamy.
6. For more on the topic of the immorality of racial fetishization, see Zheng (2016).

7. See Wynne 2020.
8. See Brown 2018.
9. See Paul et al. 2010; Wade and Harper 2020.
10. See Brennan et al. 2013.
11. See Wade and Harper 2020 for a discussion about the negative effects of racialized sexual discrimination. Relatedly, white people are more likely to casually date or have sex with racial minorities than they are to marry them, which reinforces the harmful message that racial minorities are a "type" or an "experience" (West 2020).
12. And it's worth noting that P1 itself is problematic. While racialized intimate preferences are preferences, they may not be personal, as they often are a product of social and cultural influence.
13. See Kershnar 2018, pg. 244 for his discussion about our alleged inability to control our sexual preferences.
14. See Coleman 2011 pg. 13 for his discussion of sexual goods
15. See Laumann and Yoom 1999 for a discussion about the "intraracial network effect."
16. This point is made in Hutson et al. 2018.
17. See Huemer 2003.
18. If I am lied to by my partner, I cannot make fully autonomous romantic decisions, insofar as I am lacking information that is necessary for me to make informed decisions about my romantic activities.
19. See Regan 1983 for this example.
20. And perhaps we have a right, or even a moral obligation, to interfere with her drug use.
21. See Abbate 2020 for my prior discussion of the ethics of racialized discrimination in the dating context.
22. See Davis 2007 for his discussion of "lookism."
23. It may be permissible to harm others (i.e., it may be permissible to reduce the objective welfare level of others) when competing rights are at stake. So, sometimes harming others doesn't amount to wronging them. But because there is no such thing as a "right" to disrespect the personhood of others (such as by treating them as a mere means), disrespecting the personhood of others is always a harm that is wrong.
24. See Liu 2015.
25. See Hellman 2008.
26. Law, too, recognizes this distinction: it identifies race as a protected class but not things like hair color.
27. This point is made in Mills 1994.
28. See Gallup 2013.
29. See YouGov 2018 and Skinner and Hudac 2017.
30. This point can be found in Mills 1994.
31. This point can be found in Mills 1994.
32. See Hutson et al. 2018.
33. See Marway 2018 and Mills 1994 for further discussion of Eurocentric beauty ideals.
34. See Moreland and Beach 1992.
35. See Buss 1985, Little et al. 2006, and McPherson et al. 2001.

36. And if our segregation practices, in part, influence our racialized preferences, then we can partially control our sexualized preferences and thus can be held at least partially accountable for both having and acting on these desires.
37. See Dixon 2008 and Okantah 2016.
38. See Hergovich and Ortega 2018.
39. Intimate platforms arguably have a corresponding *legal* duty based on Public Accommodation Law. For instance, if intimate platforms continually permit users to exclude Black people, there's a sense in which these platforms are saying "I'm open to the public and provide this social good, but not to Black people." This argument, however, does not apply to intimate platforms that are open only to racial minorities, since the persons who are excluded by these platforms (white people) aren't members of a protected class.
40. I also cannot thoroughly explore the question of whether these online dating platforms should be regulated by the courts and legislators. I suspect that there is a compelling argument in favor of such a proposal.
41. See Sigelman et al. 1996.

References

Abbate, C. 2020. It's Not Just a Personal Preference: Racialized Discrimination in the Tinder Context. In *College Ethics*, ed. B. Fischer, 142–154. Oxford: Oxford University Press.

Anderson, A., S. Goel, G. Huber, N. Malhotra, and D. Watts. 2014. Political Ideology and Racial Preferences in Online Dating. *Sociological Science* 1: 28–40.

Brennan, D., K. Asakura, C. George, P. Newman, S. Giwa, T. Hart, and G. Betancourt. 2013. 'Never Reflected Anywhere': Body Image Among Ethnoracialized Gay and Bisexual Men. *Body Image* 10: 389–398.

Buss, D. 1985. Human Mate Selection. *American Scientist* 73: 47–51.

Callander, D., C. Newman, and M. Holt. 2015. Is Sexual Racism Really Racism? Distinguishing Attitudes toward Sexual Racism and Generic Racism among Gay and Bisexual Men. *Arch Sex Behav* 44 (7): 1991–2000.

Coleman, N. 2011. What? What? In the (Black) Butt. *APA newsletter on Philosophy and Lesbian, Gay, Bisexual, and Transgender issues* 11: 12–15.

Davis, A. 2007. 'Lookism', Common Schools, Respect and Democracy. *Philosophy of Education* 41 (4): 811–827.

Dixon, T. 2008. Crime News and Racialized Beliefs: Understanding the Relationship between Local News Viewing and Perceptions of African Americans and Crime. *Journal of Communication* 58: 106–125.

Gallup. 2013. In U.S., 87% Approve of Black-White Marriage, vs. 4% in 1958. Retrieved from:https://news.gallup.com/poll/163697/approve-marriage-blacks-whites.aspx (Accessed September 18, 2020).

Halwani, R. 2017. Racial Sexual Desires. In *The Philosophy of Sex*, ed. R. Halwani, A. Soble, and S. Hoffman, 7th ed. Washington, DC: Rowman and Littlefield.

Hellman, D. 2008. *When is Discrimination Wrong?* Cambridge: Harvard University Press.

Hergovich, P. and J. Ortega. 2018. The Strength of Absent Ties: Social Integration via Online Dating. Retrieved from: https://doi.org/10.2139/ssrn.3044766 (Accessed September 18, 2020).

Huemer, M. 2003. Is There a Right to Own a Gun? *Social Theory and Practice* 29 (2): 297–324.

Hutson, J., J. Taft, S. Barocas, and K. Levy. 2018. Debiasing Desire: Addressing Bias and Discrimination on Intimate Platforms. *Proc. ACM Hum.-Comput. Interact.* 2 (73): 1–18.

Kershnar, S. 2018. In Defense of Asian Romantic Preference. *International Journal of Applied Philosophy* 32 (2): 243–256.

Laumann, E., and Y. Yoom. 1999. Racial/Ethnic Group Differences in the Prevalence of Sexually Transmitted Diseases in the United States: A Network Explanation. *Sexually Transmitted Diseases* 26 (5): 250–261.

Little, A., D. Burt, and D. Perrett. 2006. Assortative Mating for Perceived Facial Personality Traits. *Personality and Individual Differences* 40: 973–984.

Liu, X. 2015. No Fats, Femmes, or Asians. *Moral Philosophy and Politics* 2 (2): 1.

McPherson, M., L. Smith-Lovin, and J. Cook. 2001. Birds of a Feather: Homophily in Social Networks. *Annual Review of Sociology* 27 (1): 415–444.

Mills, C. 1994. Do Black Men Have a Moral Duty to Marry Black Women? *Journal of Social Philosophy* 25: 131–153.

Moreland, R., and S. Beach. 1992. Exposure Effects in the Classroom: The Development of Affinity among Students. *Journal of Experimental Social Psychology* 28 (3): 255–276.

Okantah, M. 2016. In Search of the Real Brother Man: An African Centered Approach to Black Cultural Identity. *Africology: The Journal of Pan African Studies* 9: 411–423.

Paul, J., G. Ayala, and K. Choi. 2010. Internet Sex Ads for MSM and Partner Selection Criteria: The Potency of Race/Ethnicity Online. *Journal of Sex Research* 47: 528–538.

Regan, T. 1983. *The Case for Animal Rights.* Los Angeles: University of California Press.

Skinner, A., and C. Hudac. 2017. 'Yuck, You Disgust Me!' Affective Bias against Interracial Couples. *Journal of Experimental Social Psychology* 68: 68–77.

Stember, C. 1978. *Sexual Racism: The Emotional Barrier to an Integrated Society.* New York: Elsevier.

Wade, R., and G. Harper. 2020. Racialized Sexual Discrimination (RSD) in the Age of Online Sexual Networking: Are Young Black Gay/Bisexual Men (YBGBM) at Elevated Risk for Adverse Psychological Health? *Am J Community Psychol* 65 (3-4): 504–523.

West, K. 2020. Interethnic Bias in Willingness to Engage in Casual Sex Versus Committed Relationships. *The Journal of Sex Research* 57 (4): 409–420.

Wynne, G. 2020. Racial Fetishization Is A Big Problem Online. Here's What Dating Apps & Users Can Do. *Bustle.* Retrieved from: https://www.bustle.com/wellness/racial-fetishization-big-problem-online-what-dating-apps-users-can-do (Accessed October 1, 2020).

YouGov. 2018. The Economist/YouGov Poll. Retrieved from:https://d25d2506sfb94s.cloudfront.net/cumulus_uploads/document/y3tke5cxwy/econTabReport.pdf (Accessed September 18, 2020).

Zheng, R. 2016. Why Yellow Fever isn't Flattering: A Case against Racial Fetishes. *Journal of the American Philosophical Association* 2 (3): 400–419.

CHAPTER 26

Sexual Ableism: Is Sex Work the Best Solution?

Kevin Mintz

Introduction

In the provocatively titled essay "Head Nurses," disability bioethicist William Peace recounts the first time he had sex after acquiring a spinal cord injury. While in a rehabilitation program, a nurse gave Peace a blow job, teaching him how he could have sex with his changed body. Peace explains that many young men in his rehabilitation program had similar experiences with other nurses. To him, these "head nurses" taught him and others more about having sex with their disabilities than doctors ever could or would. Peace's doctor did not know if a man with a spinal cord injury could have sex when Peace asked and showed no interest in helping him discover his sexual abilities.[1]

Peace's observation about his doctor describes what scholars like Tom Shakespeare have called sexual ableism. Sexual ableism is the discriminatory manifestation of lowered sexual expectations for people with disabilities in the form of individual attitudes as well as institutional policies and procedures.[2] By not providing Peace with any kind of sexual information, this professional failed to recognize Peace's disability as part of his newly acquired sexual identity. Sexual ableism also encompasses discrimination in one's choice of sexual partner on the basis of disability. Because people with disabilities often face unique challenges in finding sexual partners, professionals like Peace's

I would like to thank Nayeli Altamirano, David Boonin, Cheryl Cohen Greene, Wil Dubree, Joanne Kalogeras, Patricia Marino, Debra Satz, Laurie Shrage, Marilyn Volker, David Wasserman, and Claire Woodward, for their helpful comments on this chapter.

K. Mintz (✉)
Stanford University School of Medicine, Palo Alto, CA, USA
e-mail: kmintz@stanford.edu

D. Boonin (ed.), *The Palgrave Handbook of Sexual Ethics*,
https://doi.org/10.1007/978-3-030-87786-6_26

"head nurse" are often portrayed as compassionate heroes who provide the disabled much-needed access to the sexual component of the human experience.

To illustrate the pervasiveness of this image, consider two examples in popular culture. The Academy Award-nominated film *The Sessions* depicts the true story of a journalist with polio named Mark O'Brien who loses his virginity to a surrogate partner therapist named Cheryl Cohen Greene.[3] Surrogate Partner Therapy (SPT) is a treatment modality originally pioneered by sex researchers William Masters and Virginia Johnson.[4] SPT involves a client forming an intimate relationship with a surrogate over multiple sessions in order to practice the interpersonal skills needed to find an intimate partner. Surrogates are trained sexual health professionals who become certified through one of a handful of professional organizations, including The International Professional Surrogates Association (IPSA).[5] SPT often, but not always, involves a surrogate and client having sex.[6] Cohen Greene taught O'Brien how his body could function sexually and gave him the confidence to independently find a sexual partner.

Presenting an LGBTQ+ perspective on such a narrative, the semiautobiographical Netflix series *Special* portrays a gay man with cerebral palsy named Ryan who has his first sexual experience with a nontherapeutic sex worker, who happens to have many clients with the same disability. As in *The Sessions*, the professional depicted in *Special* is knowledgeable and compassionate, able to successfully guide his nervous disabled client through penetrative intercourse. After they have sex, Ryan anxiously tells the sex worker, "And I'm going to have sex soon for free. Okay? Just so you know."[7] This reaction speaks to an apparent tension within Ryan and potentially some others with disabilities who utilize a sex worker's services. On one hand, the experience might provide a form of sexual liberation. On another, a client with a disability may desire a sexual relationship that is not defined in terms of payment for sexual services.

To be sure, these narratives are an important component of challenging sexual ableism because they represent people with disabilities as sexual beings. Sexual expression is not cutoff to these men because of their impairments, but is achieved with a partner who helps them discover the ways in which they can experience sexual satisfaction. Moreover, these narratives, particularly *The Sessions*, have contributed to a plethora of sexual ethicists accepting SPT and other forms of sex work as legitimate means of combatting sexual ableism. My previous work on this subject, for example, has argued that access to the services of surrogates like Cohen Greene should be a matter of fair equality of opportunity in health.[8] Relatedly, a series of articles published over the last decade in the *Journal of Medical Ethics* has debated the merits of "positive sexual rights" for the disabled.[9] By positive sexual rights, these scholars refer to the idea that people with disabilities should be entitled to targeted opportunities and resources for sexual expression, as opposed to merely being able to pursue their sexual lives without interference.[10]

The concept of positive sexual rights is ethically contentious because if someone has a positive right to sex, another person has a corresponding duty to help them realize that right. Extending this reasoning to sex, some commentators claim that placing duties on an individual to help someone else realize their sexual rights is fundamentally incompatible with freedom of sexual choice. Complicating matters even further is a more fundamental ethical question: why should sex be a site for intrusion by political and social institutions in the first place? After all, history has provided many examples, particularly with regard to LGBTQ+ concerns and women's reproductive health issues, in which institutions use sex as a means of sociopolitical oppression.

For scholars who advocate for positive sexual rights, a motivating concern is the idea that without some kind of institutionalized support, people with various impairments would be unable to meaningfully exercise their sexual capacities.[11] Critics, by contrast, point out that sex is a unique component of the human experience that, unlike other social goods, cannot be subject to redistributive schemes or other social welfare measures without placing unreasonable value judgments on an individual's sexual choices. To borrow an analogy from one of these scholars, sex is not like a sandwich that can be shared based on a fair procedure. Sex is not quite like any other aspect of the human experience in that few others are simultaneously matters of especially intimate choices and reflections of pervasive social inequalities.[12] For my purposes here, I assume that any obligations surrogates or nontherapeutic sex workers have are to a client with whom they have already consented to working.

Many authors on both sides of this debate consider whether positive sexual rights can ground an ethical exemption to legal prohibitions against nontherapeutic and therapeutic sex work. For those advocating for access to sex work for the disabled, placing a corresponding duty on professional sex workers is more easily compatible with freedom of sexual choice because sex workers are professionals who, at least in the ideal case, have control over the clients they choose to work with. A surrogate, for example, may have a general duty to not discriminate against potential clients simply on the basis of disability, but that does not mean they have to accept all potential clients or have sex with a client whom they feel threatened by.[13] For those arguing against access to sex work, the gender inequalities associated with sex work are such that any positive duty to have sex with disabled clients exacerbates the exploitation these workers confront in their jobs.[14] These debates are important because they, too, challenge sexual ableism and help clarify the circumstances under which sex work can or cannot be ethically justified.

At the same time, the linking of efforts to combat sexual ableism so closely with sex work raises important questions. Why have discussions of sexual ableism focused predominantly on access to sex work for people with disabilities? Do sexual ethicists have compelling reasons for arguing that people with impairments should have access to sexual services? Finally, are there limitations to understanding sex work as a primary solution to the ethical dilemmas raised by sexual ableism? In this chapter, I answer these questions. I suggest that facilitating access to sexual opportunities for people with disabilities provides a compelling ethical justification for access to sex work. However, arguments that favor this solution without advocating other measures as well risk denying people with disabilities access to a wide array of sexual freedoms.

Specifically, enabling people with impairments to lead fulfilling sexual lives should involve much more than opportunities to use the services of a surrogate or nontherapeutic sex worker. In particular, sexual ethicists should focus broadly on the kinds of policies and resources that will enable people with disabilities to have a similar range of opportunities for sexual expression as society affords to nondisabled people. The aim of these measures, I contend, should be to ensure that people with disabilities are recognized as individuals who have the same sexual freedoms and responsibilities as anyone else who can provide sexual consent.[15]

I begin by exploring why discussions of sexual ableism have become closely linked with ethical debates on sex work. I suggest that the two literatures have become intertwined because sex work provides access to sex in an analogous way to how other professional services facilitate social inclusion for people with disabilities. From the perspective of sexual ethicists, this notion of sexual access for those with impairments constitutes a thought-provoking test case for arguments in favor of or against sex work. I then suggest that this emphasis on sex work in scholarship on sexual ableism is limiting because it reduces the sexuality of people with disabilities to the concerns of mostly men who have disability-specific difficulty with sexual acts. Emphasizing sex work further contributes to the commodification of people with disabilities' sexualities. I then suggest that therapeutic and nontherapeutic sex work should be considered as only one potential means of enabling people with impairments to lead fulfilling sexual lives. I conclude with a brief discussion of how sexual ableism affects people with intellectual disabilities who cannot legally consent, emphasizing the need to challenge the notion that people with intellectual disabilities should be categorically shielded from any kind of sexual activity.

Why Sex Work Is Often Invoked in Discussions of Sexuality and Disability

The role of sex work in combatting sexual ableism is not intuitive. After all, many other minority groups confront sexual marginalization. In the case of these other groups, though, scholars do not advocate for sex work as the primary means of sexual liberation. In fact, ethicists concerned with sexual marginalization often conceptualize how sex work might contribute to rather than eliminate sexual oppression. While most sex worker advocates point out that sex work can challenge social stigma around sexuality, many types of sex work reinforce long-standing patterns of gender and racial inequalities.[16] Ethical analyses scrutinizing sexual oppression often focus, instead, on changing interpersonal dynamics and eliminating imbalances of power between the sexually marginalized and the sexually advantaged.

Consider Sonu Bedi's work on sexual racism. His 2015 article suggests that racial equality in opportunities for sexual intimacy can be considered a matter of justice from both Rawlsian and Capabilities Approach perspectives. In particular, sexual racism contributes to the perpetuation of racial and economic inequalities in the family and simultaneously constitutes a barrier to sexual flourishing. However, Bedi's proposed remedy for these inequalities is not to distribute therapeutic or nontherapeutic sex workers to members of marginalized racial groups. Rather, he emphasizes discouraging racial stereotyping and discrimination on online dating and hookup apps. Borrowing from Chiara Cordelli, Bedi conceptualizes these apps as comprising part of the relational distributive structure around sexual relationships. Bedi means that these semiprivate, semipublic spaces distribute resources that enable people to acquire opportunities for an important social good, specifically intimacy. Bedi's recommendations, including eliminating racial filters and discouraging racially charged language, are specifically meant to undermine the racial imbalances that shape interactions between potential sexual partners.[17] I contend that efforts to remedy sexual ableism should take a similar form. Institutions comprising the relational distributive structure shaping sexual relationships should work to discourage ableist attitudes and behaviors in an individual's sexual choices.

Sexual ableism is admittedly less blatant on these applications than sexual racism in that individuals are less likely to openly state in a profile that they will not consider someone with a disability, whereas online dating sites allow people to racially filter profiles and include phrases like, "no Blacks." Despite this difference, both sexual racism and sexual ableism involve sexual preferences that are rooted in stereotypes, prejudices, and long-standing social inequalities. Why, then, has the sexual ableism literature evolved in a way that frames sex work as a primary means of sexual empowerment for people with disabilities? Why not focus, instead, on measures that discourage ableist behavior in non-paid sexual interactions?

Part of the explanation for the association between sex work and efforts to remedy sexual ableism has to do with how narratives like O'Brien's have captured the public imagination so that any ethical writing on sexual ableism now has to address SPT or nontherapeutic sex work in some way. O'Brien's experience has been documented not only in his own writings and *The Sessions*, but also the Academy Award-winning short documentary film about his life, *Breathing Lessons: The Life and Work of Mark O'Brien*.[18] Particularly because of O'Brien's severe physical disabilities, the idea that he was deserving of Cohen Greene's therapeutic assistance likely strikes an emotional chord with liberal-minded members of the general public, not to mention many disability ethicists. After all, someone with significant physical limitations *does* require some kind of support in having sex!

Independent of the public's fascination with such narratives, however, it seems that sex work and disability have become associated with one another for a more theoretically interesting reason. Disability studies scholars tend to frame the barriers confronted by people with disabilities in terms of issues of *access*. Reasonable accommodations in the occupational context, for example, are often thought of as providing access to jobs. The typist I am dictating this chapter to is facilitating my access to the writing process, because my limited dexterity makes it substantially more difficult for me to type. Similarly, surrogate partner therapists and nontherapeutic sex workers are often thought of as a mechanism to provide *access* to sexual expression for people with disabilities. Access to sexual expression on these interpretations is narrowly construed in terms of being able to access sexual experiences.

Providing access to sexual expression for people with disabilities is a part of correcting for sexual ableism. Many people with disabilities report facing disability-related barriers to sexual expression that are either the direct result of a particular impairment or the result of ableist attitudes that presume disability always involves some form of sexual dysfunction.[19] It is therefore important that any efforts to correct for sexual ableism include some appeal to the elimination of these barriers. The elimination of these barriers is what most scholars and activists mean by access to sexual expression.[20] Fully acknowledging the sexuality of people with disabilities involves making reasonable efforts to make desired consensual sexual expression as accessible to people with disabilities as possible. Despite the ethical controversies surrounding paid sex, therapeutic and nontherapeutic sex work are, nonetheless, means of facilitating access to sexual expression.

The emphasis on sex work in the sexual ableism literature is further reinforced by the extent to which feminist ethicists use the example of disability to make cases for and against the ethical permissibility of prostitution. For example, Sibyl Schwarzenbach argues that SPT has the potential to undermine the patriarchal nature of sex work because of the considerable control the surrogate has over her relationship with a client. She imagines a society where the therapy gradually contributes to the development of a sex work industry where women are more comfortable seeking sexual services from male or female sex workers.[21]

By contrast, Sheila Jeffreys contends that even in the disability case, therapeutic sex work still enforces a male sex right to women's bodies. Using O'Brien and Cohen Greene as her main case study, Jeffreys points out that many of the sexual acts done in SPT are those that a nontherapeutic sex worker and client would engage in.[22] These two examples illustrate the contours of a contentious and ever-evolving debate over sex work and how to respond to the myriad concerns it raises. Jeffreys is correct that SPT *sometimes* involves sexual acts. However, it is worth emphasizing that a considerable portion of SPT involves body awareness exercises and communication skills building which a nontherapeutic sex worker may or may not have the training, expertise, or motivation to facilitate for a client.

Consider the experience of sex worker Xaviera Hollander. She flippantly explains in her memoir *The Happy Hooker: My Own Story*: "My method? Basically the same principle as Masters and Johnson, only they charge thousands and it's called therapy. I charge $50 and it's called prostitution."[23] While Hollander does seem to facilitate sexual maturation for some of her clients, her comparison between her interactions with clients and SPT is somewhat exaggerated. For example, she tells the story of helping a shy recent law-school graduate have his first heterosexual experience after being referred to her by a psychiatrist. Unlike a surrogate, who would not initiate sex until completion of the other body awareness and intimacy-building exercises with a client, Hollander initiated sex with this client shortly after his arrival. Her purpose was to seduce him, not to help him through the process of cultivating a sexually intimate relationship. After they finished having sex and the client was about to leave, presumably never to see Hollander again, she told him, "I think you are cured, and I'm glad. However, I was the aggressor today, but from now on it is up to you. Don't be afraid of women, just try to find the type you like, and act like a man, not like a baby. And good luck."[24]

This quote illustrates that although Hollander has substantial experience as a sex worker, she has not been trained as a sexual health professional. IPSA's guidelines as well as guidelines for other sexual health professional groups would likely frown upon claiming that a client was "cured" in such a short period of time. Even if a client were cured, it is not clear after reading Hollander's account of the situation that the man would not experience sexual challenges in other unpaid heterosexual situations. Hollander could, in all likelihood, receive training to become a surrogate, but this portion of her memoir speaks to the difference between therapeutic and nontherapeutic sex work. Therapeutic sex work is meant to help a client improve their ability to share in sexual intimacy or gain greater awareness of how their body sexually functions, presumably over an extended period of time. Nontherapeutic sex work is meant to provide a client with an enjoyable and short-term sexual experience, sometimes to encourage repeat business but other times as a single sexual interaction.

My critique of Hollander is not meant to imply that nontherapeutic sex workers cannot play a role in combatting sexual ableism. Some people with disabilities may want to hire a nontherapeutic sex worker and they should be able to in societies where sex work is legal. Rather, I present this example to show how Jeffreys and like-minded critiques of SPT underestimate the amount of professional training and expertise that goes into successfully working with someone in a therapeutic situation. For example, in order to become certified through IPSA, a surrogate must complete a 100-hour didactic and experiential learning course in human sexuality and surrogate partner therapy. They must also complete an internship program where they work with clients under the supervision of a certified surrogate. The purpose of this training is to teach the surrogate how to guide the client through exercises focused on promoting body awareness, relaxation, and communication skills.[25] While some nontherapeutic sex workers may guide clients through similar exercises, they are less likely than surrogates to receive this kind of structured clinical training.

There are many ethical complexities with regard to sex work that make the distinction between SPT and nontherapeutic sex work difficult to identify. It is beyond the scope of this chapter to resolve these issues, although I have said more about them in my previous work on sexual ableism.[26] My purpose thus far has been to provide the conceptual context in which discussions of sexual ableism and sex work have become intertwined. In a literature focused in many cases on physical access to sex, it is understandable why sex work is seen as a potential means of allowing people with disabilities to exercise their sexual capacities. In some cases, like O'Brien's, sex work can have an empowering effect on people's sexual lives. The question remains, however, whether there are limitations to emphasizing access to sex work in ethical discussions of sexual ableism.

Why Sex Work Is Not Always the Best Solution for Sexual Ableism

While various forms of sex work are an important means of combatting sexual ableism, ethicists writing on disability have compelling reasons to qualify their endorsement of this solution. While sex work is a kind of sexual expression, it is hardly the only kind of sexual expression. In societies committed to individual liberty and moral pluralism, adults without disabilities have freedom of consensual sexual choice. They can decide whether they want to have sex and with whom, whether they want to have casual sex or have sex be a part of intimate relationships and the like. By overemphasizing access to sex work for people with disabilities, sexual ethicists risk neglecting considerations of how people with various impairments can have access to the same sexual freedoms and responsibilities as any other consenting adult.

Specifically, overemphasizing sex work at the expense of other potential remedies for sexual ableism excludes the concerns of people with disabilities who have the desire to lead a sexual life without involving a sex worker. Recall from the beginning of this chapter how Ryan from *Special* assured the sex worker he hired that he would soon have "sex for free." His intention, it seems, is not to assure the sex worker, who likely has an interest in Ryan's repeat business. Instead, Ryan appears to be reassuring himself that he will eventually have sexual experiences with others that do not involve hiring someone. This goal is a reasonable one shared by many people with disabilities and one that sexual ethicists scrutinizing sexual ableism should not ignore.

Relatedly, because societies are gendered such that it is considerably more stigmatizing for women to hire sex workers, an overemphasis on sex work in the sexual ableism literature deprioritizes the sexual discrimination that women with impairments confront. For example, as much as Peace's "head nurse" narrative challenges sexual ableism, a skeptical reader could reasonably charge that if the patient were female and the nurse were male, the story would not convey sexual empowerment.[27] Instead, the image of a female patient with a recent spinal cord injury receiving oral sex from a male nurse would likely invoke concerns about sexual assault.

These concerns would be legitimate in light of the evidence that women with disabilities are more susceptible to being victims of sexual violence.[28] While women with disabilities could and should be able to access sex workers on an equal basis with others, most of the ethical literature tying sexual ableism to sex work focuses on the experience of male clients. This focus deserves further scrutiny, even though it is, in part, a reflection of the fact that the majority of clients seeking therapeutic and nontherapeutic sex work are men. Focusing on the perspective of male clients likely downplays the gendered elements of interactions between sex workers and their disabled clients.

Equally important, advocating sex work as the default solution to sexual ableism helps to commodify the sexualities of people with disabilities. Patricia Marino interviews me in her book *Philosophy of Sex and Love: An Opinionated Introduction*, where I describe my own challenges with sex as a gay man with cerebral palsy. I explain that before meeting my long-term partner, sexual health professionals would always mention my hiring a surrogate or sex worker as the first potential solution to my concerns.[29] Over time, I came to see this reaction as demeaning. Although I have no moral objections to consensual sex work, I began to feel as though these professionals were asking me to put a price tag on how much my sexuality was worth. As someone who pays for physical assistance with nearly every aspect of everyday life, from getting dressed in the morning to writing academic papers, I have never wanted my sexual life to be yet another part of my disability-related expenses. The one time I seriously considered undergoing SPT, the first session was going to cost approximately $1000. The prospect of paying such a large amount of money to explore my sexuality left me with a dehumanizing feeling, as though sexual well-being was only obtainable if someone was wealthy enough to afford it.

In her critique of sex work, Elizabeth Anderson charges that prostitution inappropriately commodifies intimacy in that money cannot adequately compensate someone for sharing their sexual capacities with a partner. The good of sexual intimacy can only, in Anderson's view, be fully exchanged as an unpaid "gift" whereby partners are motivated by a reciprocal and genuine attraction to one another.[30] In my earlier work, I criticize this objection because it does not fully acknowledge that other activities, particularly in the mental health field, involve exchanging money as part of a relationship that also involves intimacy. A psychotherapist can develop a kind of professional intimacy with their clients. In fact, this intimacy might be necessary in order for the services the therapist provides to be effective. Yet, this professional does not provide their services as a "gift," expecting their clients to reciprocate the therapeutic support provided.

Similarly, had I undergone SPT, I would likely have cultivated some kind of intimacy with my surrogate. Yet, it would have been a different kind of intimacy than that shared between romantic partners. In her TEDx talk "Sexuality & disability -- a seat at the table," Cohen Greene reflects on the intimacy she shared with O'Brien when they were able to successfully have sex. She remembers telling O'Brien, "I love you in this very moment for what we just shared with each other and when you have your future relationship it's going to be so much more fulfilling."[31] Her reflection illustrates that sex workers can experience a particular kind of intimacy with their clients. Nonetheless, she recognizes that her role is not that of a romantic partner, but a sexual health professional guiding a disabled client through a very intimate experience. Surrogates are legitimately paid for their time and expertise, which if used effectively can cultivate professional intimacy, but they are still being paid to provide therapeutic intervention. For my part, even though I objectively recognize that I would have paid my surrogate for their time, services, and expertise, I wanted my sexual life to be defined in terms of this romantic rather than professional intimacy. Professional intimacy can foster relationships that help disabled people confront individual obstacles, but it does very little to change the attitudes and other social barriers that make it harder for a disabled person to find a casual or long-term sexual partner.

Consider the highly theorized debate between the medical and social models of disability. The medical model holds that the challenges people with impairments confront are individualized to their bodies and that the aim of social interventions should be to treat or cure their condition.[32] The social model of disability, by comparison, holds that disability is created by the interaction of someone's impairment with barriers in the built environment.[33] Historically the two models have been seen as oppositional but much recent ethical scholarship has pointed to how both models can be used to justify different kinds of needed interventions that improve the lives of disabled people.[34]

Sexual ableism is a case that shows why the models should not be mutually exclusive. The challenges people with disabilities confront in having sex can be both individualized to them and indicative of their social stigmatism and isolation. On the surface, advocating access to sex work is compatible with the more radical aims of the social model. Sex workers are marginalized and stigmatized through a set of institutionalized arrangements that make it more difficult for them to safely practice their profession. Similarly, according to the social model, people with impairments are rendered disabled by institutionalized arrangements that deprioritize, if not outright ignore, their bodily limitations. Advocating for access to sex work for people with disabilities would seem to challenge societal prejudices toward both communities. In certain respects, this is the case, particularly if one is sympathetic to Schwarzenbach's hypothesis about the transformative effect SPT might be able to have on the sex work industry.

Yet, there is also a sense in which advocating for access to sex work reinforces a problematic assumption that sexual ableism is, at its core, an individualized phenomenon warranting specialized professional intervention. Particularly in societies where sex work is illegal, creating exemptions for people with disabilities frames sex work as an individualized entitlement that potentially exacerbates the differences between the disability community and others. Such measures in the absence of other social interventions might very well enable some people with impairments to have sex, but would not challenge the discriminatory assumptions that underlie all forms of sexual ableism.

Consider the following example from *Loneliness and Its Opposite*, a comparative anthropological study of sexual supports for people with impairments in Sweden and Denmark. Rasmus is a largely nonverbal gay man with cerebral palsy living in a group home in Denmark. The staff of the group home is supportive of his being able to sexually express himself, so they help him hire a sex worker who comes once a month and is paid $300 for an hour-long session. The sex worker, Peter, has learned how to work with Rasmus so that he is able to fully enjoy their hour together without becoming exhausted by their sexual activity. Peter's visits are, according to the authors, the only opportunities Rasmus has to have sex. Moreover, the authors imply that Rasmus is largely unable to physically reciprocate Peter's sexual acts.[35]

While this story is framed positively within the study, one is left to wonder whether there are adjustments that could be made in Rasmus's situation so that his sexual life was not defined in terms of his paid relationship with Peter. Moreover, could some of the money spent on Peter's services be put toward adaptive equipment to help Rasmus participate more reciprocally in his interactions with Peter or potentially other sexual partners? These are ambiguities that warrant clarification from a social model perspective. As presented, however, Rasmus's challenges with sex are analyzed primarily from his individual challenges with communication and the spasticity in his limbs. Peter is a quasi-therapeutic professional who provides Rasmus with opportunities to have sex, but it is unclear whether resources and opportunities could enable Rasmus to have a broader range of options in his sexual life.

In order to expand the range of sexual options available to individuals like Rasmus, it seems necessary to examine how Bedi's extension of the relational distributive structure could be responsive to sexual ableism concerns. The laws and professional standards governing SPT and nontherapeutic sex work are one component of this structure since they determine how, if at all, people with disabilities are able to pursue particular kinds of sexual relationships. However, making the relational distributive structure accessible to people with disabilities requires much more. It might involve giving people like Rasmus access to assistive technology that enables them to participate in reciprocal social and sexual activities. I have in mind augmentative communication devices that provide the nonverbal with a means of communicating, as well as devices that help someone with physical challenges sexually stimulate a partner or help them masturbate. Further, making the relational distributive structure accessible to people with disabilities also necessitates ensuring that bars, sex clubs, and other public spaces where people seek out sexual activity, are welcoming to customers with a wide range of impairments.

Moreover, people with recent spinal cord injuries and other acquired disabilities should have access to sex education in rehabilitation programs that is not contingent on relationships like the one Peace had with his nurse. Students with and without disabilities in public schools might receive access to age-appropriate sexuality education that combats sexual ableism by addressing how different chronic illnesses might affect sexual functioning, sexual identity, and the importance of respectful sexual communication. Independent filmmakers might be given grant funding to produce erotica that portrays people with different impairments as sexually desirable and capable of taking responsibility for their own sexual satisfaction. Hollywood filmmakers might also be given incentives to produce films that portray people with disabilities actively pursuing various forms of consensual sexual relations. These are all measures that would combat sexual ableism but fall outside the scope of access to sex work. Reforms of this nature hardly guarantee that all people with disabilities will lead fulfilling sexual lives but they would ensure that members of disability communities would have a greater opportunity to pursue a sexual life not defined by ableist assumptions about their potential as sexual partners. Such areas constitute the next frontier for ethicists scrutinizing the pervasiveness of sexual ableism across societies.

Conclusion

This chapter has provided a brief exploration not only of why sexual ableism has become so closely linked to ethical debates on sex work, but also of the promises and limitations of advocating for disabled people's access to sex work. The challenges of people with impairments in accessing sex are often framed as obstacles to sexual acts. This framing contributes to SPT and nontherapeutic sex work being considered as the primary means to engage in sexual activity. It also captures the public imagination and encourages feminists on all sides of the prostitution debate to refine their positions. SPT and nontherapeutic sex work are important albeit controversial means of challenging sexual ableism because these services provide people with disabilities access to very particular kinds of sexual activity.

At the same time, they should only be considered as individual parts of a wider societal effort to combat sexual ableism. In and of themselves, these interventions do not provide access to the vast array of sexual expressions that nondisabled consenting adults can enjoy, at least in societies committed to promoting freedom of sexual choice. They are also likely to neglect the concerns of women with disabilities because of the gender dynamics surrounding sex with a paid professional. An overemphasis on sex work further commodifies the sexualities of people with disabilities, giving some of them reason to feel as though their sexual well-being comes with a monetary value. Equally important, overemphasizing these interventions comes with the risk of understanding sexual ableism as strictly an individual problem, likely neglecting the promise of measures that challenge ableist assumptions in the sexual preferences and behaviors of others.

Because of the chapter's limited scope, I have not addressed another critical area in the ethical literature on sexual ableism. In particular, how should public policy contend with the concerns of individuals whose disabilities make it difficult or impossible to provide sexual consent? Much of this literature focuses on how best to protect people with the conditions from sexual harm at the hands of others, as well as the question of whether care providers or loved ones can ethically "facilitate" the sexual life of people with intellectual impairments.[36] Both questions warrant more consideration than I can provide here. However, given the focus of this chapter, the second question is particularly worthy of discussion. Writers who advocate for sexual facilitation for people with intellectual disabilities have made an invaluable contribution to discussions of sexuality and intellectual disability in moving legal and ethical debates beyond how best to protect those with intellectual impairments from sexual harm.[37]

At the same time, they risk painting themselves into a similar corner as those who advocate access to SPT and nontherapeutic sex work. Paid caregivers who serve as sexual facilitators for those with intellectual disabilities do not perform the same labor as surrogates or other sex workers. However, they are performing a form of labor that risks commodifying the sexuality of their clients. This risk is likely inevitable for many people with intellectual impairments who need substantial support in pursuing a sexual life safely and satisfactorily. Nonetheless, discussions of how best to enable people with intellectual disabilities to combat sexual ableism should also emphasize support for comprehensive, adaptive sexuality education that maximizes the extent to which people with intellectual disabilities can engage in thoughtful sexual decision-making.

From the perspective of sexual ethics, future scholarship needs to assess the appropriate level of sexual risk to which it is ethical to expose adults with intellectual disabilities. Addressing this question would help illuminate how educational, healthcare, and other institutions, should shape the sexual lives of these individuals. In doing so, sexual ethics would be positioned to challenge many of the discriminatory assumptions that make sexual ableism an all-too-common part of the disability experience.

Notes

1. William Peace, "Head Nurses," *Atrium: The Report of the Northwestern Medical Humanities and Bioethics Program* 12 (2014): 20-22.
2. Tom Shakespeare, Kath Gillespie-Sells, and Dominic Davies, *The Sexual Politics of Disability: Untold Desires* (London and New York: Cassell, 1996): 3-6.
3. *The Sessions*, directed by Ben Lewin (Fox Searchlight Pictures, 2012), 95min, Amazon Prime Video Streaming, https://www.amazon.com/Sessions-John-Hawkes/dp/B00AYLTMMM/ref=sr_1_1?keywords=the+sessions&qid=1557498771&s=instant-video&sr=1-1.
4. William H. Masters and Virginia E. Johnson, *Human Sexual Inadequacy* (London: J. &; A. Churchill, 1971), 1-91.
5. "Training," IPSA International Professional Surrogates Assn, accessed August 9, 2020, https://www.surrogatetherapy.org/training/.
6. "What is Surrogate Partner Therapy?," IPSA International Professional Surrogates Assn, accessed August 9, 2020 https://www.surrogatetherapy.org/what-is-surrogate-partner-therapy/what-is-surrogate-partner-therapy-2/.
7. *Special*, episode 3, "Chapter 3: Free Scones," directed by Anna Dokoza, written by Ryan O'Connell, aired April 12, 2019, on Netflix, https://www.netflix.com/title/80987458.

8. Kevin Todd Mintz, "Sexual Intimacy, Social Justice, and Severe Disabilities: Should Fair Equality of Opportunity in Health Extend to Surrogate Partner Therapy?" *The Journal of Philosophy, Science & Law: Disability Special Issue* 14, no. 3 (2014): 4-15.
9. Jacob M. Appel, "Sex rights for the disabled?" *Journal of Medical Ethics* 36, no. 3 (2010): 152-54, https://doi.org/10.1136/jme.2009.033183; Steven J. Firth, "Whither a Welfare-Funded 'Sex Doula' Programme?" *Journal of Medical Ethics* 45, no. 6 (2019): 361-64, https://doi.org/10.1136/medethics-2018-105330; Alida Liberman, "Disability, sex rights and the scope of sexual exclusion," 44, no. 4 (2017): 1-4, https://doi.org/10.1136/medethics-2017-104411; FrejKlem Thomsen, "Prostitution, disability and prohibition," *Journal of Medical Ethics* 41, no. 6 (2014): 451-59, https://doi.org/10.1136/medethics-2014-102215.
10. Isaiah Berlin, "Two Concepts of Liberty," in Four Essays on Liberty (New York: Oxford University Press, 1971), 121-22.
11. Appel, "Sex rights," 153-54.
12. Amia Srinivasan, "Does anyone have the right to sex?" *London Review of Books* 40, no. 6 (2018), https://www.lrb.co.uk/the-paper/v40/n06/amia-srinivasan/does-anyone-have-the-right-to-sex
13. Kevin Todd Mintz, "Sex positive political theory: pleasure, power, public policy and the pursuit of sexual liberation," (PhD diss., Stanford University, 2019), 241, https://searchworks.stanford.edu/view/13312967.
14. Sheila Jeffreys, "Disability and the male sex right," *Women's Studies International Forum* 31, no. 5 (2008): 327-35, https://doi.org/10.1016/j.wsif.2008.08.001.
15. Because this chapter focuses on the relationship between sexual ableism and sex work, people with intellectual disabilities incapable of sexual consent are not my focus. That being said, one of the main implications of my argument for moving the sexual ableism literature beyond its emphasis on sex work is to encourage more thoughtful discussion of how the intellectually disabled might be encouraged to navigate their sexual lives as independently as possible. For more on this issue, see the conclusion of this chapter.
16. Debra Satz, "Markets in Women's Sexual Labor," *Ethics* 106, no. 1 (1995): 65-67.
17. Sonu Bedi, "Sexual Racism: Intimacy as a Matter of Justice," *The Journal of Politics* 77, no. 4 (August 2015):1001-2.
18. *Breathing Lessons: The Life and Work of Mark O'Brien*, directed by Jessica Yu (Fanlight Productions and Icarus Films, March 1996), documentary, 35:10, https://youtu.be/JzEGZnWT5dA.
19. Shakespeare et al., *Sexual Politics,* 9-12.
20. Tom Shakespeare and Sarah Richardson, "The Sexual Politics of Disability, Twenty Years On," *Scandinavian Journal of Disability Research* 20, no. 1 (2018): 82-3.
21. Sibyl Schwarzenbach, "Contractarians and Feminists Debate Prostitution," in *Prostitution and Pornography: Philosophical Debate About the Sex Industry*, ed. Jessica Spector (Stanford: Stanford University Press, 2006), 232-36.
22. Jeffreys, "Disability and the male sex right," 333-34.
23. Xaviera Hollander, *The Happy Hooker: My Own Story*, with Robin Moore and Yvonne Dunleavey (London: Tandem, 1975), 163, EPUB.
24. Hollander, *The Happy Hooker*, 166.

25. "About IPSA Training," IPSA International Professional Surrogates Assn, accessed August 9, 2020 https://www.surrogatetherapy.org/training/about-ipsa-training/.
26. Mintz, "Sexual Intimacy."
27. To raise a related point, Peace refers to the nurses who are less willing to provide sexual assistance as the "dick police" in contrast to the willing "head nurses." This phrasing objectifies the nurses and devalues the other important roles they likely play in the rehabilitation process. Peace's writing speaks to how discussions of sexual ableism cannot be entirely separated from gender roles in a person's societal context. The challenge for any potential solution to sexual ableism is how to recognize people with impairments as sexual beings without exacerbating other existing inequalities in a particular society.
28. "Sexual Abuse," Disability Justice, accessed August 9, 2020, https://disabilityjustice.org/sexual-abuse/; Sara Beth-Plummer and Patricia A. Findley, "Women With Disabilities' Experience With Physical and Sexual Abuse: Review of the Literature and Implications for the Field," *Trauma, Violence, and Abuse* 13, no. 1 (2012): 15-29.
29. Patricia Marino, *Philosophy of Sex and Love: An Opinionated Introduction* (New York, NY: Routledge, 2019), 164.
30. Elizabeth Anderson, "The Ethical Limitations of the Market," Economics and Philosophy 6, no. 2 (October 1990): 185-192, https://doi.org/10.1017/S0266267100001218.
31. Cheryl Cohen Greene, "Sexuality & disability -- a seat at the table," filmed December 2013 at TEDxFiDiWomen, San Francisco, CA, video, 17:15, https://www.youtube.com/watch?v=gds2RvmCBKE.
32. Daniel Putnam, David Wasserman, Jeffrey Blustein, and Adrienne Asch, "Disability and Justice," The Stanford Encyclopedia of Philosophy, eds. Edward N. Zalta, Fall 2019 ed., https://plato.stanford.edu/archives/fall2019/entries/disability-justice/.
33. Putnam et al, "Disability."
34. Putnam et al, "Disability."
35. Don Kulick and Jens Rydström, *Loneliness and Its Opposite: Sex, Disability, and the Ethics of Engagement* (Durham and London: Duke University Press, 2015), 199-200.
36. Simon Foley, *Intellectual Disability and the Right to a Sexual Life: A Continuation of the Autonomy/Paternalism Debate,* Routledge Advances in Disability Studies (London and New York: Routledge), 48-50.
37. Julia Bahner, "Sexual professionalism: for whom? The case of sexual facilitation in Swedish personal assistance services," *Disability & Society* 30, no. 5 (2015): 788-801; Kulick and Rydström, *Loneliness and Its Opposite*; Shakespeare and Richardson, "The Sexual Politics," 90.

CHAPTER 27

Sexual Exclusion

Alida Liberman

Introduction

What does it mean to say that someone is sexually excluded? Should we attempt to remedy such exclusion, and if so, how? Sexual goods are often run together in academic and popular discussions of these questions, which leads to a lack of clarity at best and unhelpful misunderstandings at worst. For there are many kinds of sexual goods, and exclusion from one kind of good does not entail exclusion from another. Consequently, individuals are harmed in distinct ways by exclusion from different kinds of sexual goods, and appropriate responses to such exclusion vary greatly.

In this chapter, I clarify this conceptual space by articulating several distinct kinds of sexual exclusion: (1) lack of access to sexual pleasure; (2) lack of partnered sex; and (3) lack of the social or psychological validation that comes from being seen as a sexual being. I investigate the nature of the harm associated with each kind of exclusion, and offer proposals about what our collective responses to these harms should be. In doing so, I situate the discussion of these goods within the small but growing philosophical literature about sexual exclusion and inclusion, and weigh in on debates about whether there are rights to various kinds of sexual goods. I conclude that we ought to respond to sexual exclusion by providing mechanical assistance to those who are incapable of self-stimulation, enhancing access to sexual education for everyone, and engaging in a broad and systematic effort to change the harmful social norms, stereotypes, and cultural ideals that drive exclusion from partnered sex and that can lead those who are sexually excluded to feel socially invalidated.

A. Liberman (✉)
Southern Methodist University, Dallas, TX, USA
e-mail: aliberman@smu.edu

D. Boonin (ed.), *The Palgrave Handbook of Sexual Ethics*,
https://doi.org/10.1007/978-3-030-87786-6_27

The Scope and Framing of the Problem

Clarification about the scope and framing of the topic of sexual exclusion is in order before proceeding. First, discussions of sexual exclusion in the philosophical literature are most often about how to respond to the sexual exclusion of people with disabilities. This debate focuses on whether to give people whose disabilities lead to sexual exclusion special permission to purchase sex (in jurisdictions in which prostitution is otherwise illegal), and whether sexual facilitation services for disabled people should be publicly funded or provided via charitable organizations.[1]

However, as I argue elsewhere,[2] this narrow focus on sexual exclusion and disability is both unhelpful and potentially harmful. This is due to the fact that the potential solutions that authors address are remedies to the harms of *being sexually excluded* rather than remedies to any harms that stem directly or essentially from being disabled. For being disabled is neither necessary nor sufficient for being sexually excluded; there are many people without disabilities who are sexually excluded and many people with disabilities who are not. It is true that disability is a frequent cause of exclusion from partnered sex (and the social validation that comes with it) in our ableist society. But narrowly focusing on disability status as the feature that merits a special right to sexual inclusion risks sending a false and harmful message that disability is necessarily a cause of sexual exclusion, while unhelpfully skirting around the core issue of sexual exclusion simpliciter. We should instead focus on sexual exclusion per se.

John Danaher takes up my call to focus directly on sexual exclusion, offering the first broad treatment of sexual inclusion in the analytic philosophy literature. He argues that sexual experiences should be treated as a distributive good and that sexual inclusion should be thought of "first and foremost in terms of distributive justice—i.e. as something to be addressed through principles and ideals of equality,"[3] especially for those who are sexually excluded through unjust or unfair mechanisms (e.g., prejudice or discrimination). Treating sex as a distributive good does not automatically entail that there are any rights to sex, and Danaher emphasizes that individuals never have a positive right or entitlement to have sex with anyone. But he argues that "there are some compelling moral reasons to favour recognising a bundle of both positive and negative claim rights to sexual inclusion,"[4] and suggests that candidate rights might include the removal of legal restrictions on sexual activities along with the provision of sexual education, mechanical sex aids or toys for those who cannot obtain them on their own, and (possibly) anti-discrimination training. Danaher then defends his view against four possible objections: that it upholds misogyny, that it is impossible or impracticable, that it contributes to the stigmatization of the sexually excluded, and that it involves problematic social engineering.

This chapter elaborates and expands on Danaher's project by more thoroughly assessing what kinds of negative and positive sex rights exist and examining how we should foster sexual inclusion of various sorts even when there is no right to it. Danaher focuses on whether sexual exclusion is unjust, and frames the mechanisms of sexual exclusion in terms of the (potentially overlapping) reasons why one is excluded: (1) personal features of the excluded individual (such as shyness, awkwardness, excessive romantic idealism, or overly demanding standards for potential partners); (2) social features belonging to others (such as prejudicial attitudes against the excluded individual's group or laws and social norms that discourage or prohibit certain sexual behaviors in ways that impact the excluded individual); and possibly (3) natural "mechanisms of exclusion that result predominantly or exclusively from evolved instincts or drives that lead people to favour certain kinds of sexual partner or sexual experience over others."[5]

I adopt a different framing, and instead address what kinds of harms sexual exclusion leads to and what our responses to these harms should be. Accordingly, I focus not on the mechanisms of exclusion but on the different kinds of sexual goods from which one can be excluded. For the kind of good one is excluded from tracks more closely with questions about rights and rectification than do the personal or social features that lead to exclusion. Whether and how an instance of sexual exclusion ought to be remedied depends on what sort of sexual good is lacking, rather than on whether the exclusion is due to personal or social factors. I find part (3) of Danaher's taxonomy especially unhelpful. As Danaher notes, the extent to which our sexual preferences are products of evolution—as well as the extent to which evolved preferences can be shaped or modified by culture or individual effort—is a complex empirical question for which the evidence is mixed. More importantly, whether an act of sexual exclusion is unjust depends on how it harms people and whether these harms can be remedied, and not on whether it is grounded in a preference that is "natural" or "social."[6]

Finally, in arguing that important sexual goods exist and that exclusion from them can be harmful in ways that call out for collective and individual responses, I do not mean to assume that any particular sexual good is essential to any individual person's well-being. Individuals who choose to be celibate for religious or other reasons can presumably lead flourishing lives, as can people who are asexual and accordingly feel neutral about or averse to engaging in solo or partnered sex.[7]

Exclusion from Solo Sexual Pleasure

The most basic form of sexual exclusion is the inability to attain sexual pleasure simpliciter, by which I mean stimulation leading to sexual satisfaction. Individuals will vary greatly in both the degree of their motivation to seek pleasure and in what kinds of experiences they find pleasurable, and we should be broadly inclusive of what counts as sexual pleasure. However, for the sake of discussion, I am ruling out sadistic pleasure that is inextricably bound up in another's nonconsensual pain or degradation, and am not counting this among the sexual goods from which one might be excluded.[8]

Exclusion from solo sexual pleasure is the only sort of sexual exclusion that plausibly grounds any positive claim rights. I presume that if X is a very important good, then individuals have a prima facie positive right to obtain X. This claim is highly qualified: the "very" constraint restricts it to only those goods that are central to human flourishing, and the "prima facie" constraint accommodates the many ways in which even very important goods can be unattainable and thus not something to which we have a fulfillable right (e.g., if there is a limited quantity or no way to provide them without infringing on others' more important rights). Because solo sexual pleasure is a very important good that can be readily obtained without infringing on others' rights, there is a positive right to solo sexual pleasure.

While there are religious, cultural, and philosophical traditions that denigrate sexual pleasure, or see it is as something to be managed or avoided rather than celebrated and sought after, in most secular contexts sexual pleasure is recognized as an important part of human flourishing. For example, it is a component of major philosophical theories of well-being. Hedonist theories will include sexual pleasure as one kind of intrinsically valuable pleasure. Desire-satisfaction theories of well-being will include sexual pleasure as a component of well-being so long as most individuals in fact desire it (and on some theories, would continue to desire it were they fully informed and rational). And objective list theories of well-being often include sexual pleasure on the list of objectively good things (whether as a distinct good, or more commonly, as a species of general pleasure). Sexual pleasure can also be understood as a component of physical and mental health, and the positive right to it subsumed under a right to receive what is necessary for good health. For example, the 2014 World Declaration of Sexual Rights includes the right to "the highest attainable standard of health, including sexual health; with the possibility of pleasurable, satisfying, and safe sexual experiences."[9]

Thankfully, solo sexual pleasure is not a good for which there is a fixed quantity. Nor does the pursuit of it usually require infringing on others' more important rights; so long as there is opportunity (and in some cases, access to education, assistive technology, or other forms of facilitation), sexual pleasure can be attained by oneself.[10] Furthermore, feminist scholars and activists have long argued that framing sexual pleasure as an important good to which we have a right is important for egalitarian reasons. Women's sexual pleasure is often undervalued compared with men's, or is overlooked in sexual encounters with men. Acknowledging a positive right to sexual pleasure may help combat this trend. It follows that individuals have a prima facie, defeasible, positive right to experience solo sexual pleasure, to the extent that this is possible and desirable for them. This means that we have the right to acquire whatever tools, knowledge, and other resources are needed to enable the pursuit and attainment of sexual pleasure.

Exclusion from solo sexual pleasure can stem from multiple causes, which require different responses. These causes include (but are not limited to) (1) lacking the physical ability to masturbate; (2) lacking knowledge of sex and/or an understanding of one's own sexuality; and (3) internalized cultural or religious shame or a history of trauma that hinders sexual functioning (independently of physical abilities). It is beyond my scope to address this third source of exclusion, which is better addressed by those with training in psychology and sexology. However, it is a source of exclusion from solo sexual pleasure that is clearly unjust and problematic, and those with relevant expertise should work to develop individual and collective strategies for responding to it.

Lack of Physical Ability and Mechanical Sexual Assistance

An inability to provide oneself with sexual pleasure can be caused by physical disabilities or medical conditions that make self-stimulation challenging or impossible. People who lack this ability have a pro tanto positive right to mechanical sexual assistance. Such assistance is on a par with other kinds of assistive services, and it should be provided for relevantly disabled people in whatever way healthcare and accessibility more generally are provided. Although I cannot argue for this claim at present, I assume that the state has a moral obligation to provide its citizens with comprehensive healthcare services, including meeting the accessibility needs of people with disabilities. If this is correct, it follows that mechanical sexual assistance ought to be state-funded as well. If this is not correct, then mechanical sexual assistance ought to be provided via private insurance schemes or whatever other means by which other healthcare services are provided.

Mechanical sexual assistance might involve specialized masturbation devices used by the excluded individual alone or with the assistance of a medical worker or caretaker. Adaptive technology exists to help people with disabilities move, cook, eat, engage in personal grooming, get dressed, use a computer, and more. Adaptive technologies might assist with sexual self-stimulation in a similar way.[11] Mechanical sexual assistance might also include assisting with the mechanics of partnered sex, such as placing someone with limited mobility into a certain position, setting up specialized furniture,[12] or assisting with putting on a condom.

In general, positive rights are limited insofar as they infringe on the rights of others, and the same is true of sexual rights. It follows that caretakers have a right to refuse to engage in sexual facilitation as part of their right to sexual liberty. Sexual behavior is considered especially intimate and private by many people in a way that other physical processes (such as bathing or using a toilet) are not. Moreover, the boundaries of what counts as participating in a sex act are not always clear, and certain types of sexual assistance might make the caretaker feel like they are personally involved in the sexual act or are themselves being treated as an object of sexual desire. No caretaker who does not wish to do so should be required or otherwise pressured to engage in sexual assistance.

Two solutions would enable excluded people to attain sexual pleasure without violating the sexual liberty rights of caretakers. First, caretakers could choose to join a registry of those willing to perform sexual assistance services.

They might receive additional training in this area, and perhaps be entitled to additional compensation.[13] There could also be specialized caretakers who provide this service by referral for clients whose usual caretakers are unwilling to do so.

Second, in the near future, we might develop machines that can provide sexual pleasure without the involvement of another person. Existing sex "robots" are essentially interactive sex dolls. While they use artificial intelligence to engage in basic conversation with users and display facial reactions, they are fundamentally passive and are not able to proactively engage in any sexual behavior.[14] More advanced sex robots might be developed that could provide sexual stimulation in a more autonomous way. This would be one way to satisfy the positive right to solo sexual pleasure without imposing on anyone else.[15] However, we need to tread extremely cautiously in order to avoid two major potential problems with sex robots. First, we cannot presume (as many authors writing on the topic do) that a sufficiently advanced sex robot would not itself be a moral person or a bearer or rights.[16] We must accordingly take great pains to monitor the moral status of such robots and to avoid potentially harming them or violating their rights.

Second, most current sex robots are designed to have compliant and controllable personalities, with unrealistic and highly exaggerated female bodies that are modeled on pornography and meant to appeal to a stereotypical male gaze. It is easy to imagine how their use might encourage or exacerbate the objectification and dehumanization of actual women.[17] But sex robots that are constructed to provide solo sexual pleasure as a matter of rights satisfaction need not be sexist in these ways. They might be designed in a non-humanoid way to avoid contributing to women's sexual objectification.[18] Danaher suggests drawing on various models of sex-positive feminist pornography to think about how we might construct "feminist sexbots"—for example, by including more women in their design and production and building robots that are more diverse in their physical representations and behavioral scripts.[19]

Lack of Knowledge and Provision of Sexual Education

Lack of access to sexual pleasure can also stem from a lack of knowledge, as pleasurable solo sex may not be available to individuals who do not have access to accurate and comprehensive information about how sexual pleasure works and the myriad ways in which human sexuality can be expressed. Lack of sex education can also lead to bad, dangerous, or frustrating partnered sex, which makes the moral imperative to provide comprehensive sex education overdetermined; people need knowledge both to pleasure themselves and to safely and ethically pursue the kinds of partnered sex that they desire. The current state of sex education in the United States is severely lacking.[20] Twenty states do not require sex education to be provided at all. Only twenty states (and D.C.) require the provision of information about contraception, and only nine states (and D.C.) require emphasizing the importance of consent. While eleven states (and D.C.) require the provision of inclusive content about sexual orientation, six states require that homosexuality be presented in only a negative light. Even people who receive comprehensive and inclusive sexual education in school are subjected to a myriad of false and harmful myths about sexual pleasure stemming from pornography, media, and popular culture. Women's sexual pleasure is especially likely to be misunderstood, with the false assumption that most women can and should orgasm quickly through vaginal sex alone being among the most persistent and pernicious myths.[21] LGBT+ people are especially likely to lack access to appropriate and adequate sex education,[22] as are people with disabilities.[23]

In general, we have a right to obtain the knowledge necessary for accessing the important goods to which we are entitled, as a corollary of our positive right to these goods. It follows that there is a positive right to receive adequate sexual education. This is endorsed by the U.N. Special Rapporteur on the right to education in a 2010 report.[24] While some parents provide their children with comprehensive sexual education, many others avoid the topic completely or teach their children false, restrictive, or harshly judgmental information. Sexual education should therefore be provided by the state via public schools, as well as free educational websites or printed materials accessible to those who do not attend public school or who are no longer enrolled in school. Special care should be taken to provide education for people whose sexualities are under-represented in popular media and mainstream discussions.

Whether parents should be permitted to exempt their young children from sexual education in public schools on religious or other grounds is a fraught question that is beyond the scope of this chapter. However, it is important that children are able to choose for themselves whether they wish to access sexual education when they are older; parents should not be able to exempt adolescents who do not wish to be exempted, or to prevent them from accessing educational materials on their own.

Exclusion from Partnered Sex

Most discussions of sexual exclusion focus on exclusion from partnered sex. Partnered sex is frequently connected to (and sometimes conflated with) the goods of intimacy, romance, and love. While these goods are extremely important, I am focusing only on partnered sex itself, and not on the other relational goods that it can facilitate. For partnered sex is plausibly a distinctive good even apart from these relational goods. It can provide pleasure of a different sort or degree than can solo sex; as Jacob Appel notes, "the act of physical contact with another human being appears to provide a degree of pleasure and fulfilment entirely distinct from mere masturbation, even when that contact lacks emotional intimacy."[25] Even apart from the ways in which partnered sex can bolster romance and emotional intimacy, it is frequently intertwined with other goods, such as taking pleasure in making others happy, enjoying being made happy by others, and the satisfaction of participating in a mutually beneficial project.

The Negative Right to Pursue Partnered Sex

Although partnered sex is an important good, there is no positive right (of even a prima facie sort) to receive partnered sex on demand or from any particular person. Such a right is unfulfillable: it would inevitably trample on others' negative rights to bodily autonomy and self-determination, since any putative partner has the right to refuse. Appeals to such a right also display a morally indefensible sense of entitlement and disrespect for the sexual autonomy of others.

However, it is plausible that there are negative rights to pursue consensual partnered sex. This right may be overridden in cases that involve other types of serious harm.[26] Possible examples include consensual incest (which risks harm to family structures and potential offspring) and consensual relationships where there are serious power differentials and/or conflicts of interest (e.g., between bosses and employees, faculty members and their students, therapists and their clients, politicians and lobbyists, etc.). If, as some feminists have argued, some or all forms of commercial sex are in fact seriously harmful—whether to sex workers themselves, or to women in general, or both—the negative right to buy or sell sex would be ruled out as well. Barring such cases, it is rights-violating and morally impermissible to enact policies or laws that prohibit individuals from pursuing nonharmful sex in private with consenting partners.

Obvious violations of the negative right to pursue partnered sex include laws that restrict sexual activity on the basis of race, sex, gender, religion, ethnicity, caste or class, and the like. Less obvious violations include policies against sex in assisted living facilities, nursing homes, and other residential institutions. While most institutions lack formal policies about sexual behavior between residents, many informally prohibit or discourage residents from engaging in consensual sex with each other, including by displaying negative and patronizing attitudes, depriving residents of privacy (e.g., with unlocked door policies and regular room checks), and deferring to the preferences of a resident's family about whether sexual behavior should be permitted.[27] Such policies are rights-violating and accordingly should not be permitted, unless they are needed to prevent serious harm or abuse. (For example, there are complicated and important questions about whether people with dementia or certain forms of intellectual disability have the capacity to genuinely consent to sex, which I cannot answer here.)

Similar questions arise for young adults living in residential schools and colleges. Some private institutions (such as conservative Christian colleges) prohibit students from engaging in partnered sex as part of their codes of conduct. However, this is presumably morally permissible so long as students voluntarily undertake these restrictions (e.g., they know about them when they enroll) and are able to escape from them if they so choose (e.g., they're permitted to quit or transfer if they want to). A trickier question is whether boarding schools for high school students should prohibit students from engaging in consensual partnered sex. Students who are not legal adults may not have voluntarily accepted these restrictions, as their parents may be determining which school they attend. But if the potential for coercion or abuse is high enough, this could justify putting restrictions in place.

Even trickier is the question of whether it is morally permissible for prisons to prohibit consensual sex between inmates. Many U.S. prisons forbid inmates from engaging in masturbation,[28] and sex between inmates is illegal in all U.S. states and federal prisons. (Sex between inmates and guards is also illegal, and rightfully so, as the potential for power abuses is very likely to lead to serious harm.) All else being equal, inmates should be permitted to pursue both solo sexual pleasure and consensual partnered sex, and policies forbidding this are problematically rights-violating. However, all else is likely not equal, as rates of rape, sexual abuse, and sexual coercion in jails and prisons are very high. Prison dynamics can also make it challenging to know whether sex is truly consensual, while a strong prohibition against "snitching" strongly discourages reporting violations. Whether these harms are serious enough to merit restricting inmates' right to pursue partnered sex is a complicated empirical and normative question worth further investigation.

Fostering Inclusion in Partnered Sex

Not everyone who has the right to pursue partnered sex will be able to successfully find a partner, for a variety of reasons. Geographic remoteness might limit access to suitable partners. Social anxiety might make it challenging to connect with new people. And our cultural standards of sexual desirability strongly prioritize people who are young, light-skinned, thin (for women) or muscular (for men), cisgender, nondisabled, and conventionally attractive, which can make it more challenging for people who do not fit these categories to find partners. As Don Kulick and Jens Rydstöm note, "lots of people, many of whom have no physical or intellectual impairments to speak of at all, are hindered in their search for erotic fulfilment and love by other people's prejudices, their own insecurities, and by their lack of access (because they are the wrong race, age, class, etc.) to social arenas where they conceivably might meet an erotic partner."[29]

Because partnered sex is an important good, and exclusion from it can be very painful, it would be better if more people who desired sex partners were able to find them. In light of this, we should think about how we might foster greater sexual inclusion. A crucial constraint in doing so is avoiding dangerous rhetoric about sexual entitlement. A small subset of men who are unable to find a sex partner identify as "incels" or "involuntary celibates" in online forums where they commiserate and complain about the fact that they are not having sex.[30] Their posts range from sad and sincere confessions of loneliness to overtly misogynistic tirades. A recurrent theme among incels is fierce anger—frequently accompanied by moralized resentment and sometimes by disgust—at women whom they perceive as causing their unhappiness by withholding sex from them. Discussion posts on incel forums are frequently sexist (e.g., categorizing women into hierarchies based on sexual desirability, engaging in slut-shaming) and dehumanizing (e.g., referring to women as "femoids" (short for "female humanoids"), using racial slurs).[31] They also contain morally horrific proposals about how to respond to sexual exclusion, including legalizing rape and "redistributing" sex through cash incentives or state-funded prostitution.[32] These ideas sometimes legitimize or encourage real-world violence (as with the killing sprees of Elliot Rodger in 2014 and Alek Minassian in 2018).[33]

How can we respond to exclusion from partnered sex in ways that do not lend any fuel or credibility to these sorts of narratives? One simple strategy is to develop tools that enable people who are sexually excluded due to common prejudices or conventional beauty and body norms to find each other. Dozens of specialty dating apps exist to help people with niche interests, needs, and preferences find each other, as do in-person dating clubs and groups.[34] A more challenging—but also potentially more fruitful—strategy is to think about how we should critically investigate our own sexual preferences and the cultural scripts that cement them, with the aim of potentially encouraging ourselves and others to be more open-minded or inclusive in who counts as a viable sexual partner.[35]

Amia Srinivasan articulates the importance (and risks) of such a project in a powerful and incisive essay. She argues that even though sex is clearly not a good of the sort that can or should be redistributed, and there is no positive right to sex, we can still ask questions about whether our existing sexual preferences are fair or just:

> Yet simply to say to a trans woman, or a disabled woman, or an Asian man, "No one is required to have sex with you," is to skate over something crucial. There is no entitlement to sex, and everyone is entitled to want what they want, but personal preferences—NO DICKS, NO FEMS, NO FATS, NO BLACKS, NO ARABS, NO RICE NO SPICE, MASC-FOR-MASC—are never just personal.[36]

Srinivasan suggests that we move away from an uncritical sex-positivity that focuses exclusively on consent without ever scrutinizing whether our preferences are grounded in problematic stereotypes or discriminatory attitudes:

> Consider the supreme fuckability of "hot blonde sluts" and East Asian women, the comparative unfuckability of black women and Asian men, the fetishisation and fear of black male sexuality, the sexual disgust expressed towards disabled, trans and fat bodies. These too are political facts, which a truly intersectional feminism should demand that we take seriously.... [T]he sex-positive gaze risks covering not only for misogyny, but for racism, ableism, transphobia, and every other oppressive system that makes its way into the bedroom through the seemingly innocuous mechanism of "personal preference."

How might we do this? One starting point is engaging in personal reflection about what your own sexual preferences are and why you hold them. Do your preferences reflect the oppressive systems that shape our broader social attitudes? If you use dating apps, are you setting the parameters of who your potential partners are so narrowly that you're excluding people you might be compatible with, perhaps on explicitly or implicitly biased grounds? Although fully assessing the empirical data about how our sexual preferences are formed and the extent to which they are mutable is beyond the scope of this chapter, there is reason to think that it might be possible to broaden our sexual horizons through wider exposure to more kinds of people and cultivating an attitude of open-mindedness.[37] We should also think about how the media we consume frames sex, sexuality, and desire, and consider deliberately consuming a more diverse range of stories and perspectives. As Danaher puts it in his argument for greater sexual inclusion, this "doesn't mean forcing people into sexual encounters against their will, but it does mean encouraging them to take a more critical, reflective attitude toward their own sexual psyches and not treating them as sacred, axiomatic truths."[38]

How Not to Foster Inclusion in Partnered Sex

Crucially, we should not be engaging in any sort of attempt to force people to change their sexual preferences (e.g., through brainwashing or "conversion therapy"), to shame or marginalize people for the preferences that they have, or to demand that people be attracted to those to whom they are not attracted. Practically speaking, such approaches are unlikely to succeed at fostering greater sexual inclusion. More importantly, they are ethically problematic; they violate liberty and sexual autonomy and risk unfairly criticizing people for things that are beyond their direct control. We also want to be on guard against encouraging fetishization; the line between choosing a sex partner of another race or "type" because you're open to expanding your preferences and choosing a sex partner of another race or "type" because you are exoticizing them and making false or stereotypical assumptions about them can be a thin one.

Fostering sexual inclusion by changing underlying cultural norms and beauty ideals is likely to be a long-term, slow-moving project with no guarantee of success. There are, however, two ways to immediately increase access to partnered sex for excluded people in the short term that I believe we should reject. First is the proposal that excluded people attempt to change something about themselves so that their bodies are more in line with dominant norms of sexual attractiveness. Danaher argues that

> In general, if someone has a body shape/property that is widely (though not absolutely) perceived as sexually unattractive there will be three things we can do to address their sexual exclusion: (i) we can do something to modify their body shape or property; (ii) we can match them to those who find their body shape attractive (or give them access to other outlets for sexual expression that are non-partnered, e.g., sex toys and so on); or (iii) we can try to change societal preferences to make them more sexually included.[39]

Options (ii) and (iii) are in line with what I have been suggesting. Danaher states that these strategies are most likely to be just, as they do not "compound the injustice done to an already marginalized/excluded population by making the problem something that they need to 'fix'"; however, he argues that (i) is sometimes justified "due to reasons of cost-effectiveness or triviality, consent, and the fact that other moral ends are satisfied in the process."[40] Danaher offers an example of an obese man whose body size entails that he cannot find a sex partner, arguing that the facilitation of sexual inclusion could be part of the justification (along with promoting personal health or well-being) for providing him with gastric bypass surgery for weight loss.

I disagree that advocating for permanent body modification is a morally appropriate way to attempt to increase access to partnered sex. This is most obvious for serious and potentially dangerous interventions such as gastric bypass surgery, which rearranges the internal organs, requires lifelong highly restrictive changes to eating habits, and carries a high risk of serious side effects.[41] To justify such a surgery by pointing out that it will bring someone's body more in line with narrow mainstream standards of sexual attractiveness is to give these standards a great amount of power. And encouraging another person to undertake *any* bodily change—even a comparatively safe and low-cost modification—in order to increase their access to partnered sex is to at least tacitly defer to or support these standards. Doing so implies that there is some problem with the excluded person, rather than locating the real source of the problem in the exclusionary and harmful standards to which the person does not adhere.[42] If a feature that contributes to sexual exclusion *also* hinders someone's flourishing independently of how others react to them, it makes sense to encourage them to attempt to alter that feature. But I presume that not meeting restrictive mainstream beauty standards is not *intrinsically* damaging to one's flourishing, and is harmful only because of others' discriminatory attitudes and behavior in light of these standards.

Although I think that advocating for individuals to change themselves to better conform to arbitrary and exclusionary social norms is generally ethically problematic, I also think that we should refrain from blaming or criticizing people who engage in such attempts. It will often be prudentially justified for someone who runs afoul of unjust standards to change themselves to better fit those standards. And it is not fair to blame marginalized or excluded individuals for making their own lives better by ascending a problematic hierarchy, even if this does nothing to dismantle (and perhaps further entrenches) the hierarchy itself. We should also recognize that there are many types of changes to bodies and behaviors that people might pursue in an effort to attract sexual partners, from applying makeup, to women feigning incompetence to appear unthreatening or men feigning aggressiveness to appear confident, to getting extensive cosmetic surgery.[43] Not all such efforts will contribute equally to upholding restrictive beauty ideals or sexist gender norms, and it will be challenging to determine which types of modifications sustain such problematic norms.

The second problematic way to foster immediate sexual inclusion is granting special rights to people who are sexually excluded to purchase sex, in virtue of their excluded status. (As argued in "The Scope and Framing of the Problem" section, the right way to approach this proposal is by focusing on sexual exclusion per se, and not on features that are sometimes correlated with sexual exclusion, such as disability status.) Putting this proposal into practice would require us to address messy, morally fraught, and likely impossible-to-answer questions about who is "really" sexually excluded. Is someone excluded from partnered sex only if they cannot find *any* partners, or if they cannot find partners that they find personally appealing? If the latter, does it matter how unrealistic or discriminatory their standards for a partner are? What if a heterosexual man cannot find a partner because he behaves in misogynistic ways that are off-putting to all the women he encounters? Does it matter if someone is proactively trying to find a partner, and if not, whether their lack of effort stems from restrictive gender norms (e.g., demanding that women refrain from making the first move) or laziness?

Nor should we sidestep these questions by granting access to commercial sex to everyone on the grounds that this enables those who are sexually excluded to find partners. In our current cultural environment, most sellers of sex are women and most purchasers are men. Many women who sell sex must present themselves in ways that cater to the male gaze to attract clients. They must also frequently simulate the non-purchasable goods of romantic intimacy and sexual validation, since this is what clients are often seeking. Given this, legalizing commercial sex *because* it enables people who are sexually excluded to engage in partnered sex risks further entrenching false and damaging messages about male entitlement to sex with compliant, appealing women.

I am not suggesting that commercial sex always sends a problematic social message or supports male entitlement to sex. I find it plausible that at least some forms of commercial sex are morally permissible, and that commercial sex should be legally permitted (although assessing this claim in detail is beyond the scope of this chapter). But this is for independent reasons having to do with sexual autonomy (and perhaps, in the case of legalization, with harm reduction), rather than as a remedy for sexual exclusion. When justified as a matter of exercising sexual liberty, commercial sex conveys the message that all people should be free to pursue their negative rights to engage in consensual sexual interactions. A harmful social message arises only when the justification for this is the sexually excluded status of the client, and when sexually excluded clients are overwhelmingly men purchasing sex from women who must cater to their whims. This sends the message that partnered sex is so important for men they are entitled to a special right to purchase it if they cannot get it elsewhere—not as a matter of liberty, but simply *because they cannot get it elsewhere.*

The same concerns presumably do not rule out surrogate partner therapy (or SPT, also sometimes called "sexual surrogacy"). In SPT, a client works under the direction of a licensed therapist with a surrogate partner who "participates with the client in structured and unstructured experiences that are designed to build client self-awareness and skills in the areas of physical and emotional intimacy."[44] Contra sometimes sensationalized coverage in news media, surrogates focus on helping clients learn "how to develop healthy relationships, how to touch and to receive touch, and how to be more accepting of one's body and sexuality"; "genital-genital contact may or may not be therapeutically indicated" and is "often a minor part of the therapy."[45] The aim of SPT is teaching better sexual functioning that might enable the client to be less excluded in the long-term, rather than simply providing an excluded client with a sex partner. So long as SPT does not convey a sense of male entitlement to sex—and whether it risks doing so is a complex empirical question—it should be legal and de-stigmatized as a way to promote holistic sexual health and well-being.[46]

Exclusion from Sexual Validation

The final form of exclusion that I will address is not a direct exclusion from a sexual good but an indirect negative consequence of such exclusion. Given deeply problematic background assumptions about the nature and importance of sex and about masculinity and femininity, being excluded from partnered sex can potentially lead to social stigma or feelings of personal invalidation in at least three ways.

First, adults—especially men—who have not experienced sexual intercourse (and who are not choosing celibacy for religious or other principled reasons) are often seen as pathetic or sad. In surveys of college students, male virgins report higher rates of negative emotions like anxiety, guilt, or embarrassment than do female virgins.[47] Men are also more likely to view virginity in a stigmatizing way and to see "their status as virgins as something negative and embarrassing that they needed to shed as soon as possible."[48] Men's lack of sexual experience confers lack of social status because we tend to assess masculinity in part in terms of how much (or how little) sex someone has had; unsurprisingly, stigmatizing beliefs about male virginity are positively correlated with beliefs in traditional gender roles and sexual double standards.[49] While male virginity is mocked in comedy films like *The 40-Year Old Virgin*, 40-year old virgin women are erased from the cultural landscape, even as many young women are simultaneously told that their moral worth lies in their status as "pure" virgins.[50] Sexually excluded women who try to find support online are shunned from most incel forums and other male-dominated support groups, whose members have little sympathy for the reasons why women might be excluded from partnered sex, insisting that any woman who struggles to find a sex partner simply has too-high standards.[51]

Second, our cultural conception of what counts as "real" sex is narrowly focused on male ejaculation during heterosexual intercourse, which can stigmatize or invalidate people who express their sexuality in other ways. Most obviously, this completely excludes lesbian sex from counting as "real," and erases the myriad of sexual encounters that do not involve penises penetrating vaginas. Marilyn Frye discusses how the framing of "having sex" in discrete, countable instances is a phallocentric concept that is not applicable to lesbian sex, since it implicitly assumes that male orgasm is the act that "counts" as having "had sex."[52] This narrow conception of what "counts" as sex also excludes men who cannot sustain erections (e.g., due to spinal cord injuries). For example, the documentary *Sex on Wheels* profiles Carl, a 32-year-old man who became paralyzed from the waist down in an accident. Carl had a very active sex life before the accident, and is unable to have an erection afterward. Carl assumes that his sex life is over until he visits a therapist who specializes in non-genital routes to sexual fulfillment. Carl is skeptical at first, stating, "I want my hard-on back," because the "most important tool in my box has been stolen." However, after visiting a "sacred sexual healer" who engages in skin-to-skin contact and erotic massage that is not focused on the penis, Carl becomes optimistic that he can have a fulfilling sexual life without an erection.

Third, engaging in sexual activity is often seen as an essential mark of adulthood, and people who lack sexual experience are sometimes invalidated by being infantilized. For example, Mark O'Brien, a writer with polio whose essay about hiring Cheryl Cohen-Greene for SPT inspired the movie *The Sessions*, grew up in a conservative family that never discussed sex. After he loses his virginity to Cheryl, he feels emboldened to talk to his personal attendant about sex for the first time, and subsequently feels "admitted to something from which [he] had always felt excluded: the world of adults."[53] Similarly, *Sex On Wheels* profiles a woman who hires a sex worker to help her 26-year-old intellectually disabled son lose his virginity. Anticipating the event, she asks her son: "Are you happy, John? Are you going to be a man?" At first, John responds "I am a man; I've been a man since I was 18." However, after John has sex for the first time, the sex worker reports that he "tells me he feels like a proper man now." Similarly, in a survey about the attitudes of women with intellectual disabilities toward sex, researchers noted that "many of the participants did not identify themselves as women. They identified themselves as girls. One woman said she felt like a 'grown teenager,' even though she was 56 years old."[54] In all of these cases, being sexually active seems to be a problematic prerequisite for being treated as an adult.

Responses to Sexual Invalidation

The proper response to exclusion from sexual pleasure and partnered sex is to attempt to increase access to these goods. But the proper response to sexual invalidation is not to try to provide more people with such validation, for validation of this sort is not a good worth pursuing (let alone something that anyone has a right to). The norms that form the basis for sexual validation constitute a harmful and patriarchal constraint on our flourishing. Lack of sexual experience confers lack of social status on men only because of the problematic way our society assesses both masculinity and adulthood in terms of how much (or little) sex someone has had, and presumes that "real" sex can only involve a penis penetrating a vagina. Being denied such social status is indeed painful. However, we should attempt to remedy this pain not by helping sexually invalidated men to have normative sexual experiences—thereby upholding the pernicious norms that generate their pain in the first place—but by attempting to remove the stigma associated with lacking sexual experience or having nonnormative sexual experiences.

As with the project of critically assessing and potentially expanding the horizons of sexual desirability, removing this stigma is likely to be a long-term, slow-moving project without any guarantee of success. How might this be done? For starters, we could advocate for a broader range of depictions of sexual inexperience in popular culture, and think twice about how we are framing discussions about sexual inexperience in our dating app profiles or in conversations with our friends. We should rethink the underlying gender norms that sustain the idea that men are valuable insofar as they display sexual prowess. And we should work to acknowledge the value and importance of friendships and other nonsexual relationships, and to recognize that while partnered sex is an important good, it is far from the *only* important good.[55]

Notes

1. Appel (2010) assumes that there is a right to sexual pleasure in the form of "mutual contact, not merely self-stimulation" (152) and advocates on this basis for an exemption to the prohibition of prostitution for disabled clients, proposing that these services (construed as "sexual surrogacy") be publicly funded. I offer a critique of Appel's rights-based argument in Liberman (2018). Di Nucci denies a positive right to sex, as "universal positive sexual rights are incompatible with universal negative sexual rights" (2011: 159); he proposes instead that "the sexual interests and needs of the severely disabled be met by charitable non-profit organisations whose members would voluntarily and freely provide sexual pleasure to the severely disabled" (2011: 158) or (in a later paper) by sophisticated sex robots (2017). Thomsen (2015) also argues for a legal exemption to the prohibition of prostitution, but on the grounds of beneficence and luck egalitarianism rather than on the basis of rights. By contrast, Jeffreys (2008) raises a feminist critique of the proposal that disabled people be permitted to purchase sex to avoid sexual exclusion on the grounds that this promotes a deeply problematic vision of male sexuality that oppresses and dominates women. Tracy de Boer (2015) replies that disabled men who purchase sex are in many cases "seeking the goods of sex, such as intimacy, connection, and sexual inclusion, as opposed to male domination" (66). See also Marino (2019: Ch. 11) for an excellent overview of philosophical perspectives on sex, love, and disability.
2. Liberman (2018).
3. Danaher (2020: 476).
4. Ibid.: 477.
5. Ibid.: 473.
6. Assuming that morality cannot demand the impossible of us, the existence of firm and unchangeable limits on our sexual desires—whether resulting from

evolution, social conditioning, or a combination of both—would entail that we cannot be obligated as a matter of justice to desire any particular individual or type of individual. However, demanding that someone desire another violates their sexual autonomy and is morally inappropriate on independent grounds, even if such desires are possible. For this reason, the remedies to sexual exclusion that I will address are at the level of social policy, and do not require that any individual change any particular desire. See also the "How Not to Foster Inclusion in Partnered Sex" section.

7. Experiences of asexuality vary. Some people who identify as asexual—understood as experiencing little to no sexual attraction—experience sexual desire and will masturbate or have partnered sex, while others do not. Similarly, some (although not all) people who are asexual experience romantic (as distinct from sexual) attraction. While asexuality per se does not hinder flourishing or well-being, discrimination against or misunderstanding of asexual people may lead to decreases in well-being (as with other non-normative sexual orientations). See Carrigan (2015); see also the Asexual Visibility and Education Network (https://www.asexuality.org/).
8. In doing so, I remain neutral about whether sadistic pleasure counts as a kind of valuable pleasure in its own right or whether sadism defeats or undermines the value of the pleasure.
9. The Declaration was first proclaimed by the World Association for Sexual Health (an umbrella organization representing over 100 national and international sexological organizations) in 1997 and updated in 2014. What the right to sexual health consists of is framed vaguely, as requiring "the availability, accessibility, acceptability of quality health services and access to the conditions that influence and determine health including sexual health." See https://worldsexualhealth.net/wp-content/uploads/2013/08/Declaration-of-Sexual-Rights-2014-plain-text.pdf.
10. An exception to this is sexual pleasure-seeking that violates the rights of others to avoid witnessing sexual situations without consent, such as public masturbation, to which there is no positive right.
11. Assistive masturbation devices include "vibrating cushions, remotely controlled masturbation devices, and vibrators with long handles for people who could not otherwise reach" (Quarmby 2015, citing Owens and de Than 2014).
12. For example, see the furniture available at www.intimaterider.com.
13. For this to be truly non-coercive, however, the base pay would have to be a living wage, and social services would have to meet workers' basic needs.
14. See Sharkey et al. (2017) for an overview of what current sex robots are like.
15. For an argument to this effect, see Di Nucci (2017).
16. In a review of an anthology on robot sex, Marcus Arvan (2019) worries about the fact that most of the volume's contributors presume without argument that sex robots are mere objects and are not persons.
17. See Kathleen Richardson's feminist Campaign Against Sex Robots (https://campaignagainstsexrobots.org/), along with Richardson (2016). See Danaher (2017) for an account of the possible negative symbolic consequences of using sex robots, and a proposal that these consequences be carefully assessed via an experimental approach.
18. A challenge with this proposal is that people's sexual desires may themselves be sexist, and a non-humanoid robot that provided sexual stimulation may not

satisfy these desires. The proposals I make in the "Fostering Inclusion in Partnered Sex" section about eventual reassessment of cultural standards of sexual desirability might do something to alleviate this concern in the long-term, although I am not optimistic that this is very likely.

19. Danaher (2019).
20. See https://www.guttmacher.org/state-policy/explore/sex-and-hiv-education for an overview of current regulations by state.
21. See Nagoski (2015) for a dismantling of common myths about women's sexuality, including the idea that physiological arousal (i.e., genital lubrication) and subjective sexual arousal (i.e., feeling turned on) always occur together when in many cases they do not, and the idea that "proper" sexual desire is spontaneous when many women primarily experience responsive desire in reaction to ongoing sexual stimuli.
22. A 2019 school climate survey from GLSEN found that only 8.2% of American students reported receiving LGBTQ-inclusive sex education. See https://www.glsen.org/sites/default/files/2020-10/NSCS-2019-Full-Report_0.pdf.
23. People with disabilities are often completely sheltered from learning about sex and sexuality. This might be from a misguided sense of paternalism, because caretakers do not want or know how to deal with the reality of disabled people's sexual lives, or because disabled people are falsely presumed to be asexual or incapable of having sex. For example, Tim Rose, who has cerebral palsy and uses a wheelchair, reports that in high school he "used to get sent out for sex education classes all the time, because it was seen as, 'Tim, you don't need to know this stuff'" (Verstraten 2014). Medical professionals are also sometimes hesitant to provide sexual education. For example, a 20-year-old woman reports that when she was using a wheelchair, "even my doctors were hesitant to talk about my reproductive health or contraceptives when 'I had so many other things to worry about'" (Hills 2015).
24. Muñoz (2010).
25. Appel (2010: 152).
26. See Liberman (2018) for further discussion of this point.
27. See Frankowski and Clark (2009) for a discussion of "the realities of sex and intimacy in assisted living from the perspectives of residents, families, managers, and staff, exploring the discourse of sexuality, the impact of institutional structure and the role of oversight on sexual attitudes and behaviors, and the relationship of assisted living industry values to residents' sexual expression" (25).
28. Prison overcrowding means that many prisoners lack privacy, and "public" masturbation in prison can result in anything from fines to loss of commissary privileges to being sent to solitary confinement; see Dold (2017). A major potential source of harm (which could, if serious enough, justify restricting the prima facie positive right to pursue sexual pleasure) is the way in which male inmates engage in "exhibitionist" masturbation in the presence of female guards as a form of sexual harassment; see Patterson (2017).
29. Kulick and Rydström (2015: 20).
30. See Ging (2019) for an overview of the broader online "manosphere" of which incels are a part. Incel forums have flourished on sites like 4chan, 8chan, and incels.co. Reddit banned its most prominent incel forum (r/incels) in 2017 under a policy forbidding content that glorifies or encourages violence; the incel

subreddit that replaced it (r/braincels) was banned in 2019 for violating Reddit's harassment policy.

31. For a list of terminology, see the glossary from incels.wiki (https://incels.wiki/w/Incel_Forums_Term_Glossary) as well as https://www.timsquirrell.com/blog/2018/5/30/a-definitive-guide-to-incels-part-two-the-blackpill-and-vocabulary.
32. On April 26, 2018, economist Robin Hanson posted a blog entry about the redistribution of sex in response to Minassian's "incel revolution" van attack a few days earlier, stating that "one might plausibly argue that those with much less access to sex suffer to a similar degree as those with low income, and might similarly hope to gain from organizing around this identity, to lobby for redistribution along this axis and to at least implicitly threaten violence if their demands are not met"; he noted as an aside that "sex could be directly redistributed, or cash might be redistributed in compensation." (https://www.overcomingbias.com/2018/04/two-types-of-envy.html). The post reached a wide audience after Ross Douthat discussed it in a column in *The New York Times* on May 2 (https://www.nytimes.com/2018/05/02/opinion/incels-sex-robots-redistribution.html), and was met with a wave of criticism. See also the incel notion of "sexual Marxism" https://incels.wiki/w/Sexual_Marxism.
33. In 2014, Elliot Rodger uploaded a video to YouTube decrying his status as a virgin and threatening to punish women for the "crime" of not being attracted to him; he subsequently killed six people, wounded fourteen, and took his own life. In 2018, Alek Minassian posted on Facebook that "The Incel Rebellion has already begun!", praised Elliot Rodger, and drove his van into a crowd in Toronto, killing ten and injuring sixteen.
34. For example, the Outsiders Club in the U.K. is "peer support and dating club, run by and for socially and physically disabled people" (https://www.outsiders.org.uk/outsidersclub/), while the 50plus-club hosts in-person and online meet-ups for older singles around the U.S. (https://www.50plus-club.com/).
35. See O'Shea (Forthcoming) for an argument that "there are injustices in the distribution and character of our sexual desires and that we can be held responsible for correcting these injustices" (24). O'Shea focuses on collective, political, structural solutions rather than individual responses, including "making changes to the architecture of online dating, removing unnecessary legal obstacles to sexual intimacy, and ensuring that institutional and urban design pays sufficient attention to social mixing" (25).
36. Srinivasan (2018). For further discussion of ethical problems with racialized sexual and romantic preferences, see Marino (2019: Ch. 10), Mitchell and Wells (2018), Zheng (2016), Bedi (2015), and Coleman (2011). For an argument that racialized sexual preferences need not be racist, see Halwani (2017).
37. There is some empirical evidence that body size preferences can shift in response to media exposure—for example, rural Nicaraguans who had no exposure to television preferred larger body sizes than those who were exposed to television (Boothroyd et al. 2016). See also Season 5, Episode 5 of the podcast *Invisibilia* ("A Very Offensive Rom-Com"), in which "L" realizes that she has strong racial preferences in dating, and embarks on a deliberate "experiment" to attempt to shift these preferences (https://www.npr.org/2019/04/04/709948132/a-very-offensive-rom-com); the link contains a list of scientific research about how human sexual desires (including racial preferences) are conditioned.
38. Danaher (2020: 491).

39. Ibid.: 474.
40. Ibid.
41. This is not to claim that procedures like gastric bypass should never be performed; whether the surgery offers net benefits in a particular case is a complex question that will depend on both empirical data about expected health effects as well as personal value judgments. I am arguing simply that it is morally problematic to advocate that others undergo gastric bypass *as a remedy* for sexual exclusion.
42. I presume that the harmfulness of these standards does not depend on their ultimate source—that is, whether they are the product of mere convention, evolution, or a mixture of both.
43. See https://www.thecut.com/2019/05/incel-plastic-surgery.html/ for a story about self-identified incels (some of whom seem to have severe body dysmorphia) who pursue extensive plastic surgery in an attempt to turn themselves into sexually desirable alpha males or "Chads."
44. International Professional Surrogates Association, "Surrogate Partner Therapy": https://www.surrogatetherapy.org/what-is-surrogate-partner-therapy/.
45. Ibid.
46. See Mintz (2014) for an argument defending SPT as a legitimate healthcare treatment for people with severe disabilities in fair equality of opportunity framework.
47. See Sprecher and Regan (1996) and Sprecher and Treger (2015). Although virginity is a social construct without a clear medical definition, for ease of discussion I follow the common convention of using "virgin" to mean "someone who has not had penetrative intercourse." See Valenti (2010) for discussion of the ways in which conceptualizing virginity in this way is problematic.
48. Eriksson and Humphreys (2014: 108); see also Carpenter (2005).
49. Eriksson and Humphreys (2014).
50. Valenti (2010).
51. See Chester (2018). Ironically, the term "incel" originates with a website started by a queer woman named Alana in 1993 as a support group for people who were struggling to find a romantic connection. Alana left the site after a few years, and did not know that the term had sparked a misogynistic Internet subculture until she read about Elliot Rodger's killing spree in 2014. See episode 120 of the podcast *Reply All* (https://gimletmedia.com/shows/reply-all/76h59o/).
52. Frye (1988). The presumption that sex must involve penile-vaginal penetration is so strong that someone who regularly engages in both oral and anal sex describes herself as a virgin in the following interview: https://www.thehairpin.com/2013/01/interview-with-a-virgin-scarlet/.
53. O'Brien (1990: 3).
54. Fitzgerald and Withers (2011: 9).
55. Thanks to David Boonin, Josh Crabill, and Raja Halwani for helpful commentary on this paper. Thanks also to Veromi Arsiradam, Nicole Fice, Carolyn McLeod, Emma Ryman, and Megan Winsby for feedback on earlier work that inspired this paper.

References

Appel, Jacob M. 2010. Sex Rights for the Disabled? *Journal of Medical Ethics* 36 (3): 152–154.

Arvan, Marcus. 2019. Reflections on a Future of Sex with Robots. *Metascience* 28: 163–166.

Bedi, Sonu. 2015. Sexual Racism: Intimacy as a Matter of Justice. *The Journal of Politics* 77 (4): 998–1011.

Boothroyd, Lynda G., Jean-Luc Jucker, Tracey Thornborrow, et al. 2016. Television Exposure Predicts Body Size Ideals in Rural Nicaragua. *British Journal of Psychology* 107 (4): 752–767.

Carpenter, Laura M. 2005. *Virginity Lost: An Intimate Portrait of First Sexual Experiences*. New York: New York University Press.

Carrigan, Mark. 2015. Asexuality. In *The Palgrave Handbook of the Psychology of Sexuality and Gender*, ed. Christina Richards and Meg John Barker, 7–23. London: Palgrave Macmillan.

Chester, Nick. 2018. Meet the Women of the Incel Movement. *Huck*, December 5. https://www.huckmag.com/art-and-culture/tech/meet-the-women-of-the-incel-movement/.

Coleman, Nathanial Adam Tobias. 2011. What? What? In the (Black) Butt. *APA Newsletter on Philosophy and Lesbian, Gay, Bisexual, and Transgender Issues* 11 (1): 12–15.

Danaher, John. 2017. The Symbolic-Consequences Argument in the Sex Robot Debate. In *Robot Sex: Social and Ethical Implications*, ed. John Danaher and Neil McArthur, 103–132. Cambridge, MA: MIT Press.

———. 2019. Building Better Sex Robots: Lessons from Feminist Pornography. In *AI Love You*, ed. Yuefang Zhou and Martin H. Fischer, 133–147. Cham: Springer.

———. 2020. A Defence of Sexual Inclusion. *Social Theory and Practice* 46 (3): 467–496.

De Boer, Tracy. 2015. Disability and Sexual Inclusion. *Hypatia* 30 (1): 66–81.

Di Nucci, Ezio. 2011. Sexual Rights and Disability. *Journal of Medical Ethics* 37 (3): 158–161.

———. 2017. Sex Robots and the Rights of the Disabled. In *Robot Sex: Social and Ethical Implications*, ed. John Danaher and Neil McArthur, 73–88. Cambridge, MA: MIT Press.

Dold, Kristen. 2017. Why It's Illegal for Prisoners to Masturbate. *Vice*, January 6. https://www.vice.com/en/article/ypvk7g/why-its-illegal-for-prisoners-to-masturbate.

Eriksson, Jonas, and Terry P. Humphreys. 2014. Development of the Virginity Beliefs Scale. *Journal of Sex Research* 51 (1): 107–120.

Fitzgerald, Christina, and Paul Withers. 2011. 'I Don't Know What a Proper Woman Means': What Women with Intellectual Disabilities Think About Sex, Sexuality and Themselves. *British Journal of Learning Disabilities* 41: 5–12.

Frankowski, Ann Christine, and Leanne J. Clark. 2009. Sexuality and Intimacy in Assisted Living: Residents' Perspectives and Experiences. *Sexuality Research and Social Policy* 6 (4): 25–37.

Frye, Marilyn. 1988. Lesbian 'Sex'. *Sinister Wisdom* 35: 46–54.

Ging, Debbie. 2019. Alphas, Betas, and Incels: Theorizing the Masculinities of the Manosphere. *Men and Masculinities* 22 (4): 638–657.

Halwani, Raja. 2017. Racial Sexual Desires. In *The Philosophy of Sex: Contemporary Readings*, ed. Raja Halwani, Alan Soble, Sarah Hoffman, and Jacob M. Held, 181–198. Lantham, MD: Rowman & Littlefield.

Hills, Rachel. 2015. Sex Talk Realness: How I Have Sex with a Disability. *Cosmopolitan*, April 16. http://www.cosmopolitan.com/sex-love/news/a39179/sex-talk-realness-how-i-have-sex-with-a-disability/.

Jeffreys, Sheila. 2008. Disability and the Male Sex Right. *Women's Studies International Forum* 31 (5): 327–335.

Kulick, Don, and Jens Rydström. 2015. *Loneliness and its Opposite: Sex, Disability, and the Ethics of Engagement*. Durham, NC: Duke University Press.

Liberman, Alida. 2018. Disability, Sex Rights, and the Scope of Sexual Exclusion. *Journal of Medical Ethics* 44 (4): 253–256.

Marino, Patricia. 2019. *Philosophy of Sex and Love: An Opinionated Introduction*. New York, NY: Routledge.

Mintz, Kevin Todd. 2014. Sexual Intimacy, Social Justice, and Severe Disabilities: Should Fair Equality of Opportunity in Health Extend to Surrogate Partner Therapy? *The Journal of Philosophy, Science, and Law* 14 (3): 4–15.

Mitchell, Megan, and Mark Wells. 2018. Race, Romantic Attraction, and Dating. *Ethical Theory and Moral Practice* 21 (4): 945–961.

Muñoz, Victor. 2010. *Report of the United Nations Special Rapporteur on the Right to Education: Sexual Education*. A/65/162, July 23.

Nagoski, Emily. 2015. *Come as You Are: The Surprising New Science That Will Transform Your Sex Life*. New York, NY: Simon & Schuster.

O'Brien, Mark. 1990. On Seeing a Sex Surrogate. *The Sun Magazine*. http://thesunmagazine.org/issues/174/on_seeing_a_sex_surrogate.

O'Shea, Tom. (Forthcoming). Sexual Desire and Structural Injustice. *Journal of Social Philosophy*.

Owens, Tuppy, with Claire de Than. 2014. *Supporting Disabled People with Their Sexual Lives: A Guide for Health and Social Care Professionals*. London: Jessica Kingsley Publishers.

Patterson, Brandon E. 2017. Inmates Are Using Masturbation as a Weapon. Female Guards Have Had Enough. *Mother Jones*, November 16. https://www.motherjones.com/crime-justice/2017/11/inmates-are-using-masturbation-as-a-weapon-female-guards-have-had-enough/.

Quarmby, Katherine. 2015. Disabled and Fighting for a Sex Life. *Mosaic*, 3 March. http://mosaicscience.com/story/sex-disability.

Richardson, Kathleen. 2016. Sex Robot Matters: Slavery, the Prostituted and the Rights of Machines. *IEEE Society and Technology Magazine* 35 (2): 46–53.

Sex on Wheels. 2013. Dir. Emma Young. Firecracker Films.

Sharkey, Noel, Aimee van Wynsberghe, Scott Robbins, and Eleanor Hancock. 2017. Our Sexual Future with Robots (Consultation Report). *Foundation for Responsible Robotics*, 5 July. https://responsiblerobotics.org/2017/07/05/frr-report-our-sexual-future-with-robots/.

Sprecher, Susan, and Pamela C. Regan. 1996. College Virgins: How Men and Women Perceive Their Sexual Status. *The Journal of Sex Research* 3 (10): 3–15.

Sprecher, Susan, and Stanislav Treger. 2015. Virgin College Students' Reasons for and Reactions to Their Abstinence From Sex: Results From a 23-Year Study at a Midwestern U.S. University. *Journal of Sex Research* 52 (8): 936–948.

Srinivasan, Amia. 2018. Does Anyone Have the Right to Sex? *London Review of Books* 40 (6): 5–10. https://www.lrb.co.uk/the-paper/v40/n06/amia-srinivasan/does-anyone-have-the-right-to-sex.

Thomsen, Frej Klem. 2015. Prostitution, Disability, and Prohibition. *Journal of Medical Ethics* 41: 451–459.

Valenti, Jessica. 2010. *The Purity Myth: How America's Obsession with Virginity is Hurting Young Women*. Berkeley, CA: Seal Press.

Verstraten, Katelyn. 2014. Surprise! Disabled People Have Sex. *The Star*, October 5. https://www.thestar.com/news/insight/2014/10/05/surprise_disabled_people_have_sex.html.

Zheng, Robin. 2016. Why Yellow Fever Isn't Flattering: A Case Against Racial Fetishes. *The Journal of the American Philosophical Association* 2 (3): 400–419.

PART V

Sex and Digital Technology

CHAPTER 28

Sex and Technology: From Tinder to Robot Sex

Neil McArthur

Nearly everyone now depends on technology to mediate their intimate relationships in some fashion. And there are reasons why we should see this as a good thing. Technology can benefit our sex lives and our relationships in several ways. It allows people to form connections they could not otherwise have formed, to maintain these connections over long distances, and to have experiences that are qualitatively new and different from those that were previously possible. However, the technology also presents risks. It can exacerbate social divisions and inequalities. And it is vulnerable to various forms of restriction, control, and surveillance.

This chapter argues that we should welcome the fact that sexual technology is now central to our intimate lives, and that we should equally welcome the development of new sexual technologies, such as sex robots and sexually explicit virtual reality environments. Sex tech on balance increases both the overall happiness and autonomy of users. However, this chapter further argues that we must take an active role in managing the environment in which the technology operates to ensure that the benefits are fully realized and are equally distributed.

The Technology Landscape

This is not the place to attempt a comprehensive catalog of all the technologies that can be used for sex and intimacy. Apart from the space that would be required to do so, any such catalog would date very rapidly. However, it is

N. McArthur (✉)
University of Manitoba, Winnipeg, MB, Canada
e-mail: Neil.McArthur@umanitoba.ca

D. Boonin (ed.), *The Palgrave Handbook of Sexual Ethics*,
https://doi.org/10.1007/978-3-030-87786-6_28

useful to at least briefly survey some of the main technologies that exist now, and some that we can foresee emerging in the near future, in order to better understand the philosophical issues they raise.

First, there are what I call "first wave" sexual technologies.[1] These are technologies that allow us to form connections with other people. There are various forms of text and video communication. These include asynchronous messaging applications such as iMessage and Snapchat, that allow us to send texts, videos, and photos. They also include synchronous communication tools such as Skype and Facetime. There are sites that distribute various kinds of pornography. Some of these sites, such as Chaturbate and OnlyFans, allow users to interact, either through messaging or in real time, with the performers. There are also sites and apps that can connect us with potential partners—for instance, social networks such as Facebook and Instagram, as well technologies meant explicitly for dating and casual encounters, such as Tinder and Grindr. There are also multiuser environments that allow us to interact with others, including complete strangers as well as artificial agents, inside of virtual worlds.

In addition to these technologies, most of which are now familiar, there are what I call "second wave" sexual technologies, which are at an early stage of development. These second-wave technologies create an immersive sexual experience that does not necessarily involve a human partner. The second-wave technologies that are most prominent in popular culture are sex robots, though in fact sexbots are currently quite rudimentary, and are developing fairly slowly. Virtual reality is developing more rapidly, and VR is being used to create immersive pornography, as well as immersive virtual worlds that provide erotic experiences. We are also seeing the rapid development of "teledildonics"—sex toys that are connected to a network to provide sexual stimulation in response to remote inputs. These can be integrated with virtual reality to increase the latter's immersivity.

The Benefits of Sexual Technology

I take it as a premise that sex and intimacy are sources of satisfaction and happiness for most people. I further assume that in general people prefer to have more control rather than less over their intimate lives. Other things being equal, they want the power to make decisions about their sexual and romantic behavior both free from control by other people or by the state, and free from surveillance or harassment.

First-wave technologies promote happiness and autonomy in various ways. They increase our opportunities to meet partners. Technology allows us to form connections with people we would never meet in real life, and thus to form relationships with people outside our social circles. For instance, one survey found that two-thirds of respondents reported that online dating services have made it easier to meet potential partners of a different race or ethnicity.[2] This benefits us as individuals by increasing the size of our dating pool, but it also produces social benefits, by promoting connections between members of different social groups. For people living in circumstances where their family and community watch over or control their behavior, as is common in smaller, more traditional societies, it can allow them to pursue relationships beyond ones sanctioned by those around them. It thus gives them a tool to increase their autonomy in the sphere of their personal lives.

First-wave technologies also allow us to form and maintain relationships with people who are physically distant from us. This means that we can choose partners based on common interests and compatibility rather than on the accident of geographic proximity. We can also maintain our relationships more easily if our life circumstances require us to be separated from our partners. There are some professions, such as long-haul trucking, that are inherently disruptive to relationships. Technology can help mitigate this.

First-wave technologies also allow people with specific sexual interests and members of sexual minorities to find like-minded people more easily. For example, we can look to specialized dating services, such as Star Trek Dating, as well as to the emergence of online communities such as Fetlife. Liam Wignall's study of "pup Twitter" shows how groups with highly esoteric sexual interests are able to build a community by using existing technologies in innovative ways. Wignall concludes that social networking has allowed members of sexual subcultures to enhance communication, engagement, and visibility.[3]

Second-wave technologies, which do not require human partners, can offer unique advantages beyond those offered by more familiar first-wave technologies. Second-wave technologies such as robots and virtual worlds can provide sexual experiences to people who have trouble finding partners. People may face barriers in the dating market for a variety of reasons: for instance, because they are geographically isolated, because they are not conventionally attractive, because they have a history of trauma, or because they have mental or physical disabilities. There may even be legal barriers to finding partners, as in countries where same-sex sex is illegal. Second-wave technology may not provide everyone with what they would consider an ideal substitute for human partners, but it can at least mitigate isolation, and it may help people transition toward feeling able to seek out human companions.

Second-wave technologies also promise novel sexual experiences for users. They allow people to experiment with new identities, and to explore more fluid gender roles. Kate Devlin suggests that we could see sex robotics as a way to start moving beyond gender and race altogether. She thinks these technologies could allow us "to explore issues without the restrictions of being human," and we should look upon the machine as a "blank slate that offers us the chance to reframe our ideas."[4] She herself organizes an annual "hackathon" that brings designers together to present more imaginative possibilities for sexbots, ones that can transcend simple anthropomorphism and the problematic representations of race and gender that can come with it. Sex tech can also provide users with a safe, controlled environment in which to experiment with activities, such as ones involving domination and the infliction of pain, that they might not be comfortable engaging in real life.

Second-wave technologies could also benefit people in relationships. They can address issues of desire discrepancy when one partner wants sex more than the other. Therapists report that this is one of the most common problems couples face. The technologies can also offer people in long-term relationships the benefits of novelty, and this may decrease the likelihood that they will seek sex outside the relationship. And the technologies could allow people to engage in activities that their partner may not be interested in.

Marina Adshade has speculated that second-wave technologies may lead to a more profound change in how we view relationships. She thinks they "will disentangle the association between sexual intimacy and marriage, leading to higher quality marriages."[5] As she explains, using sex robots as her paradigm:

> Access to sexbot technology will not change the biological imperative of individuals to want to share their lives, and raise their children, with another human being. But it would make it possible for individuals to choose that human being based on characteristics other than mutual sexual desire; to disentangle the association of sexual intimacy and life as a family. For example, it is not hard to imagine two heterosexual women seeing the value in forming a household and raising children together as a married couple, but with their needs for sexual companionship met by sexbot technology. Nor is it hard to imagine a homosexual man seeing the value in forming a household and raising children with a woman, since that arrangement would significantly reduce expenses associated with reproductive technologies, but with each of their needs for sexual companionship met by sexbot technology.[6]

Concerns About Sexual Technology

People have expressed numerous concerns about first-wave technologies. Some have argued that, as online dating apps have become the predominant means to meet partners, they are leading us to focus on superficial criteria such as physical appearance. Mia Levitan says that "swiping turns us all into teenagers making snap judgments solely on the basis of looks." She cites a study of the words that, when men include them in their dating profiles, lead to the greatest number of matches. It found that claiming to be "physically fit" ups the response rate by 96%. It also found that the most advantageous word to use was "6 ft."[7]

While, as I said above, dating apps can help people meet partners of different racial or cultural backgrounds, they can also become forums for discrimination. Christian Rudder, one of the founders of the dating app OKCupid, has found, based on his extensive access to data on how people use the site in the United States, that white men receive more messages, and more replies to their messages, than members of any other group. Black women receive the least attention.[8] A study in Europe showed that, on European dating sites, people of African and Middle Eastern descent receive fewer messages and replies than do white people.[9] Members of visible minorities report that those messages they do receive are often filled with racial slurs.[10] While such racist attitudes reflect those prevalent in the offline world, the anonymity of the online environment seems in many cases to exacerbate them.

Some people suggest this reveals a larger problem: because we now meet people outside our immediate social networks, it is easier to behave badly toward those we meet or are involved with. Besides racist behavior, we can also look for instance at the prevalence of ghosting—breaking off all communication with someone suddenly and with no explanation. Around a fifth of young people report having been ghosted by a romantic partner.[11]

The number would be much higher if we were to count people who have been ghosted by someone with whom they have connected online but not yet met in person. We can hypothesize that people feel more free to behave in insensitive ways when they are with someone who has no contact with anyone else they know.

While technology allows for the creation of online communities, it can also have a damaging effect on real-life ones. Michael Hobbes has chronicled what he calls the epidemic of gay loneliness, which he attributes in part to the decline of gay bars and other social spaces that is the result of technology use. He quotes a man named Adam who says:

> It's so much easier to meet someone for a hookup on Grindr than it is to go to a bar by yourself. Especially if you've just moved to a new city, it's so easy to let the dating apps become your social life. It's harder to look for social situations where you might have to make more of an effort.[12]

Statistics show there has been a notable decline in the number of gay bars, which have long served as a linchpin of the LGBTQ community. One study found that the number of gay bars in the United States has declined by 37% since 2007.[13] The author of that study told the *New York Times* that the number of lesbian bars has dropped by over half since 2007, and by over 90% since 1986.[14] The decline is not entirely due to technology, but we can hypothesize that it plays a major role. Gays and lesbians have been early adopters of dating technology—but the straight world may not be far behind. We can see the impact of technology on in-person social networks when we see, for instance, the decline in the amount of time young people spend attending social gatherings.[15]

There is also a puzzle about our current technological landscape: for all that it facilitates connections in various ways, the net effect seems to be that people are having less sex overall. Numerous explanations have been proposed for the so-called "sex recession."[16] Some have suggested that people are dissatisfied with existing technology, even as we have become more dependent on it. Others have argued that our dependence on technology simply offers us too much choice, and this in itself leads to unhappiness. Barry Schwartz has identified what he calls "the paradox of choice": that the more options we have, the less we are satisfied with the choices we make. This has been applied to the world of modern dating to help explain the decline in sexual activity.[17]

While second-wave technologies are not yet widely available, people have already begun to raise concerns about them as well. First, there is the problem of what sex robots, as well as the virtual avatars we will encounter in VR environments, will look like. While true sexbots may not exist yet, we can look to the design of existing sex dolls in order to determine what we should expect. The sex dolls currently on the market are almost all female, and they almost all reflect the image of female beauty promoted by much pornography, and indeed by the mainstream media as well. The dolls are beautiful but bland, extremely skinny, with large breasts and hips. And they are mostly white, with a few made to look Asian. If we are faced with a future when we are surrounded by what Kate Devlin calls "pornified fembots," this could significantly exacerbate the pressure that people, especially young women, already feel to be skinny, white, and beautiful.[18] This pressure puts women at risk of suffering from psychological distress, as well as eating disorders.

We may also find that sexbots are created that reenforce racial and ethnic stereotypes. There may be demand for Japanese geisha or African-American slave robots, for instance, or robots with exaggerated ethnic characteristics. Sexbots might also be programmed to speak and behave in ways that reenforce stereotypical attitudes. And quite apart from their potential impact on the attitudes of their users, members of minorities may find it deeply offensive to know that such stereotypical depictions of their group are for sale.

Artificial agents may also behave in ways that some worry will present a problematic model for intimate interactions. Sex robots may be passive, compliant, and highly sexual—exactly how misogynists think women should be. They can also be programmed to say yes whenever they are asked for sex—if their owners bother to ask at all—or perhaps even more worryingly, can be programmed to say no if their owners wish to act out fantasies concerning nonconsensual sex. This presents a model of female consent that we do not want to endorse. As Sinziana Gutiu says:

> To the user, the sex robot looks and feels like a real woman who is programmed into submission and which functions as a tool for sexual purposes. The sex robot is an ever-consenting sexual partner and the user has full control of the robot and the sexual interaction. By circumventing any need for consent, sex robots eliminate the need for communication, mutual respect, and compromise in the sexual relationship. The use of sex robots results in the dehumanization of sex and intimacy by allowing users to physically act out rape fantasies and confirm rape myths.[19]

Some have gone further and argued that the very idea of sex robots promotes ideas of women as sexual objects. Kathleen Richardson, founder of a campaign against sex robots, believes that sexbots could help normalize the sexual objectification of women. Because men will think it is acceptable to view a sexbot, which represents a woman, as an object for their own gratification, this will lead them to think it is acceptable to view actual women in the same way.[20] Benjamin Yelle says: "It's like the ultimate distillation of objectification. It's literally an object that looks like a woman."[21]

There are also concerns that the use of second-wave technologies may lead to higher levels of social isolation and other emotional issues. If robots and virtual worlds are easily available, and able to deliver appealing sexual experiences, people may shy away from human relationships. Sherri Turkle's worries about companion robots apply *a fortiori* to sex robots and other artificial intimate companions:

> The idea of a robot companion serves as both symptom and dream. Like all psychological symptoms, it obscures a problem by "solving" it without addressing it. The robot will provide companionship and mask our fears of too-risky intimacies. As dream, robots reveal our wish for relationships we can control.[22]

Relationships with artificial creatures are easier in numerous ways than those with humans. Robots and virtual companions are never sick, they never grow old or die, they never have emotional crises or need help finding a new career. But a life made up entirely of easy, safe relationships would be less rich in many ways than the lives most of us are used to—ones where we are part of a network of human relationships, with all their difficulties and frustrations.

Our bonds with artificial agents could also lead us to become less empathetic and caring in the human relationships we do have, as we become more accustomed to relationships designed to benefit ourselves. Nicholas A. Christakis says: "As AI permeates our lives, we must confront the possibility that it will stunt our emotions and inhibit deep human connections, leaving our relationships with one another less reciprocal, or shallower, or more narcissistic."[23]

A persistent concern about new technology is that it threatens to displace human workers. And this concern applies particularly to second-wave sexual technology. Robots could potentially take the place of sex workers. Adult film performers are also at risk of being displaced by artificially generated actors. Those who think that sex work is inherently exploitative will welcome these developments. However, many people see sex work as a legitimate form of employment, and those who depend on it often have few alternatives.

Both first- and second-wave technologies raise worries about access. As we become more dependent on first-wave technology in our relationships, those who cannot afford, or cannot access, new devices and reliable network connections are at a disadvantage. This affects people in rural areas and in less developed countries in particular. The problem will become more acute with second-wave technologies, which will be expensive and will require high-capacity network connections.

Responding to Concerns

Ultimately, an ethical evaluation of sexual technology requires us to balance the costs against the benefits. When it comes to first-wave technologies, I believe that the benefits, in terms of connecting us to people outside our social networks and over long distances, outweigh the costs created by people's bad behavior online and the damage to offline communities. I also believe that many of the potential problems with second-wave technology can be addressed or at least mitigated if we take an engaged role in the design and production of this technology, rather than trying to restrict access to it or ban it outright.

For instance, when it comes to the potentially racist and misogynist implications of second-wave technology, there is no question that the problem is real, and should be taken seriously. One approach, however, is to encourage designers to produce better technologies. Kate Devlin acknowledges problems with the gendered stereotypes inherent in the current crop of sexbots, but argues that our response to this should not be to "import established prudishness" into the development of this technology. Instead, she argues that we should try to promote more positive designs.[24] John Danaher provides a template for how this can work, suggesting that the designers of sexbots could take their cues from feminist pornography. He says:

> We can work to ensure better content (i.e. depictions or representations of female (and male) sexuality in robotic form); better processes (i.e. more female voices included in the production and distribution of sexbots); and better contexts (i.e. social environments and conversations surrounding the consumption and use of sexbots).[25]

The same principles can be applied to the issue of racial stereotypes. We could work to ensure that processes are in place to ensure that fewer racist sexbots are manufactured and that there is a smaller market for any that are. In dealing with both racism and sexism in the sexbot industry, government incentives or even regulation should certainly not be ruled out as well. However, the government is much more likely to have a positive impact if it engages constructively with the producers of the technologies, rather than trying to ban them or regulate them in a heavy-handed way.

Government can also play an active role in addressing the problem of access. If we recognize that sexual technology is on balance beneficial, we should be prepared to look at ways of ensuring that it is distributed to people who would benefit from it but might not be able to afford it.

The Control and Suppression of Sexual Technology

If we are convinced that sexual technology is beneficial overall, we should pay attention to the process by which it is developed and distributed. And I submit that it is currently subject to controls that severely inhibit its progress. These controls come both from the large technology monopolies and the government, and they are ultimately rooted in the stigma that sexual technology faces among the general public.

It is often said that, at least in Western countries, we live in an era of unparalleled sexual freedom. And this is true, at least with regards to the direct legal regulation of private behavior. Over the course of the past few centuries, governments, motivated by moral views about what constitutes acceptable sexual behavior, have used criminal penalties to govern numerous private sex acts, such as same-sex sex and the consumption of materials considered obscene. Activists have fought to repeal these legal restrictions, and in Western democracies, they have largely succeeded. In the United States, this process culminated in the landmark Supreme Court decision, *Lawrence v. Texas*, which declared Texas's statute criminalizing sodomy to be unconstitutional. In its ruling, the majority declared: "Our obligation is to define the liberty of all, not to mandate our own moral code."[26]

Despite *Lawrence* and similar rulings in other jurisdictions, governments have not abandoned all efforts to use the criminal law to regulate sexual technology. Notably, in 2018 the United States enacted HR-1865, known as the Fight Online Sex Trafficking Act (FOSTA) and the Stop Enabling Sex Trafficking Act (SESTA). The law removed the protection websites previously enjoyed from liability for the content they host. Specifically, it allows for the prosecution of the owners of any websites that "promote or facilitate the prostitution of another person." The result has been that nearly all sites previously used by sex workers to advertise have been shut down.[27] And even as the power of the criminal law to govern our intimate lives has receded, new forms of influence and control have emerged. The technology landscape continues to be shaped by a few key actors, and these have shown themselves to be distinctly hostile to all forms of sexual technology.

Over the past decade, people have begun using smartphones and tablets as their primary devices. But the software environment used by these devices is highly controlled. Apps must normally come from the app stores operated by Apple and Google. And both exclude most apps with explicit content—especially those designed by small companies. An example of the sort of censorship they exercise is Happy Play Time, a sexual education game designed to "eliminate the stigma around female masturbation," and featuring an anthropomorphic vulva named Happy. Though it was intended to be informative rather than erotic, Apple's App Store banned it for containing "excessively objectionable or crude content" and "containing pornographic material."[28]

Even private communication platforms are subject to forms of surveillance and control. It has been revealed that governments sometimes track the sexual activities of individuals online.[29] And companies have reserved for themselves the right to regulate how their platforms are used. For instance, the terms of service for Microsoft's Skype service gives the company the right to ban explicit language and nudity from private calls.[30]

Those who are working to develop new sexual technologies, including second-wave ones, face a hostile investment climate that inhibits innovation. Many venture capitalists, for instance, are governed by "vice clauses," that prevent them from investing in any products deemed to be sexual in nature.[31] And sex tech entrepreneurs face other barriers, such as exclusion from major trade shows and other fora. The environment for sex tech can be highly unpredictable, as we can see from the experience of the female-led sex tech company

Lora DiCarlo at the Consumer Electronics Show. In 2019, the company's Osé personal massager, marketed as a hands-free vibrator for "the holy grail of orgasms," won an award for innovation at the annual conference. However, the Consumer Technology Association (CTA), which operates the conference, stripped the company of the award, saying that "entries deemed by CTA in their sole discretion to be immoral, obscene, indecent, profane or not in keeping with CTA's image will be disqualified." It also banned the company from exhibiting at future shows. After a public backlash, the CTA reversed its decision and apologized.[32]

Governments and tech companies are able to exert control over sexual technologies because there has been little resistance, or even awareness, shown by the public. When the public does get engaged, as the above example shows, it can have an impact. Entrepreneurs respond to market incentives. They have to know that there is a market for sexual technology, and that they will not face a backlash for investing in it. And governments, at least in democratic countries, respond to public opinion. If we believe that sexual technology promises to benefit people, we must be vocal in supporting it. We should inform ourselves about the regulatory environment, and try to ensure that it is not hostile to sex tech.

Conclusion

There may seem to be a certain inevitability about the development and adoption of sexual technology. First-wave technology is already pervasive, and second-wave technology continues to progress. However, I have argued that the ethical stance we take toward the technology will impact both the speed at which it appears and the precise form it takes. While I have argued that the benefits of sexual technology outweigh the costs, I have not dismissed the concerns raised by critics. Instead, I have suggested that we should educate ourselves about sexual technology, and be vocal about the direction we would like to see it take.

Notes

1. For more on the distinction between first- and second-wave sexual technologies, see Neil McArthur & Markie L. C. Twist, "The Rise of Digisexuality: Therapeutic Challenges and Possibilities," *Sexual and Relationship Therapy* 32 (2017): 334–344.
2. Philipp Hergovich and Josue Ortega, "The Strength of Absent Ties: Social Integration via Online Dating," (2017) https://arxiv.org/pdf/1709.10478.pdf
3. Liam Wignall, "The Sexual Use of a Social Networking Site: The Case of Pup Twitter," *Sociological Research Online* 22 (2017): 21–37.
4. Kate Devlin, "In Defence of Sex Machines: Why Trying to Ban Sex Robots Is Wrong," *The Conversation* (September 17, 2015) https://theconversation.com/in-defence-of-sex-machines-why-trying-to-ban-sex-robots-is-wrong-47641

5. Marina Adshade, "Sexbot-Induced Social Change: An Economic Perspective," in John Danaher and Neil McArthur, *Robot Sex: Social and Ethical Implications* (Cambridge, Mass.: MIT Press, 2017), 289–300, at 292.
6. Adshade, 294.
7. Mia Levitan, *The Future of Seduction* (London: Unbound [ebook], 2020); Lindsay Dodgson, "Using these Words in your Dating Profile Will Get You the Most Matches, According to Dating App Badoo," *Insider* (30 August 2018), www.insider.com/words-to-put-in-your-dating-profile-to-get-most-matches-2018-8
8. See Christian Rudder, *Dataclysm: Love, Sex, Race, and Identity—What Our Online Lives Tell Us about Our Offline Selves*, New York: Broadway Books, 2014, pp. 101–109.
9. Gina Potârcă and Melinda Mills, "Racial preferences in online dating across European countries," *European Sociological Review* 31, 2015, pp. 326–341, at p. 332.
10. Curtis M. Wong, "Watch What Happened When These Two Men Swapped Grindr Profiles," *Huffington Post* (17 September 2017). https://www.huffingtonpost.ca/entry/grindr-profile-swap-what-the-flip_n_59bd352de4b0edff971c810c?ri18n=true
11. YouGov, "Poll Results: Ghosting," (28 October 2014), https://today.yougov.com/topics/lifestyle/articles-reports/2014/10/28/poll-results-ghosting; Raúl Navarro, Elisa Larrañaga, Santiago Yubero and Beatriz Víllora, "Psychological Correlates of Ghosting and Breadcrumbing Experiences: A Preliminary Study among Adults," *International Journal of Environmental Research and Public Health* 17 (2020), 1–13.
12. Michael Hobbes, "Together Alone: The Epidemic of Gay Loneliness," *Huffington Post* (2 March 2017) https://highline.huffingtonpost.com/articles/en/gay-loneliness/
13. Greggor Mattson, "Are Gay Bars Closing? Using Business Listings to Infer Rates of Gay Bar Closure in the United States, 1977–2019," *Socius* 5 (2019), 1–2.
14. Lena Wilson, "Where Did All the Lesbian Bars Go? Increasingly, They're on TV," *New York Times* (7 May 2020), https://www.nytimes.com/2020/05/07/arts/television/lesbian-bars-vida-l-word-batwoman.html
15. Teddy Wayne, "The Death of the Party," *New York Times* (16 September 2015) https://www.nytimes.com/2015/09/17/fashion/death-of-the-party.html
16. Kate Julian, "Why Are Young People Having So Little Sex?", *The Atlantic* (December 2018), https://www.theatlantic.com/magazine/archive/2018/12/the-sex-recession/573949/
17. Elizabeth Svoboda, "The Problem with Modern Romance Is Too Much Choice," *Nautilus* (6 October 2016), http://nautil.us/issue/41/selection/the-problem-with-modern-romance-is-too-much-choice
18. See Kate Devlin, *Turned On: Science, Sex and Robots*, New York: Bloomsbury, 2019.
19. Sinziana Gutiu, "The Roboticization of Consent," in Ryan Calo, Michael Froomkin, and Ian Kerr, (eds), Robot Law, Cheltenham, U.K.: Edward Elgar Publishing, 2016, pp. 186–212, at p. 187.

20. Kathleen Richardson, "The Asymmetrical 'Relationship': Parallels between Prostitution and the Development of Sex Robots," *ACM SIGCAS Computers and Society*, January 2016.
21. Glenn McDonald, "The Sex Robots Are (Almost) Here," *Experience Magazine*, June 2019. https://expmag.com/2019/06/the-sex-robots-are-almost-here/
22. Sherry Turkle, *Alone Together: Why We Expect More from Technology and Less from Each Other*, New York: Basic Books, 2011, p. 285.
23. Nicholas A. Christakis, "How AI Will Rewire Us," *The Atlantic*, April 2019. https://www.theatlantic.com/magazine/archive/2019/04/robots-human-relationships/583204/
24. Devlin, "In Defence of Sex Machines."
25. John Danaher, "Building Better Sex Robots: Lessons from Feminist Pornography," in Yuefang Zhou and Martin H. Fischer, eds., *AI Love You: Developments in Human-Robot Intimate Relationships*, New York: Springer 2019, pp. 133–147, at p. 143.
26. Lawrence v. Texas, 02-102, 539 U.S. 558, 2003.
27. Emily Witt, "After the Closure of Backpage, Increasingly Vulnerable Sex Workers Are Demanding Their Rights, *New Yorker* (8 June 2018), https://www.newyorker.com/news/dispatch/after-the-closure-of-backpage-increasingly-vulnerable-sex-workers-are-demanding-their-rights
28. Shruti Dhapolamay, "Apple's Rejection of Happy Playtime App for Female Masturbation Is Hypocritical," First Post (22 May 2014), https://www.firstpost.com/tech/news-analysis/apples-rejection-of-happy-playtime-app-for-female-masturbation-is-hypocritical-3651863.html
29. Paul Owen, "NSA 'Spied on Porn Habits to Target Radicalisers'," Guardian (27 November 2013), https://www.theguardian.com/world/2013/nov/27/nsa-files-live-coverage-developments-reaction
30. Ms. Smith, "Microsoft to Ban 'Offensive Language' from Skype, Xbox, Office and Other Services," CS Online (26 March 2018), https://www.csoonline.com/article/3264658/microsoft-to-ban-offensive-language-from-skype-xbox-office-and-other-services.html
31. Juliyen Davis, "The Companies Venture Capital Isn't Allowed to Invest In," Super Maker, https://supermaker.com/articles/the-companies-venture-capital-isnt-allowed-to-invest-in
32. Kara Fox, "Vibrator at Center of Tech Sexism Controversy Gets its Award Back," CNN 9 May 2019, https://www.cnn.com/2019/05/09/tech/ces-sex-toy-award-controversy-scli-intl/index.html

CHAPTER 29

College Party Hook Ups: Consent, Apps, and Double Standards

James Rocha

Difficulties Obtaining Consent

Hopefully, by this point, almost everyone agrees that obtaining consent is absolutely essential for morally permissible sex. However, knowing this general fact is rarely sufficient. Unfortunately, it can be challenging to know when you have obtained sexual consent correctly. This is because not everyone understands consent well enough to know how to obtain it correctly. Further, not everyone works very hard to ensure that they have obtained sexual consent. Since consent is so important for sex, understanding consent and what it requires are hugely important, especially in circumstances that make consent more difficult to effectively obtain.

Consent can be difficult to understand because consent is a unique type of agreement that alters the moral relationship between the parties to the agreement.[1] Once two people consent, they receive new moral permissions from each other and/or take on new moral obligations to each other. If I consent to mow your lawn for $50, then I become morally obligated to mow your lawn and have a moral permission to take $50 from you when I am done, just as you become morally obligated to give me $50 and have a moral permission to expect me to mow your lawn. Due to our consent, we now stand in a different moral relationship such that we morally owe each other new things that we would not owe otherwise.

J. Rocha (✉)
Department of Philosophy, Fresno State, Fresno, CA, USA
e-mail: jamesr@csufresno.edu

D. Boonin (ed.), *The Palgrave Handbook of Sexual Ethics*,
https://doi.org/10.1007/978-3-030-87786-6_29

As we can see, consent is a morally loaded term: consent changes moral relations between people, and so only agreements made in the right way can count as consent.[2] Agreeing in the right way involves whatever it takes to create the relevant moral transformation. Often, that transformation involves both new moral permissions and obligations. Sometimes, consent is more about obtaining new moral permissions because the consent can be withdrawn without violating any new duty. For instance, in a boxing match, both sides gain permissions to punch, but you are not obligated to take punches throughout the match as you have the right to withdraw your consent and leave the match.

If consent were not made correctly (to the extent that there really would be no consent, but only mere agreement), then neither permissions nor obligations would be created. If you hold a gun at my head and demand I mow your lawn, my affirmative response counts as an agreement, but it isn't consent—coercion undermines consent.[3] Similarly, consent must be made with adequate information since an agreement wouldn't morally obligate a misinformed person.[4] I am here assuming that the misinformed person isn't lacking information through a reason where we would fault them. It is not enough though that you are not deceiving the other person—consent requires an active duty to fill the other person in on the information they reasonably would require to make the agreement. If you aren't filling them in and it is reasonable for them to need the information (which you often would be in a better position to know since you have the information and they may not know that you do), then you would be manipulating their (reasonable) ignorance to trick them into a deal, which would create neither a moral permission for you nor a moral obligation for them. Finally, the person must be competent to understand what they are agreeing to, which means they must be mature enough and sober enough to agree.[5] While there may be other conditions, almost all philosophers and legal experts agree on these three (consent must be free, informed, and competent), and these conditions will work for our current purposes.

When sexual consent is well made, it creates new permissions to have sex with the other person, but there is no obligation to continue since sexual consent can be withdrawn at any time. Yet, sexual consent is often quite complicated because there is so much at stake (since it is so important to ensure you have received consent prior to sex), yet it can be very uncomfortable to discuss sex, lots of people think about sex in different and unique ways, and we do not necessarily seek out sex in the circumstances that are most suited for effectively obtaining consent without much trouble. In other contexts, such as in business and healthcare, we have worked hard to make consent routine and robust: we negotiate terms, we make sure everyone understands those terms, and we tend to write things down so all parties can read the terms. In sexual contexts, there is much less formalization and much more risk of things going wrong, especially in less than ideal circumstances.

The standard college party involves circumstances that are much less than ideal for effectively obtaining sexual consent. Students are often looking to hook up with someone they don't know well, the music is loud, and most people are drinking copious amounts of alcohol. When put in the context of consent requirements, all of these standard party routines raise worries: the partiers don't know each other well enough to know for sure what they are consenting to, overcoming the loud music often requires increased nonverbal communication that endangers the information required for consent, and the alcohol risks the competency required for consent.

This paper examines how college students party according to social scientists and relates these findings to the requirements for sexual consent. It is important to point out at the start that we will be concentrating on heterosexual college party hook ups partially because, unfortunately, that is the main focus of the social science research on hooking up in general and college party sex in particular.[6] I will argue that college parties endanger consent in a morally problematic fashion, even if most college students fortunately could end up hooking up with someone who would otherwise grant consent. I will end by considering how we could use technological means, such as a hypothetical phone app, to try to improve the situation for hook up consent at college parties—though we will also need to flag potential limits of such a remedy.

Hooking Up at College Parties

I will use "hooking up" to refer to two individuals, who are not in a committed relationship, engaging in casual sexual activity with each other.[7] The ways in which many college students typically hook up raise a series of significant moral problems, including ones related to consent.[8] When partying, students often head out in gendered groups (men together and women together), and then they find someone they wish to hook up with at the party. They are likely to become inebriated prior to initiating the hookup process.[9] Once they find someone they are attracted to, they are likely to use nonverbal cues (especially given the loud music) to indicate their interest.[10] People will likely hookup with near strangers: people who they know at some distance—perhaps they don't even know each other's names, but they share a friend, a class, live in the same building, have seen each other in the dining hall, etc.[11]

By just looking at college parties in these general terms, we already see routine factors that jeopardize consent. The participants probably do not know each other well, they are likely communicating through nonverbal means, and they have been drinking alcohol. They are likely failing the competence requirement (due to the alcohol), the information requirement (due to nonverbal communication), and are unlikely to have a good system to ensure that they are nonetheless consenting (since they do not know each other well).

Having noted that difficulty, I do wish to acknowledge that most of these young people are likely to be fortunate in how things develop. Even if they do not meet the requirements for consent, it may very well turn out that both people went to the party hoping to hook up and they will end up being at least content that they did so. They could end up lucky in the sense that neither party to the hook up ends up feeling violated afterward. Some philosophers argue that this kind of sexual scenario, where odds are good that the results will be favorable, is sufficient for consent since both persons would have consented if they could have, and both are happy with how things actually worked out.[12]

In spite of this position, I do not believe the term "consent" is being used in the case described above in the morally transformative sense I have laid out here. If people do not actually give their agreement in an informed and competent way, then the consent cannot create the moral transformation that is being sought. If the requirements for consent, such as having adequate information and being competent at the point of the agreement, are not met, then it is not in fact consent. It might be that things luckily work out so that no one wishes to blame the other person. That means that luck is the morally operative factor: both parties are simply lucky that it worked out. It does not mean that we can retroactively say they consented, even if they would have if they could have. It just means that they will not likely be blamed.

I do not even think we can conclude in such a situation that they are not worthy of moral blame—sometimes we get lucky and do not receive the moral blame we deserve. In the type of cases under discussion, people did not do everything necessary to ensure they obtained consent prior to a sexual hook up. They deserve moral blame for that since they took a large risk with another person's sexuality. They are simply fortunate that the risk didn't turn out the wrong way. If it had, they would deserve blame, and they did nothing to avoid that blame. And so, really, they deserve that blame just the same if they happen to be fortunate with no one actually blaming them. They *deserve blame*, but they won't *be blamed*.

On a contrary view, blame could derive from consequences: someone could argue that you don't count as deserving blame if no one actually feels the need to blame you. This chapter is not the right place to argue whether blame is based on the agent's intentions or the consequences of the agent's actions. It is enough here to point out that even if people think blame should depend more on consequences, it would still be true that consent was not actually obtained since the consent requirements were not met.

Sexual consent at college parties then is problematic even in the general description (loud music, alcohol, participants who don't necessarily know each other well). That is enough to tell us that consent is not easily and securely obtained at most college parties. This general result tells us that we ought to work on better promoting consent among college students. Yet, things get even worse when we look at some more specific party trends.

For instance, consider parties where students engage in a very popular style of dancing: grinding. Sociologists Shelly Ronen and Lisa Wade separately describe how grinding leads to hooking up at college parties.[13] Grinding is where the man dances in an upright position while a woman backs her buttocks into his groin area. Often, the grinding will begin with the man coming from behind and then he will suddenly press his groin against the woman's buttocks, quite possibly without her having ever seen him.[14] Wade then explains that the woman would be thought to have agreed to hook up with the man if she simply turns to see his face.[15] Thus, women often ask their friends—who can physically see the man's face—to indicate whether the man is attractive before they decide whether to turn around.[16]

This common practice makes the situation worse for consent's information requirement. The loud music makes verbal communication much harder, and so it is difficult to convey information with too much precision in general. Yet, when people are choosing hook up partners through grinding, they often choose without even having a good look at the other person—or any look at all. These hook up partners then have not even met the minimal requirement that they at least give each other a sense of their looks. The idea behind the information requirement for consent is that you must provide the other person with what any reasonable person is likely going to need to know to make their decision to agree. If they don't provide the information the other person would reasonably need, they are not doing the morally required work to create consent. In this case, they are missing the bare minimum by not even allowing the other person to have a good view of what the person they would hook up with looks like: the grinding style undermines a key piece of information that most people are seeking when deciding to hook up. Hence, this grinding style makes a mockery of the robust information required for consent. Thus, this style of choosing a hook up partner makes it very unlikely that the partners are actually giving consent when they agree to hook up.

We then see that hooking up at college parties is fraught with problems that make consent at best difficult to achieve. In some typical cases, as with grinding, it is hard to imagine there is any consent because they are not even trying to meet the information requirement. As we will see, there are other trends in college parties that make consent difficult to effectively obtain, such as the gendered double standard.

The Gendered Double Standard at College Parties

Let's consider another problematic aspect of college parties: the well-known gendered double standard, where women are routinely treated worse than men.[17] The particular issue here is that this double standard leads to confusion about what counts as consent. Because of the double standard, men tend to think it is okay for them to take women's consent as implied, even though sexual consent should be obtained in an express and affirmative manner, or so I will argue.

One indication that college parties display a double standard comes in their very organization. Sociologists Elizabeth Armstrong and Laura Hamilton describe college party themes they came across in their research, which included: Victoria's Secret, Golf Pro/Ho, and School Teacher/Sexy Student.[18] Armstrong and Hamilton note that these parties, which fraternities organized, depend on a double standard where it is only women who dress seductively. The complaint here is not about the choice to dress seductively, which can be autonomously made. It is that men in fraternities organize parties where they dictate how women must dress, while not subjecting themselves to similar styles of provocative clothing. As Armstrong and Hamilton explain: "Fraternity men choose party themes, distribute alcohol to engineer social interactions in their favor, decide who can enter and who can leave parties, and generally dictate the social lives of the campus's youngest and most vulnerable residents."[19] The double standard is not merely that men and women are being treated differently, but that certain men are dictating the terms of party engagement. And this dictating the terms of the party can create a power differential that may make it more difficult to negotiate consent fairly. We will see this point more clearly by examining a particular party theme next.

Sociologist Lisa Wade describes one party theme where we will see that the double standard creates problems for consent: the Traffic Jam or stoplight party.[20] The stoplight party involves students wearing green, yellow, or red, or holding something, such as a cup, in those colors. While the popular websites Wikipedia[21] or Urban Dictionary[22] describe this color code as indicating relationship status (green means single, red means in a relationship, and yellow implies some kind of confusion or complication), Wade's research indicates that students use the colors to imply sexual availability: "students were instructed to wear green if they were 'down to fuck,' red if they were in a relationship, and yellow if they weren't sure what they wanted."[23]

Wade's research indicates that the problems, which can be quite serious, that are created by this sexualization of the stoplight colors tend to fall more on women than men. Hence, we will see this double standard as an issue that complicates consent insofar as it is more likely that the colors will lead men to take women's consent for granted. With stoplight parties, individuals, who are usually men according to the research, take consent to be implied by the manner in which other people are dressed or by what cup those individuals hold. Looking at the specific colors, if someone has on a red top, then they have made their dissent clear, and there is no issue. If someone has on a yellow top, then they have made clear that they would like to make an independent decision and determine whether to consent when someone asks. There is also no issue there. The issue then occurs with the person who has on a green top. The question then is whether that person is actually giving implied consent by wearing green.

We should stop and consider the distinction between express consent and implied consent. Express consent involves the parties to the agreement making their consent abundantly clear to each other, such as through an affirmative response. Implied consent involves a type of agreement that is not made explicitly to the other party, but instead certain types of actions and/or the lack of other types of actions are taken to imply their consent.

Let's note from the beginning that implied consent is contentious and any given instance of supposed implied consent could be confused or mistaken. If someone didn't explicitly agree, then it is risky to assume that they have actually consented. Implied consent is especially problematic since it is the consent that creates the moral permissions in the first place. Let's consider an example.

Martial artists are not typically allowed to hurt other people in the streets. Yet, once two consent to fight, they gain permission to try to hurt each other during the arranged combat. Implied consent would be dangerous in such a violent situation. Suppose we said that stepping on the mat where the fight was taking place counted as implied consent to allow others to fight you. That would mean someone who was confused or didn't understand this rule could be severely harmed just by walking on the mat. Yet, when that confused person walked on the mat, there was no real consent since the confused person clearly did not meet the information requirement for consent. Thus, we should be suspicious of the whole idea of implied consent, and we should likely rule it out from circumstances where it could be dangerous.

Having said that, we at least seem to allow for implied consent in ordinary life, especially in business encounters. For example, when I am in a store, the store has provided me implied consent to touch the objects for sale so as to imagine them in my home. The store has given that implied consent by allowing me in, not placing signs that tell me not to touch, etc. Yet, the store can withdraw this consent if they decide I'm touching the objects in a way that they do not like. So perhaps implied consent works in certain (presumably safe) scenarios.

I believe that sexual consent, though, should always be given expressly and not assumed to be implied. In terms of the stoplight party, I believe that even the people wearing green have not provided implied consent to hook up. The problem here is that, even in the best of cases, acts of consent are open to interpretation.[24] And by the best cases, I mean ones where words are spoken with clear explanations of each person's intentions. But even then, there is room for misinterpretation because words can convey the wrong things, people may interpret the same sentence in different ways, and we certainly do not say everything we are thinking (nor should we). So, even when people provide their express, affirmative consent, there may be differences in how the parties to the consent interpret what they have agreed upon.[25]

This problem becomes clear with the green top at the stoplight party. What has the person with the green top consented to? Let's assume, for the sake of argument, that the party organizers have attempted to make it clear that wearing a green top counts as consenting to hook up. As long as there is still room for multiple, reasonable answers, we will have a plethora of situations where there can be reasonable confusion over the supposed implied consent.

Before arguing for this reasonable confusion, it is worth noting that things would still be problematic even if the confusion were not reasonable. If someone had a strange idea in their mind about what they were consenting to by wearing green, then it may be reasonable to believe that they consented to sex but the fact of the matter is that they did not *actually* consent to sex because they do not have sex on their mind. Consent won't transfer over moral permissions to do something if you never intentionally consented to allow that thing. Consent creates no moral transformation if someone is not intentionally consenting. Consent simply cannot occur unintentionally, by definition.

The moral problem then is even graver if there are multiple reasonable interpretations of wearing a green top. While there may be an expectation that the person wearing green is down to have sex with whomever (as the party organizers intended and attempted to advertise), I believe there are reasonable alternative interpretations. For instance, a person could believe wearing green means they are prepared to have sex with somebody, but they still can be choosy about whom they have sex with. Suppose the party's flyer said something like, "Green top = DTF" (which the relevant audience for the flyer would surely know means down to have sex). This person though takes it to be a reasonable position that wearing green would not entail that they are opening themselves up to have just anyone start having sex with them out of the blue. They rule out that interpretation as simply ridiculous. While it so happens that the seemingly ridiculous interpretation is what the party organizers have in mind, I want to suggest that it is reasonable to believe that the organizers wouldn't mean that interpretation since *it is fairly ridiculous*. Maybe some of the partygoers will know in fact the organizers mean the ridiculous interpretation. On the other hand, it is fair to question such an interpretation since most people both wish to be at least somewhat selective about their sexual partners and also expect others to know this selectivity is incredibly common. Since the party organizers hold an outlier view, it is reasonable that others are ruling their view out because it really is fairly ridiculous.

Suppose instead the flyer said, "Green = Ready to Hook Up with Whomever." Now, it has become more explicit on the "whomever," but the slogan is still vague in that the phrase "hook up" is ambiguous.[26] As mentioned earlier, the definition of hooking up refers to sexual activity, but that could include anything from making out to sex. So, someone could think wearing green is a way to agree in advance to hook up, but they do not necessarily believe "hooking up" has a singular, coital definition. Such a person can reasonably wear green while only being ready to make out, but not being willing to have sex.

Suppose the flyer attempts to be even more explicit: "Green = DTF with Anyone." This time, the person, who still thinks that most people wouldn't take that claim completely literally, interprets the statement to mean that they are down to *consider* sex with anyone, but they have not consented to anything (other than listening to the advances). I will argue that this interpretation too would be reasonable.

Let's imagine an analogy where someone goes to a bar. You could argue that they have given their implied consent to be around people drinking alcohol. They certainly would be acting in a problematic fashion if they attempted to register a complaint and asked the manager to get everyone else to stop drinking. So, it is fair to say they have given some implied consent to be around alcoholic drinking just by entering the bar. Suppose the bartender then repeatedly asked them to buy a drink themselves. The patron would be in the right if they insist they do not wish to buy a drink (provided that the bar has no clear rules that require drinking upon entry). They would not be in the right to insist that the bartender remain silent. I believe this nuanced position is reasonable for the following reason: by entering the bar, the patron has signaled that they are open to buying a drink. Yet, the patron has not consented to buy a drink. Perhaps they are hoping someone else will buy them a drink if they are patient. Perhaps, upon arrival, they see the bar has no drink that interests them. Perhaps they are meeting a friend who will purchase drinks later. There are numerous reasonable ways in which we can sit in a bar and not buy a drink, assuming the bar doesn't have a prior rule that requires a minimum drink purchase. By entering the bar, there is implied consent to listen to the bartender's sales pitch, but no implied consent to buy a drink. Thus, it is a situation where you may have to consider buying a drink, but you haven't consented to buy a drink, nor should you be forced to do so.

I believe the person wearing green could end up in a similar position. By wearing green, they indicate that they want to have sex. Yet, they could be like the patron who just can't find anything worth obtaining. They wanted to have sex, but the party disappoints them and there is no person approaching them with whom they are actually willing to have sex. Of course, as long as they are still at the party wearing green, they might be wrong to insist no one make sexual advances. Regardless, they have certainly not given any sexual consent, but at most consent to listen to sexual advances (or leave if they no longer wish to hear any).

Given these various possibilities, it is reasonable that a person could wear green and not believe they have granted implied consent to have sex with just anyone. Since it is reasonable for someone to believe they have not given implied consent, it is not morally acceptable for others to assume that they have been given implied consent when in fact there is no consent. The whole idea behind implied consent is that others can assume you consented without having to ask. But in this case, the mere wearing of a green top doesn't do the work of allowing others to assume there is sexual consent. They should still ask to make sure the other person wishes to have sex, regardless of the green top. And so this is not a case of implied consent. Therefore, initiating sex with someone with a green top without asking would remain a heinous act.

Let's briefly note that the situation is better with the colored cups. The improvement lies in the fact that the partygoers can put down their cups whenever they wish to do so. It is a lower cost of signaling dissent, which allows more room for drawing assumptions from someone's behavior. Since the stoplight cup party allows more room to easily and safely withdraw consent, they are preferable to the stoplight party where clothes are alleged to signal consent.

While there is clear improvement here, it still does not respond to the problem that consent is opaque. It is still uncertain what the person holding the green cup wishes to signal by doing so. There are reasonable alternatives other than that they have granted implied consent to have sex with just anyone. Hence, one still needs to ask them for express consent, and so there is no implied consent in the stoplight cup party either. The cup improves the situation, but it doesn't change the fact that consent should be established through actual, clear signals, and sexual consent should never be taken for granted due to a top, a cup, or anything else.

While the stoplight party is a particularly clear example where the double standard leads to morally problematic confusion about consent, there is reason to believe this confusion is widespread. As Lisa Wade explains from her research, the stoplight party is "particularly intense," but her research subjects "affirm that these sorts of experiences are part and parcel of going to college parties. 'Disrespectful' was a word that came up frequently, alongside the idea that men disregard women's 'boundaries.'"[27] Wade is quite explicit that men at college parties are very poor judges of consent: "Some men act instead of asking, counting on women to acquiesce or object after the fact. Because if 'just being' at a college party opens a woman up to being fondled or touched without her permission, 'just going' to his room might suggest to some men that equally sneaky, manipulative, or coercive measures are allowed."[28] And these are quite grave moral transgressions. Wade's research indicates that these men are assuming implied consent to grope women and take sexual advantage of them simply because they are at a certain type of college party together.[29] These extreme violations of consent at college parties are clearly quite serious. And so it is important to acknowledge the violations of consent that occur at college parties are dangerous.[30]

Partial Technological Solutions

As noted earlier, it is important to see that the worst-case scenarios, which are significantly problematic and should not be ignored, are not necessarily more common than situations where people are happy or at least content with their college party sexual experiences. So, there is reason to want to improve. There is reason to want to ensure consent is being safely obtained in hook ups at college parties (and everywhere else, of course). To conclude, it will be worth considering the type of changes that would help improve this situation.

There should be clearer ways to signal sexual interest at college parties. Unfortunately, wearing a green top or holding a green cup allows way too much room for confusion. Instead, interest needs to be signaled in a fashion that is particular (consent is given to specific people with whom one wishes to hook up, and not in the open to just anyone) and detailed (what exactly is being consented to). In my previous work (*The Ethics of Hooking Up*), I briefly considered what an app would have to be like to help with this sort of situation.[31] There, I argued that the imaginary moral hook up app would allow users to indicate their sexual interests, as well as to facilitate a pre-hook up conversation about how the users wished to hook up. While I acknowledged in the book that this possible app would face difficulties and shouldn't be relied upon, the general idea is a good start for using technology to try to improve the situation.[32]

In concluding here, I would like to expand on the idea of the imagined hook up app. The goal would be to develop a system that allowed for clear communication in an environment that made communication very difficult. The stoplight colors are insufficient because there is way too much room for confusion. But partygoers could develop a series of brief, prewritten responses. It would be important that they could make choices in the moment since the stoplight colors are not particularly directed at specific people but are presented to everyone equally—which is not how most people tend to hook up. Very few people enter a party and want to hook up with just anyone there. People like to make choices in the moment that allow them to be at least somewhat selective. And they will respond to different people in different ways. Thus, an app that allowed for sending a number of select prewritten responses would allow for people to choose how to respond to the person with whom they are currently interacting. If the person wishes to have different kinds of hook ups with different people, they could choose the prewritten response for the type of hook up they wish to have, and show that to the other person to check if they will consent to it.

The app would also need to allow for people to express their desires as well as their limits and reservations. Another troubling aspect of the gendered double standard for hook ups is that men are more likely to reach their sexual goals, such as achieving orgasms, than women.[33] It can be very uncomfortable to discuss sexual interests, and this conversation is unlikely to occur at a loud and crowded college party. Thus, the app can allow people to share as much as they feel comfortable sharing in an attempt to improve the physical activity itself.

Yet, the app can also allow people to express clearer limits. There are definitely some things that people will not be comfortable doing with near strangers, and they can share those things as they send their prewritten messages to each other. It would be useful to have shorthand methods to ensure people are getting these messages, and that would allow them to convey what they are interested in and what they will not be willing to do. Thus, the app allows them to facilitate a discussion while in an environment that is not conducive to communicating verbally.

It is still important to note that technology cannot solve all the problems of sexual communication, and technological solutions probably shouldn't even be the primary method for responding to these problems. For instance, were an app made to signal both consent and possibly the removal of consent as well, it could not be assumed that consent could only be removed by the app. Instead, both participants must be on alert for the other person attempting to signal that they wish to revoke their consent in any fashion that they are able to communicate it. Further, there would always be room to continue negotiation beyond the app. It cannot be taken for granted that consent can easily be granted on the app. In some cases, people may wish to alter their positions based on reading the other person's prewritten statements. And, they should always be free to make such alterations since consent should always be a fluid process that is susceptible to change, especially where change is necessary to make someone more comfortable. So, an app can facilitate a greater likelihood to achieve consent than nonverbal communication in a loud party is likely to allow for, but an app doesn't overcome all consensual issues that can arise.

The problems outlined here were related to how our society views sex. We do not take sexual consent nearly seriously enough. Men especially, according to the available research, have a very worrisome tendency to assume that they have obtained consent when they have not. As mentioned, they may often get lucky and hook up with someone who wishes to hook up with them most of the time, in spite of operating from misconceptions about consent. Yet, when people, of either gender (but especially men according to social patterns), continue to take consent for granted and do not put in the work necessary to ensure they have obtained consent, they risk eventually committing a grave moral transgression, which may amount to rape. It is of the utmost importance that we all take sexual consent much more seriously.

An app is a small step toward improving the situation. It would be better to take large steps and change our sexual culture in general and change how people hook up at parties more specifically. There should be more attention to securing consent in sexual situations. Perhaps we do not wish to go as far as business or medical consent and demand things to be put into writing. But the imagined app allows us to at least get things in writing even while it wouldn't be an official contract. The goal is to improve communication and try to ensure that people are more certain they are obtaining consent prior to hooking up. Yet, while this is a worthy goal, it is insufficient, and it is much more important that we work to make sure that we all value consent and are committed to perform the serious labor it takes to ensure there is consent prior to sex.

Notes

1. Heidi M. Hurd, "The Moral Magic of Consent," *Legal Theory* 2, no. 2 (1996), 121, 123–124, 137; David Archard, *Sexual Consent* (Boulder: Westview Press, 1998), 3, 5–6; Alan Wertheimer, "Consent and Sexual Relations," in *The Philosophy of Sex: Contemporary Readings*, eds. Alan Soble and Nicholas Power (New York: Rowman and Littlefield Publishers, 2008), 291; John Kleinig, "The Nature of Consent," in *The Ethics of Consent*, eds. Franklin G. Miller and Alan Wertheimer (New York: Oxford University Press, 2010), 4; Tom Dougherty, "Sex, Lies, and Consent," *Ethics* 123 (July 2013), 722; Tom Dougherty, "Yes Means Yes: Consent as Communication," *Philosophy & Public Affairs* 43.3 (2015), 226, 232–233.
2. There is some disagreement on this point. For views that disagree with me and think of consent as not being a moral term, see: Joel Feinberg, "Victims' Excuses: The Case of Fraudulently Procured Consent," *Ethics* 96, no. 2 (January 1986), 333–334; Thomas Mappes, "Sexual Morality and the Concept of Using Another Person," in *The Philosophy of Sex: Contemporary Readings*, eds. Alan Soble and Nicholas Power (New York: Rowman and Littlefield Publishers, 2008), 235; Dougherty, "Yes Means Yes," 226. For views that do see consent as morally loaded, see: Hurd, "The Moral Magic of Consent"; Wertheimer, "Consent and Sexual Relations," 291; Kleinig, "The Nature of Consent," 4;

Alan Wertheimer, *Consent to Sexual Relations* (Cambridge: Cambridge University Press, 2003), 119–120; Leo Katz, "Choice, Consent, and Cycling: The Hidden Limitations of Consent," *Michigan Law Review* 104, no. 4 (February 2006), 628; Robin West, "Sex, Law, and Consent," in *The Ethics of Consent*, eds. Franklin G. Miller and Alan Wertheimer (New York: Oxford University Press, 2010), 221.

3. Wertheimer, "Consent and Sexual Relations," 295, 302–304; Archard, *Sexual Consent*, 50–53; Mappes, "Sexual Morality," 231–232, 235–241; Kleinig, "The Nature of Consent," 13–16; Tom L. Beauchamp, "Autonomy and Consent," in *The Ethics of Consent*, eds. Franklin G. Miller and Alan Wertheimer (New York: Oxford University Press, 2010).
4. For more on the information requirement, see: Mappes, "Sexual Morality," 231–232; Wertheimer, "Consent and Sexual Relations," 295, 304–305; Archard, *Sexual Consent*, 46–50; Wertheimer, *Consent to Sexual Relations*, 193–214; Neil Manson and Onora O'Neill, *Rethinking Informed Consent in Bioethics* (Cambridge: Cambridge University Press, 2007); Kleinig, "The Nature of Consent," 16–17; Dougherty, "Sex, Lies, and Consent."
5. For more on the competence requirement, see: Mappes, "Sexual Morality," 233; Hurd, "Moral Magic," 140–142; Wertheimer, "Consent and Sexual Relations," 295, 305–306; Archard, *Sexual Consent*, 44–46; Wertheimer, *Consent to Sexual Relations*, 215–257; Kleinig, "The Nature of Consent," 13–14.
6. For studies that do examine same-sex party sex, see: Christian Grov, Jackeline Cruz, and Jeffrey T. Parsons, "Men Who Have Sex with Men's Attitudes Toward Using Color-Coded Wristbands to Facilitate Sexual Communication at Sex Parties," *Sex Research Social Policy* 11 (2014); Étienne Meunier, "Social Interaction and Safer Sex at Sex Parties: Collective and Individual Norms at Gay Group Sex Venues in NYC," *Sex Research Social Policy* 15 (2008).
7. See discussions such as Elizabeth L. Paul, Brian McManus, and Allison Hayes, "'Hookups': Characteristics and Correlates of College Students' Spontaneous and Anonymous Sexual Experiences," *Journal of Sex Research* 37, no.1 (2000); Norval Glenn and Elizabeth Marquardt, *Hooking Up, Hanging Out and Hoping for Mr. Right: College Women on Dating and Mating Today*, An Institute for American Values Report to the Independent Women's Forum (2001); Tracy A. Lambert, et al., "Pluralistic Ignorance and Hooking Up," *The Journal of Sex Research* 40, no.2 (2002); Kathleen Bogle, *Hooking Up: Sex, Dating, and Relationships on Campus* (New York: New York University Press, 2008); Teresa M. Downing-Matibag and Brandi Geisinger, "Hooking Up and Sexual Risk Taking Among College Students: A Health Belief Model Perspective," *Qualitative Health Research* 19, no. 9 (September 2009); Carolyn Bradshaw et al., "To Hook Up or Date: Which Gender Benefits?" *Sex Roles* 62 (2010), 661; Elizabeth Armstrong et al., "Accounting for Women's Orgasm and Sexual Enjoyment in College Hookups and Relationships," *American Sociological Review* 77, no. 3 (2012), 435; Melissa A. Lewis et al., "What is Hooking Up? Examining Definitions of Hooking Up in Relation to Behavior and Normative Perceptions," *Journal of Sex Research* 50, no. 8 (2013), 757; Herbert W. Helm Jr. et al., "Hook-Up Culture among College Students: A Comparison of Attitudes toward Hooking-Up Based on Ethnicity and Gender," *North American Journal of Psychology* 17, no. 2 (2015): 221–232; Spencer B. Olmstead et al., "Hooking Up and Risk Behaviors Among First Semester College Men:

What is the Role of Precollege Experience?" *The Journal of Sex Research* 52.2 (2015), 187; Lisa Wade, *American Hookup: The New Culture of Sex on Campus* (New York: W. W. Norton & Company, 2017), 19.

8. I have written about these issues at length at James Rocha, *The Ethics of Hooking Up: Casual Sex and Moral Philosophy on Campus* (New York: Routledge, 2020). Unfortunately, these studies, along with related studies, examine same-sex sex at sex parties, as opposed to more standard college parties, which is our focus here.
9. Paul et al., "'Hookups': Characteristics and Correlates of College Students' Spontaneous and Anonymous Sexual Experiences," 83–85; Glenn and Marquardt, *Hanging Out,* 15–16; Lambert et al., "Pluralistic Ignorance,"129; Bogle, *Hooking Up*, 30, 47, 63–64, 166–168; Downing-Matibag and Geisinger, "Sexual Risk," 1196, 1203; Lisa Menegatos et al., "Friends Don't Let Jane Hook Up Drunk: A Qualitative Analysis of Participation in a Simulation of College Drinking-Related Decisions," *Communication Education* 59.3 (July 2010); Cynthia K. Buettner, Atika Khuarna, Natasha Slesnick, "Drinking at College Parties: Examining the Influence of Student Host-Status and Party-Location," *Addictive Behaviors* 36 (2011); Jesse Owen et al., "Short-Term Prospective Study of Hooking Up Among College Students," *Archives of Sexual Behavior* 40 (2011), 337, 338–339; Julie A. Reid et al., "Casual Hookups to Formal Dates: Refining the Boundaries of the Sexual Double Standard," *Gender & Society* 25, no. 5 (October 2011), 556; Melina M. Bersamin et al., "Young Adults and Casual Sex: The Relevance of College Drinking Settings," *Journal of Sex Research* 49, no. 2–3 (2012); Lewis et al., "What is Hooking Up?," 764–765; Allyson L. Dir et al., "From the Bar to the Bed via Mobile Phone: A First Test of the Role of Problematic Alcohol Use, Sexting, and Impulsivity-Related Traits in Sexual Hookups," *Computers in Human Behavior* 29 (2013); Megan Manthos et al., "A New Perspective on Hooking Up among College Students: Sexual Behavior as a Function of Distinct Groups," *Journal of Social and Personal Relationships* 31, no. 6 (2014), 824–825; Olmstead et al., "Hooking Up and Risk Behaviors," 188; Patricia N. E. Roberson et al., "Hooking Up during the College Years: Is there a Pattern?" *Culture, Health & Sexuality* 17.5 (2015), 578, 586, 587; Tara E. Sutton and Leslie Gordon Simons, "Sexual Assault Among College Students: Family of Origin Hostility, Attachment, and the Hook-Up Culture as Risk Factors," *Journal of Child and Family Studies* 24, no. 10 (October 2015); Wade, *American Hookup*, 28–31, 43–45, 70–91.
10. Monica Moore, "Courtship Signaling and Adolescents: 'Girls Just Wanna Have Fun'?" *The Journal of Sex Research* 32, no. 4 (1995); Douglas N. Husak and George C. Thomas III, "Rapes without Rapists: Consent and Reasonable Mistakes," *Philosophical Issues: Social, Political, and Legal Philosophy* 11 (2001), 88, 93–94, 98; Lambert et al., "Pluralistic Ignorance," 129; Bogle, *Hooking Up*, 33–39; Shelly Ronen, "Grinding on the Dance Floor: Gendered Scripts and Sexualized Dancing at College Parties," *Gender and Society* 24.3 (June 2010), 367–373.
11. Rocha, *Ethics of Hooking Up*, 7.
12. See: Wertheimer, *Consent to Sexual Relations*, 152–155; Husak and Thomas III, "Rapes without Rapists," 92–103.
13. Ronen "Grinding"; Wade, *American Hookup*, 31–39.
14. Wade, *American Hookup*, 33. Also see Ronen "Grinding," 361, 366–369.

15. Wade, *American Hookup*, 38. Also see Ronen "Grinding," 361–362, 369–370.
16. Wade, *American Hookup*, 33–37.
17. Mary Crawford and Danielle Popp, "Sexual Double Standards: A Review and Methodological Critique of Two Decades of Research," *The Journal of Sex Research* 40, no. 1 (2003); Bogle, *Hooking Up*, 96–127; Glenn and Marquardt, *Hanging Out*, 21–22; John Draeger, "Can Girls Go Wild With Self-Respect?" in *College Sex: Philosophy for Everyone*, eds. Michael Bruce, Robert M. Stewart, and Fritz Allhoff (West Sussex: Wiley-Blackwell, 2010), 198–208; Reid et al., "Casual Hookups,"; Owen et al., "Short Term," 338–339; Laura D. Kovac and Dawn E. Trussell, "'Classy and Never Trashy': Young Women's Experiences of Nightclubs and the Construction of Gender and Sexuality," *Leisure Sciences* 37 (2015), 201–204; Heather Kettrey, "What's Gender Got to Do With It? Sexual Double Standards and Power in Heterosexual College Hookups," *The Journal of Sex Research* 53.7 (2016); Allison B. Wolf, "A Hookup of Her Own," *International Journal of Applied Philosophy* 30.2 (2017), 195–196.
18. Elizabeth A. Armstrong and Laura T. Hamilton, *Paying for the Party: How College Maintains Inequality* (Cambridge: Harvard University Press, 2013), 51.
19. Ibid., 53.
20. Wade, *American Hookup*, 207–210.
21. See https://en.wikipedia.org/wiki/Stop_light_party (accessed 29 July 2020).
22. See https://www.urbandictionary.com/define.php?term=stoplight%20party (accessed 29 July 2020).
23. Wade, *American Hookup*, 207–208.
24. Manson and O'Neill, *Rethinking Informed Consent*, 42–48.
25. Onora O'Neill, "Between Consenting Adults," *Philosophy and Public Affairs* 14, no. 3 (Summer 1985), 255–256; David M. Adams, "Date Rape and Erotic Discourse," in *Date Rape: Feminism, Philosophy, and the Law*, ed. Leslie Francis (University Park: The Pennsylvania State University Press,1996), 38–39; Manson and O'Neill, *Rethinking Informed Consent*, 45–48; Patricia Marino, "The Ethics of Sexual Objectification," *Inquiry* 51, no. 4 (August 2008), 355, 357; Laura Backstrom et al., "Women's Negotiation of Cunnilingus in College Hookups and Relationships" *Journal of Sex Research* 49, no. 1 (2012), 6–9.
26. Glenn and Marquardt, *Hanging Out*, 5, 13, 22; Bogle, *Hooking Up*, 28–29, 85–88; Lewis et al., "What is Hooking Up?"; Armstrong et al., "Women's Orgasm," 435; Wade, *American Hookup*, 40; Wolf, "A Hookup of Her Own," 191.
27. Wade, *American Hookup*, 210.
28. Ibid.
29. Ibid., 206.
30. For more on these problematic trends, see: Paul et al., "'Hookups,'" 81; Lambert et al., "Pluralistic Ignorance," 130; William F. Flack, Jr. et al., "Risk Factors and Consequences of Unwanted Sex Among University Students: Hooking Up, Alcohol, and Stress Response," *Journal of Interpersonal Violence* 22.2 (February 2007), esp. 150–152; Robyn L. Fielder et al., "Sexual Hookups and Adverse Health Outcomes: A Longitudinal Study of First-Year College Women," *Journal of Sex Research* 51, no. 2 (2014), 140–141; Sutton and Simons, "Sexual Assault,"; Kimberly A. Tyler, "Alcohol Expectancy, Drinking Behavior, and Sexual Victimization among Female and Male College Students,"

Journal of Interpersonal Violence 37.15 (2015), 2301 and 2314; Wendasha Jenkins Hall and Amanda E. Tanner, "US Black College Women's Sexual Health in Hookup Culture: Intersections of Race and Gender," *Culture, Health & Sexuality* 18.11 (2016), 1265, 1268; Kaitlin M. Boyle and Lisa Slattery Walker, "The Neutralization and Denial of Sexual Violence in College Party Subcultures," *Deviant Behavior* 37.12 (2016).

31. Rocha, *Ethics of Hooking Up*, 154–155.
32. Ibid.
33. Laura Backstrom et al., "Women's Negotiation of Cunnilingus"; Wade, *American Hookup*, 157–179.

CHAPTER 30

#MeToo and the Ethics of Doxing Sexual Transgressors

Peter Brian Barry

Catherine McKinnon opens an opinion piece by praising the #MeToo movement for "accomplishing what sexual harassment law to date has not," adding that:

> [T]oday's survivors are initiating consequences none of them could have gotten through any lawsuit—in part because the laws do not permit relief against individual perpetrators, but more because they are being believed and valued as the law seldom has. Women have been saying these things forever. It is the response to them that has changed.[1]

If #MeToo has succeeded in altering public responses to allegations of sexual misconduct, that result is worth noting. Equally worth noting is that those results are sometimes secured via extralegal mechanisms. For example, in October of 2018, Moira Donegan created a Google Spreadsheet titled "Shitty Media Men," an anonymous crowdsourced document recording and detailing allegations of sexual misconduct perpetrated by men in the publishing industry. To her surprise, the document went viral and at least some alleged perpetrators were investigated and fired.[2] One of Harvey Weinstein's early accusers, Rose McGowan, tweeted in October of 2016 using #WhyWomenDontReport "because it's been an open secret in Hollywood/Media & they shamed me while adulating my rapist,"[3] later identifying Weinstein by name. Examples can

P. B. Barry (✉)
Saginaw Valley State University, University Center, MI, USA
e-mail: pbbarry@svsu.edu

D. Boonin (ed.), *The Palgrave Handbook of Sexual Ethics*,
https://doi.org/10.1007/978-3-030-87786-6_30

be multiplied of individuals using social media and the Internet to identify sexual transgressors. The identified individuals are thereby doxed.

By most accounts, "dox" and "doxing" (alternatively: "doxx" and "doxxing") refer to revenge tactics from hacker culture that reveal a target's identity, thereby exposing them to undesirable attention from both legal and extralegal entities. There are strong taboos against doxing operative in online communities and suspension or expulsion is a common punishment as are reprisals. Among persons deeply immersed in online culture "Doxing has only negative connotations"[4] and it is often described in moralized terms: as a "perilous form of justice,"[5] a "downright dangerous" tactic that "makes data into a weapon,"[6] and "cyberbullying."[7] Doxing is little discussed by professional philosophers but when discussed it is ominously described as a "spectre... haunting the Internet."[8] Generally, I understand doxing to involve the intentional public release onto the Internet of personally identifying information by a third party, often, but not necessarily, with the intent to humiliate, threaten, intimidate, or punish.[9] Doxing involves, at least, the publication of personally identifying information about a subject—that is, the doxed individual—that is purportedly connected to other personally identifying information. This non-moralized definition is preferable to moralized definitions of "dox" and its variants as moralized definitions rule out morally permissible doxing by fiat. I contend that, at least sometimes, doxing is morally permissible and praiseworthy.

In this chapter, I defend the thesis that doxing sexual transgressors is sometimes a morally permissible response to their sexual misconduct. I consider several rationales for thinking that doxing sexual transgressors can be justified. I first consider the possibility that doxing sexual transgressors is justified in virtue of warning prospective victims of sexual misconduct. I then consider the related possibilities that sexual transgressors are liable to be punished in virtue of their misconduct and that doxing them can realize some goods of just punishment. I also consider the possibility that doxing engages subjects in a sense that needs to be explained. I also consider and defang a series of objections along the way. What emerges is, I think, a fairly strong case that doxing sexual transgressors is sometimes morally permissible.

A caveat: the meaning and extension of "sexual transgressor" is unclear. In health care ethics, sexual transgression is "A form of sexual misconduct consisting of inappropriate and sexualized touching of a patient that stops short of overt sexual relations,"[10] a definition that suggests that having sexual relations with a patient is not a sexual transgression, and thus that a clinician who has sex with a patient is not a sexual transgressor. I stipulate that I understand "sexual transgressor" to denote anyone responsible for transgressing moral norms governing appropriate sexual conduct. Accordingly, rapists are sexual transgressors, but domestic abusers might not be if their misconduct violates other, nonsexual norms. Sexual harassers are sexual transgressors even if their victims fail to make the case that a reasonable person could not operate in the workspace—if, say, their actual target is not easily rattled—just so long as their conduct does violate sexual norms. So understood, intent is not a necessary element for being a sexual transgressor: someone who unknowingly or unwittingly violates moral norms about sex counts. Accordingly, I do not argue that all sexual transgressors may be permissibly doxed. The case for doxing sexual transgressors is strongest when subjects are clearly culpable and their transgression a serious one—that is, when it causes or places a victim at risk of serious harm and does not merely cause or risk disliked mental states like offense or discomfort. While doxing is fairly often categorically dismissed as wrongful or unjust, I contend that it is sometimes a morally permissible response to sexual transgressors, but especially when extant legal mechanisms fail to protect or secure justice for victims of sexual misconduct.

Doxing as Warning

One obvious reason to dox a sexual transgressor is because they are prone to causing harm to innocents or saddling them with unjustifiable risks. Doxing can thus be understood as a means for warning others to ensure that they can reconsider future interactions, revise current arrangements, and so forth. Doxing is quicker than other, more familiar legal mechanisms for warning potential victims: criminal investigations and trials are slow processes if they occur at all; their operations are largely shielded from the public eye; nondisclosure agreements prevent victims from alerting others, and so forth. By contrast, doxes can spread rapidly among social media users thus expediting the process. Doxing can also help to disseminate information to potential victims who might otherwise go uninformed. The Shitty Media Men document is illustrative: various "whisper networks" alerted some potential victims to the risks posed by various sexual transgressors, but they reached only a limited number of persons and missed some especially vulnerable ones.[11] Further, the tools of social media allow for commentary and follow-up to clarify and elicit information: direct messages can be sent via social media, tweets replied to, and so forth.

Understood as a mechanism for warning, doxing resembles other, morally justifiable mechanisms used to warn potential victims of misconduct. For example, doxing sexual transgressors is arguably analogous to publishing sex offender registries: the purpose of each is to provide information about a public threat so that measures can be taken to prevent or minimize the risk of future harm. This analogy is suspect, however. Crucially, individuals placed on sex offender registries have been convicted and not merely accused, however credibly, of misconduct. The typical dox of a sexual transgressor comes much earlier in the process and identifies the subject as a sexual transgressor prior to being found legally responsible and may occur precisely because the sexual transgressor is or is believed to be unlikely to be found legally responsible. And since convicted sex offenders have gone through the process of criminal conviction they have been given due notice and afforded the opportunity to defend themselves. Doxed subjects usually get no such opportunity prior to publication. Relatedly, websites that track sex offenders are the product of and maintained by public entities: for example, the National Sex Offender Public Website is hosted and maintained by a network of federal and state government agencies, including the United States Department of Justice. Doxes, like whisper networks, are usually private affairs even when organized and thus less amenable to accountability and review, among other concerns.

A better analogy is available. Doxing sexual transgressors is arguably more like whistleblowing.[12] While different conceptions of whistleblowing are available, it is commonly supposed that a whistleblower (1) has "insider status" within an institution, (2) discloses nonpublic information, (3) brings attention to illegal or immoral activity, (4) avoids of the usual chain of command of their institution, and (5) intends to stop that activity.[13] The analogy is strengthened if the doxer of a sexual transgressor is "on the inside" of an institution. Donegan, for example, was employed in the media industry and is thus able to dish about its inner workings but also avoided reporting to, say, a director of human resources or an equity officer who would normally take receipt of complaints about sexual misconduct, suggesting that (1) and (4) are met. Similarly, a graduate student might dox the adviser who harassed him thereby revealing the tawdry secrets of his department instead of reporting to his Title IX Coordinator. If the sexual transgressor's misconduct is purely a private affair performed outside the confines of what we would normally understand as an institution, there may be no one with "insider" status and no chain of command to avoid, although the semantics of "institution" can be stretched a bit: a woman harassed by construction workers in her neighborhood might post their photos in a community Facebook group instead of reporting to law enforcement, for example. In any case, some insiders (or "insider-like persons") will also meet conditions (2), (3), and especially (5) by doxing sexual transgressors, just as some whistleblowers do.

Another point about doxing and whistleblowing: some, but not all, cases of whistleblowing are properly regarded as supererogatory conduct—that is, as conduct that is "above and beyond the call of duty" or, more precisely, conduct that is not morally required of us but is of great moral value when performed. The same can be said of doxing. Whistleblowing is sometimes claimed to be grounded in a general moral duty—say, to prevent harm[14] or wrongdoing.[15] If whistleblowers have a duty to prevent harm, then the whistleblower simply does what morality requires of her and, by definition, she does not supererogate. But whistleblowing often imposes significant costs and liabilities on whistleblowers, perhaps more than morality can require us to take on. If whistleblowing is not actually morally obligatory because it is too risky, then the whistleblower supererogates. Now, for all that has been said, in some cases in which doxing is the best or only means to warn and thereby prevent harm, doxing is morally obligatory and someone who doxes in these conditions does not supererogate. But sometimes credibility demands revealing personal details, even if a dox is anonymous, which increases risks of reprisal and retaliation, perhaps more than morality can require. In those cases, doxers, like whistleblowers, supererogate, another point in favor of the analogy between doxing and whistleblowing.

This analogy does not entail that doxers may take any steps they regard as necessary to alert the public and familiar moral constraints remain operative. A whistleblower who alerts the public to governmental malfeasance but carelessly or needlessly unmasks and endangers the lives of innocent foreign operatives may well act wrongly even if acting from duty. Doxers too must sometimes take steps to minimize harms done or posed to innocents, a concern that leads to an objection to doxing sexual transgressors. Undeniably, some innocents have been wrongly doxed, suffering unjustly as a result, and undoing the damage done is challenging. For example, Dr. Kyle Quinn, Assistant Professor of Engineering at the University of Arkansas, bears a slight resemblance to Andrew Dodson who was photographed carrying a Tiki torch and wearing an "Arkansas Engineering" t-shirt at the Charlottesville "Unite the Right" rally in 2017. Amateur investigators confused Quinn and Dodson, and Quinn endured a public campaign for his firing and came to fear for his safety.[16] There are also a handful of high-profile cases in which sexual transgressors were publicly and almost certainly wrongly identified, including the 2014 case of an alleged rape at a fraternity party at the University of Virginia reported in *Rolling Stone* magazine. The accused were cleared in both instances, but only after significant damage and at great cost.

There is little chance of denying that doxing sexual transgressors can imperil innocents, but so can many legal and social practices that we tolerate, including legal punishment. More to the point, the current objection is less an objection to doxing sexual transgressors in particular than to irresponsible doxing generally: the mistaken identification of Quinn probably wouldn't have held up with more careful checking and the offending *Rolling Stone* article is a famous example of poor journalism. The methodology of some doxers probably needs to be seriously altered; in particular, the use of crowdsourcing, a famously unreliable method of identification, should probably be abandoned. But doxers could instead utilize or at least approximate the use of professional techniques and standards routinely utilized by professional journalists. Revising current legal norms can also help to ensure responsible doxing. For example, doxers who publish would be open to civil and perhaps criminal liability if they cannot establish that their allegations were properly vetted, and online platforms might also be open to civil and criminal liability for publishing the work product of irresponsible doxers much as newspapers can be held liable for their publications. In the United States, this would require revising or amending §230 of the Communications Decency Act (CDA) 47 U.S.C. which provides that "No provider or user of an interactive computer service shall be treated as the publisher or speaker of any information provided by another information content provider," a statute widely thought to immunize online operators from liability for publishing what others say. This revision is controversial but endorsed by some privacy scholars[17] and it has analogs in copyright law: online platforms can be held liable under a contributory liability theory if they intentionally induce or encourage direct infringement, and they can be held vicariously liable if they are given due notice and fail to remove the offending content.[18] Similarly, online publishers could be held liable if they either intentionally induce or encourage irresponsible doxing or fail to remove such doxes when given notice. Holding online operators liable may have some costs but those costs should not be blown out of proportion: the mere existence of the copyright infringement tort has not quelled development and innovation just as the existence of the defamation tort has not silenced online expression, so it is far from clear why holding online operators liable must unacceptably chill online speech and expression.

In sum, there is value in doxing sexual transgressors to warn others of possible harms, and the risks that doxing poses to innocents can probably be tolerably reduced. If all this is right, then there is at least one clear rationale for doxing sexual transgressors.

Doxing and Punishment

Another rationale that purports to justify doxing sexual transgressors begins with the observation that sexual transgressors often do not merely transgress social or sexual norms; they violate the moral rights of their victims. The rights violated will vary depending on the nature of the misconduct, but they include rights to bodily and sexual autonomy, privacy, life, and health, in addition to negative rights to be free from intimidation and harassment, among others. These moral rights are sometimes codified into civil and criminal codes, but even when they are not their violation amounts to serious wrongdoing. On a familiar theory of punishment, wrongdoers forfeit the rights that they violate, including rights that would otherwise make punishing them wrong. On this theory—that is, the forfeiture theory—since persons normally have a moral right not to be punished, the moral permissibility of punishment is centrally a question about rights.[19]

To be sure, the forfeiture theorist has some explaining to do. For example, she needs to explain why engaging in criminal behavior should justify the loss of moral rights since not all criminal behavior is morally wrong. But, according to forfeiture theory, punishment is most plausible when a criminal violates another person's moral rights, not when she engages in criminal conduct per se. Since victimless crimes do not involve the violation of another person's moral rights, the forfeiture theorist can deny that agents of victimless crimes forfeit any rights that would otherwise make punishing them wrong.[20] The forfeiture theorist will also want to explain just how violated rights and forfeited rights are related in an independently plausible way; she will probably want to deny that violated and forfeited rights must be identical lest she be stuck with the result that rapists can be justly raped and torturers justly tortured, a result that at least many find counterintuitive. The forfeiture theorist should also deny that those who misbehave only a little may be justifiably subjected to especially harsh treatment suggesting that violated and forfeited rights must be proportional in some interesting sense. So, forfeiture theorists should probably endorse something like the following:

> *The Forfeiture Claim* (FC): If *P* violates *Q*'s moral right to *X*, then *P* forfeits *P*'s own moral right to *X* (or some equivalent right or set of moral rights).[21]

Note that equivalence here does not require identity; FC only demands that the forfeited right is morally on a par with the right that is violated in terms of its weight and importance. If incarcerating murders and rapists is just, then their rights of free movement and association must be on a par with the rights that they violate if not necessarily identical. Also note that, as stated, FC does not require punishment: even if wrongdoers forfeit moral rights that would otherwise make punishing them wrong, there might be still other reasons to refrain from punishing them. All FC implies is that moral wrongdoers by virtue of their own conduct are no longer protected by a range of moral rights such that some range of conduct that was previously wrongful is now morally permissible, or, at least, not wrongful in virtue of violating those forfeited rights.

Talk of rights forfeiture is relevant in this context since doxing can violate or threaten important moral rights; subjects can be harassed and intimidated if a dox elicits a torrent of unwanted and hostile communications, especially if the doxer intends to bring about such a result; it increases the risk of physical and financial harms; and so forth. But if a subject forfeits their own rights to be free from harassment and intimidation, for example, doxing them need not be wrongful. For example, some racist subjects have clearly harassed and intimidated their victims and so, by FC, forfeit their negative rights to be free from such treatment. In that case, doxing them is not wrongful even if they are thereby harassed or intimidated, or at least, not wrongful in virtue of violating those rights.[22] Similarly, if sexual transgressors violate their victims' rights to be free from harassment and intimidation, they too cannot complain if they are harassed or intimidated as a consequence of having been doxed, or at least, doxing them is not wrongful in virtue of violating those rights.

Reflection on a real-life example helps. The website AutoAdmit bills itself as "The most prestigious law school discussion board forum in the world," but was a tool for pseudonymous users to, among other things, troll female students at Yale Law School, offering up hundreds of comments that were variously threatening and sexually explicit.[23] Martha Nussbaum details the way that the two female students suffered as a result: they were harassed and intimidated, but also objectified in a way that imperiled their career prospects, among other tangible harms.[24] Suppose that the pseudonymous individuals were doxed: there is a strong case that doxing them is morally permissible since, aside from whatever poetic justice there is in having their own weapons used against them, FC implies that they forfeit their own rights to be free from such harassment and intimidation and objectification. Sexual transgressors responsible for more serious sexual misconduct might also be justly doxed given FC: a rapist violates his victim's right to bodily autonomy, but that is not all he does. Even though the right to be free from harassment is not equivalent to the right to bodily autonomy, the fact that the rapist exposes his victim to a range of psychological harms akin to the psychological harms associated with harassment, including fear and anxiety, suggests that he has also violated her right to be free from those harms as well. Accordingly, he forfeits at least those rights that would otherwise make doxing him wrong.

With some exceptions, the typical forfeiture theorist only claims that forfeiture of those rights that would otherwise make punishment wrongful is a necessary condition for just punishment, not a sufficient condition.[25] Arguably, punishment must also serve some further morally worthy goal to be justified. The case for doxing sexual transgressors would be stronger if doxing can be linked to some goals of just punishment, but this isn't terribly hard. I have no particular theory of just punishment to defend here and aligning with any one theory would weaken the present argument, but it should suffice to note that doxing can realize multiple goals of just punishment suggested by competing theories. I note some such goals in the remainder of this section.

One influential theory of punishment, retributivism, entails that the very fact that sexual transgressors have engaged in wrongdoing suffices to justify their punishment. While the literature on retributivism is voluminous, retributivists tend to agree about a cluster of claims, including the claims that certain wrongdoers deserve to be punished and that their punishment is intrinsically good.[26] On an especially strong statement of the view, retributivism just is "the view that punishment is justified by the desert of the offender."[27] Without necessarily endorsing that strong view, the nerve of any retributivist rationale for doxing sexual transgressors is going to appeal to the fact that transgressors deserve to be treated in some way that we do not normally think that persons should be treated: their own misconduct warrants harsh treatment that would not otherwise be permissible. Note that while retributivism might be understood narrowly as a defense of legal punishment, the appeal to desert purports to justify punishment generally, including private acts of punishment administered by nonstate actors. So even if doxers are not state agents empowered to impose criminal punishment, doxing might still be justified by appealing to the claim that it gives subjects what they deserve. And the retributivist case for doxing is especially strong if the familiar legal mechanisms of civil and criminal law are not up to the task of giving sexual transgressors what they deserve.

Retributivism will not justify doxing any and all sexual transgressors either because some sexual transgressors do not deserve to be punished—say, because they have a valid excuse or because there is some generally good reason to exempt them from punishment—or because doxing them will cause them to suffer more than they deserve. Just as doxing is not justified if it results in the harassment or intimidation of a sexual transgressor who has not forfeited their right to be free from harassment and intimidation, doxing will not be justified if it causes them to suffer undeservedly or excessively. One especially powerful challenge to a defense of doxing will begin with the observations that doxing can quickly result in a torrent of unwanted hostile communication and that it is difficult to bring such communications to an end once the torrent begins. Horror stories in which subjects have been the victims of online bullying and cyber-harassment are common enough that, arguably, a general policy of doxing sexual transgressors could only be adopted by a seriously negligent or reckless agent. Still, it does not follow that there are no cases in which sexual transgressors deserve the especially harsh treatment sometimes caused by doxing; some clearly do.

Doxing sexual transgressors can secure other goods highlighted by competing theories of just punishment. For example, doxing might be justified on the grounds that it has an adequate deterrent effect, either by making it less likely that the subject will engage in future wrongdoing or by making other persons less likely to engage in future wrongdoing presumably because the subject's punishment disinclines them to replicate such behavior out of a fear that they will be treated similarly. It is an open question whether doxing sexual transgressors must deter future wrongdoing generally or sexual misconduct in particular to be justified, but arguably any deterrent effect is of moral significance. Any deterrence theory of punishment will face the familiar objection that deterrence can be secured by punishing the innocent so long as it is generally believed that the subject of punishment is guilty.[28] However, the current rationale for doxing sexual transgressors pairs with the claim that some sexual transgressor has forfeited whatever rights that would otherwise make punishing him wrongful. Insofar as the innocent have forfeited no relevant right, this familiar objection is blocked.

Independent of any deterrent effect, it is widely thought that part of what makes something an instance of punishment is the condemnatory character of the measure taken. Some penalties, such as parking tickets, deter crime but don't socially condemn a wrongdoer's actions in the way that a full-dress criminal trial replete with conviction and sentencing does.[29] Arguably, another good of just punishment is its tendency to express condemnation of a wrongdoer or her actions. But doxing can realize this good as well: the doxing of various "Unite the Right" participants was, at least in part, an effort to publicly condemn their perceived racism in an equally public manner.[30] True, in cases of criminal punishment an agent of the state ensures that some wrongdoer is condemned, while doxers are usually private individuals expressing their personal condemnation and enjoining others to do the same. But, by permitting some subclass of sexual transgressors—again, those sexual transgressors who have forfeited the requisite rights—to be doxed, a state is arguably acting as a proxy or as implicitly approving of their condemnation. So here too, doxing sexual transgressors can realize a goal of just punishment.

Understanding doxing as a means of punishing sexual transgressors prompts another objection ignored thus far. I have suggested that doxing sexual transgressors can realize at least some legitimate goals of just punishment, but doxing is invariably an action performed by private individuals and not the state—at least, not state actors acting in that capacity—such that doxing is bound to be ungoverned by safeguards that regulate legal punishment. Most notably, due process is secured in just legal punishment while subjects of doxing receive little of it: subjects are not usually given advance notice that their personal information will be published, allegations are not necessarily vetted prior to doxing, there is no appeals process or formal opportunity to respond to allegations, no opportunity to confront and cross-examine an accuser, no independent tribunal to oversee things, and so forth. While some legal officials, including those at the highest levels, have singled out sexual transgressors as being especially in need of guarantees of due process[31] others worry that the use of extralegal mechanisms, like doxing, actually undermines the goals of #MeToo by making it more likely that victims of sexual misconduct are not taken seriously[32] and that legitimate fact-finding will be hindered.[33]

Here too, this objection is not directed at doxing sexual transgressors in particular but to doxing generally, although there might be some variants of the objection that are specific to doxing sexual transgressors. For example, if the objection is that doxed sexual transgressors might lose their jobs absent due process, it is worth noting that, at least in the United States, due process is not generally guaranteed in the context of employment law, and the general perception that workers cannot be fired but for good cause is widespread.[34] Carving out employment protections for doxed sexual transgressors, usually male, thus has the feel of patriarchal exceptionalism unless the call for due process in the context of employment is suitably general.

Bracketing that concern, there are at least two responses to the due process objection. First, there arguably currently exists "a set of emerging procedural norms for making and evaluating public reports of sexual assault, harassment, and misconduct against high-profile leaders in business, media, and government"[35] that will tend to secure the guarantees of due process. Indeed, "in most instances in which high-profile leaders have lost their positions, the allegations have been vetted and scrutinized by media, and decision-makers have required some form of corroborative evidence or conducted their own investigations before taking action."[36] True, this response is cold comfort to doxed subjects who are not "high-profile leaders" but, and this is the second response, the existence of these extralegal norms governing cases involving high-profile sexual transgressors suggests the possibility of legal norms governing the practice of doxing sexual transgressors generally. More to the point: if there are informal norms governing #MeToo, there can surely be legal norms governing doxing sexual transgressors which would establish legal side-constraints adequate to guarantee due process. The proposed reforms to promote responsible doxing suggested above will also tend to ensure that something approximating due process will be realized.

Doxing as Public Reason

The condemnatory effect of doxing sexual transgressors, highlighted in the previous section, is closely related to another rationale for doxing sexual transgressors. While doxing is often described in positively illiberal terms—as cyber-harassment, as incitement, as vigilantism—there is a case to be made that doxing exemplifies a liberal ideal in action.

Doxing sexual transgressors has the feel of a speech act that should be familiar to faithful liberals. According to public reason liberalism, laws are legitimate only if they can be justified by distinctively public reasons—that is, reasons that are in some sense acceptable to, accessible to, or shareable by all reasonable citizens,[37] reasons that would be reasonable for any citizen to assent to.[38] So understood, public reason liberalism flows from a theory of social and political morality according to which right actions are those that can be justified to others on grounds that they cannot reasonably reject.[39] Different public reason liberals have different views about just what stands in need of justification. Rawls initially demands that "the limits imposed by public reason do not apply to all political questions but only to those involving constitutional essentials and questions of basic justice."[40] Other public reason liberals hold that coercive laws generally need public justification.[41] If coercive laws generally require justification by public reason, it is presumably because they are, well, coercive. But it is not just laws that can be coercive; John Stuart Mill identified the "moral coercion of public opinion" as a means of social control in need of justification. The existence of either legal or social norms enabling doxing probably needs to be subjected to the gauntlet of public reason to garner liberal justification as either threatens the liberty of doxed subjects.

I contend that doxing can be a mechanism for *engaging* one's fellow citizens—that is, for eliciting public reasons from them in defense of their actions and associations. Engagement involves demanding public reasons from a subject on the one hand and providing a means for sharing them on the other. So understood, doxing does not merely condemn a subject; it invites her into a public arena and functions as the opening salvo of a dialogue about the matter at hand. The use of social media to elicit a response from a subject is by its very nature a social exercise as is the opportunity to provide public reasons in one's defense. Engagement, so understood, is not only consistent with liberal values and commitments, it is the ideal of public reason in practice: as Rawls puts things, "The ideal of public reason does hold for citizens when they engage in political advocacy in the political forum."[42] True, many subjects will not be responsive to attempts to engage them via dox. But this is not fatal to understanding doxing as a mechanism for engaging. Just as defying the values of a moral community is evidence of a capacity for moral communication with them,[43] deliberately refusing engagement is evidence of a capacity to proffer public reasons. So, in doxing, a subject is treated as the sort of individual capable of supplying public reasons whether or not they respond.

The problem with at least many mechanisms used for doxing is that there is little opportunity for a doxed subject to respond and provide the requested justification: while an audience may see the initial tweet in which a subject is doxed, they may not have the patience or opportunity to wade through a stream of retweets and replies where a subject's reply emerges. For this rationale to succeed, doxers need to be genuinely sincere in their demands for public reasons and subjects given a genuine opportunity to offer up reasons in their defense: if doxers get what they said they want, their complaint should be withdrawn and the veracity of the subject's reply acknowledged in as public a manner as the initial challenge. Doxers unwilling to take these steps cannot sincerely avail themselves of this justification and legal penalties should deal with those who do so. Social penalties might be appropriate as well: doxers who dox unjustly would thereby forfeit at least some rights that would tend to make shaming or ostracizing them wrongful.

A rationale for doxing sexual transgressors maps onto this general defense of doxing: doxing sexual transgressors is justified to the extent that it brings them into the public arena and elicits public reasons from them. To illustrate, consider the case of Aziz Ansari, a popular comedian publicly identified online as allegedly having violated social norms governing dating and sex, allegations which fell short of accusing him of criminal sexual misconduct.[44] Insofar as the information published online identified Ansari as a sexual transgressor, he was doxed. What is notable is that Ansari did not categorically dismiss the allegations and did not devolve into victim blaming, and if he did not outright apologize he at least expressed remorse and publicly reflected on the matter in a way that can reasonably be understood as promoting the public discussion of sexual norms.[45] Some commentators were unimpressed with Ansari's response, but justification demands not that anyone actually endorses the offered justification, only that public reasons are, in some sense, made *available* to them.[46] To the extent that Ansari was successfully engaged, doxing him was both justified and of social value.

To be sure, real-life examples of doxing exemplifying public reason are hardly legion, and the typical dox falls well short of this ideal. No surprise, then, that doxing is often understood as a shaming mechanism, a deliberately objectionable characterization. For example, doxers are sometimes described as "shame-slinging vigilantes"[47] and doxing as "shaming in cyber-mobs."[48] Some kinds of doxing are especially prone to this sort of objection. In one of the few philosophical discussions of doxing, David M. Douglas helpfully distinguishes deanonymizing doxing in which "personal information establishing the identity of a formerly anonymous individual is released" from *targeting doxing* in which a doxer "discloses personal information that reveals specific details of an individual's circumstances that are usually private, obscure, or obfuscated," since anonymity and obscurity are not the same thing. Both are distinct from delegitimizing doxing in which a doxer "reveals intimate personal information that damages the credibility of that individual."[49] Delegitimizing doxing is not always wrongful. In some cases, a subject arguably deserves to have their credibility damaged—if, say, they have willfully and wrongly misrepresented their credentials or if damaging their credibility is a necessary means for advancing an important public interest. Still, Douglas often characterizes delegitimizing doxing in morally dubious ways: as instances in which a doxer "attempts to shame and humiliate the subject,"[50] "vilifies the subject,"[51] and "objectifies its subject" by portraying her "as unworthy of personhood, without legitimate interests of her own that should also be recognized."[52] Since doxing sexual transgressors does sometimes humiliate, vilify, and objectify, the charge that doxing generally and doxing sexual transgressors in particular amounts to shaming seems to stick. This result should be especially disquieting to faithful liberals given Martha Nussbaum's argument that there is an "apparent tension" between liberalism's commitment to recognizing and respecting the equal dignity of citizens and shaming penalties that undermine recognition of that dignity.[53] If doxers really are shame-slinging vigilantes, then doxing sexual transgressors threatens their right to be treated with equal dignity and doxing them is seriously wrongful.

There are multiple responses to this sort of objection. First, even if doxing sexual transgressors does shame them and thereby fails to adequately respect their dignity, the liberal commitment to respecting the equal dignity of citizens might be trumped by other, weightier considerations of justice. Unless the faithful liberal is willing to regard the commitment to respecting the dignity of citizens as an absolute requirement of justice, then that commitment will not absolutely prohibit shaming. Second, one could recall FC above and maintain that sexual transgressors have forfeited their right to be treated with dignity in virtue of their sexual misconduct. This is hardly an implausible assumption as at least very many sexual transgressors will fail to respect the dignity of their victims, among other moral failings. But if that right is forfeited, shaming them need not be wrongful. Third, one could deny that shaming penalties are illiberal. Despite Nussbaum's powerful rejection of shaming penalties, some faithful liberals have defended shaming albeit within restrictions.[54] Fourth, the liberal demand prohibiting shaming penalties seems to impose an obligation on the state, not private citizens, which cannot demand that individual citizens refrain from shaming behavior without fairly stringent impositions on individual exercises of free expression and association. In this vein, Mill allows, apparently, that the exercise of individual liberty might result in fairly substantial setbacks to individual persons via the exercise of the moral coercion of public opinion, even if the state would be unjustified in imposing those same setbacks.[55] So, even if sexual transgressors cannot be targeted by the state for humiliation or stigmatization, they have no right to demand that the liberal state restrict other persons from acting in ways that ensures that those sexual transgressors suffer severe penalties when others choose not to associate with them and encourage others to similarly refrain from association, for example. Fifth, noting that most punishments will have some kind of shaming effect, the defender of doxing sexual transgressors could allow that punishing with the intention of shaming is clearly prohibited; for that reason, pillories and the stockade are out. But she might still insist that it is permissible to intend to punish for some other reason—say, to engage a subject—even if she merely foresees that the subject will be shamed, so long as the doxer takes at least some steps to ensure that the intended goal is realized and the foreseeable consequence is minimized.

These responses help to blunt the objection that doxing amounts to illiberal shaming. The liberal state need not undermine the equal dignity of its citizens by permitting them to be doxed. Rather, it affirms their equal dignity by acknowledging that they are equally capable of proffering and providing public reasons in their defense and thereby acknowledging that they are the kind of person who can be engaged. Of course, doxers who sincerely intend to engage their subject are committed to allowing their subject to defend herself; engagement demands fair play and permitting a subject to provide the very justification that you have sought. Doxers who fail to permit subjects the opportunity to offer public reasons in their defense may be rightly subjected to both legal and social punishments, and, once again, the actual practice of doxing will probably have to be seriously reformed. All this is consistent with putting the liberal ideal of public reason into action and permitting sexual transgressors to be doxed.

Conclusion

I have identified multiple rationales for doxing sexual transgressors: the use of an extralegal mechanism, like doxing, can ensure that potential victims of sexual misconduct are warned; it can help to secure some of the goods of punishment that have often gone unrealized for victims of sexual misconduct via familiar legal mechanisms; it can facilitate public reason in action. This justification of doxing sexual transgressors does not entail that doxing generally is justified, partly because victims of other kinds of transgressions may be better situated to find relief in extant legal processes. Still, if doxing is ever justified, then the usual categorical dismissal of doxing is mistaken. For at least some sexual transgressors, being doxed is an altogether suitable response to serious wrongdoing.

Notes

1. Catharine A. MacKinnon, "#MeToo Has Done What the Law Could Not," *New York Times* (Feb. 4, 2018), https://www.nytimes.com/2018/02/04/opinion/metoo-law-legal-system.html (Last accessed: 28: September 2020).
2. Moria Donegan, "I Started the Media Men List. My Name is Moira Donegan," *The Cut* published by New York (10 January 2018): https://www.thecut.com/2018/01/moira-donegan-i-started-the-media-men-list.html (Last accessed: 28 September 2020).
3. https://twitter.com/rosemcgowan/status/786723360550035460.
4. Mat Honan, "What is Doxing?" *Wired*. 6 March 2014. https://www.wired.com/2014/03/doxing/ (Last accessed: 25 June 2018).
5. Emma Grey Ellis, "Whatever Your Side, Doxing is a Perilous Form of Justice," *Wired*. 17 August 2017. https://www.wired.com/story/doxing-charlottesville/ (Last accessed: 29 March 2018).

6. Jasmine McNealy, "What is Doxxing, and Why is It So Scary?," *The Conversation* 16 May 2018. https://theconversation.com/amp/what-is-doxxing-and-why-is-it-so-scary-95848 (Last accessed: 22 August 2018).
7. Laura Sydell, "Some are Troubled by Online Shaming of Charlottesville Rally Participants," *All Tech Considered*. National Public Radio, 15 August 2017. https://www.npr.org/sections/alltechconsidered/2017/08/15/543566757/twitter-account-names-and-shames-far-right-activists-at-charlottesville (Last accessed 29 March 2018).
8. David M. Douglas, "Doxing: a Conceptual Analysis," *Ethics of Information Technology* 18 (2016), p. 199.
9. Peter Brian Barry, "Doxing Racists," *The Journal of Value Inquiry* (2020): https://doi.org/10.1007/s10790-020-09747-0.
10. "Sexual Transgression," *Segen's Medical Dictionary*, https://medical-dictionary.thefreedictionary.com/Sexual+Transgression (Last accessed: 19 October 2020).
11. Jenna Wortham, "We Were Left Out," *The New York Times Magazine* (13 December 2017): https://www.nytimes.com/interactive/2017/12/13/magazine/the-reckoning-women-and-power-in-the-workplace.html (Last accessed: 30 June 2020).
12. Douglas, "Doxing," p. 206.
13. See Jeffrey Moriarty, "Business Ethics," *The Stanford Encyclopedia of Philosophy* (Fall 2017 Edition), Edward N. Zalta (ed.), URL = <https://plato.stanford.edu/archives/fall2017/entries/ethics-business/>.
14. Richard T. De George, *Business Ethics*, Upper Saddle River, NJ: Pearson, 2003), p. 299.
15. Michael Davis, "Whistleblowing," in Hugh LaFollette (ed.), *Oxford Handbook of Practical Ethics* (New York: Oxford University Press, 2003), pp. 539–563.
16. Daniel Victor, "Amateur Sleuths Aim to Identify Charlottesville Marchers, but Sometimes Misfire," *New York Times* 14 August 2017. https://www.nytimes.com/2017/08/14/us/charlottesville-doxxing.html (Last accessed: 12 October 2020).
17. Daniel Solove, "Speech, Privacy, and Reputation on the Internet," in *The Offensive Internet*, p. 23.
18. Natalia Hornchick, "Reaching Through the 'Ghost Doxer': An Argument for Imposing Secondary Liability on Online Intermediaries," *76 Washington and Lee Law Review* 1307 (2019), pp. 1314–1315.
19. Christopher Heath Wellman, *Rights Forfeiture and Punishment* (Oxford: Oxford University Press, 2017), p. 2.
20. Wellman, ibid., pp. 148–171.
21. Boonin, *The Problem of Punishment*, p. 105. See also Alan H. Goldman "The Paradox of Punishment," *Philosophy & Public Affairs* 9 (1979), p. 45. Boonin articulates but does not accept (FT).
22. See Barry, "Doxing Racists."
23. Danielle Citron, *Cyber Civil Rights*, 89 *Boston University Law Review* 61 (2009), pp. 71–72.
24. Martha Nussbaum, "Objectification and Internet Misogyny," in *The Offensive Internet: Speech, Privacy, and Reputation*, Saul Levmore and Martha C. Nussbaum, eds. (Cambridge: Harvard University Press, 2010), pp. 73–75.
25. Christopher Morris, "Punishment and the Loss of Moral Standing," *Canadian Journal of Philosophy* 21:1 (1991), p. 63. For a minority view, see Christopher

Heath Wellman, "The Rights Forfeiture Theory of Punishment," *Ethics* 122: 2 (January 2012), p. 371.

26. Walen, Alec, "Retributive Justice," *The Stanford Encyclopedia of Philosophy* (Fall 2020 Edition), Edward N. Zalta (ed.), URL = <https://plato.stanford.edu/archives/fall2020/entries/justice-retributive/>.
27. Michael Moore, "The Moral Worth of Retribution," in *Responsibility, Character, and the Emotions: New Essays in Moral Psychology*, Ferdinand Schoeman, ed. (Cambridge: Cambridge University Press, 1987), p. 87.
28. For discussion, see Boonin, *The Problem of Punishment*, esp. pp. 41–52.
29. Joel Feinberg, "The Expressive Function of Punishment," *The Monist*, 49 (1965), pp. 397–423.
30. Camila Domonoske, "On the Internet, Everyone Knows 'You're Racist': Twitter Account IDs Marchers," *The Two-Way*. National Public Radio, 14 August 2017. https://www.npr.org/sections/thetwo-way/2017/08/14/543418271/on-the-internet-everyone-knows-you-re-a-racist-twitter-account-ids-marchers (Last accessed: 20 October 2020).
31. See Emily Stewart, Trump Wants "Due Process" for Abuse Allegations. I Asked 8 Legal Experts What That Means, *VOX* 12 February 2018. https://www.vox.com/policy-and-politics/2018/2/11/16999466/what-is-due-process-trump (Last accessed: 12 October 2020).
32. Emma Grey Ellis, "How #HimToo Became the Anti #MeToo of the Kavanaugh Hearings," *WIRED* 27 September 2018, https://www.wired.com/story/brett-kavanaugh-hearings-himtoo-metoo-christine-blasey-ford/ (Last accessed: 12 October 2020).
33. Shira A. Scheindlin & Joel Cohen, "After #MeToo, We Can't Ditch Due Process," *The Guardian* 8 January 2018, https://www.theguardian.com/commentisfree/2018/jan/08/metoo-due-process-televictions (Last accessed: 12 October 2020).
34. Pauline T. Kim, "Bargaining with Imperfect Information: A Study of Worker Perceptions of Legal Protection in an At-Will World," *Cornell Law Review* 83: 105 (1997), p. 110.
35. Jessica A. Clarke, "The Rules of #MeToo," *University of Chicago Legal Forum*, Vol. 2019, Article 3 (2019), pp. 39–40.
36. Ibid., pp. 40–41.
37. Gerald, Gaus, *Contemporary Theories of Liberalism* (London: Sage Publishing, 2003), p. 209.
38. Andrew Jason Cohen, "Contemporary Liberalism and Toleration," in *The Cambridge Companion to Liberalism*, ed. by Steven Wall (Cambridge: Cambridge University Press, 2015), p. 192.
39. Thomas Scanlon, *What We Owe to Each Other* (Cambridge: Harvard University Press, 2008).
40. John Rawls, *Political Liberalism*, 2nd ed. (New York: Columbia University Press, 2005), p. 214.
41. Gerald Gaus, *The Order of Public Reasons* (New York: Cambridge University Press, 2011), pp. 495–497; Jonathan Quong, *Liberalism without Perfection* (New York: Oxford University Press, 2011), pp. 233–250; Kevin Vallier, *Liberal Politics and Public Faith: Beyond Separation* (New York: Routledge, 2014), p. 24.
42. Rawls, *Political Liberalism*, p. 215.

43. Gary Watson, "Responsibility and the Limits of Evil: Variations on a Strawsonian Theme," in *Perspectives on Moral Responsibility*, ed. By John Martin Fischer and Mark Ravizza (Ithaca: Cornell University Press, 1994), pp. 119–148.
44. Katie Way, "I Went on a Date with Aziz Ansari. It Turned into the Worst Night of My Life." Babe.net: https://babe.net/2018/01/13/aziz-ansari-28355 (Last accessed 20 October 2020).
45. Anna North, "Aziz Ansari has addressed his sexual misconduct allegation. But he hasn't publicly apologized." *Vox.com.* 12 July 2019: https://www.vox.com/identities/2019/7/12/20690303/aziz-ansari-sexual-misconduct-accusation-right-now (Last accessed 20 October 2020).
46. Vallier, *Liberal Politics and Public Faith*, pp. 28–29.
47. Ellis, "Whatever Your Side, Doxing is a Perilous Form of Justice."
48. Decca Muldowney, "Inside the Left's Online Efforts to Out White Supremacists."
49. Douglas, Doxing," p. 199.
50. Ibid., p. 205.
51. Ibid., p. 206.
52. Ibid., p. 209.
53. Martha Nussbaum, *Hiding from Humanity: Disgust, Shame, and the Law* (Princeton: Princeton University Press, 2004), p. 4.
54. See, for example, Thom Brooks, "Shame on You, Shame on Me? Nussbaum on Shame Punishment," *Journal of Applied Philosophy* 25:4 (2008), pp. 322–334.
55. Which is not to say that there is no problem for Mill. See Dan Threet, "Mill's Social Pressure Puzzle," *Social Theory and Practice* 44:4 (2018), pp. 539–565 for a recent discussion.

CHAPTER 31

Naughty Fantasies (With a New Postscript Including Sex Robots)

John Corvino

Is it wrong to eroticize activities that are themselves wrong? Consider the following scenarios:

(1) Sally is a feminist who enjoys light sadomasochism, including spanking, whipping, and slave-master roleplaying. Sometimes she is the "top" in such activities, and sometimes her boyfriend is, but in all cases the activities are consensual and mutually pleasurable.
(2) Fred is a gay man who finds himself aroused by stories of fraternity hazing, particularly those involving people being forcibly stripped or humiliated. He is against such activity in "real life," and indeed he discourages college-age friends from joining fraternities because of these and other troubling practices. Nevertheless, he occasionally visits websites that contain such stories for the purpose of erotic stimulation.
(3) Ramona occasionally fantasizes about rape while masturbating or having sex with partners. Given her serious opposition to rape and to exploitation of women more generally, she feels guilty about such fantasies. Nevertheless, she finds that they intensify her sexual experiences.

This paper was originally published in 2002 in the *Southwest Philosophy Review* [Vol. 18, Number 1 (January 2002), pp. 213–220]. I've made a few minor amendments to the original text for clarity and added a new postscript.
Content note: This paper includes discussion of rape and sexual humiliation.

J. Corvino (✉)
Wayne State University, Detroit, MI, USA
e-mail: j.corvino@wayne.edu

D. Boonin (ed.), *The Palgrave Handbook of Sexual Ethics*,
https://doi.org/10.1007/978-3-030-87786-6_31

Some (notably certain feminists) have argued that actively entertaining the sorts of fantasies described above is always wrong, insofar as these fantasies depend on models of domination and exploitation. Others (including other feminists) have defended such practices on grounds of personal freedom and privacy. In this chapter, I would like to revisit this debate by clarifying and evaluating one of the core arguments within it. Although my conclusions will be somewhat tentative, I hope at least to elucidate some of the ethical and conceptual issues raised by these practices, which are not merely controversial but also rather common. As I note in the postscript, the advent of highly sophisticated artificial-intelligence "sex robots," including some with a resistance setting that allows users to simulate rape or sexual assault, gives the issue new urgency.

For lack of a better term, I shall group these practices under the rubric "naughty fantasies." In using the term "naughty" I do not intend to diminish the seriousness of the topic. The alternative "guilty fantasies" is unacceptable because it prejudges the issue. "Sadomasochism" is too narrow, though much of the prior feminist debate on this topic has focused on sadomasochism. "Violent fantasies" is both too narrow and too broad: Not all such fantasies are *violent* in any but the loosest sense (cf. Fred's fraternity humiliation scenes), and not all violent fantasies are sexual. By "naughty fantasies" I mean any sexual fantasy, *either imagined or acted out*, that appears to eroticize an activity that is itself morally wrong. (The "appears to" is key: Later I will question whether the paradigmatic cases actually do this.) By "eroticization," I mean actively regarding the activity with sexual desire—although as I explain below, I think the phenomenon of eroticization is itself in need of better analysis. The argument I wish to clarify and evaluate is as follows:

(1) It is wrong to eroticize activities that are themselves wrong.
(2) Naughty fantasies eroticize activities that are themselves wrong.
Therefore, it is wrong to pursue naughty fantasies.

Let us call this the "naughty-fantasies argument." Note that the first premise is intended *a priori.* One could, of course, make the consequentialist argument that those who pursue naughty fantasies are more likely to engage in *actual* exploitation than those who do not, or that they are more likely to be violent or callous or incapable of intimacy—but this is not the argument under consideration here. Rather, the question is whether it is wrong *in itself* to fantasize about wrongful activities.

One way to rebut the argument is to attack the second premise. Patrick Hopkins takes this approach in his defense of sadomasochism.[1] According to Hopkins, sadomasochism does not eroticize a wrongful activity, it eroticizes the *simulation* of a wrongful activity:

> In the case of SM [sadomasochism]... it should not be assumed that SM participants actually find pleasure in the torture of slaves, nor in the cries of a rape victim, nor in the humiliation of women, nor in the relentless assault of an attacker. In fact, it is a central ethical and political value of those SMists who also profess

> to be feminists that such events are indeed evil, deplorable, and repugnant. At the same time, however, it is possible to desire the *simulation* of those events, to lust after the context of a negotiated and consensual "submission" or "domination."… [T]he sadomasochist can *desire the simulation itself*, not as an inferior copy of the real thing, not as a copy of anything at all, but as simulation qua simulation.[2]

In one sense, Hopkins is correct when he claims that SM participants do not find pleasure in the torture of slaves and so on: The typical naughty fantasizer is as horrified by *actual* atrocities as anyone else. But during fantasy, the distinction between the fictional and the actual is not so clear. When Fred reads a fraternity-hazing story on the internet, does it matter for erotic purposes whether the story is true or not? Suppose Fred discovers that a favorite story that he had believed to be "fictional" is actually a biographical account. He might feel guilty about continuing to fantasize about the story, but will he find it less erotic? Might he not even find it *more* erotic?

The answers to these questions depend on Fred's particular situation. But whatever the answers, the question remains whether it is psychologically possible for Fred (or anyone else) to eroticize a *simulation qua simulation*. In order to understand this question, put aside the issue of eroticization and consider first what it means to *attend to* a simulation qua simulation. Take barn facades, for example. One can attend to these qua barns, in which case one might notice their charming architecture, their idyllic settings, their bright red siding. Or one can attend to them qua simulations, in which case one might notice their apparent three-dimensionality, their realistic stature, or their propensity for appearing in Gettier-type counterexamples. Note that the features noticed when attending to the barn facade qua simulation are largely exclusive of those noticed when attending to it qua barn: Indeed, when attending to it qua simulation, one is mindful of precisely those features that make the object *not* a real barn.

The case is somewhat different in naughty fantasies. True, SM participants frequently attend to the pleasure of their partners, and to that extent, they are mindful of features that distinguish their activities from actual violence. But they are also mindful of features that occur in the "real" case—the spanking, the quickened heartbeat, the gasps, and groans. When they eroticize these features, SM participants (and other naughty fantasizers) seem to be eroticizing not simulations qua simulations, but domination and its manifestations. The simulation is not the object of arousal; rather, it is the vehicle for the object of arousal.

To see this point more clearly, let us suppose that Fred is watching a porn film depicting fraternity hazing. If Fred were to focus on the film qua simulation, what features would he notice? Presumably, he would attend to things like the convincingness (or lack thereof) of the actors' humiliation, the realistic nature of the dorm-room sets, or the level of detail in the storyline. Such attention would be more appropriate to the film critic than the sexual participant.

Indeed, it is not even clear what it would mean to *eroticize* such features (although someone who did would possess a most interesting fetish). What turns people on in naughty fantasies is not depiction, but rather what is being depicted.

This is not to say that the object of a naughty fantasy—that is, the part that "turns on" the fantasizer—is typically clear or well-defined. Fred may not know why he enjoys fraternity-hazing scenes, and he will likely find that over-analysis of this issue serves merely to spoil his enjoyment. Lacking a developed "phenomenology of arousal," we may find that it is harder to explain what the object of arousal *is* than to explain what it is *not*. In any case, we must reject Hopkins' claim that the object is simulation qua simulation and consider a different approach to the naughty-fantasies argument.

Perhaps it would be more promising to grant the second premise of the naughty-fantasies argument and instead attack the first, which states that it is always wrong to eroticize wrongful activities. Hopkins employs this approach briefly when he writes,

> In fact, SM scenes gut the behaviors they simulate of their violent, patriarchal, defining features. What makes events like rape, kidnapping, slavery and bondage evil in the first place is the fact that they cause harm, limit freedom, terrify, scar, destroy, and coerce. But in SM there is attraction, negotiation, the power to halt the activity, the power to switch roles, and attention to safety. Like a Shakespearean duel on stage, with blunted blades and actors' training, violence is simulated, but is not replicated.[3]

Hopkins is correct that the central wrong-making features of rape are absent from rape fantasies. Moreover, one could argue that rape fantasies not only lack the wrong-making features but also add a good-making one, namely the pleasure of the fantasizer. But this argument misreads the first premise of the naughty-fantasies argument as being consequentialist in nature. Recall that proponents of the argument are not (or at least, not *here*) concerned with whether such fantasies have good or bad consequences. The fact that people take pleasure in such fantasies is precisely the problem, according to this argument: Some pleasures are bad because their objects are inappropriate as objects of pleasure.

There are at least two non-consequentialist ways to understand the first premise of the naughty-fantasies argument. One might adopt a virtue-ethics approach and argue that naughty fantasies are incompatible with good character, particularly if they are pursued habitually. The idea here is not that regular pursuit of such fantasies might cause one to be more violent, or callous, or indifferent to human suffering; that argument would be consequentialist. Rather, the idea is that naughty fantasies are incompatible with some virtue or set of virtues that have some *noninstrumental* value in a well-lived life.[4] Someone lacking such virtues would be missing something, even if neither she nor anyone else is less happy as a result. Alternatively, one might argue that actively entertaining naughty fantasies is wrong in itself, apart from any connection with virtue. I will focus on this latter claim, though much of what I say will apply to the virtue-ethics version as well.[5]

The intuition that drives this claim is that any seriously wrongful activity merits an attitude of disapproval, and eroticization of such an activity is inconsistent with this attitude. I suggest that the best way to defend this intuition is dialectically, through the use of further examples.

Consider Raymond, who collects and studies newspaper accounts of *actual* rapes in order to enhance his erotic life. Suppose that he, like Ramona, is deeply against rape and would never commit it. Now one might have trouble accepting this supposition, mainly because it seems incompatible with Raymond's enjoyment of real-life rape stories. In other words, there seems to be some incoherence involved in being against rape while deriving sexual pleasure from reading about real rapes. Yet notice that on consequentialist grounds, Raymond is in the clear. (We are assuming that Raymond is sincere in his opposition and that he is no more inclined to be violent or callous because of his predilections.) His pleasure in reading such accounts results in a net gain in utility.

Or consider Charlie, married with grown children, who sometimes fantasizes about having sex with underage girls. He is deeply opposed to child abuse and would never engage in such activity himself. On a few occasions he has purchased magazines with erotic images of females who are in fact over eighteen but who appear to be as young as twelve, and he has privately begun writing (and via the internet, sharing) fictional stories about forcible sex with these girls. His relationships with his wife and others appear unaffected by this activity, which occupies very little of his time and which he would abandon entirely were it not for the sexual charge it brings him. Or if Charlie's case does not trouble you, consider Chester, who is like Charlie in every respect except that the stories he shares are *actual accounts* of adult-child rape. Consequentialist concerns aside, there is a strong intuition that in Chester's case (if not in Charlie's as well) the eroticization of child abuse represents a failure to exhibit the appropriate attitude toward a deeply wrongful activity.

One might object in the cases of Raymond and Chester, who seek real life accounts rather than fictional ones, on the grounds that their predilections are *causally dependent* on the existence of actual atrocities. The problem is that the same seems true in the case of Ramona, Fred, and Charlie. Presumably, naughty fantasies give them a sexual charge in large part because of the intense emotions associated with *actual* domination and humiliation. Thus, if causal dependence on real atrocities tarnishes Raymond's and Chester's fantasies, it tarnishes all naughty fantasies. (Indeed, a major thrust of feminist ethics is that human actions cannot be evaluated apart from the social context that imbues them with meaning.)

The point about inappropriate attitudes can be illustrated with non-sexual examples as well. Some years ago I attended a large Southern university where one of the local fraternities annually held an "Old South Ball." The fraternity, which was notorious for its white-only membership, would hire Black students to pose as "slaves" at the ball for the sake of verisimilitude. Needless to say, this event regularly provoked a serious outcry within the campus community. While some defended the fraternity on the grounds that the Black actors were willingly (though, to many minds, inexplicably) participating, most thought that the event involved a serious failure on the part of all participants to adopt an appropriate attitude toward slavery. The fact that these actors were paid well was beside the point.

This is not to say that any and all depictions of wrongful practices are bad. Compare the Old South Ball to the mini-series *Roots* and the difference is immediately apparent: Those who watch *Roots* should (and typically do) sympathize with the slaves; those fraternity members who participated in the ball simply enjoyed themselves at the "slaves'" expense. On the other hand, if someone were to watch *Roots* and smile or cheer during a lynching scene, others might reasonably object. Similar points can be made about war movies. A person who watches a World War II film, for example, and fails to sympathize with the Jews is missing something—a point that was made humorously during a *Seinfeld* episode, when Jerry was chided mercilessly by his parents after being caught making out with his girlfriend during *Schindler's List*.

Video games may provide an even better analogy because of their more clearly participatory nature (and see the postscript for discussion of sex robots). Suppose Victor enjoys games in which he assumes the role of a fighter pilot bombing various military targets. Such games are quite common and are widely regarded as harmless (the familiar objections to violent toys notwithstanding). Now suppose the game is based on an *actual* war—say, the civil war in Bosnia—and Victor assumes the role of a Serb targeting Muslim civilians. Further, suppose that although Victor is opposed to the "ethnic cleansing" in Bosnia, he really enjoys his video game and cheers loudly every time a Slavic Muslim in the game is killed. Even if we were convinced that Victor would never behave this way toward ethnic minorities in "real life," would we not find his behavior during the games problematic?

The analogy is, needless to say, imperfect. For one thing, eroticizing something is not the same as cheering for it (more on this below). But the example does capture the concern that anti-SM feminists and others have about naughty fantasies: Rape, humiliation, and exploitation are real and all too-common evils. To seek out their depictions for the purpose of pleasuring oneself is to fail to acknowledge that evil properly.

Having done my best to defend this intuition, let me conclude by noting a few points that militate against it. (I acknowledged at the outset that my conclusions in this chapter are tentative.) The naughty-fantasies argument (in particular, the first premise) depends upon the assumption that eroticization involves a kind of pro-attitude which, as such, is incompatible with condemnation of the depicted activity. Yet the truth of that assumption is by no means clear, for several reasons. First, as noted above, the object of eroticization is often nonspecific. Just as one can enjoy a movie (say, *Schindler's List*) without endorsing particular characters, perhaps one can eroticize a wrongful behavior while remaining fully cognizant of its wrongfulness. This again points to the need for a more developed phenomenology of eroticization.

Second, eroticization may not be fully voluntary. If Ramona stumbles upon a rape account in the newspaper and she subsequently becomes aroused, has she really adopted an attitude? Isn't it more accurate to say that she has experienced a *reaction* to a stimulus (and an uninvited stimulus at that)? Insofar as the reaction is involuntary, Ramona cannot be held morally accountable for it. Of course, the matter is somewhat different when she is not merely reading the newspaper but instead perusing her local adult bookstore. There seem to be various possible degrees of participation in erotic arousal, ranging from the involuntary to the deliberate. It is toward the latter end of the spectrum that one begins to suspect a pro-attitude.

Yet even if Ramona's arousal is deliberate, there is still another defense she can marshal—especially if she herself is a rape survivor. Is it not possible that in eroticizing wrongful activities, those who pursue naughty fantasies thus rob these activities of their wrongfulness? I have in mind the kind of transformation intended when, for example, LGBT persons use the word "queer" or Jews make jokes about the Holocaust—Mel Brooks's musical comedy *The Producers* being a brilliant example of the latter. Of course, these sorts of reclamations raise troubling questions of their own. But they also raise the helpful suggestion that the solution to the naughty-fantasies dilemma lies in areas having very little to do with sex.

One final point: Although I have tried to focus on a non-consequentialist argument here, one might wonder whether consequentialism is lurking behind the intuitions driving that argument. For it is possible that people believe that it is wrong to eroticize wrongful activities because they cannot shake the suspicion that those who do so are more likely to engage in wrongful behavior, stipulations to the contrary notwithstanding.[6] Such is one of the pitfalls of philosophical thought experiments. Moreover, it should be noted that there *are* important consequentialist arguments surrounding this issue, on both sides. (One could, for example, point to the negative effects of suggesting to people that it would be morally wrong for them to dwell on fantasies that are otherwise largely involuntary.) A fuller treatment of naughty fantasies would need to address these concerns. In this chapter, I have attempted to lay some groundwork for further philosophical investigation.[7]

Postscript (2021)

A friend of mine once told me that when you don't know what to think about a topic, go out for a long walk. This paper, written in the early 2000s, is the written equivalent of a long, meandering walk. When colleagues learned that I was writing a paper entitled "Naughty Fantasies," they naturally asked, "What's your overall conclusion?" I told them quite frankly that I didn't have one. I still don't, really.

For that reason, and also for reasons of time and interest, I resisted the temptation to write a new paper: "Naughty Fantasies Revisited." Instead, I decided to make some minor edits correcting ambiguities in the original paper and to add this brief postscript.

Two of those ambiguities are worth noting. First, the original paper defined naughty fantasies as fantasies that eroticize activities that are themselves wrong. The problem with that definition is that it made nonsense of my raising objections to the second premise of the naughty-fantasies argument: One cannot reasonably question whether naughty fantasies eroticize wrongful activities if one has defined naughty fantasies as fantasies that do exactly that. (Rookie philosophical mistake.) So here I define naughty fantasies as fantasies that *appear to* do this, allowing me later to question whether they in fact do so.

Second, in the original paper I mentioned that the fictional Ramona had herself been raped. Friends of mine who taught the paper later informed me that this detail tended to muddy the waters: Quite understandably, many harbor the intuition that rape victims are entitled to whatever they need in their fantasy lives to reclaim their sexual agency. So I eliminated that detail here. Still, the case raises the following related question: Does it make a difference whether the fantasizer imagines themselves in the position of victim or perpetrator? If I were to write "Naughty Fantasies Revisited" I might explore that question; for now, I leave it to readers.

In recent years the paper's central question has been given new salience by the invention of lifelike sex robots with artificial intelligence, which can be programmed to "resist" the user's advances: "Oooooh, you're being bad! Stop it!" Presumably, these allow users the opportunity better to simulate rape and sexual assault. Perhaps even more concerning, it is possible to design these sex robots to look and sound like real-life persons (such as famous actresses). If we bracket the consequentialist concerns, is there something wrong with the use of such "resisting" robots? If so, is it any different from pursuing the same scenarios through pure fantasy, unassisted by technology? The "fantasy" path is more likely for most people, as these sex robots are currently quite expensive—although that of course could change.

This leads me to a final question: Can we, *should* we, bracket the consequentialist concerns? While I gesture in the original paper at "the pitfalls of philosophical thought experiments," I increasingly worry that such bracketing is lazy at best. If ever I were to write "Naughty Fantasies Revisited," I would want to examine the empirical research more deeply. In particular, I worry about the kinds of attitudes toward women encouraged by such fantasies and by the media and technology (such as sex robots) that enable them. For now, I hope the paper continues to provide some useful intuition pumps for others who explore these questions.

Notes

1. Patrick Hopkins, "Rethinking Sadomasochism: Feminism, Interpretation, and Simulation," *Hypatia* 9: I (1994), pp. 116–141.
2. Ibid., p. 198.
3. Ibid., p. 197.
4. For a defense of a non-instrumentalist understanding of the virtues see Walter Schaller, "Are Virtues No More Than Dispositions to Obey Moral Rules?" *Philosophia* 20 (July 1990).
5. Notice that if one adopts the virtue-ethics understanding of the first premise it is no longer true *a priori,* since the virtues are learned through experience.
6. I am indebted to Alastair Norcross for raising this point.
7. I am indebted to Robert Card, Alessandro Giovannelli, Herbert Granger, Mark Huston, Bruce Russell, Robert Titiev, Susan Vineberg, and various participants in the Sixty-third Annual Meeting of the Southwestern Philosophical Society for helpful suggestions on earlier drafts of this paper. I am particularly grateful to David Boonin for inviting the new version.

Index[1]

NUMBERS AND SYMBOLS

[1] Note: Page numbers followed by 'n' refer to notes.

D. Boonin (ed.), *The Palgrave Handbook of Sexual Ethics*,
https://doi.org/10.1007/978-3-030-87786-6

D

The manufacturer's authorised representative in the EU is Springer
Nature Customer Service Centre GmbH, Europaplatz 3, 69115 Heidelberg,
Germany. If you have any concerns regarding our products, please
contact ProductSafety@springernature.com

Printed and bound by CPI Group (UK) Ltd, Croydon, CR0 4YY
30/09/2026
02225490-0020